Algorithms for Intelligent Systems

This book series publishes research on the analysis and development of algorithms for intelligent systems with their applications to various real world problems. It covers research related to autonomous agents, multi-agent systems, behavioral modeling, reinforcement learning, game theory, mechanism design, machine learning, meta-heuristic search, optimization, planning and scheduling, artificial neural networks, evolutionary computation, swarm intelligence and other algorithms for intelligent systems.

The book series includes recent advancements, modification and applications of the artificial neural networks, evolutionary computation, swarm intelligence, artificial immune systems, fuzzy system, autonomous and multi agent systems, machine learning and other intelligent systems related areas. The material will be beneficial for the graduate students, post-graduate students as well as the researchers who want a broader view of advances in algorithms for intelligent systems. The contents will also be useful to the researchers from other fields who have no knowledge of the power of intelligent systems, e.g. the researchers in the field of bioinformatics, biochemists, mechanical and chemical engineers, economists, musicians and medical practitioners.

The series publishes monographs, edited volumes, advanced textbooks and selected proceedings.

More information about this series at http://www.springer.com/series/16171

Nikhil Marriwala · C. C Tripathi · Shruti Jain ·
Shivakumar Mathapathi
Editors

Soft Computing for Intelligent Systems

Proceedings of ICSCIS 2020

Springer

Editors
Nikhil Marriwala
Department of Electronics
and Communication Engineering
Kurukshetra University
Kurukshetra, Haryana, India

C. C Tripathi
University Institute of Engineering
and Technology (UIET)
Kurukshetra University
Kurukshetra, Haryana, India

Shruti Jain
Department of Electronics
and Communication Engineering
Jaypee University of Information
Technology
Waknaghat, Himachal Pradesh, India

Shivakumar Mathapathi
Co-Founder of Dew Mobility
Santa Clara University
Santa Clara, CA, USA

ISSN 2524-7565 ISSN 2524-7573 (electronic)
Algorithms for Intelligent Systems
ISBN 978-981-16-1047-9 ISBN 978-981-16-1048-6 (eBook)
https://doi.org/10.1007/978-981-16-1048-6

This Springer imprint is published by the registered company Springer Nature Singapore Pte Ltd.
The registered company address is: 152 Beach Road, #21-01/04 Gateway East, Singapore 189721, Singapore

Preface

The aim of this conference is to serve for educators, researchers, and developers working in the area of recent advances and upcoming technologies utilizing computational sciences in signal processing, imaging, computing, instrumentation, artificial intelligence, and their applications. This conference will provide support and aid to the researchers involved in designing latest advancements in communication and intelligent systems that will permit the societal acceptance of ambient intelligence. The proceedings of the conference "Soft Computing for Intelligent Systems" (SCIS-2020) encompass all branches of Artificial Intelligence, Computational Sciences, and Machine Learning which are based on computation at some level such as AI-based Internet of Things, Sensor Networks, Robotics, Intelligent Diabetic Retinopathy, Intelligent Cancer Genes Analysis using Computer Vision, Evolutionary Algorithms, Fuzzy Systems, Medical Automatic Identification Intelligence System and Applications in Agriculture, Healthcare, Smart Grid, Instrumentation Systems, etc. It presents the latest research being conducted on diverse topics in Soft Computing with the goal of advancing knowledge and applications in this rapidly evolving field. Authors were invited to submit papers presenting novel technical studies as well as position and vision papers comprising hypothetical/speculative scenarios. It presents the latest research being conducted on diverse topics in intelligent technologies to advance knowledge and applications in this rapidly evolving field. The conference is seen as a turning point in developing the quality of human life and performance in the future, and therefore it has been identified as the theme of the conference. The overall goal of this conference is to present the latest snapshot of the ongoing research as well as to shed further light on future directions in the area of soft computing and artificial intelligence. The aim of this conference is to serve for researchers, educators, and developers working in the area of upcoming technologies in the field of Computational Systems; Deep Learning; IOE; Sensor Networks; Robotics; Intelligent Body Area Network for Healthcare; Embedded Systems; and Intelligent Computing Techniques, Modeling, and Simulations.

The overall goal of this conference is to present the latest snapshot of the ongoing research as well as to shed further light on future directions in this space. Authors are invited to submit papers presenting novel technical studies as well as position and vision papers comprising hypothetical/speculative scenarios.

For the proper review of each manuscript, every received manuscript was first checked for plagiarism and then the manuscript was sent to three reviewers. In this process, the committee members were involved and the whole process was monitored and coordinated by the General Chair. The Technical Program Committee involved senior academicians and researchers from various reputed institutes. The members were from India as well as abroad. The technical program mainly involves the review of the paper. A total of **240** research papers were received, out of which 50 papers were accepted and registered and presented during the 3-day conference, acceptance ratio is **20.83%**.

An overwhelming response was received from the researchers, academicians, and industry from all over the globe. The papers were received from pan-India with states **Tamil Nadu, Karnataka, Maharashtra, Haryana, Telangana, Jammu And Kashmir, Assam, Kerala, Uttar Pradesh, Punjab, Rajasthan, Chandigarh, Andhra Pradesh, Delhi, Himachal Pradesh, Madhya Pradesh,** and not to mention our neighboring states. The authors are from premium institutes IITs, NITs, Central Universities, NSIT, PU, and many other reputed institutes.

Organizers of SCIS-2020 are thankful to University Institute of Engineering & Technology (UIET) which was established by Kurukshetra University in 2004 to develop as a "Centre of Excellence" and offer quality technical education and to undertake research in engineering and technology. The ultimate aim of the UIET is to become a role model for engineering and technology education not only for the state of Haryana but also for the world over to meet the challenges of the twenty-first century.

The editors would like to express their sincere gratitude to general chairs, plenary speakers, invited speakers, reviewers, Technical Programme Committee members, International Advisory Committee members, and Local Organizing Committee members of SCIS-2020, without whose support the quality and standards of the conference could not be maintained. Special thanks to Springer and its team for this valuable publication. Over and above, we would like to express our deepest sense of gratitude to UIET, Kurukshetra University, Kurukshetra, for hosting this conference. We are thankful to Technical Education Quality Improvement Programme (TEQP-III) for sponsoring the International Conference SCIS-2020 event.

Kurukshetra, India — Nikhil Marriwala
Kurukshetra, India — C. C Tripathi
Waknaghat, India — Shruti Jain
Santa Clara, USA — Shivakumar Mathapathi

Contents

About the Editors

Dr. Nikhil Marriwala (B.Tech., M.Tech. and Ph.D. in Engineering and Technology) is working as Assistant Professor and Head of the Department Electronics and Communication Engineering Department, University Institute of Engineering and Technology, Kurukshetra University, Kurukshetra. He did his Ph.D. from National Institute of Technology (NIT), Kurukshetra in the department of Electronics and Communication Engineering. He did his post-graduation (M.Tech.) in Electronics and Communication Engineering from IASE University and did his B.Tech. in Electronics and Instrumentation from MMEC, Mullana, Kurukshetra University, Kurukshetra. He has more than 18 years of experience teaching graduate and postgraduate students. More than 31 students have completed their M.Tech. dissertation under his guidance. His areas of interests are Software-Defined Radios, Cognitive Radios, Soft Computing, Wireless Communications, Wireless Sensor Networks, Fuzzy system design, and Advanced Microprocessors. He has published more than 5 book chapters in different International books, has authored more than 10-books with Pearson, Wiley, etc. and has more than 30 publications to his credit in reputed International Journals and 20 papers in International/National conferences. He also has 04 patents published to his credit with 02 International Patents. He has been Chairman of Special Sessions in more than 6 International/National Conferences and has delivered a keynote address at more than 3 International conferences. He has also acted as organizing secretary for more than 3 International conferences. He is having an additional charge of Training and Placement Officer, UIET, Kurukshetra University, Kurukshetra and heading the T&P cell for more than 10 years now. He is the single point of contact (SPOC) and head of the local chapter of SWAYAM NPTEL Local Chapter of UIET, KUK. He is the SPOC for Infosys campus connect program for UIET, KUK. He is also the He is a reviewer for many reputed journals such as the *International Journal of Communication Systems*, Wiley, *IEEE Signal Processing Letters*, *International Journal of Measurement Technologies* and *Journal of Organizational and End User Computing* (JOEUC), *Egyptian Informatics Journal*—Elsevier, *Instrumentation Engineering* (IJMTIE), *International Journal of Interactive Communication Systems and Technologies* (IJICST), *Current Journal of Applied Science and Technology*, UK. He was awarded as the **"Career Guru of the Month"** award by Aspiring Minds.

C. C Tripathi did his Ph.D. (Electronics) from Kurukshetra University, Kurukshetra. Since 2016, he is working as Director, University Institute of Engineering Technology (an autonomous institute), Kurukshetra University, Kurukshetra and Dean Faculty of Engineering and Technology. As Director, he is also heading institute academic bodies like the board of studies, academic council with four UG 8 PG programs and spearheading research in various engineering and applied sciences departments in the institute. Microelectronics, RF MEMS for Communication and Industrial Consultancy are his specialization areas. He has developed Micro-fabrication R&D Lab and RF MEMS R&D Lab. He is a member of more than 14 Professional Bodies. He has published more than 81 papers in reputed journals and more than 44 papers in National/International conferences. He has also filed four patents. He has implemented TEQIP-II and TEQIP-III grants of '10.00 Crores and '7.00 Crores respectively by preparing Institution Development Plan (IDP). He has also guided 7 Ph.D.s and more than 20 M.Tech. (PG) students in Engineering and Technology. He is a senior member of IEEE society and reviewer of a number of prestigious journals. He has successfully conducted more than 40 FDPs and 02 International conferences and a number of National Conferences.

Shruti Jain is Associate Professor in the Department of Electronics and Communication Engineering at Jaypee University of Information Technology, Waknaghat, Himachal Pradesh, India, and has received her Doctor of Science (D.Sc.) in Electronics and Communication Engineering. She has a teaching experience of around 15 years. She has filed 06 patents out of which 01 patent is granted and one is published. She has published more than 15 book chapters, and 100 research papers in reputed indexed journals and in international conferences. She has also published six books. She has completed two government-sponsored projects. She has guided 06 Ph.D. students and now has 01 registered student. She has also guided 11 M.Tech. scholars and more than 90 B.Tech. undergraduates. Her research interests are image and signal processing, soft computing, bio-inspired computing and computer-aided design of FPGA and VLSI circuits. She is a senior member of IEEE, a life member and Editor-in-Chief of Biomedical Engineering Society of India and a member of IAENG. She is a member of the Editorial Board of many reputed journals. She is also a reviewer of many journals and a member of TPC of different conferences. She was awarded by Nation Builder Award in 2018–2019.

Prof. Shivakumar Mathapathi is an adjunct faculty member at multiple universities including the University of California San Diego, Sonoma State University, and Santa Clara University, and he is also teaching at regional community academia including ethnically diverse institutions such as Ohlone and De-Anza College in California. Shivakumar has been a passionate educator for nearly 25 years of industry experience with a strong desire to help students recognize the connection between learning and industry needs. He is an accomplished researcher and a co-founder of Dew Mobility Incorporated and Xtrans Solutions (P) Limited. His teaching and research areas include the Internet of things, machine learning, artificial intelligence, cybersecurity and blockchain. He has also authored numerous articles published in prestigious journals, including IEEE, and various conferences, such as NIST Smart City projects. He is the co-founder of Xtrans, which is a leading company in establishing the centre of excellence (CoE) to assist higher education to educate their students about the next generation Internet research and development (R&D) programs. Shivakumar is also a team lead of the education cluster at the Global City Team Challenge project in collaboration with National Institute of Standards and Technology (NIST), under the U.S. Department of Commerce. He is the key contributor in creating tutorials, workshop and practice use cases for smart cities. The workshop also discusses IoT standards and protocol which would help community partners and city/municipality staff to get familiar with national as well as international IoT standards.

Chapter 1
Object Detection from the Seabed Imaging Data Using Soft Computing Techniques

U. Anitha, G. D. Anbarasi Jebaselvi, R. Narmadha, Vishnu Vardhan, and Sri Pavan

1 Introduction

Nearly all the underwater practices are conducted by remote-operated vehicles (ROV) guided by human hands. The most important tasks that these ROV vehicles should be able to perform freely are operations such as visual check of man-made structures like pipeline and offshore structures object detection of mines, obstacle avoidance, etc. [1–5]. To compact with underwater image processing, consider primary physics of sunshine propagation within the water medium [6, 7]. Underwater region causes degradation effects, which are not present in an image captured in air. Sonar images are with less visibility since light gets attenuated while travelling through the water, so the pictures result in poor contrast and hazy. Light gets attenuated and restricts the viewing distance up to 20 m in natural water and less than 5 m in muddy spot. The light in water causes attenuation by absorption that eliminates a portion of light energy and scattering, which changes sunshine path direction. The attenuation, absorption and scattering activities of light rays in water impact the functioning of underwater imaging systems [8, 9]. Forward scattering that randomly diverged light on its way from an object to the camera generally results in blurring of the image features. Backward scattering superimposes the event on the image and hides the original scene [10, 11]. The subsistence of the gliding particles referred to 'marine snow' (extremely variable in a similar way and strength) boost both absorbing and scatter effects. In summary, pictures we have a vested interest can suffer by more consequent problems: limited range visibility, low distinction, non-costume lighting, blurring, bright colour lessened (bluish appearance) and noise. In image applications, it contains object boundaries and object shadows and noise. The image processing restoration processes planned to improve a corrupted image employing a model of the degradation [12, 13]. The sonar image contains mainly speckle noise which is a source by the intrusion of the sea-floor sediment along with multi-path reverberation.

U. Anitha (✉) · G. D. Anbarasi Jebaselvi · R. Narmadha · V. Vardhan · S. Pavan
Sathyabama Institute of Science and Technology, Chennai, India

N. Marriwala et al. (eds.), *Soft Computing for Intelligent Systems*, Algorithms for Intelligent Systems, https://doi.org/10.1007/978-981-16-1048-6_1

It is confirmed that filtering techniques like median, SRAD and FROST are more appropriate for removing speckle noise in acoustic images [14]. Image enhancement technique is a pre-processing step for segmentation, object detection and recognition in computer vision applications [15, 16]. Segmentation has been done by fuzzy C-means algorithm based on spatial neighbour information, new results indicate that this algorithm be able to get the good segmentation but spends much time [17, 18]. The existing system faces problems like difficult to predict the K-value in clustering segmentation method. Initial seed value (K-value) in clustering method has a tough impact on image results. The region growing segmentation process is not chosen for its reduced functions and automated features come about are not having exact values. Pre-processing steps are needed to seek out which sort of filtering is getting to be more helpful. A desirable area is selected from the segmental image to calculate its volume [19]. Volume level of the preferred area is larger than the original area. The region growing algorithm will portion image not just the area but also the non-area with high-intensity ratio.

2 Methodology

In this proposal, each of the resulting images is given as a convolution between the image and a 2D filter, where the width of the filter increases with the parameter. Adaptive mean adjustment (AMA) is an enhancement technique used to enhance contrast in images. The adaptive method calculates multiple histograms which correspond to definite section of the image and improve the local contrast of a picture to bring out the core details. However, it is a bent to amplify noise more in comparatively uniform regions of a picture. A CLAHE (contrast limited adaptive histogram equalization) prevents it by limiting the amplification. In this module, modified adaptive mean shift (M-AMS) algorithm is going to be implemented to beat the matter of existing system of segmentation mishandling by initializing standard membership values in order that standard segmentation result for each image, number of clusters and initial standard intensity values was obtained. Next to the achievement of preliminary clustering process, the clusters were successively updating the cluster weights and membership degree. Vector membership values are distributed identically to all members in the cluster for maintaining cluster updating. The following high-level features are going to be extricated in the research work: autocorrelation, cluster prominence, entropy, cluster shade, dissimilarity energy, maximum probability, variance, entropy, information measure of correlation and homogeneity. The time-domain relationship of pixels in co- occurrence matrix (GLCM) is mentioned as grey-level spatial dependence matrix. Every image element I, J in GLCM is the summation of number of times that the I-pixel value along with the J-pixel value. Rearrange all above features into the matrix and store it as a database. The intention of feature extraction technique is to imply the image in its efficient and special kind of single values or a matrix vector. Feature extraction is executed automatically from a picture. The two-dimensional discrete wavelet transform (2D-DWT) is used here for image feature extraction.

3 Proposed System

From Fig. 1 showcases the step-by-step process of the proposed system; initially, input image which was taken from the database is fed into double conversion process and it will double the precision with great intensity, then forwarded to next step and once calculation of the size of image is done, the image undergo through filtering process by hybrid median filter and later proceeds to enhancement process. Here the image is enhanced by the adaptive median filter followed by clustering method by FCM algorithm, which is part of segmentation process. Next the segmented image is fed to GLCM algorithm in feature extraction procedure. In next step to the feature extraction, classification of extracted image by using SVM is done. Hence, the image is tested with the image of the database and transmitted using OFDM to receiver. Totally there are 65 images in the database all related to flight parts such as Boeing's window, tail, wings, engine and spinner. In these datasets, there are nearly 15 images on each of the flight part, and on this 10 images were used for training and 5 images have been tested for each part. This has been implemented using MATLAB software.

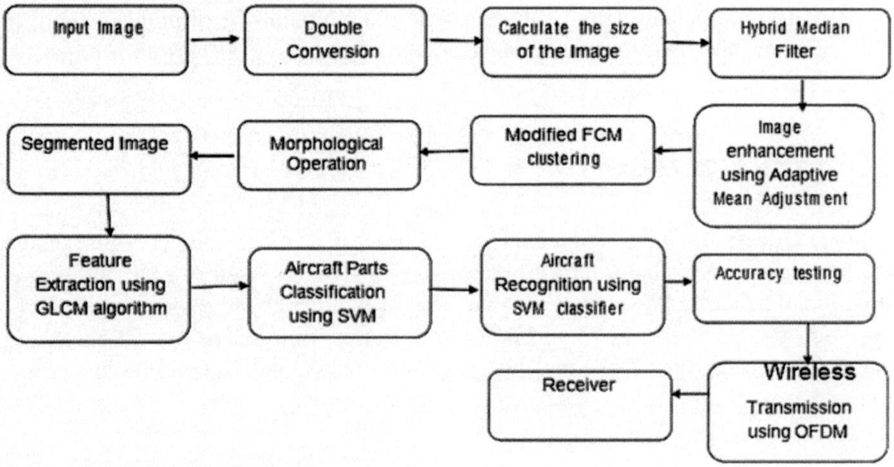

Fig. 1 Proposed system block diagram

3.1 Double Conversion

In MATLAB, im2double takes a picture as input and returns a picture of sophistication double. If the given image is of sophistication double, the output image is just like it. If the given image is not double, im2double sends the consequent image of sophistication double, rescaling or compensating the information as necessary. It converts binary image BW to a double precision intensity image.

3.2 Hybrid Median Filter

Hybrid median filter is the superior version of median filter, which can remove the noise well against median type. Image filters produce a replacement image from an indigenous by working on the pixel values. Filters are to enhance contrast, suppress noise, find edges and locate features. It would like to reinforce the standard of images.

3.3 Image Enhancement Using Adaptive Mean Adjustment

An entropy adaptive sub-histogram equalization (EASHE) is proposed for the research. The proposed method splits the histogram by entropy of input image into four segments, and therefore the dynamic range of every sub-histogram is adjusted.

3.4 Segmented Image

In image processing, image segmentation is a process of subdividing a digital image into many segments, also referred to an image object. The goal of segmentation is depiction of a picture into something that is more significant and easier to analyse. In this work, it uses hybrid FCM and morphological operation. The hybrid clustering segmentation integrates K-means clustering and fuzzy C-means (FCM) clustering.

3.5 FCM Clustering

Fuzzy C-means (FCM) is a method of grouping which is derived from fuzzy pure mathematics, and allows piece of knowledge to belong of two or more clusters. FCM is popularly used for soft segmentations but more sensitive to noise to overcome this problem, a modified FCM method is proposed for this work.

3.6 Morphological Operation

It is an image processing operation that process on shape of the objects in an image with respect to its neighbouring pixels. The shape of the objects is traced and identified.

3.7 Feature Extraction (GLCM)

Texture analysis has been done using grey-level co-occurrence matrix (GLCM). Create a GLCM, and then extract statistical processes from this matrix. The following are the GLCM features extracted in this research work such as autocorrelation, cluster prominence, entropy, cluster shade, dissimilarity energy, maximum probability, variance, entropy, information measure of correlation and homogeneity.

3.8 SVM Classifier

An algorithm intuitively works on creating linear decision boundaries to classify multiple classes. The ability of a hypothesis to correctly classify dataset is known as generalization. In this case, SVM performs better than that of the ordinary neural network in such a way that the results are convered fastly.

3.9 Wireless Transmission Using OFDM

A picture is inserted as input and is processed by break down into pixel levels, coding it and orthogonal frequency division multiplexing transmission through an AWGN channel. Receiving is additionally done at the top with an ofdm.bmp image file.

4 Experimental Results

Figure 2 is the image taken as input image, which is 2D image. This image is taken under water captured by SONAR. It is one of the pictures available in database. Figure 3 is the output image of median filter; it is a nonlinear digital filtering technique.

The pre-processing steps helps to improve the results of further image processing techniques. Figures 4 and 5 are the output image of the hybrid median filter. It is a superior version of the median filter; eliminate noise better than the original

Fig. 2 Input image

Fig. 3 Median filter output

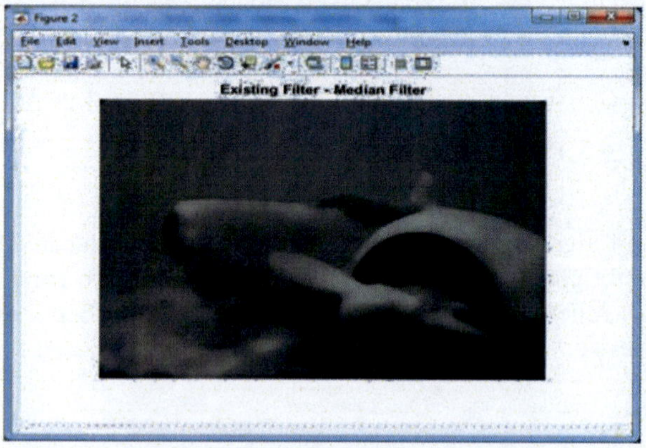

Fig. 4 Hybrid median filter

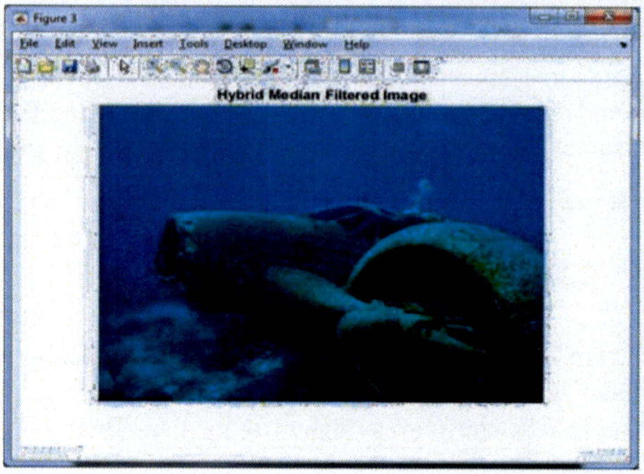

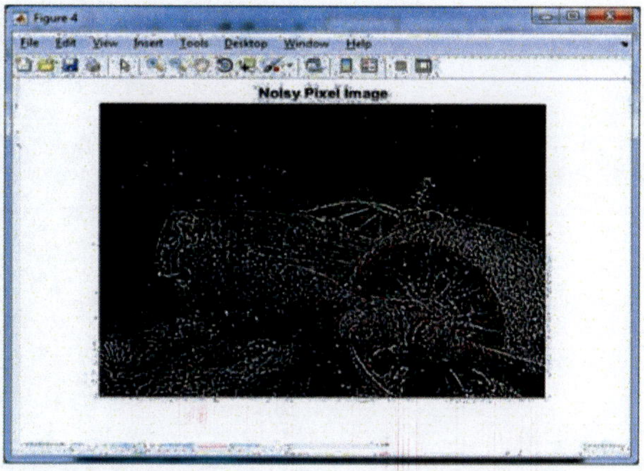

Fig. 5 Noisy pixel image

median filter. It generates new image from the original by working it on the pixel values. The obtained image is suppressed noise, enhanced contrast, to find edges and with located features and Fig. 5 is the output of the difference between the original and hybrid image. Figure 6 is the NTSC (National Television Standard Committee) colour image. It is an analog encoding system which results ground color to the sample image.

Then Fig. 7 is an output of image enhancement process. This process is used to make the image features more distinct by making use of optimal colours available on the output device. It is done by changing the range of an image intensity values to increase contrast. Figure 8 is obtained as an output of image segmentation process. Image segmentation is a method used for dividing digital image into multiple segments.

This process is also helpful for the analysis of the image easily and Fig. 9 shows detection of object shape that is present in input image. By this image we came to

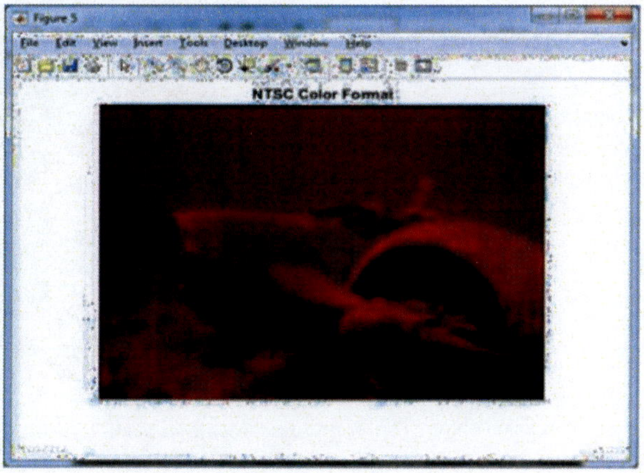

Fig. 6 NTSC image

Fig. 7 Contrast enhanced image

Fig. 8 Segmented image

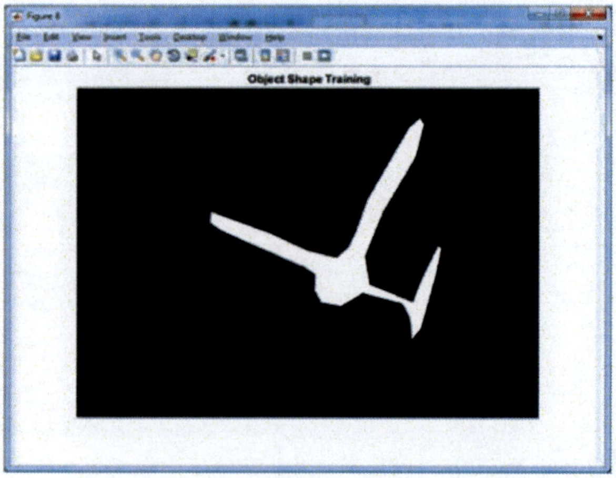

Fig. 9 Object detection

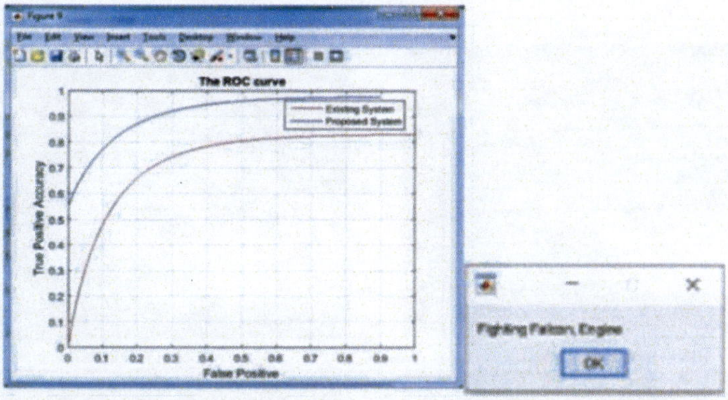

Fig. 10 Training accuracy curve and object identified as fighting falcon engine

know that some object is present in the image. After this step, the object shape is detected.

Figure 10 shows the training window for the neural network using an optimization algorithm to get a set of weights to map inputs with outputs. This training dataset is to create a good mapping of inputs to outputs and this window shows the name of the part or segment that is identified through output. The output of the corresponding input image is identified as fighting falcon engine.

Figure 11 is another example for this work for the detection of Boeing window in the flight.

5 Conclusion

It concludes that the overall efficiency of the proposed system is 91% by developing and introducing the new algorithms to the existing system. Here the hybrid median filter has been used instead of median filter, which helped to great extent in image enhancement process by removing noises. In the classification process to obtain better efficiency, support vector machine (SVM) is used. Algorithms in the proposed system result in satisfactory outputs. In future, by increasing the number of images in the database training, detection of many parts of the flight can be carried out.

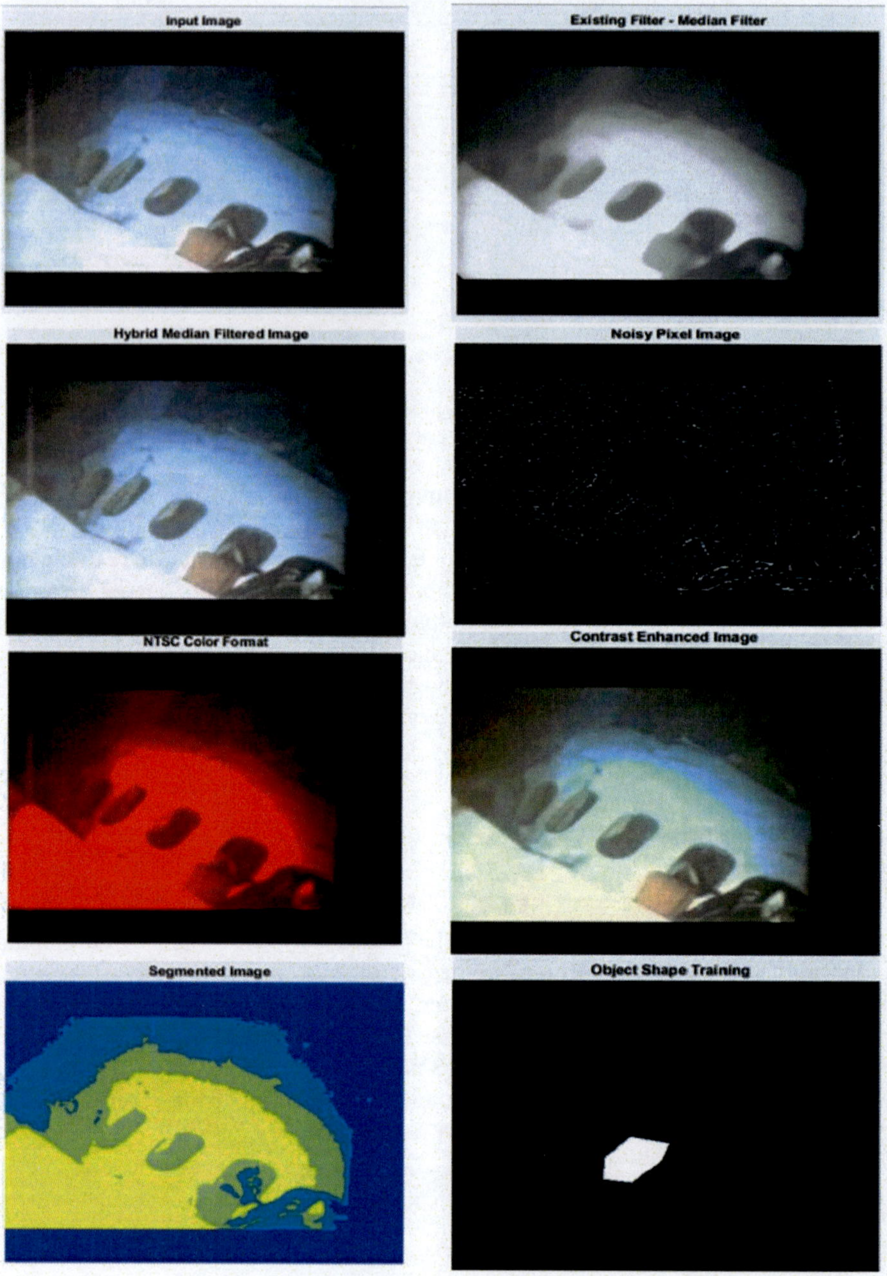

Fig. 11 Detection of Boeing window

References

1. Alagar VS, Thiel LH (1981) Algorithms for detecting M-dimensional objects in N-dimensional spaces. IEEE Trans. Pattern Anal. Mach. Intell. 3:245–256
2. Balasuriya BAAP, Fujii T, Ura T A vision based interactive system for underwater robots. In: Proceedings of IEEE IROS '95, Pennsylvania, pp 561–566 (1995)
3. Ripley BD (1996) Pattern recognition and neural networks, Cambridge University Press
4. Balasuriya BAAP, Fujii T, Ura T (1996) Underwater pattern observation for positioning and communication of AUVS. Proc IEEE IROS '96, 193–20
5. Branca A, Stella E, Distante A (1998) Autonomous navigation of underwater vehicles. In: Proceedings Oceans '98, Nice, France, pp 61–65
6. Duntley SQ Light in the sea. J Op Soc Am (1963)
7. Davies RS (1990) Remote visual inspection in nuclear, pipeline and underwater industries. Mater Evol 48:797–803
8. Anitha U, Malarkkan S Underwater object identification and recognition with sonar images using so ft computing techniques. Indian J. Geo-Marine Sci. (2018)
9. Lane DM, Stoner JP (1994) Automatic interpretation of sonar imagery using qualitative feature matching. IEEE J Ocean Eng 19:391–405
10. Anitha U et al (2019) Sonar image segmentation and quality assessment using prominent image processing techniques. Appl Acoust
11. Buckingham MJ, Berkout BV, Glegg AAL (1992) Imaging the ocean ambient noise. Nature 356:327–329
12. Foresti GL, Gentili S, Zampato M Autonomous underwater vehicle guidance by integrating neural networks and geometrical reasoning. Int J Imaging Syst Technol (in press)
13. Nguyen H-T et al Study on the classification performance of underwater sonar image classification based on convolutional neural networks for detecting a submerged human body. Sensors (2020)
14. Yan J, Meng J, Zhao J Real-time bottom tracking using side scan sonar data through one-dimensional convolutional neural networks. Remote Sens (2019)
15. Fleischer SD, Rock SM (1998) Experimental validation of a real-time vision sensor and navigation system for intelligent underwater vehicles. In: Proceedings of IEEE Conference on Intelligent Vehicles. Stuttgart, Germany
16. Shi Hong, et.al.: An underwater ship fault detection method based on Sonar image processing, Journal of physics, (2016).
17. Foresti GL, Murino V, Regazzoni CS, Trucco A (1997) A voting-based approach for fast object recognition in underwater acoustic images. IEEE J Ocean Eng 22:57–65
18. Mandhouj I et al Sonar Image Processing for Underwater Object Detection Based on High Resolution System (2012)
19. Zhu J et al Underwater Object Recognition Using Transformable Template Matching Based on Prior Knowledge (2019)

Chapter 2
Enhanced Voltage Regulation of 16-Bus Micro-Grid System Using Sliding Mode Controller

Akhib Khan Bahamani and G. Srinivasulu Reddy

1 Introduction

"Examination of 4 switch 3 phase inverter system utilizing FLC" was exhibited by Zaky [1]. This exertion proposed delicate changing strategy to decrease the exchanging misfortunes and an alternate delicate exchanging system was recommended to lessen exchanging pressure. "Various information aggregation techniques for smart-grid" were given by Chen [2]. This work manages brilliant meter utilized for estimating power culmination and the complete fulfillment of the whole buyer was got by utilizing know-it-all meter [3]. "PR-controller-based synchronizing framework associated inverter" was given by Seifi [4]. This exertion managed the straightforward technique to execute PR controller in A-B-C outline. This exertion analyzed the aftereffects of matrix-converter framework with PI and PR controllers. "Examination and usage of 2 level PV inverter framework with PR controller" was given by Dey [5]. This work utilized recurrence bolted loop to improve the conformance of PV inverter framework. "Structure of discrete current controller the PWM converter" was given by Doncker [6]. Changed PR controller was utilized to improve the exhibition of PWM converter energy-conversion-system (W.E.C.S). "For the grid-side and rotor-side converters placed in the rotor circuit of the D.F.I.G," power procedures are revealed [7]. "Displaying, control, and simulation" of a photovoltaic power system for grid-associated and stand-alone applications are done. It proposes a hybrid system comprising of a photovoltaic (PV) exhibit and rechargeable battery incorporated to the distribution grid with the plan to perform "load-sharing with the distribution grid." The PV array and the battery are associated with the DC side of the voltage-source inverter (VSI) through a boost converter and buck-boost converter, respectively [7–9].

A. K. Bahamani (✉)
Department of EEE, Narayana Engineering College, Nellore, Andhra Pradesh, India

G. Srinivasulu Reddy
Narayana Engineering College, Nellore, Andhra Pradesh, India

N. Marriwala et al. (eds.), *Soft Computing for Intelligent Systems*, Algorithms for Intelligent Systems, https://doi.org/10.1007/978-981-16-1048-6_2

P.E. (power electronic) systems for the "grid-integration of RES (renewable-energy-sources): A survey": "The consumption of allocated energy assets is gradually being sought after as an increment." The purpose of a "power-electronic interface" is responsible to necessities related to the sustainable power source itself as well as to its impacts on the power system task [10, 11]. A valuable current for "PV-sun-oriented power generator is included with the grid. It exhibits the itemized plan and displaying of grid incorporated with the photovoltaic solar power generator.

2 Research Gap

Enhancement of time response of SBMG with closed-loop PR and SM controllers is not present in the above literature. Hence, the proposed work compares the responses of 16-bus micro-grid. Systems with PR and SM controllers under closed-loop conditions. Simulation results of 16-bus micro-grid with the overhead stated controllers are included along with the time-domain parameters. This work proposes SMC for the control of SBMG.

3 Proposed System

16-bus micro-grid with PRC and SM controllers is shown in Fig. 1. Voltage of load bus is sensed and it is compared with the reference voltage. The voltage error is applied to PR/SMC. The output of controller adjusts the pulse width applied to DPFC.

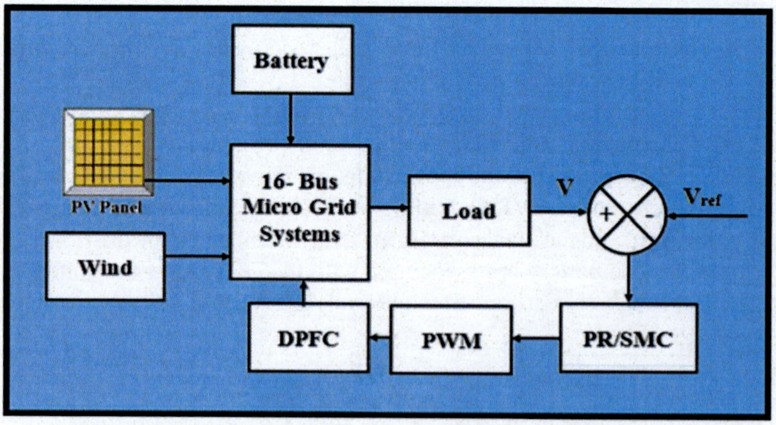

Fig. 1 Block diagram of 16-bus micro-grid with PRC and SM controllers

Modeling of PV system: The equation for the solar system is as follows:

$$I = I_{pv} - I_D \tag{1}$$

where

I output current,
I_{pv} current supplied by PV, and
I_D diode current.

$$I_{pv} = I_0\left(e^{kv_T} - 1\right) \tag{2}$$

where

I_0 leakage current,
V_T threshold voltage, and
K constant.

Modeling of wind System

$$P = 0.5\rho A\,V3$$

where

P power output,
ρ air density, and
V velocity.

Control operation of wind-based MGS: The controlled rectifier at the output of the wind generator produces variable DC. Thus, the DC yield of the rectifier can be controlling the pulse width applied to the devices of the rectifier.

$$V_0 = 1.35V_L\cos\alpha \tag{3}$$

where

V_L line voltage of the wind generator and
α firing angle.

4 Controller Technology

4.1 Current Control by Using Proportional Resonant Control

In order to normalize the grid current, a single-phase feedback current loop is employed. The model of the current control and the plant is illustrated in Fig. 2. As space vector theory cannot be useful to the single-phase voltage-source inverter, the controller design and modeling of the system cannot be finished in dq-frame. Thus, controller should be able to follow single-phase sinusoidal current reference directly.

The transfer function of the LCL filter is simply represented by plant $G_f(s)$, which is given as follows:

$$G_f(s) = \frac{sC_f R_d + 1}{s^3 L_I L_g C_f + s^2 C_f R_d (L_I + L_g) + s(L_I + L_g)} \tag{4}$$

Alternatively, a proportional-resonant (P-R) control which is based on “internal model principle” recommended by Francis and Wonham has an infinite gain. The PR-$G_i(s)$ is in the following form:

$$G_i(s) = K_p + \frac{K_i s}{S^2 + 2\delta\omega_0 s + \omega_0^2} \tag{5}$$

Here K_p and K_i are the proportional and integral gain correspondingly, "δ" is the damping factor, and ω_0 is power frequency of the grid voltage. The infinite gain of P-R control is reduced by damping factor δ to increase the bandwidth and thus dynamics of the system remains stable.

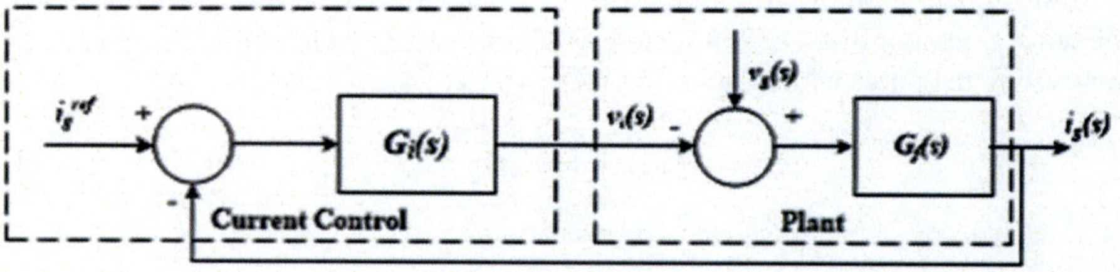

Fig. 2 Block diagram of current controller

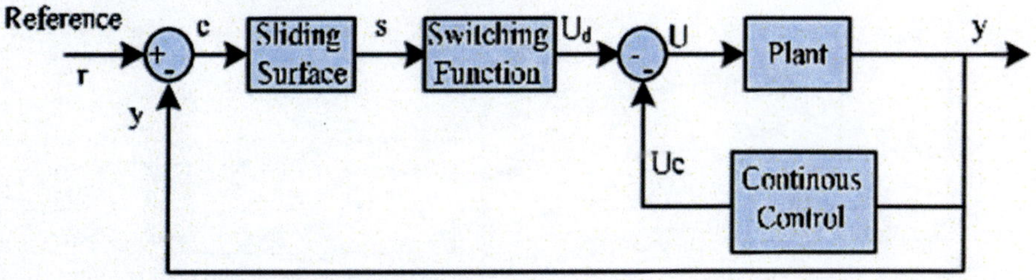

Fig. 3 Block diagram of sliding mode controller

4.2 *Slide Mode Controller*

The block diagram of "sliding mode controller" is shown in Fig. 3. In sliding mode, the majority of the controllers comprise mistake of single or numerous states of the system in the sliding surface (for example, inductor current or capacitor voltage).

Additionally, a few controller comprises error and both of the time derivatives and the integral of the error in the sliding surface to stabilize the system. In this case, the sliding surface can be symbolized as a second-order differential equation for which broad numerical study is obliged to guarantee system steadiness. Another surface is characterized for the enhancement in the steady-state error and settling time, which comprises voltage error and square of the capacitor current.

5 Simulation Results

5.1 *Open-Loop DPFC 16-Bus Mircro-grid with Load Disturbance*

Circuit diagram of open-loop SBMG with load disturbance is delineated in Fig. 1. There are three WGs and two PV systems. Load disturbance occurs at bus 6. Voltage at bus 6 of SBMG with load disturbance is shown in Fig. 2 and its value is 0.6×10^4 V. RMS voltage at bus 6 of SBMG with load disturbance is shown in Fig. 3 and its value is 4000 V. The reduction in voltage at bus 6 is due to increase in load. Current at bus 6 of SBMG with load disturbance is shown in Fig. 4 and its value is 56 A. Real power at bus 6 of SBMG with load disturbance is shown in Fig. 5 and its value is 1.4×10^5 W. Reactive power at bus 6 of SBMG with load disturbance is shown in Fig. 6 and its value is 3.3×10^4 VAR.

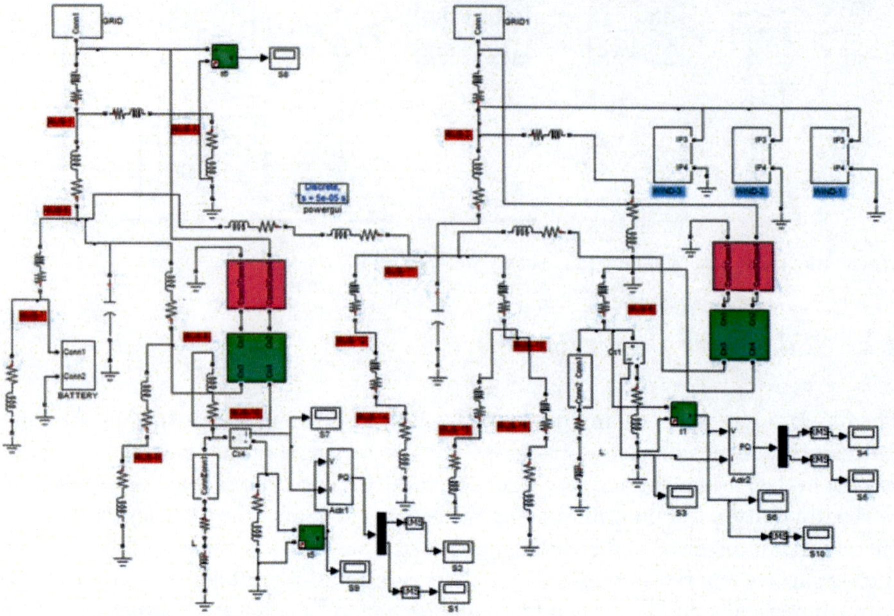

Fig. 4 Circuit diagram of open-loop SBMG with load disturbance

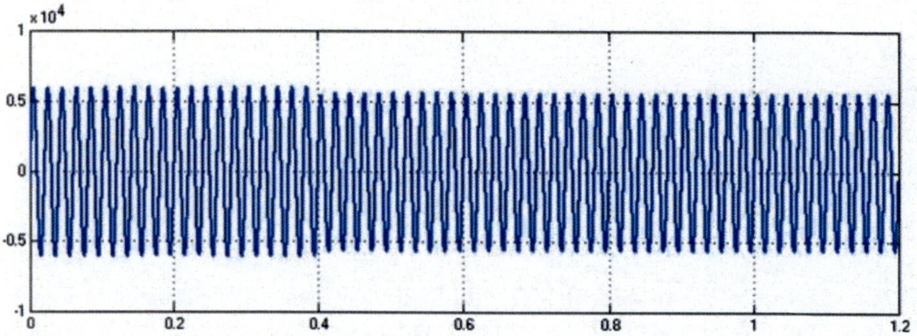

Fig. 5 Voltage at bus 6 SBMG with load disturbance

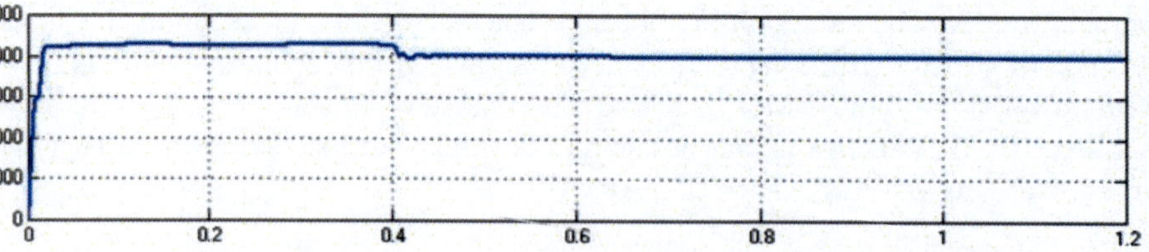

Fig. 6 RMS voltage at bus 6 SBMG with load disturbance

5.2 Closed-Loop SBMG with PR Controller

Circuit diagram of closed-loop SBMG with PR controller is delineated in Fig. 7. Voltage at bus 6 of closed-loop SBMG with PR controller is shown in Fig. 8 and its value is 0.7×10^4 V. RMS voltage at bus 6 of closed-loop SBMG with PR controller is shown in Fig. 9 and its value is 4400 V. Current at bus 6 of closed-loop SBMG with PR controller is shown in Fig. 10 and its value is 58 Amp. Real power at bus 6 of closed-loop SBMG with PR controller is shown in Fig. 11 and its value is 1.7×10^5 W. Reactive power at bus 6 of closed-loop SBMG with PR controller is shown in Fig. 12 and its value is 3.8×10^4 VAR.

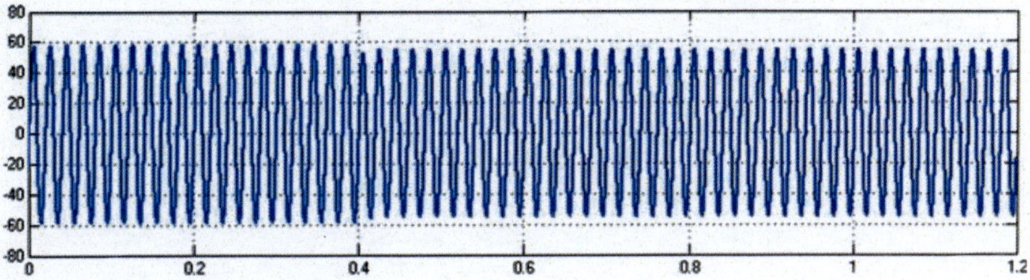

Fig. 7 Current at bus 6 SBMG with load disturbance

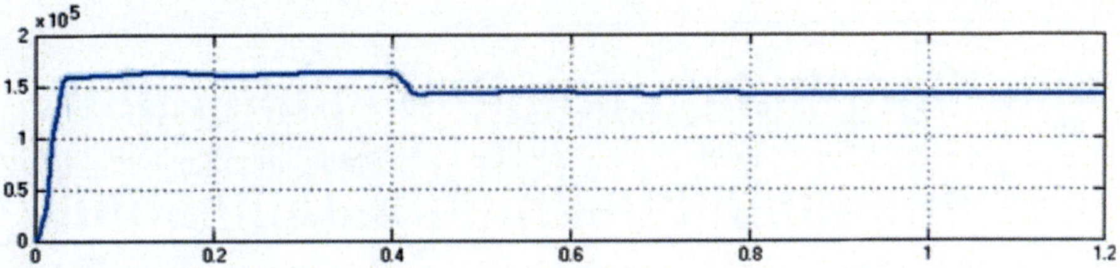

Fig. 8 Real power at bus 6 SBMG with load disturbance

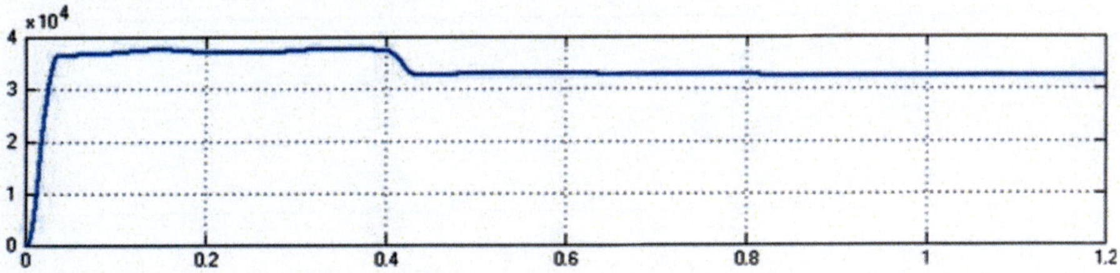

Fig. 9 Reactive power at bus 6 SBMG with load disturbance

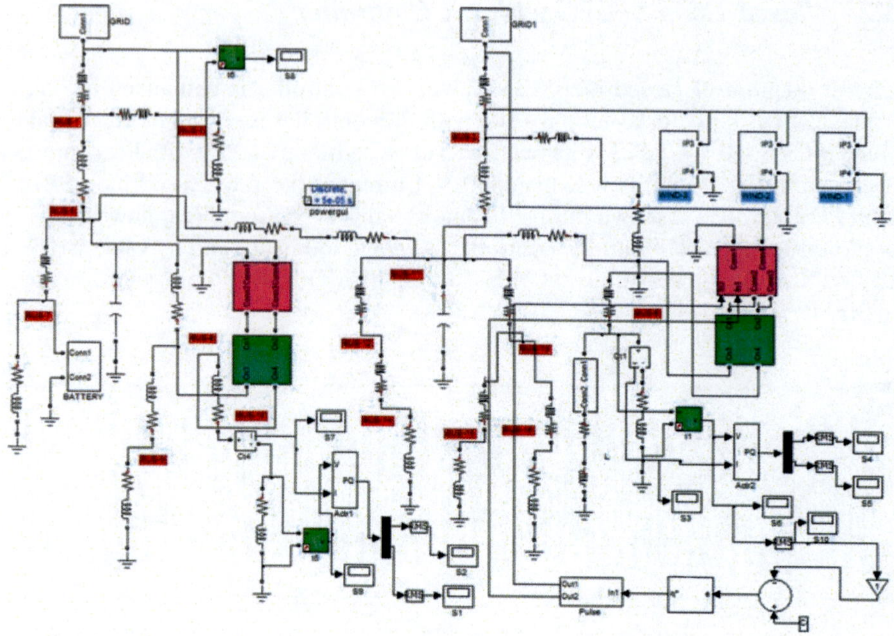

Fig. 10 Circuit diagram of closed-loop DPFC-based 16-bus mircro-grid with PR controller

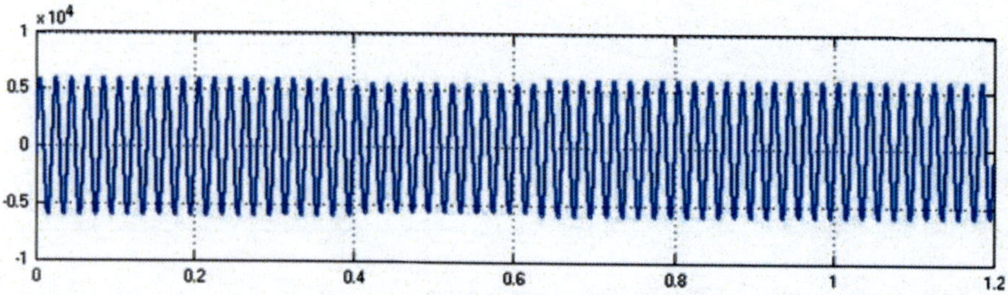

Fig. 11 Voltage at bus 6of closed-loop SBMG with PR controller

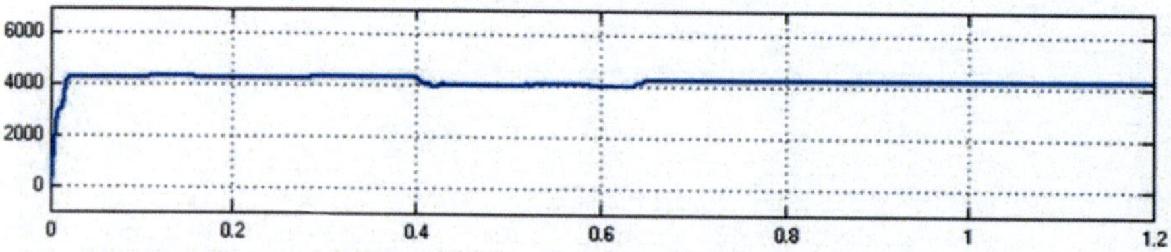

Fig. 12 RMS voltage at bus 6 of closed-loop SBMG with PR controller

5.3 Closed-Loop SBMG System with SM Controller

Circuit diagram of closed-loop DPFC-SBMG with SM controller is delineated in Fig. 13. Voltage at bus 6 of DPFC-SBMG with SM controller is shown in Fig. 14 and its value is 7000 V. RMS voltage at bus 6 of DPFC-SBMG with SM controller is shown with SM controller in Fig. 15 and its value is 4400 V. Current at bus 6 of DPFC-SBMG with SM controller is shown in Fig. 16 and its value is 58Amp. Real power at bus 6 of DPFC-SBMG with SM controller is shown in Fig. 17 and its value is 1.7×10^5 W. Reactive power at bus 6 of DPFC-SBMG with SM controller is shown in Fig. 18 and its value is 3.8×10^4 VAR.

Bar chart comparison of time-domain parameters is given in Fig. 19. Comparison of time-domain parameters using PRC and SMC is given in Table 1. By using SMC, the rise time is reduced from 0.42 to 0.41 s; peak time is reduced from 0.58 to 0.43 s;

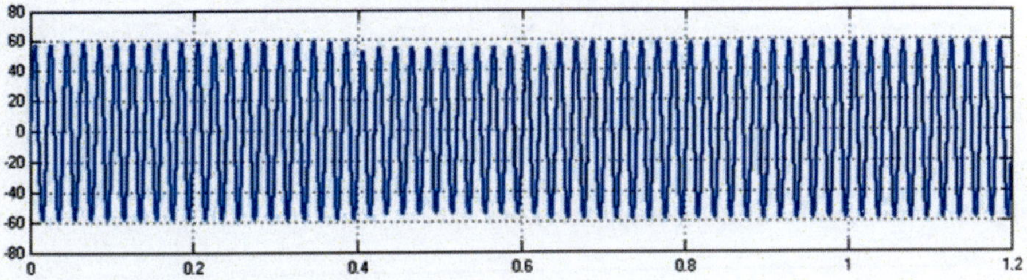

Fig. 13 Current at bus 6 of closed-loop SBMG with PR controller

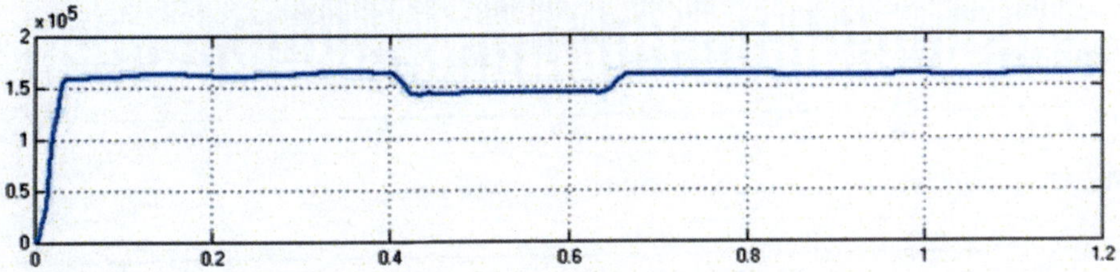

Fig. 14 Real power at bus 6 of closed-loop SBMG with PR controller

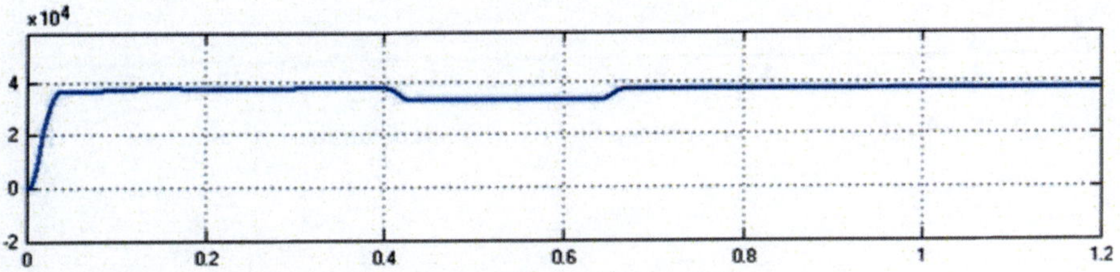

Fig. 15 Reactive power at bus 6 of closed-loop SBMG with PR controller

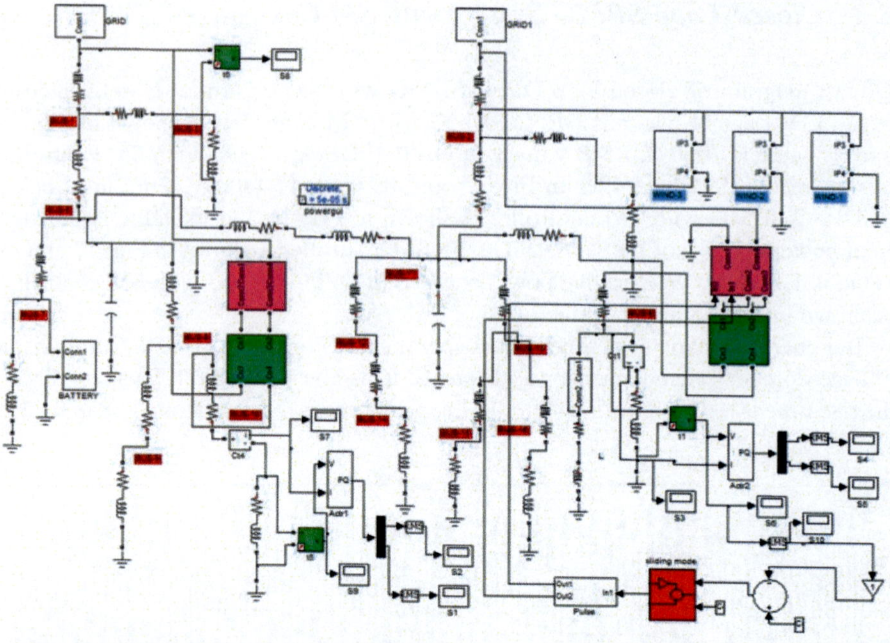

Fig. 16 Circuit diagram of closed-loop SBMG with SM controller

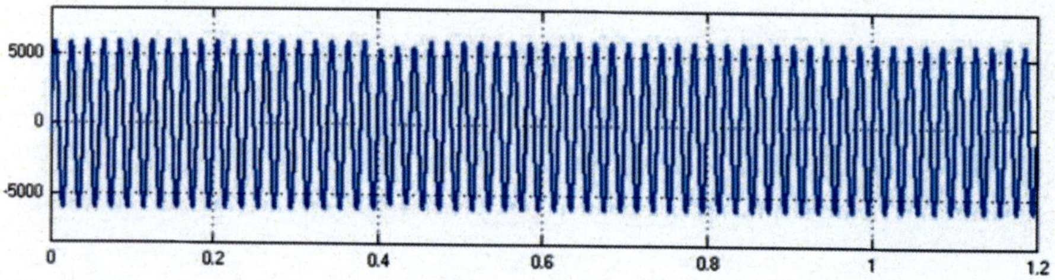

Fig. 17 Voltage at bus 6 of closed-loop SBMG with SM controller

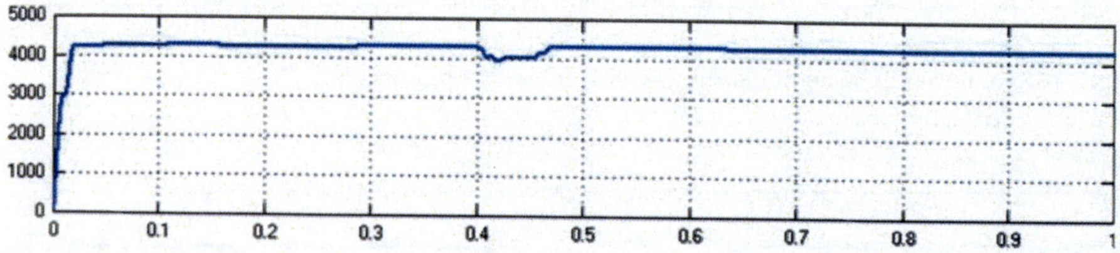

Fig. 18 RMS voltage at bus 6 of closed-loop SBMG with SM controller

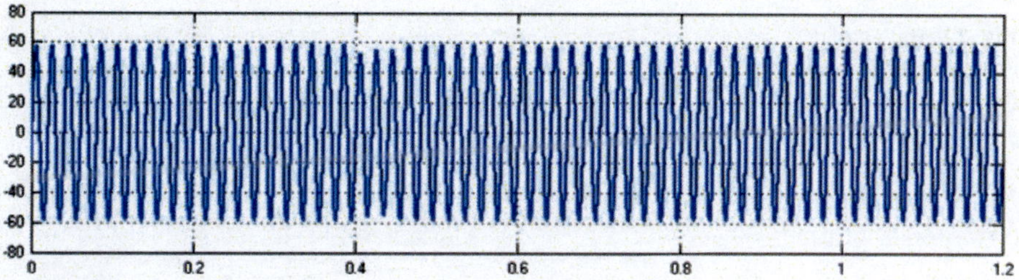

Fig. 19 Current at bus 6 of closed-loop SBMG with SM controller

Table 1 Comparison of time-domain parameters using PRC and SMC

Type of controller	$T_{r(SEC)}$	$T_{p(SEC)}$	$T_{s(SEC)}$	$E_{ss(V)}$
PRC	0.42	0.58	0.66	1.9
SMC	0.41	0.43	0.45	0.8

settling time is reduced from 0.66 to 0.45 s; and steady-state error is reduced from 1.9 to 0.8 V. Hence, the outcome represents that the closed-loop SBMG with SM controller is superior to closed-loop DPFC 16-bus mircro-grid with PR controller. Real, Reactive power at bus 6 of closed loop SBMG with SM controller are shown in the Figs. 20 and 21 respectively. Figure 22 shows bar chart comparison of Time domain parameter with PRC and SMC.

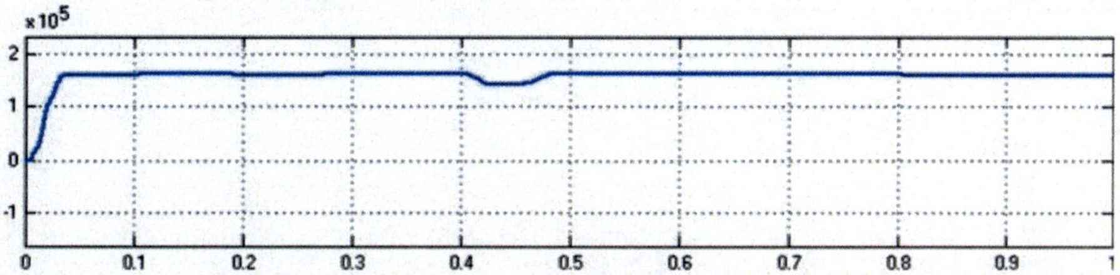

Fig. 20 Real power at bus 6 of closed-loop SBMG with SM controller

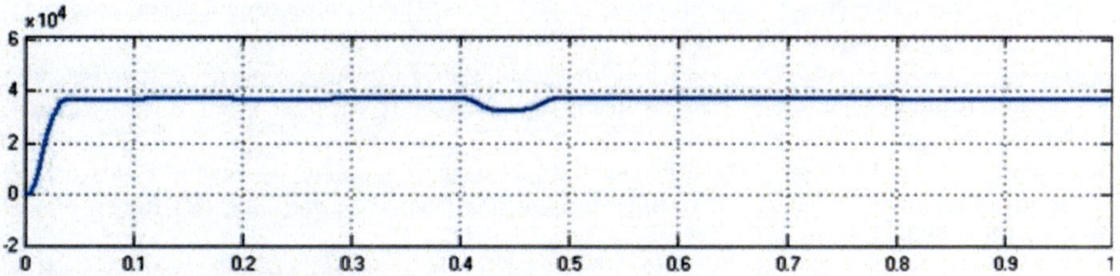

Fig. 21 Reactive power at bus 6 of closed-loop SBMG with SM controller

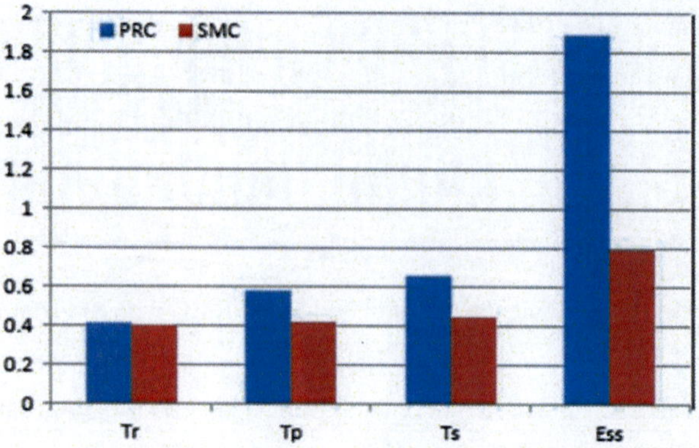

Fig. 22 Bar chart comparison of time-domain parameters

6 Conclusion

Closed-loop DPFC-based 16-bus micro-grid systems with PR and SM controllers are simulated and outcomes are presented. By using SMC, the rise time is reduced from 0.42 to 0.41 s; peak time is reduced from 0.58 to 0.43 s; settling time is reduced from 0.66 to 0.45 s; and steady-state error is reduced from 1.9 to 0.8 V. Hence, the outcome represents that the closed-loop SBMG with SM controller is superior to closed-loop SBMG with PR controller.

The present work deals with the simulation of closed-loop DPFC-based SBMG with PR and SM controllers. Closed-loop DPFC-based SBMG with MPC can be done in future.

References

1. Seifi K, Moallem M (2019) An-adaptive-PR-controller-for-synchronizing-grid-connected-inverters. IEEE Trans IE 66(3):2034–2043
2. Zaky MS, Metwaly MK (2017) A-performance-investigation-of-a-4switch3-phase -inverter-fed-IM-drives-at-low-speeds-using-fuzzy-logic & PI-controllers. IEEE Trans PE 32(5):3741–3753
3. Chen Y, Martínez-Ortega J, Castillejo P, López L (2019) A-homomorphic-based –multiple-data-aggregation-scheme-for-smart-grid. IEEE Sens J 19(10):3921–3929
4. Zhang J, Li L, Dorrell DG, Guo Y (2019) Modified-PI-controller-with-improved-steady-state-performance-and-comparison-with-PR-controller-on-direct-matrix- converters. Chinese J Electric Eng, 5(1):53–66
5. Kumar N, Saha TK, Dey J (2018) Control, implementation, and analysis of a dual two level photovoltaic inverter based on modified proportional resonant controller. IET-Renew Power Gen 12(5):598–604
6. VanDerBroeck CH, Richter SA, Bloh JV, DeDoncker RW (2018) Methodology for analysis and design of discrete time current controllers for three phase PWM converters. CPSS Tran Power-Electron Appl 3(3):254–264
7. Geng H, Xi X, Yang G (2017) Small signal stability of power system integrated with ancillary controlled large-scale DFIG based wind farm. IET-Renew Power Gen 11(8):1191–1198

8. Israr M, PandeyAK (2017) Modeling & control of utility grid connected solar photovoltaic array integrated system using MATLAB. In: 2017-Intererrnational, Conference, on Computation of Power, Energy Information & Communication (ICC-PEIC). Melmaruvathur, pp 500–505
9. Guimarães JS, RicardodeAlmeida B, Tofoli FL, deSouzaOliveira D (2018) Three phase grid connected WECS with mechanical power control. IEEE Trans Sustain Energy 9(4):1508–1517
10. Harrabi N, Souissi M, Aitouche A, Chaabane M (2018) Intelligent control of grid connected AC–DC–AC converters for a WECS based on T–S fuzzy interconnected systems modelling. IET P.E 11(9):1507–1518
11. Bahamani AK, Sreerama Reddy GM, Ganesh V (2016) Power quality Improvement in fourteen bus system using non conventional source based ANN controlled DPFC. Indonesian J Electric Eng Comput Sci 4(3)

Chapter 3
Comparative Review of MAC Architectures

Purra Dinesh, Kishore Sanapala, Grande Naga Jyothi, and R. Sakthivel

1 Introduction

In the recent times, there is rapid advancements in the computing platforms due to the evolutionary development occurring in microelectronics. Large-capacity data processing or image/video processing, audio signal processing are some real-time signal processing applications which have great demand [1, 2].

Area, delay, and energy utilization are the key performance aspects for all the microelectronic circuits/architectures used in the digital signal processors (DSP) [2, 3]. Some of the crucial steps involved in DSP are convolution, filtering of the inner products. MAC is a fundamental elementary unit in every digital processor. Multiplier plays a prominent role in developing MAC units. Repeated multiplication and accumulation processes are performed by the MAC unit to accomplish complicated operations in DSP [4–7]. The basic MAC structure with adder, multiplier, and the accumulator as the key constituent elements is shown in Fig. 1 [1]. In this, firstly, the partial products are produced by the multiplier and the response of the multiplier is delivered to the adder. The results of the multiplier and the previously accumulated results are added. Some of the adders like ripple carry adder and carry-save adder are generally used in implementation of DSPs. In order to test different MAC units, efficient design of multipliers and adders is necessary.

The rest of the paper is organized as follows. Section 2 presents the overview of fundamental arithmetic units used in the MAC architecture. Section 3 gives the comparative review on various MAC models. Conclusions from the comparisons are made in Sect. 4.

P. Dinesh · K. Sanapala (✉)
Department of ECE, Marri Laxman Reddy Institute of Technology and Management, Hyderabad, India

G. N. Jyothi · R. Sakthivel
School of Electronics Engineering, Vellore Institute of Technology, Vellore, India

N. Marriwala et al. (eds.), *Soft Computing for Intelligent Systems*, Algorithms for Intelligent Systems, https://doi.org/10.1007/978-981-16-1048-6_3

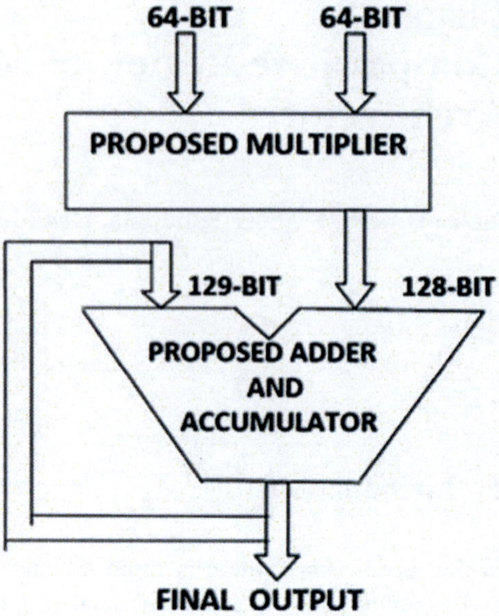

Fig. 1 Basic MAC architecture

2 Arithmetic Units of MAC

2.1 Adder

In digital signal processing computations, adder plays the essential role also as an analytical part of the MAC. Therefore, while choosing an adder, least delay, moderate power, and area efficiency are the key aspects. Many adders are proposed in the literature. Ripple carry adder is the basic fast adder obtained by cascading the 1-bit full adder blocks; the numbers with large bit length can be added by using this adder [1, 6, 8]. But the disadvantage of ripple carry adder is its more carry propagation delay. Carry bypass adder is another type of adder where the full adders are separated into groups. Carry skip adder is another type of adder where the name itself indicates that this adder is used to skip the logic. This adder contains a simple ripple carry adder. In order to command the latency which is assembled by the rippling effect, a carry look ahead adder is designed. Advantage of this adder is that it is used to remove the propagation time or transit time which exists in aligned adders. By adding 3 bits at the same instant of time, carry-save adder can be used. Delay can also be minimized using this structure.

2.2 Multiplier

A multiplier is a bottleneck used to examine the performance of a MAC at different stages. Figure 2 shows the basic 2 × 2 multiplier block used in MAC design. Multipliers are categorized into two types which are also known as clocked multipliers and array multipliers. To replicate the shift and multiplicand algorithm, the serial data multiplier is used. Commercially many series multiplier chips are accessible, which can be cascaded using many 1-bit multipliers. In this case, multipliers are arranged in a parallel manner. Serially the multiplicand bits are introduced. These are used in scaling by constant-type applications whereas serial processing is employed in correlation applications which require a serial multiplier. In simpler VLSI implementations, array multipliers are preferred for smaller size of operands. In array multipliers, the final output is achieved without registering the partial products. Hence, array multiplier output can be achieved quickly. Owing to some disadvantages like high complexity and testability in parallel multipliers, serial data multipliers are preferred in DSP.

Ripple carry array multiplier technique is similar to array multiplier technique. But here based on the obtained carry value, the partial product stages/levels go round to the next level from the previous level. In microprocessor circuits, Wallace multiplier, which is one of the finest tree multipliers, is used. Different algorithms can be operated by using these multipliers. Dadda multiplier is the modified form of the Wallace multiplier. Booth multiplier is another type of multiplier which is used in reducing the number of repetition steps. Modified Booth algorithm (MBA) is a type of multiplier which is also used in many MAC architectures. Since booth multipliers consume more power, these are not preferred for low power applications. Baugh-Wooley (BW) multiplier is mostly preferred for its higher energy efficiency [9, 10].

Fig. 2 Basic 2 × 2 multiplier used in MAC unit

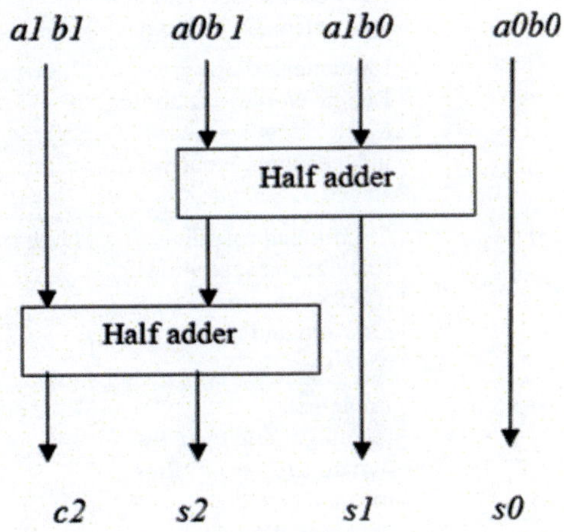

3 Comparison of MAC Models

In order to provide short and precise information, the comparison of various MAC models is shown in Table 1 with the methodology and their advantages and disadvantages. Table 2 shows the performance comparison of different MAC units with respect to power, delay, and area. From the comparisons, it is evident that all the designs compromise with one of the performance metrics in order to show the improvements with either of the metrics—power, area, and delay.

4 Conclusion

This paper presents the short and concise comparative review of different MAC architectures with respect to different performance aspects, advantages, and disadvantages. However, it is evident that all the proposed MAC models compromise with one of the performance parameters in order to show the improvements with either of the parameter—power, delay, and area. It is observed that for implementing the MAC models for advanced applications like machine learning requires high performance with acceptable energy consumption. In this case, the CNN-algorithm-based MAC units could be of key choice. The other category of applications where the processors consume more standby power needs the low-power MAC models with acceptable

Table 1 Comparative review of different MAC models

MAC design	Methodology	Advantages	Disadvantages
[1]	Final adder of the multiplier is replaced by a carry-save adder with new sign extension technique	High-speed area and energy efficient	Operating mode flexibility is limited
[4]	Implemented using Baugh-Wooley multiplier by clock gating the independent pipeline stages	Low power	Large area and more critical delay paths
[5]	Conditional rounding, scaling, and saturation (CRSS) portion is removed and involves re-pipelining	High speed	More number of gates required
[6]	Implements reconfigurable logic by dynamically controlling the data bits and component modules	Low power	Requires high clock frequency

(continued)

Table 1 (continued)

MAC design	Methodology	Advantages	Disadvantages
[7]	Flare outs of the barrel shifter control lines are half of the IBM scheme	Low power consumption	Delay is more
[8]	Two prototypes of an 8-bit,16-bit, and 32-bit MAC unit were introduced as FPGAs	Less area and low power consumption	Limited operating frequency
[9]	The modified Booth algorithm and three-dimensional technique are used in solving the problems of delay	Low power consumption	Large area
[10]	Single instruction multiple data (SIMD) and the multiply with implicit-accumulate (MIA) features	High performance Low power	Number of 16-bit MACs per cycle are more when compared to SA-110 MACs
[11]	A fast half-carry-save Montgomery modular multiplication algorithm with dual-core multiplier accumulator	High speed Less area Low power consumption	Additional hardware requirement
[12]	Fixed/floating-point merged architecture to support 16-bit half precision floating-point multiplication	Less area, Low power consumption	Delay is high
[13]	Using weight sharing, the full range of values in a trained convolution neural network (CNN) are put in MAC	Low complexity, area, and power consumption	More CNNs are required
[14]	Approximation is introduced to both the multiplication and accumulation stages	Reduced area power product	Inaccurate for some input combinations
[15]	Custom 32-bit floating-point data format is used to change the number of computations	Less complex	Requires high operating frequency
[16]	Packing two MAC operations into one DSP block	High performance	Area is doubled

Table 2 Performance comparison of different MAC models

MAC design	Technology	Frequency (MHz)	Bit size	Power (mW)	Delay (ns)	Area
[1]	65 nm, 1.1 V	–	32	15.62	2.06	43,397 μm^2
[4]	TSMC 65 nm	584.8	16	0.780	1.71	10,730 μm^2
[5]	0.5 μm, 3.3 V	–	256	–	12.7	21.154 K gates
[6]	–	16.93	512	118.25	–	–
[8]	XSA-100 V1.1 board	–	8	–	101.85	100 CLB slices
			16		234.37	374 CLB slices
			32		453.92	1385 CLB slices
[10]	180 nm, 1.3 V	600	32	86	–	655,200 μm^2
[11]	TSMC 90 nm	369	32	9.76	2.76	47.2 K gates
[12]	STM-90 nm	–	32	14.07	0.8	42,710.90 μm^2
[13]	PDK 45 nm ASIC	100	16	~100	–	~100 Kgates
[14]	TSMC 65 nm, 1 V	–	8	0.2609	–	483 μm^2
[16]	Xilinx Virtex7 485 T FPGA	280	8, 16	–	–	–

performance. In this case, the approximate computation-based MAC units are the better choice for the designer.

References

1. Hoang TT, Själander M, Larsson-Edefors P (2010) A high-speed, energy-efficient two-cycle multiply-accumulate (MAC) architecture and its application to a double-throughput MAC unit. IEEE Trans Circuits Syst I 57(12):3073–3081
2. Jyothi GN, Sanapala K, Vijayalakshmi A (2020) ASIC implementation of distributed arithmetic based FIR filter using RNS for high speed DSP systems. Int J Speech Technol 23:259–264
3. Sanapala K, Yeo SS (2018) Schmitt trigger-based single-ended 7T SRAM cell for internet of things (IoT) applications. J Supercomput 74:4613–4622
4. Warrier R, Vun CH, Zhang W (2014) A low-power pipelined MAC architecture using Baugh-Wooley based multiplier. In: IEEE 3rd Global Conference on Consumer Electronics (GCCE),

Tokyo, pp 505–506
5. Smith SC (2005) Development of a large word-width high-speed asynchronous multiply and accumulate unit. Integr VLSI J 39(1):12–28
6. Tatas K (2007) Architecture design of a coarse-grain reconfigurable multiply-accumulate unit for data-intensive applications. Integr VLSI J 40(2):74–93
7. Pillai RVK, Al-Khalili D, Al-Khalili AJ (2000) Low power architecture for floating point MAC fusion. IEE Proc Comput Digital Tech 147(4):288–296
8. Lee S, Kim D, Nguyen D, Lee J (2019) Double MAC on a DSP: boosting the performance of convolutional neural networks on FPGAs. IEEE Trans Comput Des Integr Circuits Syst 38(5):888–897
9. NagaJyothi G, SriDevi S (2017) Distributed arithmetic architectures for fir filters—a comparative review. In: IEEE International conference on wireless communications, signal processing and networking (WiSPNET), Chennai
10. Chandralekha V et al Design of 8 bit and 16 bit Reversible ALU for low power applications. In: IEEE 5th international conference on computing communication and automation (ICCCA). Noida, India (2020)
11. Abdelgawad A, Bayoumi M (2007) High speed and area-efficient multiply accumulate (MAC) unit for digital signal processing applications. In: IEEE international symposium on circuits and systems. New Orleans, LA, pp 3199–3202
12. Farooqui AA, Oklobdzija VG (1998) General data-path organization of a MAC unit for VLSI implementation of DSP processors. In: IEEE international symposium on circuits and systems (ISCAS). Monterey, CA, pp 260–263
13. Liao Y, Roberts DB (2002) A high-performance and low-power 32-bit multiply-accumulate unit with single-instruction-multiple-data (SIMD) feature. IEEE J Solid-State Circuits 37(7):926–931
14. Ding JH, Wang D, Tan H (2018) A high-performance RSA coprocessor based on half-carry-save and dual-core MAC architecture. Chinese J Electron 27(1):70–75
15. Zhang H, Lee HJ, Ko S (2018) Efficient fixed/floating-point merged mixed-precision multiply-accumulate unit for deep learning processors. In: IEEE international symposium on circuits and systems (ISCAS). Florence, pp 1–5
16. Garland J, Gregg D (2017) Low complexity multiply accumulate unit for weight-sharing convolutional neural networks. IEEE Comput Archit Lett 16(2):132–135
17. Adams E, Venkatachalam S, Ko S (2019) Energy-efficient approximate MAC unit. In: IEEE international symposium on circuits and systems (ISCAS). Sapporo, Japan, pp 1–4
18. Lv A, Wang C, Hou L, Zeng Z, Guo J, Jiang N (2018) An arithmetic unit and multiplying accumulation unit of a custom floating point data format. In: IEEE 3rd international conference on integrated circuits and microsystems (ICICM)

Chapter 4
Proposal of ASLR for Voice Disorders

Ravi Gorli, Ch. Demudu Naidu, and G. Pandit Samuel

1 Introduction

Kevin Ashton (1999) gave a novel-paradigm framing as Internet-of-Things (IoT) [1]. The basic motto of IoT is 'the pervasive presence around us for a variety of things may be objects which, through by unique addressing schemes, which are able to interact with each other and cooperate with neighbours to reach common goals' [2]. Internet of Things (IoT) is a network intelligence which connects different things via Internet on the need of information exchange and communicates with each other through the information sensing devices in accordance with agreed protocols. It achieves the goal of intelligent tracking, monitoring, identifying, locating and managing things [3]. IoT has three characteristics: comprehensive perception, reliable transmission and intelligent processing [4]. The IoT is creating huge billion and trillion things to the network for communicating with each other.

IoT offers a means of things which become context aware and they were able to sense, interact, communicate and exchange data, information and knowledge [5]. The world is laying steps towards smart technologies where it has entered into maximum fields such as smart health care, manufacturing, smart grids, smart city, smart agriculture, smart vehicles and smart waste management and as of now many other fields are also getting equipped with IoT [6].

In healthcare monitoring, a lot of research is going in the implementation of smart heathcare applications, smart wearables are introduced where different parameters such as ECG, photoplethysmography (PPG), HR, BP, body temperature and galvanic

R. Gorli (✉) · Ch. Demudu Naidu · G. Pandit Samuel
Anil Neerukonda Institute of Technology and Sciences, Visakhapatnam, India
e-mail: ravi.it@anits.edu.in

Ch. Demudu Naidu
e-mail: chdnaidu.it@anits.edu.in

G. Pandit Samuel
e-mail: samuel.it@anits.edu.in

N. Marriwala et al. (eds.), *Soft Computing for Intelligent Systems*, Algorithms for Intelligent Systems, https://doi.org/10.1007/978-981-16-1048-6_4

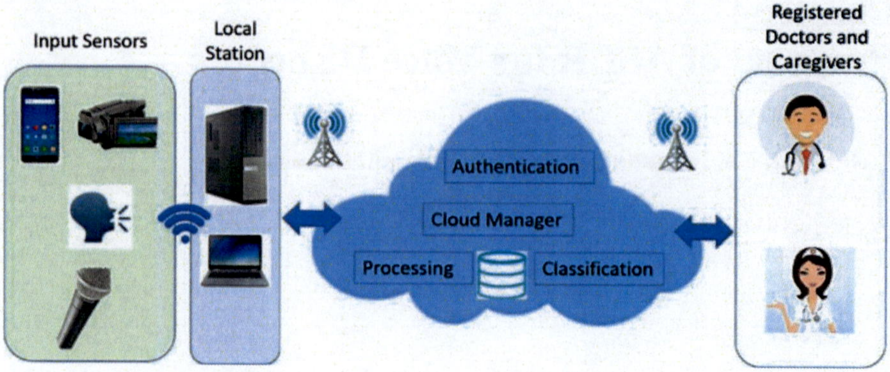

Fig. 1 Model for smart voice disorder [1]

skin response (GSR) are measured using a vest connected to the human [7]. Clinical applications for chronic illness detection, including cardiac arrhythmia, hypertension, heart failure, sleep disorder, voice disorder emotional problem, cognitive impairment and functional decline, are also monitored using the smart wearables connected to the human, where they continuously monitor [7] (Fig. 1).

1.1 Voice Signals

Voice signals produce sound with pressure of air vibrations ex-hauled from lungs and shaped and modulated with vibrations of vocal-fold and vocal-tract resonance. Production of voice is leaded by different structures such as respiratory system which is intensity influenced and the larynx production of voice cornerstone, the vibration of vocal folds determines and the vocal tract constitutes the pharynx, nasal and paranasal cavities, timbre change, sonority responsible and laryngeal sound resonance [8].

1.2 Automatic Speech Recognition

Automatic speech recognition (ASR) is the process of extracting the text from the words spoken. It was still a task of challenged atmosphere where the spoken persons may pronounce differently with different accents and dialects with styles of different rates with a nature of several emotions, and due to these it leads to a variability with noise in environment of several microphones and reverberation of different devices used for recording of voice signals. The architecture of ASR is given in Fig. 2 where the speech signal is followed by feature extraction with MFCC and the

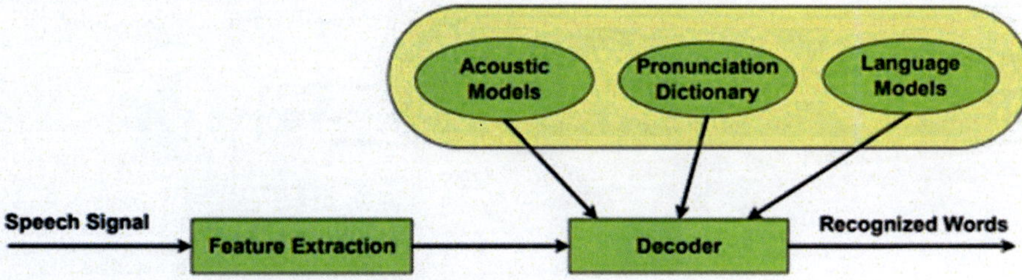

Fig. 2 Architecture of automatic speech recognition [1]

decor which consists of acoustic models, pronunciation dictionary, language models and the recognized words saved in the database [9].

1.2.1 Deep Learning for IoT

Deep learning is considered as a main stream for Internet-of-Things applications. Over machine learning, deep learning results in a better performance and the iot operations are done using huge high end data so the deep learning processing makes it more comfortable when compared with ML techniques. Automatic extraction of new features is done in DL. The performance of ML is based on feature accuracy which is extracted after identification. High-level feature learning is the process of identifying human's faces from images and wording language from speech voices, and multimedia processing information is done more effectively with DL [9].

1.2.2 MFCC

Mel-frequency central coefficient (MFCC) is common base feature extraction technique for ASR, parts of human's production-ed speech and perception-ed speech are mimiced with MFCC, logarithmic perception of pitch and loudness of human's auditory system and tries to eliminate speaker-dependent characteristics by excluding the harmonics and fundamental frequency. MFCC utilizes the change in the feature vector as a feature for dynamic nature representation of speech [9] (Fig. 3).

1.3 GMM Classifier

Gaussian mixture model (GMM) is a probability density function for parametric which are represented with weighted sum of Gaussian density components. GMM are used widely in models of parametric and biometric system feature extraction, and special features of speaker recognition system of vocal tracts are represents with GMM. The equation represents the GMM model [10],

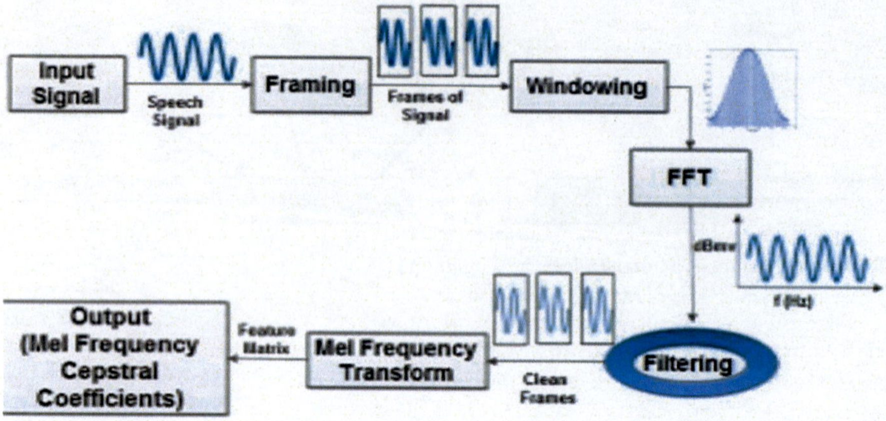

Fig. 3 MFCC [1]

$$P(\frac{x}{\lambda}) = wig(\frac{x}{\mu i \sum i}) \tag{1}$$

$$P(x) = \frac{1}{\sigma\sqrt{2\pi}} e^{-(x-\mu)^2/2\sigma^2} \tag{2}$$

With mean vector μi and co-variance matrix Σi, the mixture weights satisfy the constraint that wi M $i = 1 = 1$. The mean vectors, co-variance matrices and mixture weights from all component densities parameterize the complete GMM. The notation collectively represents these parameters.

$$\lambda = wi, \mu i, \sum i; i = 1, \ldots, M \tag{3}$$

MFCCs GMM distribution is as shown in Fig. 6. Hence, D-dimensional MFCCs can be modelled by GMM model of M-mixtures [11].

1.4 Evaluation Methods:

For evaluation of performance, we use P—precision, R—recall, A—accuracy, F1-measure, true positive rate (TPR) and false positive rate (FPR) [11]:

$$\text{P} = \text{TP}/(\text{TP} + \text{FP}) \tag{4}$$

$$\text{R} = \text{TP}/(\text{TP} + \text{FN}) \tag{5}$$

$$\text{A} = (\text{TP} + \text{TN})/(\text{TP} + \text{FP} + \text{TN} + \text{FN}) \tag{6}$$

$$\text{F1 measure} == (2 * \text{Precision} * \text{Recall})/(\text{Precision} + \text{Recall}) \tag{7}$$

$$\text{TPR} = \text{TP}/(\text{TP} + \text{FN}) \tag{8}$$

$$\text{FPR} = \text{FP}/(\text{FP} + \text{TN}) \tag{9}$$

1.5 Database

In our basic research, we have worked with databases such as SVD database [12] and MEEI database [13]. The SVD database contains four pitch intonations of normal, low, high and medium of three vowels a, i, e [14] and the MEEI consists vowels of a and with two different frequencies from 25 to 50 kHz [15].

2 Proposed Framework

2.1 Brief Framework

In the present world, a lot of diseases are dragging the human into many illness disorders which may affect reducing their life span, infact the humans are affected internally or externally from head to toe. Identifying the cause and stage of disease plays a key role in diagnosis of the disease. As per the ancient technologies, doctors are not up to mark of identifying the cause and stage of the disease. As the treatment goes with the face-to-face interaction, identifying the problem is difficult for identifying as the patient does not have enough knowledge of symptoms.

We have taken voice disorder identification as a main motto of our research problem along with implementing of a new model known as automatic speech linguistic recogniser (ASLR) along with analysis for voice disorder identification and analysis. The ASLR plays a major role for the identification of the voice disorder in which, in fact, the manual face-to-face interaction of patient with doctor is replaced with the smart technologies involvement. So the patient can be anywhere in the world or can be busy with his work but his treatment goes on.

As the voice device connected to the human recognises the voice and forwards the signals to the server where the speech linguistic model identifies whether the patient is facing with any voice disorder, the voice signal identifies a disorder signal and then the model forwards interacting the specialists related to it for the diagnosis of the disorder and the prescription given by the doctor is forwarded to the human. So the treatment goes on and the model continuously monitors the voice disorder of person at every stage and in counter identifies and diagnose the disease carried on until the person gets relief.

We have proposed a model for one of the applications of smart health, i.e. implementation of voice disorder identification with smart sensors. Figure 4 describes the model with different stages which are interconnected to each other. The voice signals are captured from the person with connection of voice sensors, they will record the voice signals within a span of intervals and the signals are forwarded to the cloud where the signal processing will be done using the signal processing algorithm and the pre-processed signals are saved in the cloud where the initial identification whether the person is facing with any disorder or not, if facing then the signals are forwarded to the main section where the classification of disorder will be done using the Gaussian mixture model and then depending on the type of disorder the vice sample goes to the particular specialist along with a notification to the person. In return, the doctor's prescription with treatment details will be received from the doctor and these are saved along with compassion of past treatments and these prescriptions will be forwarded to the patient.

- Stage 1: The voice samples are received from person using any smartphone or any voice recording smart devices such as Amazon Echo, Google Assistant, Apple Siri, IBM Assistant Recording voice or any wireless nanosmart sensors connected to the human in wireless to the mobile phone. Samples from the person can be either forwarded from the human to the cloud or automatically using smart sensors connected to the human which can also send the signals to the cloud without any intervention to the human.

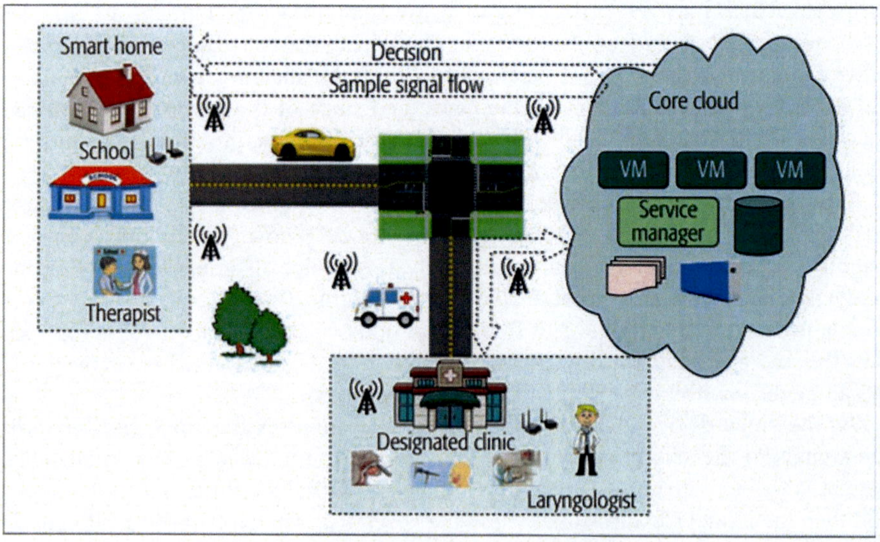

Fig. 4 The proposed smart healthcare framework [1]

- Stage 2: The pre-processing of signals is done where the correction of signals is done along with differentiation of the voice, unvoiced or silent. Here the speech signals are classified based on the characteristics of the signals with different correction algorithms.
- Stage 3: Identification of voice disorder is done, where the main stream of voice disorder identification is done using different machine learning algorithms with steps such as firstly identification of voice disorder whether the person is facing any voice disordered or not with initial classification algorithms and then deep classification of finding the exact voice disorder and vice versa specialist proposal.
- Stage 4: Proposal of medication with past information and also consulting the specialist automatically and vice versa to the patient's prescription.

Flow representation of proposed framework is shown in Fig. 5. The flow representation shows different stages starting from the request service followed by registration and the request grants from the server after which the service is accepted and the voice signals are uploaded and different tasks are performed on the cloud server and the results are forwarded to the client and then the SLP comes into contact via the server and later the following consultation occurs depending on the disease and the patient via contact with doctors.

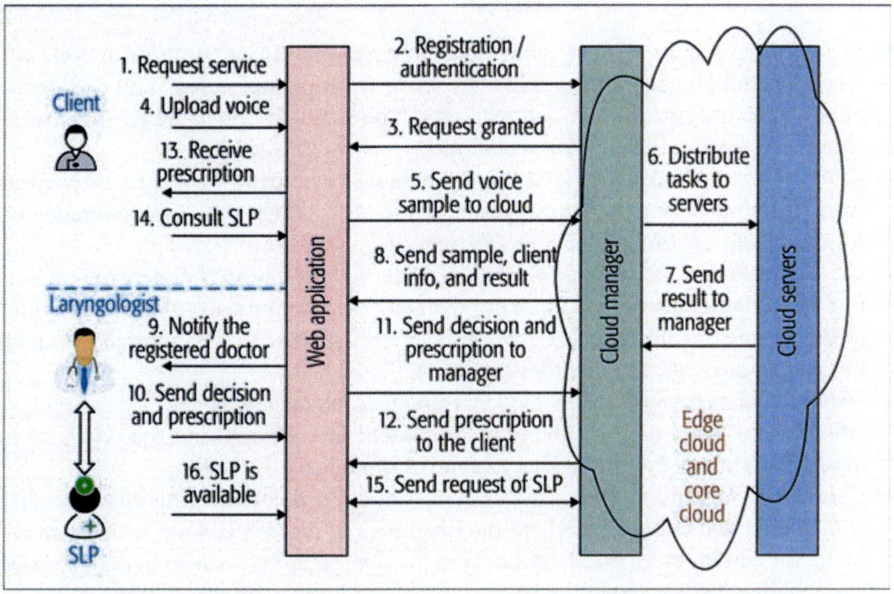

Fig. 5 Task flow of the proposed voice disorder detection and treatment framework [1]

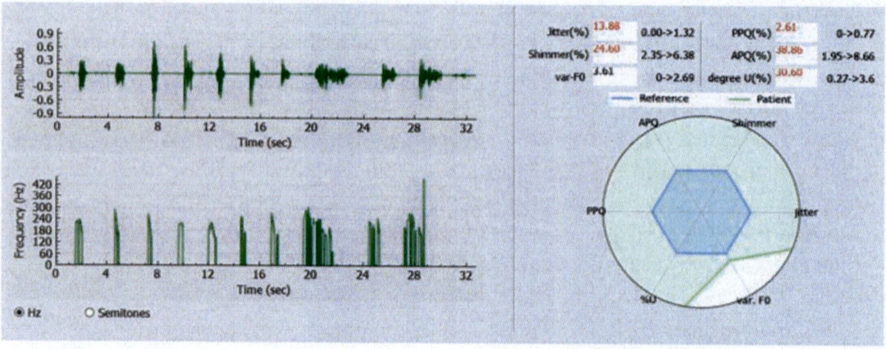

Fig. 6 Phonation analysis [1]

2.2 *Software Development*

A new software with the name ASLR is proposed for incorporating all the stage operations from the starting task of receiving of signals from sensors or through an application to the end of the classification and diagnosis. The development of software is proposed in different stages.

- In the initial stage, a smart application is developed in an Android mobile for receiving the signals, either uploaded voice from person along with additional option of recording the voice automatically. The received signals are pre-processed using different pre-processing techniques.
- Second stage is the development of voice database where the voice signals received from different sensors and the application are all stored in the central storage in a format of segregation of uid, sex and age.
- On this database, the classification of disorder identification is done where it will find out whether the patient is facing with any disorder or not and then according to the classified information database is again stored with the columns, namely, uid, name, sex, age and disorder (yes/not).
- Using the above database the deep learning techniques are applied for exact classification of what type of disorder the patient is suffering and the database is modified in a new column added as type of disorder.
- Using the disorder data, which depends on the patient disorder, particular specialist is allocated and connected where the diagnosis of disease is done with a request to doctor and in return prescription from the doctor and it is saved in the database and which is forwarded to the particular patient to take the meditation.

2.3 Creation of Database

We have a proposal for creation of new voice database with cloud. The database consists of voice signals collected among people around different age groups, genders, disorders and so on which is collected via smart sensors connected to people and also data collected from the smart application through smartphones and tabs. The collected data is stored in an ordered format after pre-processing of voice signals. The complete database is used for the classification of disorder identification with different machine learning techniques applied to the data and after the classification, again the classified data is stored back into the voice data.

3 Results and Discussion

In our experiment, we performed on SVD database which is standardized, feature extraction with MFCC and classification with Gaussian mixture model (GMM). The sample voice is divided into training and test databases where the GMM classifier uses 12 MFCCs, energy, and their deviates. Use of different number of Gaussian (power of 2). We have taken 101 persons' date from SVD database which consists of 38 pathological and 63 healthy voice data and the features extracted from MFCC, Shimmer and Jitter and we have checked which type of mixture we have to use from 3 to 15 where we have got better performance for training set with 5 mixture and test data with 11 mixture so we have chosen best mixture with calculation of their performance comparing with each other. Then the average classified rate is calculated, we got 98.4 with train data and 95.2 with test data and the same process is applied with HMM model where we got 96 for train data and 93 with test data and then we have applied with ANN model where we got 90 with train data and 92 with test data. So comparing the three models, we got better performance with GMM model, so we will choose GMM model. GMM classification for test data gives a correct higher classification for healthy and pathological voices. The false negative = 2.4, i.e. pathological speech data was misclassified and as is healthy the case, true negative rate (TN) = 97.6 denotes more correctly pathological were classified. The above models are applied on a sample data and the actual model is implemented in future with comparison with GMM and SVM using the complete SVD database and our own database.

3.1 Graphical Comparisons

We have identified the disorders in another method finding the features of healthy and disorder persons with comparison of phonation, prosody, articulation, intelligibility and DDK.

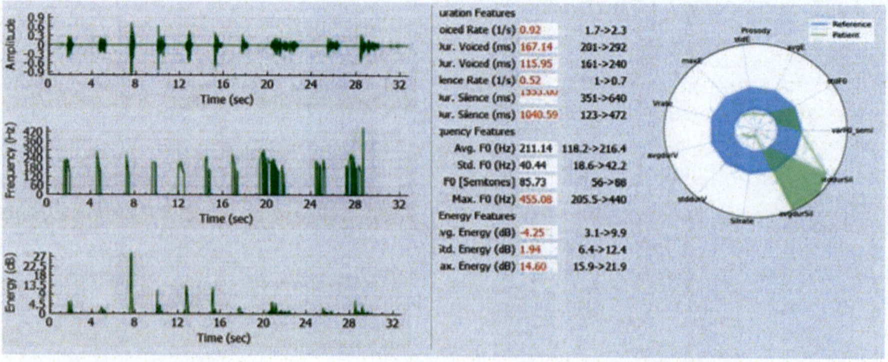

Fig. 7 Prosody analysis [2]

3.1.1 Phonation Analysis

The phonation analysis has been extracted by comparison of healthy person's and disorder person's signals where the graph shows the difference in the circle with the hexagon and the boundary values and the results are shown in Fig. 6.

3.1.2 Prosody Analysis

The Prosody analysis has been extracted by comparison of healthy and disorder persons' signals where the graph shows the difference in the circle with the hexagon and the boundary values and the results are shown in Fig. 7.

3.1.3 Articulation Analysis

The articulation analysis has been extracted by comparison of healthy and disorder persons' signals where the graph shows the difference in the circle with the hexagon and the boundary values and the results are shown in Fig. 8.

3.1.4 Intelligibility Analysis

The intelligibility analysis has been extracted by comparison of healthy and disorder persons' signals where the graph shows the difference in the circle with the hexagon and the boundary values and the results are shown in Fig. 9.

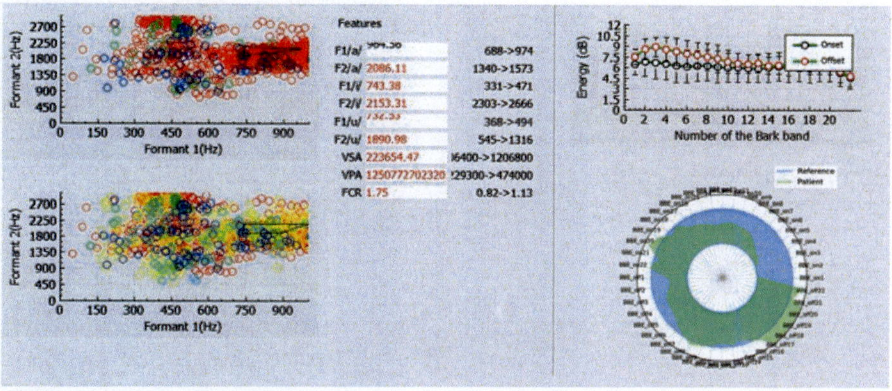

Fig. 8 Articulation analysis [3]

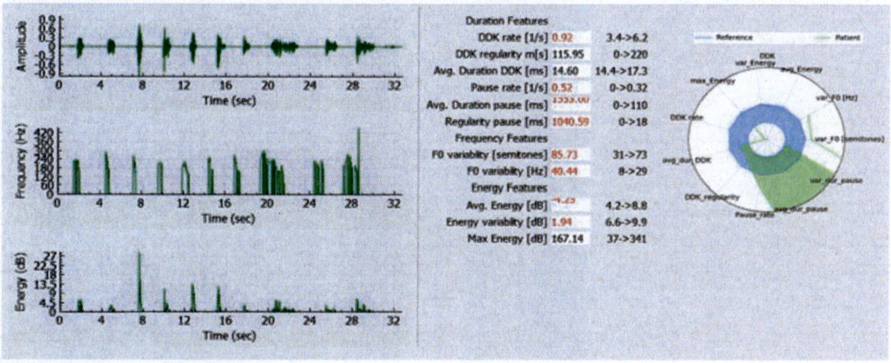

Fig. 9 Intelligibility analysis [4]

3.1.5 DDK Analysis

The DDK analysis has been extracted by comparison of healthy and disorder persons' signals where the graph shows the difference in the circle with the hexagon and the boundary values and the results are shown in Fig. 10.

3.1.6 Evaluation Analysis

The evaluation analysis has been extracted by comparison of healthy and disorder persons' signals where the graph shows the difference in the circle with the hexagon and the boundary values and the results are shown in Fig. 11.

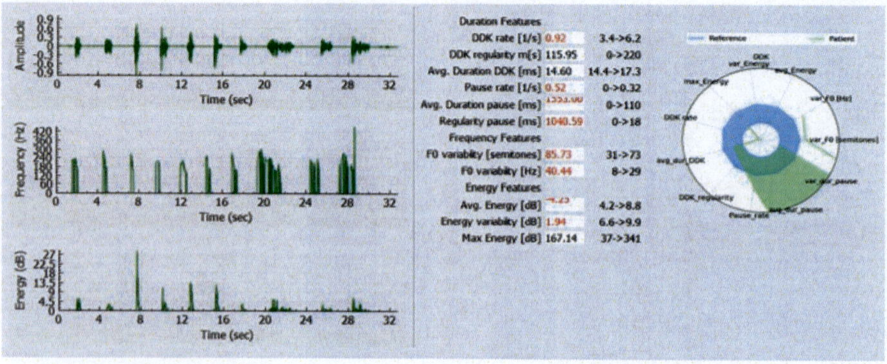

Fig. 10 DDK analysis [5]

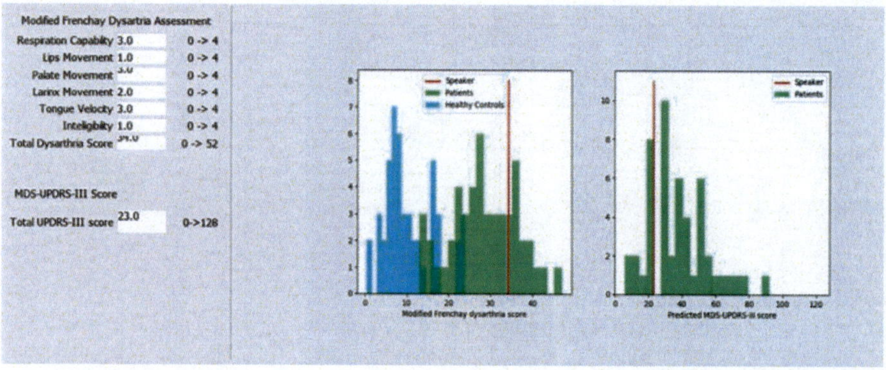

Fig. 11 Evaluation analysis [6]

3.2 Smart Healthcare Framework

A smart healthcare framework for voice disorder with automation is proposed for implementation of a environment where the smart sensors and smart devices in contact with the person will automatically detect the voice signals and the pre-processing of voice signals is done at the initial stage. In this model, two stages of voice disorder identification are proposed. First stage is the initial stage, whether person is facing with any voice disorder or not. Followed by the deep classification of actual disorder at the second stage vice versa, connected environment of doctors.

4 Conclusion

A smart healthcare framework for voice disorder with automation is proposed for implementation of an environment where the smart sensors and smart devices in contact with the person will automatically detect the voice signals and the pre-processing of voice signals is done at the initial stage. In this model, two stages of voice disorder identification are proposed. First stage is the initial stage, whether person is facing with any voice disorder or not. Followed by the deep classification of actual disorder at the second stage vice versa, connected environment of doctors and patients with cloud is equipped. In the result section, we have done with a sample data in which we have taken 101 samples from the SVD database where we got GMM performance well compared with ANN and HMM models. The actual proposed model with implementation is followed in the next publication with GMM model and SVM with full database and our database will be on future.

References

1. Agrawal S, Das ML (2011) Internet of things—a paradigm shift of future Internet applications. In: 2011 Nirma University International Conference on Engineering, Ahmedabad, Gujarat, pp 1–7
2. Iera A, Floerkemeier C, Mitsugi J, Morabito G (2010) The internet of things [Guest Editorial]. IEEE Wirel Commun 17(6):8–9
3. Stankovic JA (2014) Research directions for the internet of things. IEEE Internet Things J 1(1):3–9
4. Liu T, Lu D (2012) The application and development of IoT. In: Proceedings of International Symposium Information Technology in Medical Education (ITME), vol 2, pp 991–994
5. Chen S, Xu H, Liu D, Hu B, Wang H (2014) A vision of IoT: applications, challenges, and opportunities with China perspective. IEEE Internet Things J 1:349–359
6. Su K, Li J, Fu H (2011) Smart city and the applications. In: 2011 International Conference on Electronics, Communications and Control (ICECC), Ningbo, pp 1028–1031
7. Sundaravadivel P, Kougianos E, Mohanty SP, Ganapathiraju MK (2018) Everything you wanted to know about smart health care: evaluating the different technologies and components of the internet of things for better health. IEEE Consumer Electron Mag 7(1):18–28
8. Cesari U, De Pietro G, Marciano E, Niri C, Sannino G, Verde L (2018) Voice disorder detection via an m-health system: design and results of a clinical study to evaluate Vox4Health. BioMed Res Int Article ID 8193694, 19 pages, 2018U
9. Lee T et al Automatic speech recognition for acoustical analysis and assessment of cantonese pathological voice and speech. In: 2016 IEEE International Conference on Acoustics, Speech and Signal Processing (ICASSP), Shanghai, pp 6475–6479
10. Chauhan PM, Desai NP (2014) Mel frequency cepstral coefficients (MFCC) based speaker identification in noisy environment using wiener filter. In: 2014 International Conference on Green Computing Communication and Electrical Engineering (ICGCCEE), Coimbatore, pp 1–5
11. Martinez D, Lleida E, Ortega A, Miguel A, Villalba J (2012) Voice Pathology Detection on the Saarbrücken Voice Database with Calibration and Fusion of Scores Using MultiFocal Toolkit. https://doi.org/10.1007/978-3-642-35292-811

12. Usama M, Ahmad B, Wan J, Hossain MS, Alhamid MF, Hossain MA (2018) Deep feature learning for disease risk assessment based on convolutional neural network with intra-layer recurrent connection by using hospital big data. IEEE Access 6:67927–67939
13. Massachusetts Eye and Ear Infirmary, Elemetrics Disordered Voice Database (Version 1.03), Voice Speech Lab., Boston, MA, USA (1994). Available: https://www.kayelemetrics.com/
14. Pujol P et al (2005) Comparison and combination of features in a hybrid HMM/MLP and a HMM/GMM speech recognition system. IEEE Trans Speech Audio Process 13(1):14–22
15. Barry WJ, Putzer M Saarbrucken Voice Database. Available: https://www.stimmdatenbank.coli.uni-saarland.de/. Accessed 10 May 2018

Chapter 5
Methodological Analysis with Informative Science in Bioinformatics

Sahil Jindal, Nikhil Marriwala, Archit Sharma, and Rhythm Bhatia

1 Introduction

The term bioinformatics coined in 1865 is often from the Mendel's discovery of genetic inheritance [1]. Bioinformatics is a knowledge base field that evolves methods and software package tools for understanding biological data. Bioinformatics is the application of computing within the biological field to urge information from the biological information. As a knowledge base domain of science, bioinformatics integrates computing, statistics, arithmetic and engineering technologies to review and elaborate biological information. In the current years, fast evolution in genetics and proteomics has triggered great deal of information. Bioinformatics plays a crucial role in gathering, distinguishing, analysing, storing and classifying the genetic information [1].Computational biology and bioinformatics square measures are involved with the employment of computational techniques to grasp biological phenomena and to accumulate and utilize biological knowledge, increasingly large-scale knowledge [2]. Artificial intelligence has been seeking a great deal of attention because it tries to duplicate human intelligence for analysing complicated knowledge around North American nation. The two major subsets of AI named as machine learning and deep learning have created a great deal of pleasure within the analysis community for its viable exploitation in several fields [3].

Often, complicated biological phenomena involve several biological aspects, and therefore cannot be explained in one knowledge sort. For this reason, bioinformatics involves an interlinked analysis of many totally different knowledge varieties and will

S. Jindal (✉) · A. Sharma · R. Bhatia
University Institute of Engineering and Technology, Kurukshetra University, Kurukshetra, Haryana, India

N. Marriwala
Department of Electronics and Communication Engineering, Kurukshetra University, Kurukshetra, Haryana, India

N. Marriwala et al. (eds.), *Soft Computing for Intelligent Systems*, Algorithms for Intelligent Systems, https://doi.org/10.1007/978-981-16-1048-6_5

provide a holistic understanding of sophisticated biological phenomena [4]. Evolution, particularly in humans or in alternative organisms that have large populations, may be a very tough and complicated system to model. As a result, bioinformatics can utilize the prognostic power of machine learning, biologists will peek into the longer term and see what the foremost possible population outcomes are, or what chances an explicit population's DNA can evolve into a specific version [4]. Genomics is one among the foremost necessary fields in bioinformatics where the amount of sequences on the market is increasing exponentially. These information obtained have to be compelled to be processed so as to get useful data [5].

Bioinformatics involves the process of biological knowledge mistreatment approach-supported computation and arithmetic. The biological knowledge has fully grown exponentially in recent times resulting in two problems, one issue is of economical information storage and therefore the second issue deals with the finding of helpful information using deep learning from the huge databases. And the feature learning is enabled mechanically by deep learning that represents a machine learning technique [6]. Deep learning is used to merge feature extraction and makes a lot of reconciling and is compatible with prediction and creates superior progress in handling knowledge of next-generation sequencing [7]. As the part of drug development, various resources such as clinical studies, high-resolution medical pictures, electronic medical records, genomic profiles, etc. are taken into consideration to assist the whole process. Medical researchers that have in-depth knowledge sets may be analysed by robust AI systems [8]. In short, it is optimum to say that bioinformatics may be a wealthy vein for knowledge science as a result of the huge amounts of information.

2 Artificial Intelligence in Bioinformatics

Artificial intelligence is a part of engineering science which has been specialized in handling issues by laptop scientists through the employment of heuristic and probabilistic approaches since 1950s. AI approaches handle issues wherever there's no demand for the fully demonstrably correct or best answer and a strong constraint. However, there is a necessity for a solution that is healthier than one presently famous or that is suitable among the outlined constraints or a weak constraint. As long as several issues in bioinformatics do not have a powerful constraints, there is many scope for applying AI techniques to a variety of bioinformatics issues [9]. Bioinformatics may be a major helper of the recent advancements in computer science (AI). As associated with nursing knowledge base field of science and technology, bioinformatics aims to develop strategies, tools and software package to enhance our understanding of biological information [10].

Bioinformatics is the gathering, organization and analysis of a large quantity of biological knowledge and AI that recently shows a progress within the field of bioinformatics, like molecular structure prediction. Artificial intelligence has found increasing applications within the field of bioinformatics [11]. There are basically two reasons initiated here, which define the role of artificial intelligence involved in the growth of bioinformatics, i.e. the primary one antedate to the starting of computing, when once living organisms supplied the information for the planning of process strategies that were more favoured as source of real information, as well as the tolerance of unsure and inaccurate information, auto-adaptation and learning skills and the other antedate to the start of the century when a data-intensive science field once started as biology, with the disclosure of diverse instruments scanning molecules like bacteria, viruses and even genes at the molecular level [12].

Bioinformatics conjointly has vital potential in resolution population and biological process genetic science queries. Evolution, particularly in humans or in different organisms that have huge populations, is an especially lengthy and sophisticated system to model. As a result, bioinformatics is being used to harness the prophetic power of computer science, due to which the biologists will be able to peak into the longer term and see what the foremost possible population outcome is going to be, or what possibilities an exact population's deoxyribonucleic acid can evolve into a selected version. Bioinformatics has been given a spotlight amid COVID-19 [4]. The structural bioinformatics tools have the applications of AI and have effective methodologies to style active novel compounds and drugs against medical specialty disorders like cancer and other mutated virus diseases arising in today's era through in silico means, that is, by exploiting the tools having AI applications [12].

The applications of AI in bioinformatics have the power to annotate the information towards logical conclusions [13]. The simulations of various models, annotations of biological sequences, process drug planning, virtual screening and factor prediction will expeditiously be predicted through the association of AI and bioinformatics [8]. The advancement in AI and bioinformatics has important contribution in health sciences as well as immune informatics [14] and vaccinology and ends up in the clinical bioinformatics, high-throughput screening, sickness bar and medicine. In the context of bioinformatics, AI is often helpful in facilitation of any clinical analysis as well as AI and heuristic ways will give a key solution for the new challenges that are exhibited into a data massive science by the ongoing alteration of biology [15].

3 Machine Learning in Bioinformatics

Machine learning, a subfield of technology involving the event of algorithms that learn the different possible ways to form predictions with supporting knowledge, includes a variety of rising applications within the field of bioinformatics. The goal is to use advance procedures of biology and informative science to analyse the behaviour of molecules and therefore the chance of obtaining a remedy which will benefit the society in all ways. Basically the two problems raise during exponential rise of the quantity of biological knowledge defined as economical data storage and management and the extraction of helpful data from the huge knowledge databases

[5]. Machine learning techniques in different area units of biological domain are applied for data extraction from knowledge and its utilization for benefit of society.

Machine learning has been employed in bioinformatics for prediction and discovery with the increase of high handiness and sort of molecular-level knowledge [4]. One space of chance for machine learning is within the prediction of genomic options that describe any genomic region with some annotated form, like sort of a sequence. For different bioinformatics applications, a subfield of AI known as machine learning has become a strong tool. In the medical domain, a lot of data is generated on daily basis and to extract the important features from there, machine learning has been an extensively used and successful methodology. Machine learning makes predictions that is based on the best fit model and uses the training data to discover basic patterns and build computational models that work in the area of bioinformatics. As well as, some good algorithms associated with machine learning such as random forests, support vector machines, etc. make an effort to learn the various aspects of systems biology, genomics and different other domains [16]. As considering the good execution of algorithms of machine learning, the features are the important parts that represent the type of data and used for extracting some useful information in bioinformatics [17]. So in order to identify the appropriate features that are used in specific areas, a design of features is done by domain expertise or by the human engineers.

Machine learning strategies are particularly helpful at prediction and pattern detection under massive datasets. There are a variety of rising applications of machine learning inside the bioinformatics area [18]. Machine learning could be a vastly powerful tool in bioinformatics that is very helpful in predicting and detecting specific required data from a database. And through data and training experience, machine learning provides the more useful solutions. Machine learning has several characteristics, one is employed to decrease false-positive rates, and it is the power of machine so as to extend the performance supported past information [19]. Machine learning is mainly used to predict sequences of DNA and RNA strands. Bioinformatics additionally has important potential in resolution population and organic process genetic science queries.

Machine learning is used in many domains of bioinformatics or a computed biology such as genomic sequencing, microarray examination, organic processes, etc. Within the past many decades, the immensely great deal of genetic sequence information generated has provided large knowledge banks that help the researchers to productively examine and combine and present this information with the help of different procedural ways [20]. Basically the bioinformatics goal is to enhance the various perspectives of understanding the biological data and this will be achieved by machine learning particularly for correct classification and prediction of tremendous quantity of information from dynamic setting [21]. So to explore more areas in bioinformatics, ML comprises different subsets like supervised learning, unsupervised learning, semi-supervised, hybrid, reinforcement learning, and one of the most result oriented areas, i.e. deep learning.

4 Deep Learning in Bioinformatics

Deep learning is a branch of ML where artificial neural network algorithms are designed to work like human brain and match the level of neuron architecture. The basic mechanism in deep learning has been designed from the concept of neural networks which comprises signalling process that makes a strong interconnected link in between different neurons that store and process the huge amount of data and inspired from there DL is used for computational analysis that basically tries to enhance the result status or improves the accuracy. In this aspect, various deep learning algorithms are used in the area of bioinformatics to extract features and knowledge from the biological or biomedical databases [16].

AI ways like deep neural network improve the decision capability in biological and chemical applications. Together with the bioinformatics, deep learning in order to handle huge knowledge has attained success in numerous domains. In bioinformatics, deep learning is used as a research objective that is executed in terms of input data and characterized in such a way that demonstrated the architectures of deep learning. With the succession of the large knowledge period in biology, it is predictable that deep learning can become progressively necessary within the sphere and can be integrated in large majorities of study pipelines [22]. Extracting inherent valuable information from the available massive knowledge remains as a huge downside in machine biology and in bioinformatics [23].

Deep learning improves the process of feature engineering over other machine learning algorithms and extracts the essential features on its own by using feature engineering. So, it is responsible for major developments in different areas where for many years AI community has struggled [24]. By extracting the simpler features from biological data, deep learning is initiated to learn complex features by integrating simpler features. Deep learning network is complex and based on the phenomenon of the multiple non-linear layers of artificial neural networks that increase the abstraction level in representations of data [16].

Deep learning provides many solutions to lot of basic biological queries. Deep learning strategies strived to an oversized collection which comprise patient information about their physical conditions may offer important aspects of different patients' data. Deep learning provides wide platforms for augmentation in the existing biological processes and desegregation of many sorts of omics knowledge. Deep learning is also useful for organic phenomenon prediction tasks [25].

One-dimensional CNNs and RNNs area units have similar temperament for tasks associated with DNA-binding and RNA-binding proteins, epigenomics and ribonucleic acid junction [18]. Domain consultants generally play vital roles in planning and execution of knowledge appropriately, coding the foremost salient previous data. So deep learning system provides base for many scientists working in biology field for optimizing their tasks [25]. Deep learning calculates the promising results in the field of bioinformatics.

5 Bioinformatics—An Intelligent Approach

Bioinformatics, mainly the application of computational data and techniques, is used to analyse the information linked to biomolecules at large scales, has now developed as part of molecular biology and is used in many of structural biology, genetics and genomics. Bioinformatics aids scientists to understand simple 1D to complex 3D structures of protein or other macromolecules. So, it is turning into inconceivable for degree knowledge to cipher and differentiate the entries with whole lot of databases. Thus, square measures of ML and AI techniques are used to analyse biological information sets and thus notice and identify the similarities and patterns existing in varied databases.

Bioinformatics combines biology and knowledge science, giving machine learning and computing ways a true and vital purpose. The main aim of bioinformatics is to use the facility of machine learning and knowledge science to explore biological systems and processes too complicated to be explored by hand. It links knowledge science technologies with biological tasks [4]. Gilbert et al., 2003 associate empirical comparison of Maths learning systems, rule-based learning systems and ensemble methods [13].

Bioinformatics creates heuristic approaches and sophisticated algorithms for exploitation computing and knowledge technology so as to resolve biological issues. Intelligent implication of the information will accelerate biological information discovery. Data processing, as biological intelligence, tries to seek out reliable, new, helpful and meaningful patterns in immense amounts of information [18].

Bioinformatics implicate the formation and development of algorithm exploitation techniques as well as procedure intelligence, applied math and statistics, information processing and organic chemistry to resolve biological issues sometimes. Some analysis efforts within the field embody prediction of organic phenomenon, sequence analysis, sequence finding, order annotation, molecular interactions, etc. [26]. Intelligent bioinformatics needs solely rudimentary data of biology, bioinformatics or engineering science and is geared towards interested readers notwithstanding discipline [18].

Figure 1 shows how bioinformatics is the defined nature of data and informative science with the main aspect of biology and biochemistry and further how the term introduced as computational biology system in combination of all aspects including AI, ML and DL.

6 Computational Aspect of Informative Data in Bioinformatics

Bioinformatics used potential of many domains such as biology, statistics and applied sciences to analyse biological information. It includes scripting to study biological information and their methods, likewise as a selected analysis 'pipelines' that square

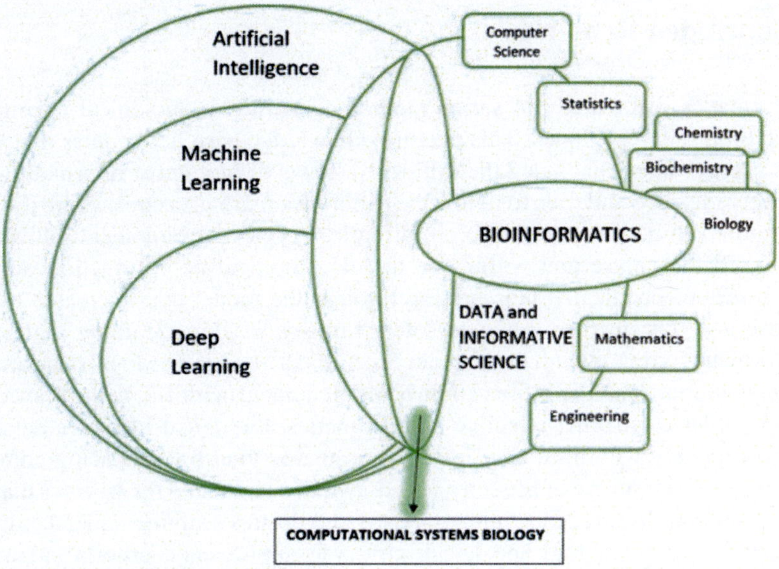

Fig. 1 Illustration of combinational view

measure continuously used, notably within the domain of genetic science [27]. Bioinformatics is the science of decoding technology by use of biological. Because of the large success in three-dimensional modelling of biomolecules, genomics and biological systems and to understand this large amount of data, a biological knowledge is required.

Bioinformatics could be a many-sided space concerned within the advancement of process, techniques, methods and software package tools for biological knowledge. Complicated analysis is required to take the results with immense quantity in biological science [26]. Computational approaches tend to draw upon skills in computer code development, information development and management and visual image ways to convey data contained inside knowledge sets. In general, the basic steps of management of biological and genetic information are to collect, store, analyse and integrate knowledge with the utilization of bioinformatics [28].

Simply put, machine biology is mainly like learning biology exploitation machine techniques that are the understanding of the science. In the COVID-19, testing is vital and pressing, the mixture of AI associates in nursing bioinformatics can play a huge vital role in attention and in chase and observance of public health [15]. Bioinformatics focuses additionally on the engineering aspect and therefore the creation of tools that can be employed with biological knowledge to resolve issues arising and for general welfare of the society especially in terms of health care.

7 Concluded Remarks

Bioinformatics is the study of varied method concerned in biological information with pc tools and techniques. This exciting knowledge base field comes at a value. In this survey paper concluded the different aspects of biological information with informative science and tries to define the different square measures where the analysis required to be processed in bioinformatics. Typically, biological information is advanced to analyse and within the fast development of information ought to be processes mechanically and learning through the model that is created by past expertise and this may be achieved solely through machine learning algorithms. Bioinformatics grew out of life science as a result, and it became transparent to rearrange and analyse the information being generated with the help of specialist skills. In order to conduct effective bioinformatics, biological literature databases still grow quickly with necessary information that is vital for enhancing computational growth. The sphere of life science is drawn as customized for AI ways that data and information discovery technique is hard and desires extreme understanding of every computational biology and data discovery as improvement criteria and to identify problems that once augmented by data discovery algorithms and come up with a far higher judgement of biological systems. The conceptual challenges to informatics in procedure biology and in bioinformatics area unit are rousing and provide assurance to supply problems which will still operate in the event of upgraded device for intelligent systems.

References

1. Dixit P (2015) Machine Learning in Bioinformatics: A Novel Approach for DNA Sequencing. https://doi.org/10.1109/ACCT.2015.73
2. Ezziane Z (2016) Applications of artificial intelligence in bioinformatics: a review, vol 30. https://doi.org/10.1016/j.eswa.2005.09.042
3. Shreya A AI in Bioinformatics. https://towardsdatascience.com/ai-in-bioinformatics-a1acdc3cdd89. Accessed 05 Oct 2020
4. Ye A Bioinformatics: How AI Can Contribute to the Study of Life. https://towardsdatascience.com/bioinformatics-how-ai-can-contribute-to-the-study-of-life-7a67f3d62a9f. Accessed 07 Oct 2020
5. Larrañaga P, Calvo B, Santana R, Bielza C, Galdiano J, Inza I, Lozano JA, Armañanzas R, Santafé G, Pérez A, Robles V (2006) Mach Learn Bioinform 7(1):86–112. https://doi.org/10.1093/bib/bbk007
6. Aditya Shastry K Machine Learning for Bioinformatics. Part of the Algorithms for Intelligent systems book series (AIS). https://doi.org/10.1007/978-981-15-2445-5_3
7. Han H, Liu W (2019) The coming era of artificial intelligence in biological data science. BMC Bioinform 20:712. https://doi.org/10.1186/s12859-019-3225-3
8. Nagarajan N (2019) Application of Computational Biology and Artificial Intelligence Technologies in Cancer Precision Drug Discovery, ArticleID: 8427042. https://doi.org/10.1155/2019/8427042
9. Narayanan AK, Olsson EC, Björn (2002) Artificial intelligence techniques for bioinformatics. PMID: 15130837, vol 1

10. Golkarieh H How AI is shaping the future of bioinformatics. https://medium.com/optima-ai/how-ai-is-shaping-the-future-of-bioinformatics-f4aa17bce5a6. Accessed 08 Oct 2020
11. Biswas S How pairing AI with bioinformatics will be a blessing. https://www.healthworkscollective.com/how-pairing-ai-with-bioinformatics-will-be-a-blessing/. Accessed 08 Oct 2020
12. Nicolas J Artificial Intelligence and Bioinformatics. https://hal.inria.fr/hal-01850570/file/My_Ai_and_Bioinformatics. Accessed 08 Oct 2020
13. Sajda P (2006) Machine learning for detection and diagnosis of disease. Annual Rev Biomed Eng 8:537–565
14. Lee JW, Lee JB, Park M, Song SH (2005) An extensive comparison of recent classification tools applied to microarray data. Comput Stat Data Anal 48(4):9–15
15. Udacity Team. How healthcare is using Bioinformatics and AI to improve population health. https://blog.udacity.com/2020/07/how-healthcare-is-using-bioinformatics-and-ai-to-improve-population-health.html. Accessed 08 Oct 2020
16. Min S, Lee B, Yoon S (2017) Deep Learning in Bioinformatics, pp 851–869. https://doi.org/10.1093/bib/bbw068
17. Goodfellow I, Bengio Y, Courville A (2016) Deep Learning. Book in preparation for MIT Press
18. Larranaga P, Calvo B, Santana R, Bielza C, Galdiano J, Inza I, Lozano JA, Armananzas R, Santafe G, Perez A, Robles V (2006) Machine learning in bioinformatics. Brief Bioinform 7(1):86–112
19. Tyagi N Understanding Bioinformatics as the application of machine learning. https://www.analyticssteps.com/blogs/understanding-bioinformatics-application-machine-learning. Accessed 08 Oct 2020
20. Greenwood M Is Machine Learning the Future of Bioinformatics? https://www.azolifesciences.com/article/Is-Machine-Learning-the-Future-of-Bioinformatics.aspx. Accessed 08 Oct 2020
21. Jayanthi K, Mahesh C (2019) Need of Machine Learning In Bioinformatics, vol 8, Issue 11
22. Li Y, Huang C, Ding L, Li Z, Pan Y, Gao X (2019) Deep learning in bioinformatics: introduction, application, and perspective in big data era. https://doi.org/10.1101/563601
23. Tang B, Pan Z, Yin K, Khateeb A (2019) Recent Advances of Deep Learning in Bioinformatics and Computational Biology. https://doi.org/10.3389/fgene.2019.00214
24. LeCun Y, Bengio Y, Hinton G (2015) Deep learning. Nature 521(7553):436–444
25. Ching T, Himmelstein DS, Beaulieu-Jones BK, Kalinin AA, Do BT, Way GP, Ferrero E et al (2018) Opportunities and obstacles for deep learning in biology and medicine. J R Soc Interf 15. https://doi.org/10.1098/rsif.2017.0387
26. Madiajagan M, Sridhar Raj S (2019) Deep Learning and Parallel Computing Environment for Bioengineering Systems. https://doi.org/10.1016/B978-0-12-816718-2.00022-1
27. Wikipedia, the free encyclopedia. "Bioinformatics". https://en.wikipedia.org/wiki/Bioinformatics. Accessed 08 Oct 2020
28. Lacroix Z, Critchlow T (2003) Bioinformatics. https://doi.org/10.1016/B978-155860829-0/50003-6

Chapter 6
Effect of Laser Pulse in Modified TPL GN-Thermoelastic Transversely Isotropic Euler–Bernoulli Nanobeam

Iqbal Kaur, Parveen Lata, and Kulvinder Singh

Nomenclature

δ_{ij}	Kronecker delta,
c_{ijkl}	Elastic parameters,
β_{ij}	Thermal elastic coupling tensor,
T	Temperature change,
T_0	Reference temperature,
t_{ij}	Stress tensors,
e_{ij}	Strain tensors,
u_i	Displacement components,
$w(x,t)$	Lateral deflection of beam
ρ	Medium density,
C_E	Specific heat,
K_{ij}	Thermal conductivity,
τ_q	Phase lag of heat flux
τ_v	Phase lag of thermal displacement gradient
τ_t	Phase lag of temperature gradients
t	Time
K^*_{ij}	Materialistic constant
α_{ij}	Linear thermal expansion coefficient
ε	Dimensionless key number
I	Moment of inertia of cross section
$c_{11}I$	Flexural rigidity of the beam
I_0	Laser intensity
t_0	The pulse rise time

I. Kaur (✉) · P. Lata
Department of Basic and Applied Sciences, Punjabi University, Patiala, Punjab, India

K. Singh
UIET, Kurukshetra University, Kurukshetra, India

N. Marriwala et al. (eds.), *Soft Computing for Intelligent Systems*, Algorithms for Intelligent Systems, https://doi.org/10.1007/978-981-16-1048-6_6

R_a	Absorptivity of the irradiated surface
δ	The absorptive depth of heating energy
M	Flexural moment of cross section of beam
M_T	Thermal moment of inertia
$\beta_1 M_T$	Thermal moment of beam
$f(t)$	Temporal profile of laser beam

1 Introduction

In last decade, the demand for micro- and nano-materials structures is continuously increasing due to their special electronic, electrical and mechanical properties. Due to these properties, nano-materials are used in nano-beams as elementary structural components in micro-electromechanical system (MEMS)/nano-electromechanical systems (NEMS) and piezoelectric devices. Recently, the laser pulse technology have gain interest in the field of non-destructive testing and processing of material. The radiation laser light generates wave motion in the solid material. Therefore, laser pulse have important applications in Micro and nano-mechanical resonators.

Quintanilla and Racke [1] discussed the three-phase lag heat conduction models with Lyapunov function. El-Karamany and Ezzat [2] proposed a TPL micro-polar theory of thermoelasticity. Kumar et al. [3], Kumar and Kumar [4] studied the thermoelastic damping by using the TPL theory of thermoelasticity of a microbeam resonator. Codarcea-Munteanu and Marin [5] proposed a model for dipolar materials with TPL theory of thermoelasticity with voids. Abouelregal [6] proposed a novel model of TPL thermoelastic theory of higher order using fractional calculus. Paul and Mukhopadhyay [7] discussed the thick plate with diffusion with TPL thermoelasticity theory.

Abd-Elaziz and Othman [8] studied the thomson and thermal loading effect of non-gaussian laser beam on magnetothermoelastic porous medium with Green-Nagdhi theory of type-III. Zenkour [9] studied thermomechanical response of microbeams with refined two-temperature multi-phase-lags theory and the modified couple stress analysis. Kumar and Devi [10] studied the thermoelastic beam with modified couple stress theory. Zenkour and Abouelregal [11] studied axially moving microbeam due to sinusoidal pulse heating. Lata and Kaur [12] studied the thermoelastic analysis in transversely isotropic beam with GN-theory of type-II and type-III. Hobiny et al. [13] proposed a general model with TPL GN-III theory of thermoelasticity in a 2D porous medium for thermoelastic wave.

Despite of this, several researchers worked on a different theory of thermoelasticity as Abbas and Othman [14], Marin [15–17], Abbas and Youssef [18], Sharma and Grover [19], Abbas and Marin [20], Othman and Abbas [21], Kumar and Abbas [22], Riaz et al. [23], Bhatti et al. [24–26], Lata and Kaur [27–29], Kaur and Lata [30, 31], and Bhatti et al. [32].

In this study we have examined the vibration phenomenon in 2D transversely isotropic homogeneous Euler–Bernoulli nanobeam with laser pulse with new modified three-phase lag Green Naghdi (TPL GN) model. The ends of nanobeam are considered to be simply supported and have constant temperature. Temperature is assumed to vary sinusoidally. Laplace transforms are used to derive the non-dimensional expressions for lateral deflection, axial displacement, temperature distribution, axial stress and thermal moment in the transformed domain and numerical inversion techniques are used to find the expressions in the physical domain. Effect of new modified TPL GN heat transfer is represented graphically for lateral deflection, axial displacement, temperature distribution, axial stress and thermal moment using the MATLAB software. Comparisons are made with the different thermoelasticity theories. Few specific cases are also derived.

2 Basic Equations

For an anisotropic medium the constitutive relations, following Kumar et al. [33] are given by

$$t_{ij} = c_{ijkl}e_{kl} - \beta_{ij}T. \tag{1}$$

where

$$\beta_{ij} = c_{ijkl}\alpha_{ij}, \tag{2}$$

Here $c_{ijkl}(c_{ijkl} = c_{klij} = c_{jikl} = c_{ijlk})$ are elastic parameters,$\beta_{ij} = \beta_i\delta_{ij}$, $K_{ij} = K_i\delta_{ij}$, i is not summed.

Following Zenkour and Mashat [34], a modified TPL GN equation for conduction of heat is given by

$$\mathcal{L}_1 T_{,ij} = \mathcal{L}_2\big(\rho C_E T + \beta_{ij}T_0 e_{ij}\big) - \mathcal{L}_0(\rho Q_0), \tag{3}$$

where time differential operators $\mathcal{L}_i(i = 0, 1, 2)$ are given by

$$\mathcal{L}_1 = K_{ij}\left(1 + \sum_{r=1}^{n}\frac{(\tau_t)^r}{r!}\frac{\partial^r}{\partial t^r}\right)\frac{\partial}{\partial t} + \varepsilon K_{ij}^*\left(1 + \sum_{r=1}^{n}\frac{(\tau_v)^r}{r!}\frac{\partial^r}{\partial t^r}\right), \tag{4}$$

$$\mathcal{L}_0 = \left(1 + \sum_{r=1}^{n}\frac{(\tau_q)^r}{r!}\frac{\partial^r}{\partial t^r}\right)\frac{\partial}{\partial t}, \tag{5}$$

$$\mathcal{L}_2 = \mathcal{L}_0\frac{\partial}{\partial t} \tag{6}$$

where

$$\varepsilon = 0 or 1 n \geq 0 \text{ and } 0 \leq \tau_v < \tau_t < \tau_q.$$

The phonon-electron interaction or phonon scattering creates phase lag τ_t, and the fast-transient thermal inertia effects create τ_q. The phase lags τ_v, τ_q, and τ_t are intrinsic properties of the medium and are small in magnitude.

Some special cases for Three-Phase Lag (TPL), Dual Phase Lag (DPL), Single Phase Lag (SPL) and Simple for GN theories are also obtained from Eq. (2) as

(i) TPL G–N III model ($\varepsilon = 1, n \geq 1$)

$$\left[K_{ij}\left(1+\sum_{r=1}^{n}\frac{(\tau_t)^r}{r!}\frac{\partial^r}{\partial t^r}\right)\frac{\partial}{\partial t}+K_{ij}^{*}\left(1+\sum_{r=1}^{n}\frac{(\tau_v)^r}{r!}\frac{\partial^r}{\partial t^r}\right)\right]T_{,ij} = \left(1+\sum_{r=1}^{n}\frac{(\tau_q)^r}{r!}\frac{\partial^r}{\partial t^r}\right)\frac{\partial}{\partial t}\left[\frac{\partial}{\partial t}(\rho C_E T + \beta_{ij}T_0 e_{ij}) - (\rho Q_0)\right] \quad (7)$$

(ii) DPL G–N III model ($\varepsilon = 1, n \geq 1, \tau_t = 0$)

$$\left[K_{ij}\frac{\partial}{\partial t}+K_{ij}^{*}\left(1+\sum_{r=1}^{n}\frac{(\tau_v)^r}{r!}\frac{\partial^r}{\partial t^r}\right)\right]T_{,ij} = \left(1+\sum_{r=1}^{n}\frac{(\tau_q)^r}{r!}\frac{\partial^r}{\partial t^r}\right)\frac{\partial}{\partial t}\left[\frac{\partial}{\partial t}(\rho C_E T + \beta_{ij}T_0 e_{ij}) - (\rho Q_0)\right] \quad (8)$$

(iii) SPL G–N III model ($\varepsilon = 1, n \geq 1, \tau_t = \tau_v = 0$)

$$\left[K_{ij}\frac{\partial}{\partial t}+K_{ij}^{*}\right]T_{,ij} = \left(1+\sum_{r=1}^{n}\frac{(\tau_q)^r}{r!}\frac{\partial^r}{\partial t^r}\right)\frac{\partial}{\partial t}\left[\frac{\partial}{\partial t}(\rho C_E T + \beta_{ij}T_0 e_{ij}) - (\rho Q_0)\right] \quad (9)$$

(iv) DPL G–N II model ($\varepsilon = 0, n \geq 1$)

$$K_{ij}\left(1+\sum_{r=1}^{n}\frac{(\tau_t)^r}{r!}\frac{\partial^r}{\partial t^r}\right)\frac{\partial}{\partial t}T_{,ij} = \left(1+\sum_{r=1}^{n}\frac{(\tau_q)^r}{r!}\frac{\partial^r}{\partial t^r}\right)\frac{\partial}{\partial t}\left[\frac{\partial}{\partial t}(\rho C_E T + \beta_{ij}T_0 e_{ij}) - (\rho Q_0)\right] \quad (10)$$

(v) SPL G–N II model ($\varepsilon = 0, n \geq 1, \tau_t = 0$)

$$K_{ij}T_{,ij} = \left(1 + \sum_{r=1}^{n} \frac{(\tau_q)^r}{r!} \frac{\partial^r}{\partial t^r}\right)\left[\frac{\partial}{\partial t}(\rho C_E T + \beta_{ij} T_0 e_{ij}) - (\rho Q_0)\right] \quad (11)$$

(vi) Simple G–N III model ($\varepsilon = 1, \tau_t = \tau_v = \tau_q = 0$)

$$\left(K_{ij}\frac{\partial}{\partial t} + K^*_{ij}\right)T_{,ij} = \frac{\partial}{\partial t}\left[\frac{\partial}{\partial t}(\rho C_E T + \beta_{ij} T_0 e_{ij}) - (\rho Q_0)\right] \quad (12)$$

(vii) Simple G–N II model $\left(K_{ij} \to 0, \varepsilon = 1, \tau_q = 0\right)$

$$\left(K^*_{ij}\right)T_{,ij} = \frac{\partial}{\partial t}\left[\frac{\partial}{\partial t}(\rho C_E T + \beta_{ij} T_0 e_{ij}) - (\rho Q_0)\right] \quad (13)$$

(viii) Simple G–N I model ($\varepsilon = 0, \tau_t = \tau_q = 0$)

$$\left(K_{ij}\right)T_{,ij} = \left[\frac{\partial}{\partial t}(\rho C_E T + \beta_{ij} T_0 e_{ij}) - (\rho Q_0)\right] \quad (14)$$

3 Formulation of the Problem

Consider a transversely isotropic homogeneous thermoelastic nanobeam (Fig. 1) of length ($0 \leq x \leq L$), width $\left(-\frac{b}{2} \leq y \leq \frac{b}{2}\right)$ and thickness $\left(-\frac{h}{2} \leq z \leq \frac{h}{2}\right)$, where x, y and z are the cartesian axes, L denotes the length, b the width and h the thickness of the nanobeam. The x-axis is taken along the axis of the nanobeam and y–z plane is along one of the end ($x = 0$) of the nanobeam with the origin located at the centre of this end $x = 0$ of the nanobeam. The ends of the nanobeam are kept at uniform temperature T_0.

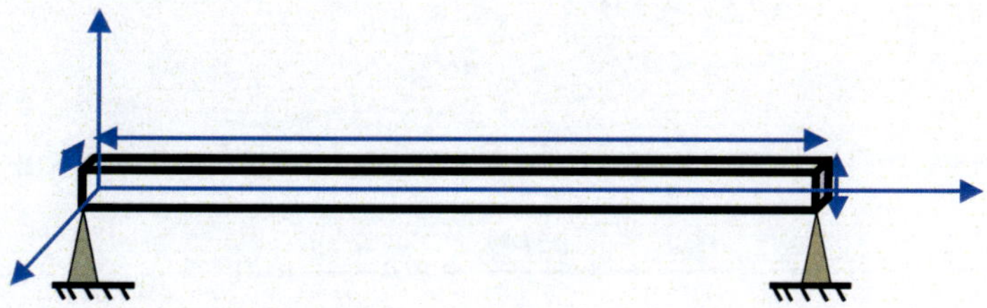

Fig. 1 Structural design of the nanobeam

Following Rao [35], the displacement components due to Euler–Bernoulli beam theory are given by

$$u_1(x, y, z, t) = -z\frac{\partial w}{\partial x}, u_2(x, y, z, t) = 0, u_3(x, y, z, t) = w(x, t) \tag{15}$$

Initially, the nanobeam is unstrained and unstressed.

$$w(x, t)|_{t=0} = \left.\frac{\partial w(x, t)}{\partial t}\right|_{t=0} = 0,\ T(x, z, t)|_{t=0} = \left.\frac{\partial T(x, z, t)}{\partial t}\right|_{t=0} = 0. \tag{16}$$

The 1D constitutive equation obtained from Eq. (1) with the help of Eq. (15) becomes

$$t_{xx} = -c_{11}z\frac{\partial^2 w}{\partial x^2} - \beta_1 T \tag{17}$$

where $\beta_1 = (c_{11} + c_{13})\alpha_1 + c_{13}\alpha_3$ and due to Euler–Bernoulli hypothesis, the thermoelastic parameter $\beta_3 = 2c_{13}\alpha_1 + c_{33}\alpha_3$ does not appear.

The flexural moment of the cross section for the transversely isotropic homogeneous nanobeam is given by

$$M(x, t) = -\int_{-\frac{h}{2}}^{\frac{h}{2}}\int_{\frac{-b}{2}}^{\frac{b}{2}} t_{xx} z\mathrm{d}z\mathrm{d}y = c_{11}I\frac{\partial^2 w}{\partial x^2} + \beta_1 M_T \tag{18}$$

where

$$M_T = b\int_{\frac{-h}{2}}^{\frac{h}{2}} Tz\mathrm{d}z \tag{19}$$

$$I = \frac{bh^3}{12}.$$

The transverse equation of motion for the free nanobeam is given by Rao [35]

$$\frac{\partial^2 M}{\partial x^2} + \rho A\frac{\partial^2 w}{\partial t^2} = 0 \tag{20}$$

where $A = bh$ is the cross-sectional area. Using Eq. (18) in Eq. (20), we get

$$c_{11}I\frac{\partial^4 w}{\partial x^4} + \beta_1\frac{\partial^2 M_T}{\partial x^2} + \rho A\frac{\partial^2 w}{\partial t^2} = 0. \tag{21}$$

According to Kumar and Vohra [38] for $t = 0$, the upper surface $z = \frac{h}{2}$, of transversely isotropic thermoelastic homogeneous nanobeam is exposed to laser pulse having

$$f(t) = I_0 \frac{t}{t_0^2} e^{-(t/t_0)} \tag{22}$$

where $t_0 = 2ps$ is the pulse rise time and therefore the energy source $Q_0(z,t)$ as

$$Q_0(z,t) = \frac{R_a}{\delta} e^{\frac{\left(z-\frac{h}{2}\right)}{\delta}} f(t) = \frac{R_a}{\delta} I_0 \frac{t}{t_0^2} e^{\left(\frac{\left(z-\frac{h}{2}\right)}{\delta} - \frac{t}{t_0}\right)}. \tag{23}$$

The heat conduction Eq. (2) using Eqs. (15) and (23) can be written as

$$\begin{aligned}
&\left[K_1\left(1+\sum_{r=1}^{n}\frac{(\tau_t)^r}{r!}\frac{\partial^r}{\partial t^r}\right)\frac{\partial}{\partial t}+\varepsilon K_1^*\left(1+\sum_{r=1}^{n}\frac{(\tau_v)^r}{r!}\frac{\partial^r}{\partial t^r}\right)\right]\frac{\partial^2 T}{\partial x^2}\\
&+\left[K_3\left(1+\sum_{r=1}^{n}\frac{(\tau_t)^r}{r!}\frac{\partial^r}{\partial t^r}\right)\frac{\partial}{\partial t}+\varepsilon K_3^*\left(1+\sum_{r=1}^{n}\frac{(\tau_v)^r}{r!}\frac{\partial^r}{\partial t^r}\right)\right]\frac{\partial^2 T}{\partial z^2}\\
&=\left(1+\sum_{r=1}^{n}\frac{(\tau_q)^r}{r!}\frac{\partial^r}{\partial t^r}\right)\left[-z\beta_1 T_0\frac{\partial^4 w}{\partial x^2\partial t^2}+\rho C_E\frac{\partial^2 T}{\partial t^2}\right.\\
&\left.-\left(\frac{\rho R_a I_0}{\delta t_0^2}\left(1-\frac{t}{t_0}\right)e^{\left(\frac{\left(z-\frac{h}{2}\right)}{\delta}-\frac{t}{t_0}\right)}\right)\right].
\end{aligned} \tag{24}$$

The dimensionless quantities are given by

$$\begin{aligned}
&x' = \frac{x}{L},\ z' = \frac{z}{L},\ w' = \frac{w}{L}, h' = \frac{h}{L}, b' = \frac{b}{L}, (t', \tau_t', \tau_v', \tau_q') = \frac{c_1}{L}(t, \tau_t, \tau_v, \tau_q),\\
&T' = \frac{T}{T_0},\quad \rho c_1^2 = c_{11},\ t_{xx}' = \frac{t_{xx}}{\beta_1 T_0}, a_1' = \frac{a_1}{L^2}, a_3' = \frac{a_3}{L^2}, M_T' = \frac{M_T}{T_0 L^3}, \delta' = \frac{\delta}{L}
\end{aligned} \tag{25}$$

Now applying the dimensionless quantities from Eq. (25) on Eqs. (21) and (24), and after suppressing the primes the equations in non-dimensional form can be expressed as

$$c_{11} I \frac{\partial^4 w}{\partial x^4} + \beta_1 T_0 \frac{\partial^2 M_T}{\partial x^2} + \rho A c_1^2 \frac{\partial^2 w}{\partial t^2} = 0, \tag{26}$$

$$\begin{aligned}
&\left[K_1\left(1+\sum_{r=1}^{n}\frac{(\tau_t)^r}{r!}\frac{\partial^r}{\partial t^r}\right)\frac{c_1}{L}\frac{\partial}{\partial t}+\varepsilon K_1^*\left(1+\sum_{r=1}^{n}\frac{(\tau_v)^r}{r!}\frac{\partial^r}{\partial t^r}\right)\right]\frac{\partial^2 T}{\partial x^2}\\
&+\left[K_3\left(1+\sum_{r=1}^{n}\frac{(\tau_t)^r}{r!}\frac{\partial^r}{\partial t^r}\right)\frac{c_1}{L}\frac{\partial}{\partial t}+\varepsilon K_3^*\left(1+\sum_{r=1}^{n}\frac{(\tau_v)^r}{r!}\frac{\partial^r}{\partial t^r}\right)\right]\frac{\partial^2 T}{\partial z^2}
\end{aligned}$$

$$= \left(1+\sum_{r=1}^{n}\frac{(\tau_q)^r}{r!}\frac{\partial^r}{\partial t^r}\right)\left[-z\beta_1 c_1^2\frac{\partial^4 w}{\partial x^2\partial t^2}+\rho C_E c_1^2\frac{\partial^2 T}{\partial t^2}\right.$$
$$\left.-\frac{L^2}{T_0}\left(\frac{\rho R_a I_0}{\delta t_0^2}\left(1-\frac{L}{c_1}\frac{t}{t_0}\right)e^{\left(\frac{\left(z-\frac{h}{2}\right)}{\delta}-\frac{L}{c_1}\frac{t}{t_0}\right)}\right)\right] \tag{27}$$

The Laplace transform is defined by

$$\mathcal{L}[f(x,z,t)] = \int_0^\infty e^{-st} f(x,z,t)\mathrm{d}t = \bar{f}\,(x,z,s). \tag{28}$$

Using Eq. (28) in Eqs. (26) and (27) we get

$$c_{11} I\frac{\partial^4\,\bar{w}}{\partial x^4}+\beta_1 T_0\frac{\partial^2 \bar{M}_T}{\partial x^2}+\rho A c_1^2 s^2\,\bar{w}=0, \tag{29}$$

$$\left[K_1\left(1+\sum_{r=1}^{n}\frac{(\tau_t s)^r}{r!}\right)\frac{sc_1}{L}+\varepsilon K_1^*\left(1+\sum_{r=1}^{n}\frac{(\tau_v s)^r}{r!}\right)\right]\frac{\partial^2\,\bar{T}}{\partial x^2}$$
$$+\left[K_3\left(1+\sum_{r=1}^{n}\frac{(\tau_t s)^r}{r!}\right)\frac{c_1 s}{L}+\varepsilon K_3^*\left(1+\sum_{r=1}^{n}\frac{(\tau_v s)^r}{r!}\right)\right]\frac{\partial^2\,\bar{T}}{\partial z^2}$$
$$=\left(1+\sum_{r=1}^{n}\frac{(\tau_q s)^r}{r!}\right)\left[-z\beta_1 c_1^2 s^2\frac{\partial^2\,\bar{w}}{\partial x^2}+\rho C_E c_1^2 s^2\,\bar{T}\right.$$
$$\left.-\left(\frac{\rho R_a I_0 L^2}{\delta t_0^2 T_0}\left(\frac{1}{\left(s+\frac{L}{t_0 c_1}\right)}-\frac{L}{t_0 c_1}\frac{1}{\left(s+\frac{L}{t_0 c_1}\right)^2}\right)e^{\frac{\left(z-\frac{h}{2}\right)}{\delta}}\right)\right]. \tag{30}$$

For a nanobeam, sinusoidally varying temperature along the z-direction can be written as

$$\bar{T}(x,z,s) = \bar{\bar{T}}(x,s)\sin\frac{\pi z}{h}. \tag{31}$$

The thermal moment of inertia using (31) becomes

$$\bar{M}_T = b\int_{\frac{-h}{2}}^{\frac{h}{2}} \bar{T} z\,\mathrm{d}z = \frac{2bh^2}{\pi^2}\bar{\bar{T}}(x,s)) \tag{32}$$

The Eq. (29) by using Eqs. (31) and (32) becomes

$$\frac{d^4\bar{w}}{dx^4} + \delta_7\frac{d^2\bar{\bar{T}}}{dx^2} + \delta_8\frac{12bs^2}{h^2}\bar{w} = 0. \tag{33}$$

Using (31) in Eq. (30) and then multiplying the resultant equation by z and integrating w.r.t z from limits $-h/2$ to $h/2$ we get

$$\delta_3\frac{d^2\bar{\bar{T}}}{\mathrm{d}x^2} + \delta_4\bar{\bar{T}} + \delta_5\frac{d^2\bar{w}}{\mathrm{d}x^2} + \delta_6 = 0 \tag{34}$$

where

$$\delta_1 = \frac{\delta h}{2}\left(1 + e^{\frac{-h}{\delta}}\right) - \delta^2\left(1 - e^{\frac{-h}{\delta}}\right),$$

$$\delta_2 = \frac{\rho R_a I_0 L^2 \delta_1}{\delta t_0^2 T_0},$$

$$\delta_3 = \frac{2h^2}{\pi^2}\left[K_1\left(1 + \sum_{r=1}^{n}\frac{(\tau_t s)^r}{r!}\right)\frac{sc_1}{L} + \varepsilon K_1^*\left(1 + \sum_{r=1}^{n}\frac{(\tau_v s)^r}{r!}\right)\right],$$

$$\delta_4 = -2\left[K_3\left(1 + \sum_{r=1}^{n}\frac{(\tau_t s)^r}{r!}\right)\frac{c_1 s}{L} + \varepsilon K_3^*\left(1 + \sum_{r=1}^{n}\frac{(\tau_v s)^r}{r!}\right)\right] - \frac{2h^2}{\pi^2}\left(1 + \sum_{r=1}^{n}\frac{(\tau_q s)^r}{r!}\right)\rho C_E c_1^2 s^2,$$

$$\delta_5 = \left(1 + \sum_{r=1}^{n}\frac{(\tau_q s)^\alpha}{r!}\right)\frac{h^3\beta_1 c_1^2 s^2}{12},$$

$$\delta_6 = \delta_2\left(1 + \sum_{r=1}^{n}\frac{(\tau_q s)^r}{r!}\right)\left(\frac{1}{\left(s + \frac{L}{t_0 c_1}\right)} - \frac{L}{t_0 c_1}\frac{1}{\left(s + \frac{L}{t_0 c_1}\right)^2}\right),$$

$$\delta_7 = \frac{24\beta_1 T_0}{\pi^2 c_{11} h},\ \delta_8 = \frac{12bs^2}{h^2}.$$

Now we have to solve the set of Eqs. (33) and (34), i.e.

$$\left(D^4 + \delta_8\right)\bar{w} + \delta_7 D^2\bar{\bar{T}} = 0, \tag{35}$$

$$\delta_5 D^2\bar{w} + \left(\delta_3 D^2 + \delta_4\right)\bar{T} = -\delta_6 \tag{36}$$

Eliminating $\bar{\bar{T}}$ in Eqs. (35) and (36) we obtain

$$\left[D^6 - pD^4 + qD^2 - r\right]\bar{w} = 0 \tag{37}$$

Eliminating $\bar{w}$ in Eqs. (35) and (36) we obtain

$$\left[D^6 - pD^4 + qD^2 - g\right]\bar{\bar{T}} = m \tag{38}$$

where $p = \frac{-\delta_4+\delta_7\delta_5}{\delta_3}, q = \delta_8, g = \frac{-\delta_4\delta_8}{\delta_3}, m = \frac{-\delta_6\delta_8}{\delta_3}$. The differential equation governing the lateral deflection $\bar{w}(x, s)$ Eq. (37) takes the form

$$\left(D^2 - \lambda_1^2\right)\left(D^2 - \lambda_2^2\right)\left(D^2 - \lambda_3^2\right)\bar{w}(x, s) = 0 \tag{39}$$

where $\pm\lambda_1, \pm\lambda_2$ and $\pm\ \lambda_3$ are the characteristic roots of the equation $\lambda^6 - p\lambda^4 + q\lambda^2 - g = 0$ and hence,

$$\lambda_1^2 + \lambda_2^2 + \lambda_3^2 = p,$$

$$\lambda_1^2\lambda_2^2 + \lambda_2^2\lambda_3^2 + \lambda_1^2\lambda_3^2 = q,$$

$$\lambda_1^2\lambda_2^2\lambda_3^2 = \mathrm{g}.$$

Assume Lateral Deflection $\bar{w}(x, s)$ as

$$\bar{w}(x, s) = \sum\nolimits_{i=1}^{3}\left[A_i e^{\lambda_i x} + B_i e^{-\lambda_i x}\right] \tag{40}$$

where A_i and $B_i, i = 1, 2, 3$ are constants and the thermal moment $\bar{T}(x, s)$ is obtained from Eq. (38) and as

$$\bar{T}(x, s) = \sum\nolimits_{i=1}^{3}\left[A_i' e^{\lambda_i x} + B_i' e^{-\lambda_i x}\right] - \frac{\delta_6}{\delta_4} \tag{41}$$

where A_i', B_i',i = 1,2, 3 are constants. Putting (41) and (40) in (36) gives

$$\begin{bmatrix} A_i' \\ B_i' \end{bmatrix} = \frac{-\delta_5\lambda_i^2}{\delta_3\lambda_i^2 + \delta_4}\begin{bmatrix} A_i \\ B_i \end{bmatrix}, i = 1, 2, 3. \tag{42}$$

Using Eq. (42) in Eq. (41) yields

$$\bar{T}(x, s) = \sum\nolimits_{i=1}^{3}\varsigma_i\left[A_i e^{\lambda_i x} + B_i e^{-\lambda_i x}\right] - \zeta_1, \tag{43}$$

where

$$\varsigma_i = \frac{-\delta_5\lambda_i^2}{\delta_3\lambda_i^2 + \delta_4},$$

$$\zeta_1 = \frac{\delta_6}{\delta_4}.$$

The axial displacement is

$$\bar{u}(x,z,s) = -z\sum\nolimits_{i=1}^{3}\left[\lambda_i A_i e^{\lambda_i x} - \lambda_i B_i e^{-\lambda_i x}\right] \tag{44}$$

Thermal moment is given by

$$\bar{M}_T = \frac{2bh^2}{\pi^2}\beta_1\left\{\sum\nolimits_{i=1}^{3}\varsigma_i\left[A_i e^{\lambda_i x} + B_i e^{-\lambda_i x}\right] - \zeta_1\right\} \tag{45}$$

Temperature distribution is given by

$$\bar{T}(x,z,s) = \left\{\sum\nolimits_{i=1}^{3}\varsigma_i\left[A_i e^{\lambda_i x} + B_i e^{-\lambda_i x}\right] - \zeta_1\right\}\sin\frac{\pi z}{h} \tag{46}$$

Axial stress is given by

$$\bar{t}_{xx}(x,z,s) = \sum\nolimits_{i=1}^{3}\left(-c_{11}z\lambda_i^2 - \beta_1\varsigma_i\sin\frac{\pi z}{h}\right)\left[A_i e^{\lambda_i x} + B_i e^{-\lambda_i x}\right] + \beta_1\zeta_1\sin\frac{\pi z}{h}. \tag{47}$$

4 Boundary Conditions

For both the simply supported ends of nanobeam which are maintained at a constant temperature T_0 and the boundary conditions can be written as

$$w(x,t)|_{x=0} = w(x,t)|_{x=L} = 0, \tag{48}$$

$$\left.\frac{\partial^2 w(x,t)}{\partial x^2}\right|_{x=0} = \left.\frac{\partial^2 w(x,t)}{\partial x^2}\right|_{x=L} = 0. \tag{49}$$

$$T(x,z,t)|_{x=0,L} = T_0\sin\frac{\pi z}{h}. \tag{50}$$

Applying dimensionless quantities and Laplace transform defined by Eqs. (25) and (28) on Eqs. (48)–(50) the dimensionless boundary conditions becomes

$$\bar{w}(x,s)|_{x=0} = \bar{w}(x,s)|_{x=1} = 0, \tag{51}$$

$$\left.\frac{\partial^2 \bar{w}(x,s)}{\partial x^2}\right|_{x=0} = \left.\frac{\partial^2 w(x,s)}{\partial x^2}\right|_{x=1} = 0. \tag{52}$$

$$\left.\bar{\bar{T}}(x,s)\right|_{x=0,1} = 1 \tag{53}$$

5 Solution

Substituting the values of $\bar{w}$ and $\bar{\bar{T}}$ from Eqs. (40) and (41) in Eqs. (51)–(53), we derive the value of A_i and B_i as

$$A_i = \frac{\Delta_i}{\Delta}, \; B_i = \frac{\Delta_{i+3}}{\Delta}, \; i = 1,2,3 \tag{54}$$

and

$$\Delta = \begin{vmatrix} 1 & 1 & 1 & 1 & 1 & 1 \\ \lambda_1^2 & \lambda_2^2 & \lambda_3^2 & \lambda_1^2 & \lambda_2^2 & \lambda_3^2 \\ e^{\lambda_1} & e^{\lambda_2} & e^{\lambda_3} & e^{-\lambda_1} & e^{-\lambda_2} & e^{-\lambda_3} \\ \lambda_1^2 e^{\lambda_1} & \lambda_2^2 e^{\lambda_2} & \lambda_3^2 e^{\lambda_3} & \lambda_1^2 e^{-\lambda_1} & \lambda_2^2 e^{-\lambda_2} & \lambda_3^2 e^{-\lambda_3} \\ \varsigma_1 & \varsigma_2 & \varsigma_3 & \varsigma_1 & \varsigma_2 & \varsigma_3 \\ \varsigma_1 e^{\lambda_1} & \varsigma_2 e^{\lambda_2} & \varsigma_3 e^{\lambda_3} & \varsigma_1 e^{-\lambda_1} & \varsigma_2 e^{-\lambda_2} & \varsigma_3 e^{-\lambda_3} \end{vmatrix}. \tag{55}$$

$\Delta_i (i = 1, 2, 3, \ldots, 6)$ are obtained by replacing the columns by $\left[\, 0,\ 0,\ 0,\ \ 0,\ 1 + \zeta_1,\ 1 + \zeta_1 \,\right]$ in Δ_i. Substituting the values from (54) in Eqs. (40), (42)–(47), we obtain the lateral deflection, axial displacement, thermal moment, the temperature distribution and axial stress of the transversely isotropic thin beam.

6 Inversion of Laplace Transform

To derive the solution in the physical domain, the transforms in Eqs. (40), (42)–(47) have to be inverted using inverse Laplace transform integral defined by

$$f(x,t) = \frac{1}{2\pi i} \int_{e^{-i\infty}}^{e^{+i\infty}} \bar{f}(x,s) e^{-st} \mathrm{d}s. \tag{55}$$

This integral is evaluated using Romberg's integration method with adaptive step size described in Press et al. [36].

7 Particular Cases

1. If we take ($\varepsilon = 1, n \geq 1$) in Eq. (2) then we can derive the dimensionless expressions for the lateral deflection, the axial displacement, the thermal moment, the temperature distribution and the axial stress from Eqs. (40), (44)–(47) for the transversely isotropic nanobeam with TPL G–N III heat transfer.
2. If we take ($\varepsilon = 1, n \geq 1, \tau_t = 0$) in Eq. (2) then we can derive the dimensionless expressions for the lateral deflection, the axial displacement, the thermal moment, the temperature distribution and the axial stress from Eqs. (40), (42)–(47) for the transversely isotropic nanobeam with DPL G–N III heat transfer.
3. If we take ($\varepsilon = 1, n \geq 1, \tau_t = \tau_v = 0$) in Eq. (2) then we can derive the dimensionless expressions for the lateral deflection, the axial displacement, the thermal moment, the temperature distribution and the axial stress from Eqs. (40), (42)–(47) for the transversely isotropic nanobeam with SPL G–N III heat transfer.
4. If we take ($\varepsilon = 0, n \geq 1$) in Eq. (2) then we can derive the dimensionless expressions for the lateral deflection, the axial displacement, the thermal moment, the temperature distribution and the axial stress from Eqs. (40), (42)–(47) for the transversely isotropic nanobeam with DPL G–N II heat transfer.
5. If we take ($\varepsilon = 0, n \geq 1, \tau_t = 0$) in Eq. (2) then we can derive the dimensionless expressions for the lateral deflection, the axial displacement, the thermal moment, the temperature distribution and the axial stress from Eqs. (40), (42)–(47) for the transversely isotropic nanobeam with SPL G–N II heat transfer.
6. If we take ($\varepsilon = 1, \tau_t = \tau_v = \tau_q = 02$) in Eq. () then we can derive the dimensionless expressions for the lateral deflection, the axial displacement, the thermal moment, the temperature distribution and the axial stress from Eqs. (40), (42)–(47) for the transversely isotropic nanobeam with Simple G–N III heat transfer.
7. If we take $\left(K_{ij} \rightarrow 0, \varepsilon = 1, \tau_q = 0\right)$ in Eq. (2) then we can derive the dimensionless expressions for the lateral deflection, the axial displacement, the thermal moment, the temperature distribution and the axial stress from Eqs. (40), (42)–(47) for the transversely isotropic nanobeam with simple G–N II heat transfer.
8. If we take ($\varepsilon = 0, \tau_t = \tau_q = 0$) in Eq. (2) then we can derive the dimensionless expressions for the lateral deflection, the axial displacement, the thermal moment, the temperature distribution and the axial stress from Eqs. (40), (42)–(47) for the transversely isotropic nanobeam with simple G–N I heat transfer.

8 Numerical Results and Discussion

The effect of different GN theories in the theoretical results is illustrated for transversely isotropic cobalt material. For a single crystal of cobalt, the physical data following Dhaliwal and Singh [37] is given by

$$\begin{aligned}
&c_{11} = 3.07 \times 10^{11}\mathrm{Nm}^{-2}, c_{12} = 1.650 \times 10^{11}\mathrm{Nm}^{-2},\\
&c_{13} = 1.027 \times 10^{10}\mathrm{Nm}^{-2}, c_{33} = 3.581 \times 10^{11}\mathrm{Nm}^{-2},\\
&c_{44} = 1.510 \times 10^{11}\mathrm{Nm}^{-2}, C_E = 4.27 \times 10^{2}\mathrm{jKg}^{-1}\mathrm{deg}^{-1},\\
&\beta_1 = 7.04 \times 10^{6}\mathrm{Nm}^{-2}\mathrm{deg}^{-1}, \beta_3 = 6.90 \times 10^{6}\mathrm{Nm}^{-2}\mathrm{deg}^{-1},
\end{aligned}$$

$K_1 = 0.690 \times 10^2\mathrm{Wm}^{-1}\mathrm{Kdeg}^{-1}$, $K_3 = 0.690 \times 10^2\mathrm{Wm}^{-1}\mathrm{K}^{-1}$, $K_1^* = 0.02 \times 10^2\mathrm{NSec}^{-2}\mathrm{deg}^{-1}$, $K_3^* = 0.04 \times 10^2\mathrm{NSec}^{-2}\mathrm{deg}^{-1}$, $L/h = 10, b/h = 0.5, \rho = 8.836 \times 10^3\mathrm{Kgm}^{-3}$.

In the Figs. 2, 3, 4, 5 and 6, the effect of GN-III theories on the lateral deflection, axial displacement, thermal moment, the temperature distribution and axial stress of the transversely isotropic thin beam is shown

- The solid line represents TPL GN-III theory.
- The dashed line represents DPL GN-III theory.
- The dotted line represents SPL GN-III theory.

Figure 2 shows the effect of TPL, DPL and SPL GN-III theories on lateral deflection w w.r.t. length x. It is found that the lateral deflection w increases for the initial range $x < 0.25$ and then decreases for TPL, DPL and SPL GN-III theories. Figure 3 illustrates the effect of TPL, DPL and SPL GN-III theories on axial displacement u w.r.t. beam length. For TPL, DPL and SPL GN-III theories the axial displacement decreases for x < 0.4 and then increases with a difference in magnitude. Figure 4

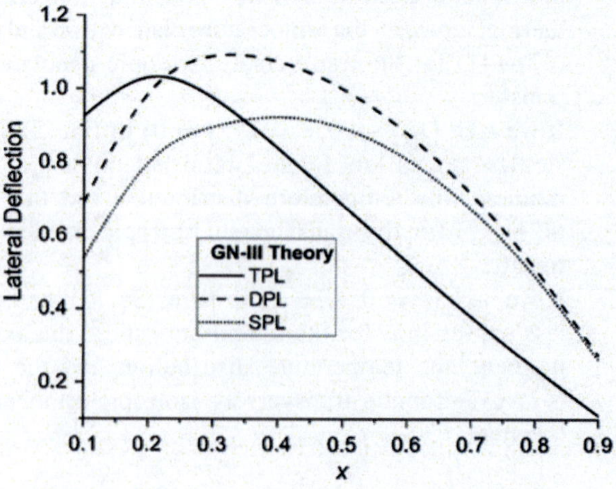

Fig. 2 Variations of lateral deflection $w(x, t)$ with length of beam

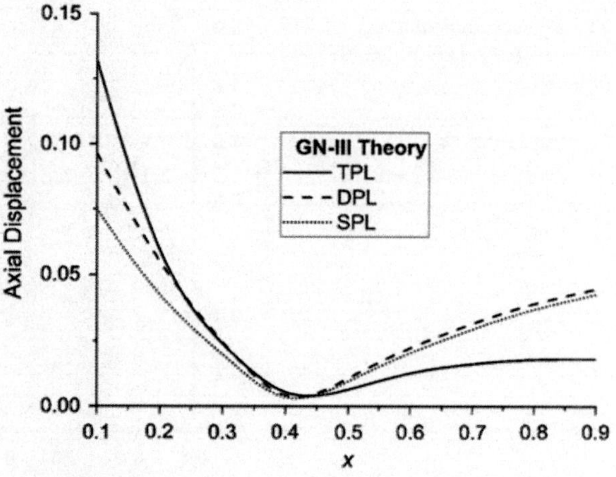

Fig. 3 Variations of axial displacement $u(x, z, t)$ with length of beam

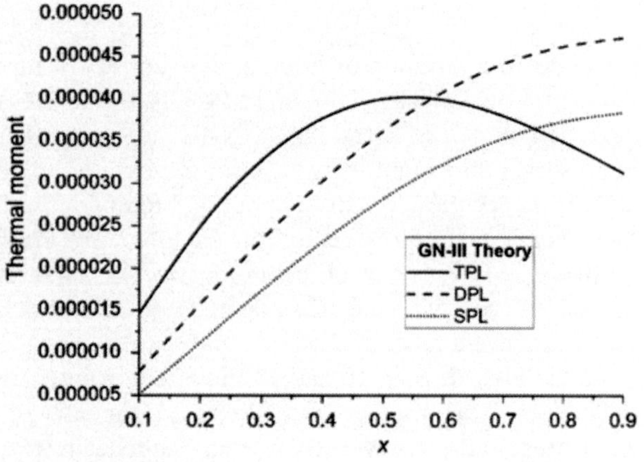

Fig. 4 Variations of thermal moment $M_T(x, t)$ with length of beam

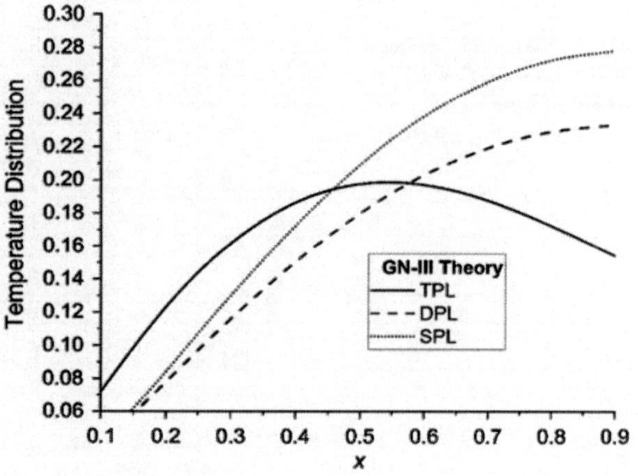

Fig. 5 Variations of temperature $T(x, z, t)$ with length of beam

Fig. 6 Variations of axial stress $t_{xx}(x, z, t)$ with length of beam

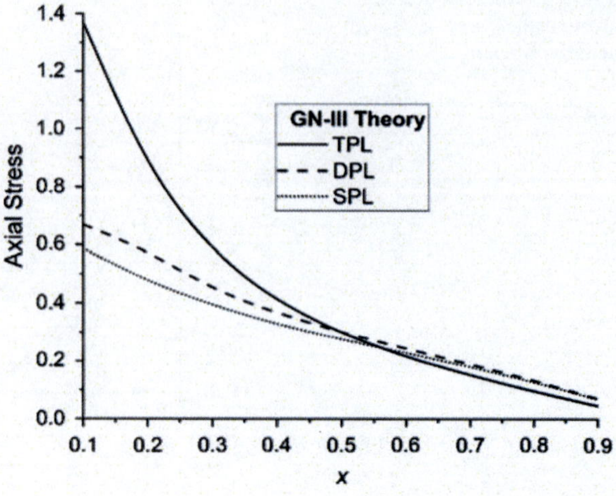

illustrates the variation of thermal moment M_T w.r.t. beam length and Fig. 5 illustrates the effect of TPL, DPL and SPL GN-III theories on temperature $T(x, z, t)$ w.r.t. beam length. For SPL and DPL GN-III theory, the thermal moment and the temperature shows an instant increase with increase in length of the beam while for TPL GN-III theory the thermal moment and temperature first increases for $x < 0.5$ and then decreases. Figure 6 represents the effect of TPL, DPL and SPL GN-III theories on the axial stress t_{xx} w.r.t. beam length. The axial stress t_{xx} decrease gradually in all GN-III cases with a difference in magnitude which becomes almost same at the end of the beam length.

In the Figs. 7, 8, 9, 10 and 11 the effect of intensity of laser pulse on the lateral deflection, axial displacement, thermal moment, the temperature distribution and axial stress of the transversely isotropic thin beam with TPL GN-III theory is shown:

Fig. 7 Variations of lateral deflection $w(x, t)$ with length of beam

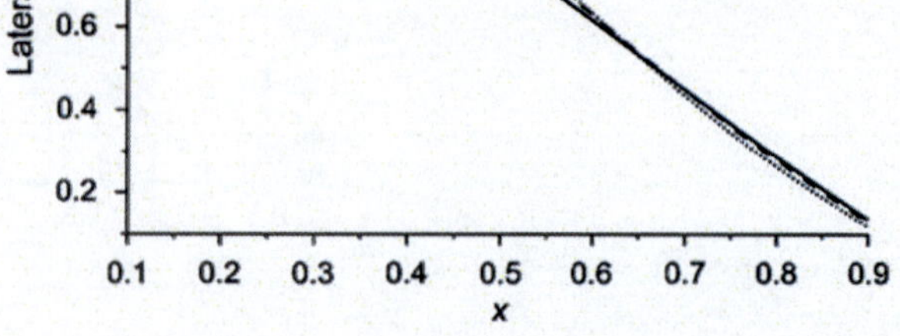

Fig. 8 Variations of axial displacement $u(x, z, t)$ with length of beam

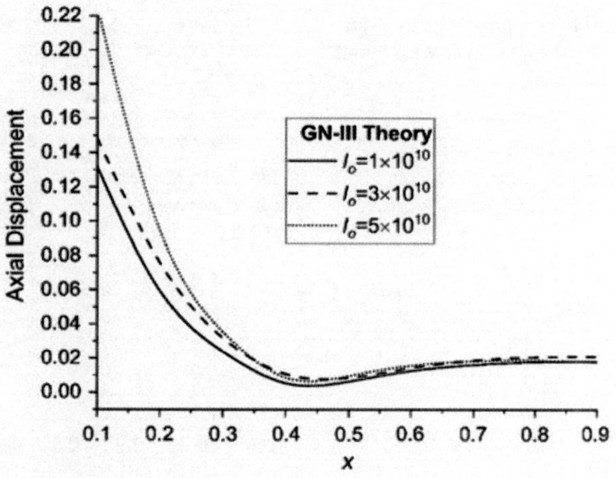

Fig. 9 Variations of thermal moment $M_T(x, t)$ with length of beam

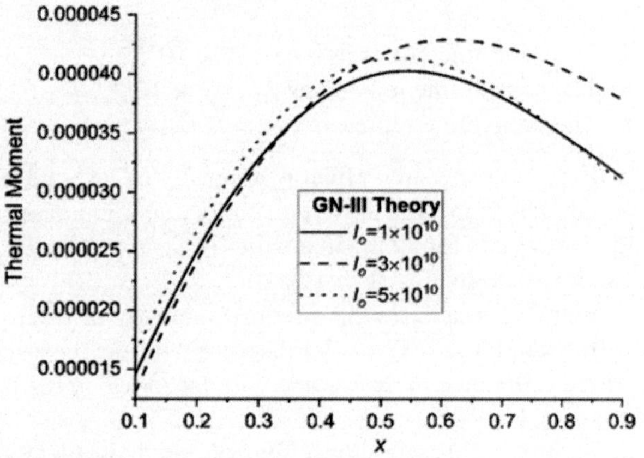

Fig. 10 Variations of temperature $T(x, z, t)$ with length of beam

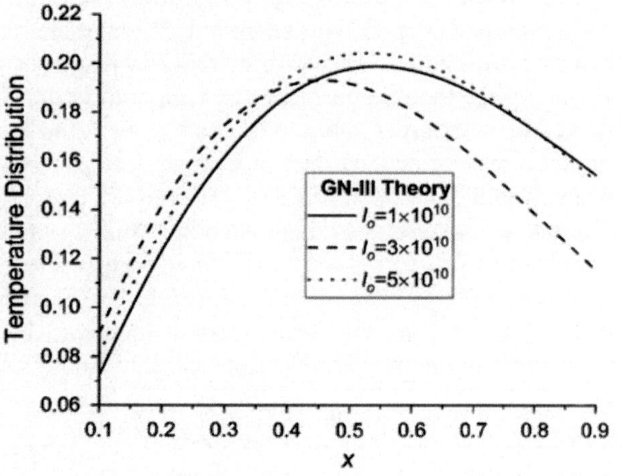

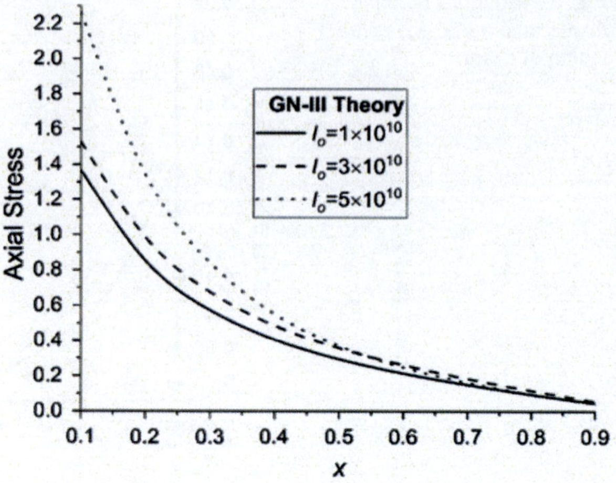

Fig. 11 Variations of axial stress $t_{xx}(x, z, t)$ with length of beam

- The solid line represents $I_0 = 1 \times 10^{10}$.
- The dashed line represents $I_0 = 3 \times 10^{10}$.
- The dotted line represents $I_0 = 5 \times 10^{10}$.

Figure 7 shows the effect of intensity of laser pulse on lateral deflection w w.r.t. length. It is found that the lateral deflection w increases for the initial range $x < 0.2$ and then decreases for all the intensities of laser pulse with larger value for $I_0 = 5 \times 10^{10}$ and lowest value for $I_0 = 1 \times 10^{10}$.

Figure 8 illustrates the effect of intensity of laser pulse on axial displacement u with beam length. The axial displacement decreases for $x < 0.4$ and then increases with a difference in magnitude with larger value for $I_0 = 5 \times 10^{10}$ and lowest value for $I_0 = 1 \times 10^{10}$.

Figure 9 illustrates the effect of intensity of laser pulse on Thermal moment M_T w.r.t beam length and Fig. 5 illustrates the effect of intensity of laser pulse on Temperature $T(x, z, t)$ with beam length. the thermal moment and the temperature shows an instant increase with increase in length of the beam while for TPL GN-III theory the thermal moment and temperature first increases for $x < 0.5$ and then decreases with larger value for $I_0 = 5 \times 10^{10}$ and lowest value for $I_0 = 1 \times 10^{10}$. Figure 6 represents the effect of intensity of laser pulse on the axial stress t_{xx} w.r.t. beam length. The axial stress t_{xx} decreases gradually in all GN-III cases with a difference in magnitude which becomes almost same at the end of the beam length with larger value for $I_0 = 5 \times 10^{10}$ and lowest value for $I_0 = 1 \times 10^{10}$.

In the Figs. 12, 13, 14, 15 and 16, the effect of GN-III theories on the lateral deflection, axial displacement, thermal moment, the temperature distribution and axial stress of the transversely isotropic thin beam is shown

- The solid line represents TPL GN-III theory when $\varepsilon = 1$
- The dashed line represents DPL GN-II theory when $\varepsilon = 0$.
- The dotted line represents Simple GN-I theory when $\varepsilon = 0$, $\tau_t = \tau_q = 0$.

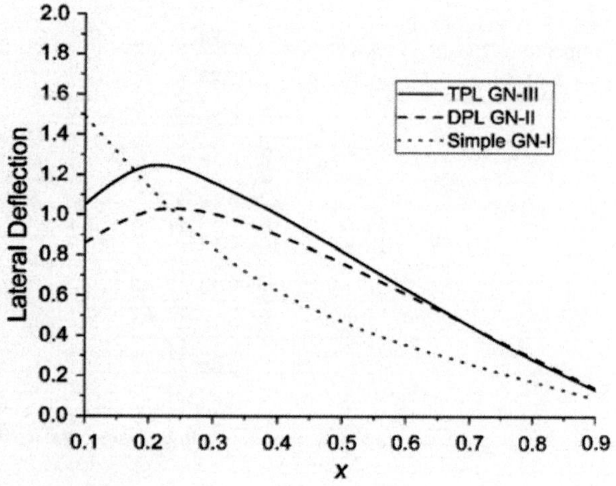

Fig. 12 Variations of lateral deflection with length of beam

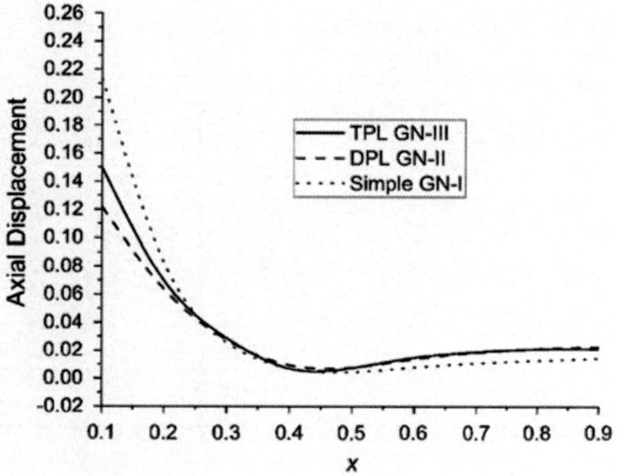

Fig. 13 Variations of axial displacement $u(x, z, t)$ with length of beam

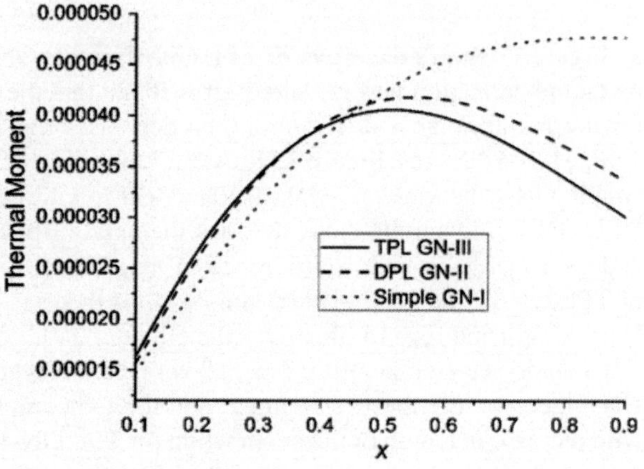

Fig. 14 Variations of thermal moment $M_T(x, t)$ with length of beam

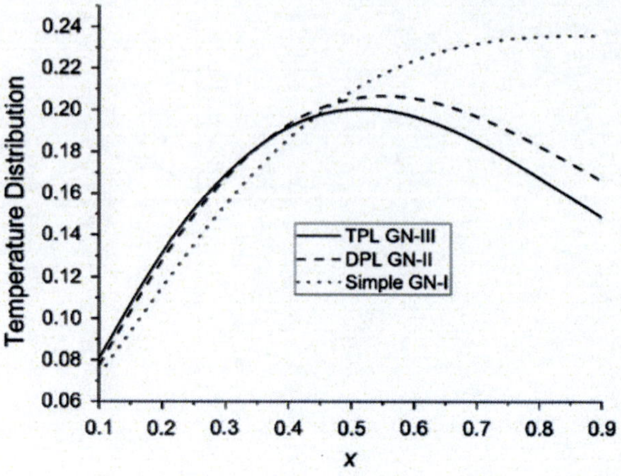

Fig. 15 Variations of temperature $T(x, z, t)$ with length of beam

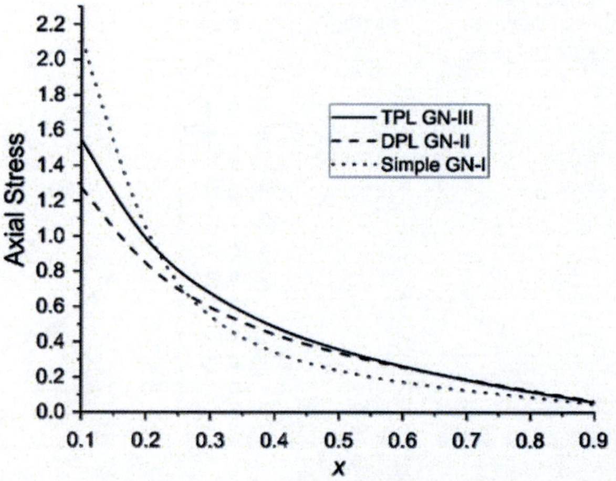

Fig. 16 Variations of axial stress $t_{xx}(x, z, t)$ with length of beam

Figure 12 shows the effect of TPL GN-III, DPL GN-II and simple GN-I theories on lateral deflection w w.r.t. length. It is found that the lateral deflection w increases for the initial range $x < 0.25$ and then decreases for TPL GN-III, DPL GN-II and simple GN-I theories. Figure 13 illustrates the effect of TPL GN-III, DPL GN-II and simple GN-I theories on axial displacement u with beam length. For TPL GN-III, DPL GN-II and simple GN-I theories the axial displacement decreases for $x < 0.4$ and then increases with a difference in magnitude. Figure 14 illustrates the effect of TPL GN-III, DPL GN-II and simple GN-I theories on thermal moment M_T with beam length and Fig. 15 illustrates the effect of TPL GN-III, DPL GN-II and simple GN-I theories on temperature $T(x, z, t)$ w.r.t. beam length. For DPL GN-II and simple GN-I theories, the thermal moment and the temperature shows an instant increase with increase in length of the beam while for TPL GN-III theory the thermal moment and temperature first increases for $x < 0.5$ and then decreases.

Figure 16 represents the effect of TPL GN-III, DPL GN-II and simple GN-I theories on the axial stress t_{xx} w.r.t. beam length. The axial stress t_{xx} decrease gradually in all cases with a difference in magnitude which becomes almost same at the end of the beam length.

9 Conclusions

- A new mathematic model is used to study the vibration phenomenon in transversely isotropic thermoelastic simply supported nanobeam in the context of new modified generalised three-phase lag Green Naghdi (TPL GN) model with multi thermal relaxation times subjected to laser pulse using Euler–Bernoulli theory have been investigated.
- Laplace transform technique has been utilised to obtain expressions for lateral deflection, axial displacement, temperature distribution, axial stress and thermal moment in the transformed domain and numerical inversion techniques are used to find the expressions in the physical domain and are shown graphically.
- It is observed that there is out of TPL, DPL, SPL theories, TPL theory has significant impact on lateral deflection, axial displacement, thermal moment, temperature distribution and axial stress of the transversely isotropic thermoelastic beam and shows the better results.
- The results exhibit the effect of intensity of laser pulse on the behaviour and variation of lateral deflection, axial displacement, thermal moment, temperature distribution and axial stress of the transversely isotropic thermoelastic beam. As the intensity f the laser pulse increased, higher is the variations in lateral deflection, axial displacement, thermal moment, temperature distribution and axial stress.
- This research helps in the design and construction of MEMS/NEMS, beam-type accelerometers, sensors, resonators, continuum mechanics, experimental wave propagation and other branches of engineering.

References

1. Quintanilla R, Racke R (2008) A note on stability in three-phase-lag heat conduction. Int J Heat Mass Transf 51(1–2):24–29
2. El-Karamany A, Ezzat MA (2013) On the three-phase-lag linear micropolar thermoelasticity theory. Eur J Mech a Solids 40:198–208
3. Kumar R, Kumar R, Kumar H (2018) Effects of phase-lag on thermoelastic damping in micromechanical resonators. J Therm Stresses 41:1115–1124
4. Kumar R, Kumar R (2019) Effects of phase lags on thermoelastic damping in micro-beam resonators. Int J Struct Stab Dyn 19(9):1971005
5. Codarcea-Munteanu L, Marin M (2019) A study on the thermoelasticity of three-phase-lag dipolar materials with voids. Bound. Boundary Value Problems

6. Abouelregal AE (2019) Three-phase-lag thermoelastic heat conduction model with higher-order time-fractional derivatives. Indian J Phys
7. Paul K, Mukhopadhyay B (2019) Two-dimensional generalized magnetothermoelastic diffusion problem for a thick plate under laser pulse heating with three-phase lag effect. J Eng Phys Thermophys 92:516–527
8. Abd-Elaziz EM, Othman MIA (2019) Effect of Thomson and thermal loading due to laser pulse in a magneto-thermo-elastic porous medium with energy dissipation. J Appl Math Mech 99(8)
9. Zenkour AM (2018) Refined two-temperature multi-phase-lags theory for thermomechanical response of microbeams using the modified couple stress analysis. Acta Mech 229(9):3671–3692
10. Kumar R, Devi S (2017) Interactions of thermoelastic beam in modified couple stress theory. Appl Appl Math Int J (AAM) 12(2):910–923
11. Zenkour AM, Abouelregal AE (2015) Thermoelastic vibration of an axially moving microbeam subjected to sinusoidal pulse heating. Int J Struct Stab Dyn 15(6):1–15
12. Lata P, Kaur I (2019) A study of transversely isotropic thermoelastic beam with green-Naghdi Type-II and Type-III theories of thermoelasticity. Appl Appl Math Int J (AAM) 14(1):270–283
13. Hobiny A, Alzahrani FS, Abbas I (2020) Three-phase lag model of thermo-elastic interaction in a 2D porous material due to pulse heat flux. Int J Numer Methods Heat Fluid Flow ahead-of-print:ahead-of-p
14. Abbas IA, Othman MI (2012) Generalized thermoelastic interaction in a fiber-reinforced anisotropic half-space under hydrostatic initial stress. J Vib Control 18(2):175–182
15. Marin M (1994) The Lagrange identity method in thermoelasticity of bodies with microstructure. Int J Eng Sci 32(8):1229–1240
16. Marin M (2010) Some estimates on vibrations in thermoelasticity of dipolar bodies. J Vib Control SAGE J 16(1):33–47
17. Marin M (2010) A partition of energy in thermoelasticity of microstretch bodies. Nonlinear Anal Real World Appl 11(4):2436–2447
18. Abbas IA, Youssef HM (2012) A nonlinear generalized thermoelasticity model of temperature-dependent materials using finite element method. Int J Thermophys 33:1302–1313
19. Sharma JN, Grover D (2011) Thermoelastic vibrations in micro-/nano-scale beam resonators with voids. J Sound Vib 330(12):2964–2977
20. Abbas I, Marin M (2017) Analytical solution of thermoelastic interaction in a half-space by pulsed laser heating. Phys E-Low-Dimension Syst Nanostruct 87:254–260
21. Othman MIA, Abbas IA (2011) Effect of rotation on plane waves at the free surface of a fibre-reinforced thermoelastic half-space using the finite element method. Meccanica 46:413–421
22. Kumar R, Abbas IA (2013) Deformation due to thermal source in micropolar thermoelastic media with thermal and conductive temperatures. J Comput Theor Nanosci 10(9):2241–2247
23. Riaz A, Ellahi R, Bhatti MM, Marin M (2019) Study of heat and mass transfer in the Eyring-Powell model of fluid propagating peristaltically through a rectangular compliant channel. Heat Transf Res 50(16):1539–1560
24. Bhatti MM, Ellahi R, Zeeshan A, Marin M, Ijaz N (2019) Numerical study of heat transfer and Hall current impact on peristaltic propulsion of particle-fluid suspension with compliant wall properties. Modern Phys Lett B 35(35)
25. Bhatti MM, Yousif MA, Mishra SR, Shahid A (2019) Simultaneous influence of thermo-diffusion and diffusion-thermo on non-Newtonian hyperbolic tangent magnetised nanofluid with Hall current through a nonlinear stretching surface. Pramana 93(6):88
26. Bhatti MM, Elelamy AF, Sait SM, Ellahi R (2020) Hydrodynamics interactions of metachronal waves on particulate-liquid motion through a ciliated annulus: application of bio-engineering in blood clotting and endoscopy. Symmetry 12(4):532–547
27. Lata P, Kaur I (2019) Plane wave propagation in transversely isotropic magnetothermoelastic rotating medium with fractional order generalized heat transfer. Struct Monitor Mainten 6(3):191–218

28. Lata P, Kaur I (2019) Transversely isotropic thermoelastic thin circular plate with time harmonic sources. Geomech Eng 19(1):29–36
29. Lata P, Kaur I (2019) Effect of inclined load on transversely isotropic magneto thermoelastic rotating solid with time harmonic source. Adv Mater Res 8(2):83–102
30. Kaur I, Lata P (2019) Transversely isotropic thermoelastic thin circular plate with constant and periodically varying load and heat source. Int J Mech Mater Eng 14(10):1–13
31. Lata P, Kaur I (2019) Thermomechanical interactions in a transversely isotropic magneto thermoelastic solids with two temperatures and rotation due to time harmonic sources. Coupled Syst Mech 8(3):219–245
32. Bhatti MM, Shahid A, Abbas T, Alamri SZ, Ellahi R (2020) Study of activation energy on the movement of gyrotactic microorganism in a magnetized nanofluids past a porous plate. Processes 8(3):328–348
33. Kumar R, Sharma N, Lata P (2016) Effects of thermal and diffusion phase-lags in a plate with axisymmetric heat supply. Multidiscipl Model Mater Struct (Emerald) 12(2):275–290
34. Zenkour AM, Mashat DS (2020) A laser pulse impactful on a half-space using the modified TPL G-N models. Sci Rep 10(1):4417
35. Rao S (2007) Vibration of continuous systems. Wiley, New Jersey
36. Press W, Teukolshy SA, Vellerling WT, Flannery B (1986) Numerical recipes in Fortran. Cambridge University Press Cambridge
37. Dhaliwal R, Singh A (1980) Dynamic coupled thermoelasticity, Hindustan Publication Corporation, New Delhi, India
38. Kumar R, Vohra R (2019) Response of thermoelastic microbeam with double porosity structure due to pulsed laser heating. Mech Mech Eng 23(1):76–85. https://doi.org/10.2478/mme-2019-0011

Chapter 7
Designing Techniques for 4G LTE Networks with QoS-Aware RD Network Based on Radio Resource Organization Approach

T. GangaPrasad and M. S. S. Rukmini

1 Introduction

Long-Term Evolution (LTE) and improved cellular system signifies the wireless technologies which lead to develop the future mobile broadband services. The LTE provides various advantages such as higher system capacity, low cost per bit, higher spectrum efficiency and higher data rate [1]. Hence the requirements of cellular networks are satisfied by using the heterogeneous wireless networks, i.e. LTE [2] as well as advanced LTE. Due to the small and macrocells, the hetero generous networks satisfy the necessities of growing broadband mobile traffics [3]. The device to device (D2D) communication as a cellular infrastructure is also represented as LTE direct communication. This D2D communication has high spectrum and energy efficiency as well as the Peer to Peer service has the advantages of reuse gain, proximity and hop [4]. In the architecture of third Generation Partnership Project (3GPP) only one node is located among the enhanced Node B (eNB) and User equipment (UE). Accordingly, the eNB contains the radio network controller which has the responsibility for Traffic balancing, mobility and radio resource management (RRM) [5].

Additionally the self-optimization in LTE is defined as process using the performance measures of UE and eNB that used for tuning the system parameters for obtaining the optimal performance [6], the less capacity in the cell edge is considered as constraint in the LTE advanced (LTE-A) systems. The reason behind the LTE with less capacity is interference caused by the users which leads to create aggressive frequency. In addition a complete isolation among the various parallel services and signaling overhead creates the issues in the random access network of

T. GangaPrasad
ECE Department, Vignan Foundation for Science Technology and Research (Deemed to be University), Guntur, AP, India

M. S. S. Rukmini (✉)
Vignan Foundation for Science Technology and Research (Deemed to be University), Hyderabad, Telangana, India

N. Marriwala et al. (eds.), *Soft Computing for Intelligent Systems*, Algorithms for Intelligent Systems, https://doi.org/10.1007/978-981-16-1048-6_7

LTE [7]. The key issue in the RRM is the interdependency of various OSI stack layers since the stack layer needs the existence of suitable cross-layer optimization techniques for achieving better performance. Additionally the existence of various access methods with multiple interfaces requires the deployment of various RRM modules as the same equipment [8]. Some of the existing RRM methods are given as hybrid approach [9] carrier aggregation based RRM [10] and path loss threshold-based component carrier and cluster configuration algorithm [11]. The major contribution of this research paper is given as follows.

The binary version of CSA is used in the 5G network for allocating an optimal bandwidth and power to the desired UEs since an effective searching process and less cooperation complexity are leads to improve the RRM using the BCSA.

The fifth generation (5G) and Long-term evolution-advanced (LTE-A) networks have turn out to be the most promising communication technologies; the credit goes to the aggregation of carrier components [1]. These technologies offers superior bandwidth with advanced throughput, maximize the number of admitted users, meets the increasing anxiety of mobile traffic in the network. The divergence necessities from various vertical industries have determined the standard shift in the next-generation (i.e. 5G) movable networks. In these technologies, network slicing has emerged as a major standard for this purpose by sharing and isolating resources over the same 5G physical infrastructure. To accurately fulfill the requirements of different quality-of-service (QoS) imposed by various network slices for different vertical applications, it is necessary to initiate a QoS-aware programmable data plane that is configurable to implement the QoS commitments [2].

As the wireless networks have constrained lifetime due to their limitations. In previous techniques, for example, in the wireless mobile communication system in the first generation, it uses analog technology for communication instead of a digital one. The design of the structure is based on the technology used for the transmission of the data packets [3]. These technologies have the goals for making a standard infrastructure that should be organized in a way to sustain the obtainable as well as the prospective services. These all generation styles require the transportation of the information should be designed in an organized way that it can develop itself as the technology changes [4]. On the other hand, all these development should be done exclusively to compromise the existing services of the current network. The well-known network, i.e., Wireless LAN, with predetermined internet and maintaining the mobile internet wirelessly since the matching QoS as predetermined Internet. Finally, the fifth-generation technique is predicated on fourth-generation technology only for making some extended uprising to the 5G technology. These issues should be resolved that are wider exposure and improvement from one to another technology for making a difference between the technologies [5]. Certainly, the throughput delays and blocked calls are some of the most important measurements of the level of QoS in the network. In this paper, a comprehensive view is provided upon various aspects of RROA, its challenges, and existing schemes. The existing RROA schemes are presented with their respective architecture which has a significant impact on the approaches. The problem of RROA is multi-dimensional and different dimensions are presented with their respective solutions such as interference or energy management.

Therefore, bandwidth increasing stills to be an insufficient solution for maintaining a good QoS level in the whole network, especially during congested situations, unless if it deals with the access management of network resources [6].

This paper is prearranged as follows. Section 2 contains the literature survey; it explains how to advance the scheduling organization system based on radio resource organization approach with QoS-Aware algorithm in 5G networks and LTE-A. Section 3 shows the proposed algorithm applied for LTE-A and 5G networks. The result comparison is given in Sect. 4. Finally, the conclusion of the work is done with a conclusion.

2 Literature Survey

The Long-Term Evolution (LTE) networks can be considered as other names of the fourth-generation technique due to its flying features. The systems of these generations can provide a speed of 10 M to 1Gbps, with good security as well as enhanced battery lifetime, and it will charge less per bit cost with some complicated design structure. Finally, the fifthgeneration mobile systems were begins at the end of 2010. This generation of mobile systems can provide large area coverage with good connectivity. The most important thing about this technology is it can work on Wireless World Wide Web and come as a complete wireless network system exclusive of restrictions. The fifth–generation system has as discussed extremely manageable, it gives very high transmission speed with elevated capacity, the unique feature is it provides a very bulky data propagation capacity of Gbps, the processing capacity s also very fast in contrast with the preceding generation technologies, and as compared to other preceding techniques it has more effectiveness with attraction [7].

The statements through the mobile has to turn out to be more fashionable in most recent years due to its exceptional growth starting from various generations of first to fifth generation in mobile machinery. It happens only due to the extremely sky-scraping augment in the number of users as well as the obligation of examination friendly broadcasting skills. The word generation refers to the revolutionization in the broadcasting of the information data with attuned bands of frequencies [5]. Since 1980, the mobile infrastructure has undergone significant variations from that time only this technology knowledgeable an enormous development. The first-generation systems are the earliest movable phones to be used and had an introduction in 1982 till 1990. The best example of the first-generation mobile systems is the advanced mobile phone system that is utilized through frequency-modulated technology. The basic features of first-generation systems are, the 2.4 kbps is the speed of the system, it uses an analog signal, the quality of the audio is poor, the lifetime of the battery is also poor with a large size, and it also faces security issues because it provides very poor security.

To transmit multiple streaming, in mp4form, from a sink to mobile clients various authors examined the scheduling algorithms. Several researchers have given an approach, in an epoch-by-epoch framework, to allocate the transmission slots fairly for the users in a wireless manner. Then they presented a fast lead-aware greedy scheduling algorithm. Zhiqiang et al. present a sub-channel design in dense scenarios. Various authors prove that a particular form of power allocation outperforms in contrast with the conventional schemes. Finally, they project an efficient method for jointly allocating power and a sub-channel based on the binary power allocation. Anyway, the main contribution of this generation is various technologies were introduced by the first generation only. These technologies are MTS, Advanced AMTS, IMTS, and a very famous technology is PTT.

2.1 LTE-QoS RRM

The network parameters like delay, jitter, bandwidth, and packet loss play a vital role in designing QoS-based LTE Systems. On the off chance that the system is limited in terms of resources guarantying essential QoS is significant for smooth network tasks explicitly for real-time steaming and multi-media applications like online games, Voice over IP (VoIP), and IPTV. The radio resource management (RRM) holds the power consumption of every sector in each cell. These cells are calculated depends upon the output of the antenna adjustment module. And SINR also calculated depends on the channels model, interference from the base adjacent BSs and transmission power of the BS. The limit of every user and the entire system is determined dependent on the power and SINR.

In parallel with the LTE radio access, packet core networks are also evolving into the flat SAE architecture. This new architecture is designed to optimize network performance, improve cost-efficiency, and facilitate the uptake of mass-market IP-based services. There are only two nodes in the SAE architecture user plane: the LTE base station (eNodeB) and the SAE Gateway, as shown in Fig. 1. The LTE base stations are connected to the Core Network using the Core Network–RAN interface, S1. This flat architecture reduces the number of involved nodes in the connections.

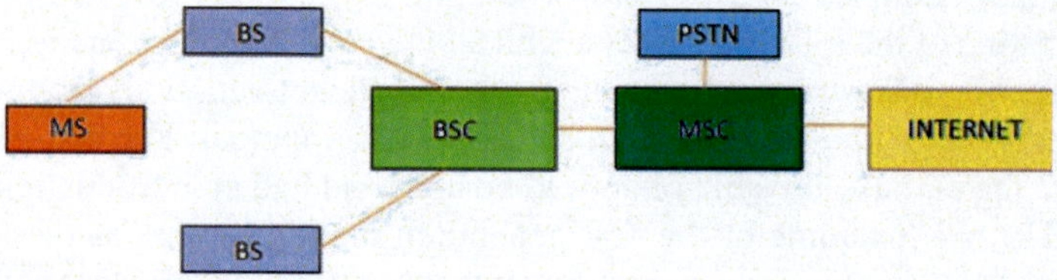

Fig. 1 Advance mobile organization

3 Proposed Work

3.1 Designing of Proposed Model

This section is divided into two parts as proposed model designing and the problem methodology. By doing so, we can allocate the available resources as well as can identify resource allocation with consequent solutions.

Methodology

The drawback of the current techniques has less improvement in downlink performance in soft-frequency reuse-based LTE. To overcome the drawback of the above-mentioned problem methods the proposed work mainly contributes in different ways as

1. CWO algorithm is proposed to share queuing criterion information, for the routing process in the network, with the neighbors used for the QoC-RRM scheme.
2. The RDNN is introduced by QoC-RRM to classify different terms based on multiple constraints, for the users, of priority and non-priority.
3. An assumption is made for the QoC that, for upcoming networks, depends on the management techniques for the available radio resource
4. For making a proper solution A hybrid algorithm is proposed, i.e., QoS-aware optimal confederation algorithm.
5. The allocated resources are directly maintained by the sinks only.

Designing of the proposed Model

The radio networks, here parallel with the LTE, with access packet core networks, are evolving into the SAE system architecture. For network performance optimization it is designed and can be used as a new architecture for improving the efficiency of the network along with IP-based service facilities for the mass-market. The SAE Gateway is the node in the SAE system architecture with e Node B as a second node for the user plane [12]. The basic structure for the Long-Term Evolution is shown in Fig. 1.

The QOC-RRM

The LTE networks are implemented by using a different technique called radio resource management for future applications. These proposed RRM schemes are designed using quality of services techniques along with aware optimal confederation technique which is nothing but the advanced method of QoS method. So the proposed scheme is designed by using advanced methods for the route-finding metrics such as Chaotic Weed Optimization (CWO) and the well-known Recurrent Deep Neural Network (RDNN) algorithms.

4 Performance Evaluation and Comparison

The evaluation of the performance is done in two different ways either by using the Sum Rate or the Guaranteed Rate. These methods, i.e., QOC-RRM performance using sum rate and QOC-RRM performance using guaranteed rate are discussed below with their outcomes.

4.1 QOC-RRM Performance Using Sum Rate

Besides, Fig. 2 as well as Fig. 3 shows the basic user performance compared with MOC-based SOC-RRM and with some other techniques. Furthermore, due to the constrained resource for the BSs, the basic CG SON RRM gives the worst performance of interference reading of RRM in the BSs. It is not an accurate illustration of the authentic downlink interference (Table 1).

Figure 4 depicts the presentation of the summation rate with diverse internal users for the indoor case. The performance shows that through this we can get the best sum rate comparison. When building distance MBS gets increased the macro base stations are utmost in the proposed work. Hence, we can say that the proposed gives an incremented 12% performance in contrast with the existing work (Tables 2, 3).

Hence, by seeing the Fig. 5th we can say that the building distance MBSs get increased the macro base stations are best in our proposed outdoor sum rate. The

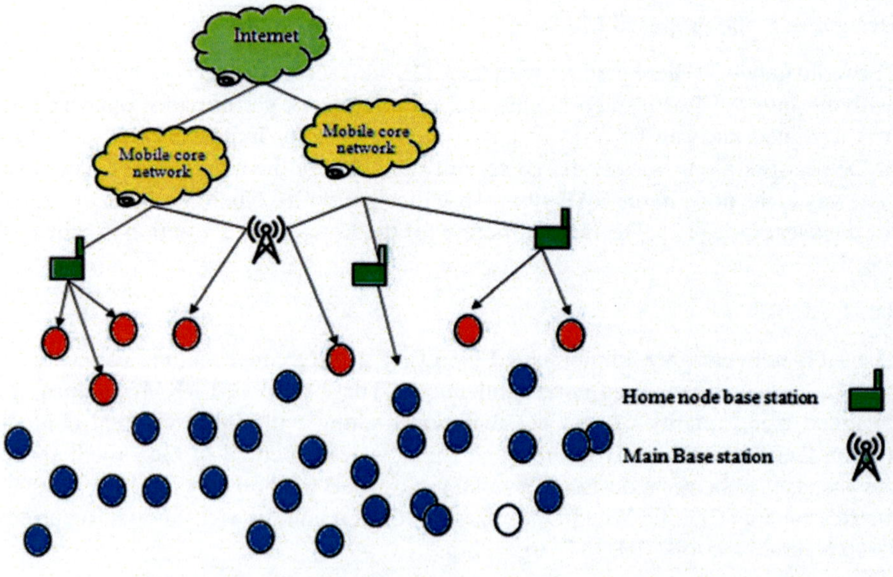

Fig. 2 System model (Proposed QoC-RRM)

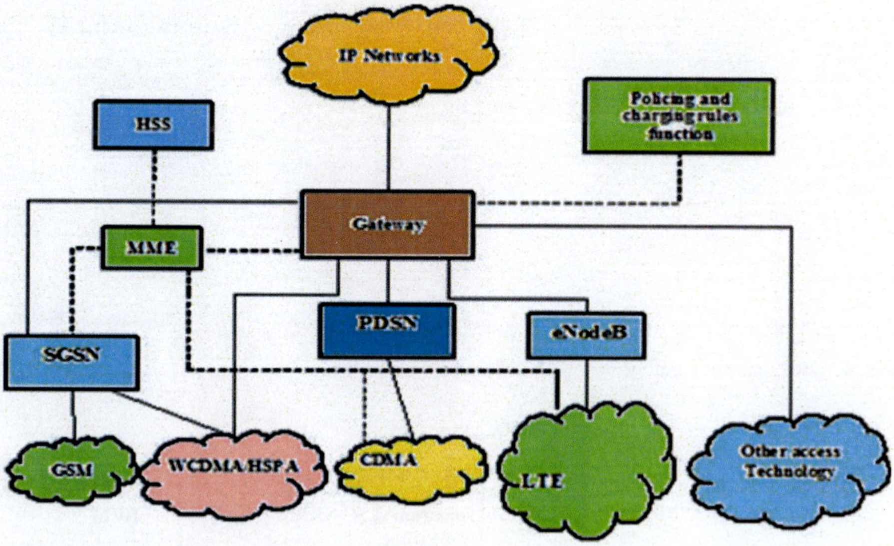

Fig. 3 The architecture of long-term evolution

Table 1 Bit rate (Indoor Users)

Sr. No	Building distance	Basic CG SONRRM	Dynamic CG SONRRM	Full spectrum utilization	Moc-based SONRRM	QOC-RRM	Self-Organization RRM
		Bit rate (MBS)	Bit rate (MBS)	Bit rate (MBS)	Bit Rate (MBS)	Bit rate (MBS)	Bit rate (MBS)
1	400	30	41	39	76	86	64
2	500	31	43	39	78	88	65
3	600	32	45	40	79	89	66
4	700	33	47	41	84	84	67
5	800	34	49	42	86	96	68
6	900	35	50	43	88	98	69
7	1000	36	53	44	90	99	71
8	1100	37	55	45	94	104	72
9	1200	38	57	46	98	120	74
10	1300	39	59	47	100	129	76
11	1400	40	60	48	110	130	77
12	1500	41	62	50	120	140	80

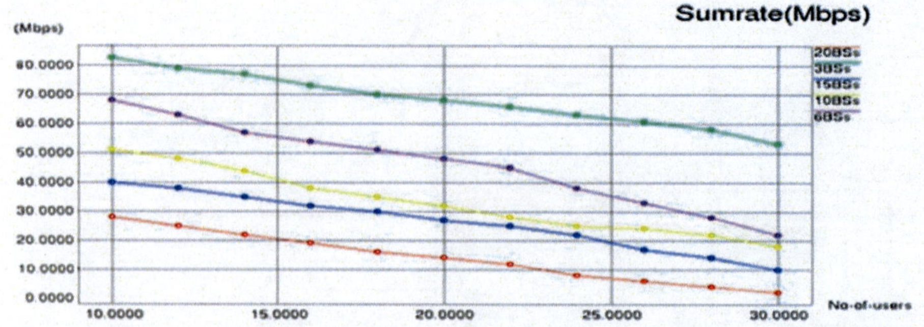

Fig. 4 User sum rate outdoor

Table 2 Outdoor sum rate

Sr. No	Building Distance	Basic CGSONRRM	Dynamic CGSONRRM	Full spectrum utilization	Moc-Based SONRRM	QOCRRM	Self-Organized RRM
		Sum rate (MBS)	Sum rate (MBS)	Sum rate (MBS)	Sum rate (MBS)	Sum rate (MBS)	Sum rate (MBS)
1	400	4	5	6	14	25	7
2	500	3	4	5	13	24	6
3	600	2	3	4	12	23	5
4	700	1	2	3	11	22	4
5	800	0.9	1	2	10	21	3
6	900	0.8	0.9	1	9	20	2
7	1000	0.7	0.8	0.9	8	19	1
8	1100	0.6	0.7	0.8	7	18	0.9
9	1200	0.5	0.6	0.7	6	17	0.8
10	1300	0.4	0.5	0.6	5	16	0.7
11	1400	0.3	0.4	0.5	4	15	0.6
12	1500	0.3	0.3	0.4	3	14	0.5

resulting outcome shows that the proposed work gives a 17% improved performance in comparison with the existing work.

4.2 *QOC-RRM Performance Using Guaranteed Rate*

Figure 6 shows that significant guaranteed rate improvement for users is given by the proposed method. Hence, we can say through the figure that the proposed technique can offer at slightest 50% QoS improvement compared to the existing works by users guaranteed rate (Tables 4, 5).

Table 3 Bit rate (Outdoors)

Sr. No	Building distance	Basic CGSONRRM	Dynamic CGSONRRM	Full spectrum utilization	Moc-Based SONRRM	QOC-RRM	Self-Organized RRM
		Bit rate (MBS)	Bit rate (MBS)	Bit rate (MBS)	Bit rate (MBS)	Bit rate (MBS)	Bit rate (MBS)
1	400	151	150	140	158	168	154
2	500	153	152	138	161	169	158
3	600	155	153	130	165	170	165
4	700	160	154	126	171	171	167
5	800	165	156	121	176	172	168
6	900	170	164	118	180	169	169
7	1000	143	168	110	174	168	171
8	1100	145	160	108	170	167	164
9	1200	147	154	106	165	166	158
10	1300	138	153	97	160	170	154
11	1400	132	152	94	151	174	152
12	1500	130	151	90	149	178	150

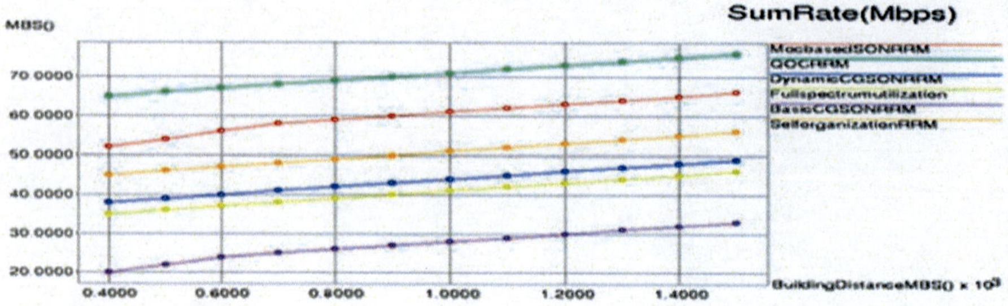

Fig. 5 Indoor sum rate

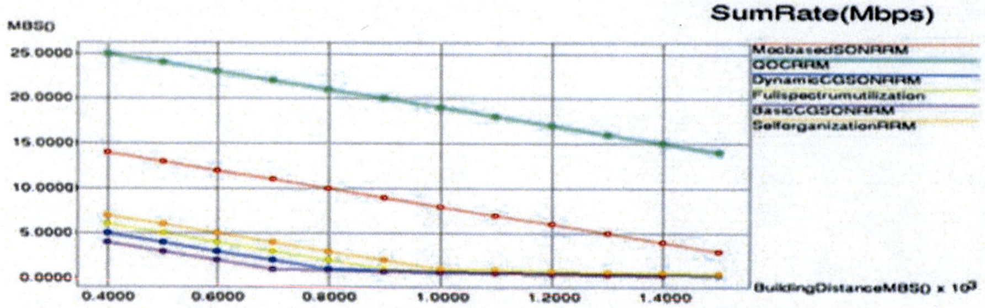

Fig. 6 Sum rate (Outdoor)

Table 4 Guaranteed rate (w.r.t Threshold)

Sr. No	Building distance	Basic CGSONRRM	Dynamic CGSONRRM	Full dpectrum utilization	Moc-Based SONRRM
1	10	410	410	310	210
2	12	444	418	318	218
3	14	467	429	329	229
4	16	480	431	331	231
5	18	493	436	336	236
6	20	523	438	338	238
7	22	559	480	380	280
8	24	580	484	384	284
9	26	597	488	388	288
10	28	625	500	400	300
11	30	640	540	440	360

Figure 7 shows a threshold presentation for the guaranteed rate that has the most excellent performance compared with the dissimilar station for the sinks. While there is no increment in the consumer number, the kbps are at its utmost rate in the

Table 5 Sum rate outcomes (w.r.t no. of Users)

Sr. No	Threshold (dB) / Sum rate (kbps)	3BSs	6BSs	11BSs	15BSs	20BSs
1	10	83	68	51	40	28
2	12	79	63	48	38	25
3	14	77	57	44	35	22
4	16	73	54	38	32	19
5	18	70	51	35	30	16
6	20	68	48	32	27	14
7	22	66	45	28	25	12
8	24	63	38	25	22	8
9	26	61	33	24	17	6
10	28	58	28	22	14	4
11	30	53	22	18	10	2

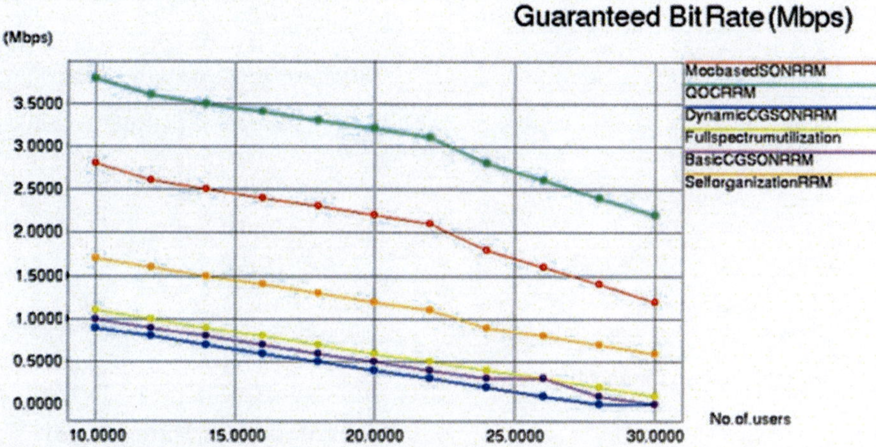

Fig. 7 Guaranteed rate performance comparison

proposed work. The projected threshold rate for guaranteed will use 15BSs that give the superlative result in contrast with the existing work (Table 6).

Figure 8 shows the Indoor performance of the guaranteed rate which gives the best sum rate compared with the different indoor users. When building distance MBS gets increased the macro base stations are maximum in our proposed indoor sum rate. Our proposed indoor sum rate compared with other existing systems 10% improved in our performance. Figure 9 presents the performance for the outdoor rate that gives the superlative outcome in contrast with dissimilar users for outdoor techniques. When building distance MBSs get increased the macro base stations are best in the proposed outdoor guaranteed method. The projected outdoor guaranteed in comparison with other existing systems 15% improved in our performance (Fig. 10 and Table 7).

Table 6 User's bit rate

Sr. No	No. of Users	Basic CGSONRRM	Dynamic CGSONRRM	Full spectrum utilization	Moc-Based SONRRM	QOCRRM	Self-Organization RRM
		Bit rate	Bit rrate	Bit rate	Bit rate	Bit rate	Bit rate
1	10	1	0.9	1.1	2.8	3.8	3.8
2	12	0.9	0.8	1	2.6	3.6	3.6
3	14	0.8	0.7	0.9	2.5	3.5	3.5
4	16	0.7	0.6	0.8	2.4	3.4	3.4
5	18	0.6	0.5	0.7	2.3	3.3	3.3
6	20	0.5	0.4	0.6	2.2	3.2	3.2
7	22	0.4	0.3	0.5	2.1	3.1	3.1
8	24	0.3	0.3	0.4	1.8	2.8	2.8
9	26	0.3	0.1	0.3	1.6	2.6	2.6
10	28	0.1	0	0.2	1.4	2.4	2.4
11	30	0	0	0.1	1.2	2.2	2.2

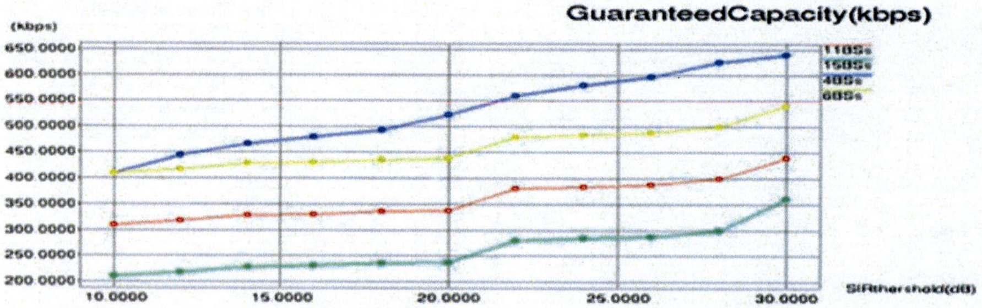

Fig. 8 Guaranteed rate (Threshold)

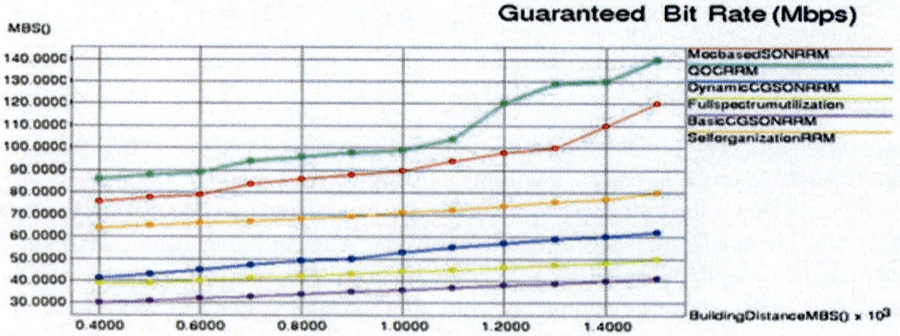

Fig. 9 Indoor guaranteed rate

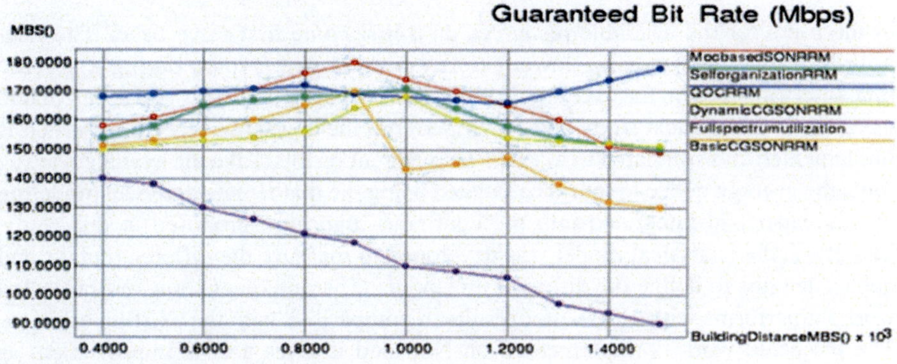

Fig. 10 Performance of outdoor Guaranteed Bit rate

Table 7 User's sum rate Calculation

Sr. No	No.of Users	Basic CGSONRRM	Dynamic CGSONRRM	Full spectrum utilization	Moc-based SONRRM	QOCRRM	Self-organization RRM
		Sum rate	Sum rate	Sum rate	Sum rate	Sum rate	Sum rate
1	10	30.5	49	31.5	45	55	42
2	12	30.6	48	31.6	46	56	43
3	14	30.7	47	31.7	47	57	44
4	16	31.2	46	32.2	48	58	45
5	18	31.5	45	32.5	49	59	46
6	20	31.7	44	32.7	50	60	47
7	22	32.1	43	33.1	51	61	48
8	24	32.4	42	33.4	52	62	49
9	26	33.3	41	34.3	53	63	50
10	28	33.8	40	34.8	54	64	51
11	30	34	39	35	55	65	52

5 Conclusion

Compassion is annoying in today's life for turning out to be completely wireless and trying to become a continuous way into the edifying data anytime and wherever through improved bandwidth, less cost, enhanced superiority, with high speed. The proposed work implements an LTE network through QoS along with radio resource management for making it LTE-A. In this case, the QOC-RRM method is used for QoS. The CWO algorithms are proposed and analyzed with other chaotic weed optimization for queuing principle with the routing process to share the information data. Hence, after receiving the data the sink, based on priority, schedules the priority of the users for the available resources on a first come first serve basis. The low-priority traffic classes, using lower priority bearers, are favored during a specific time interval based on the average waiting time for each class. This one is calculated using the mathematical study and integrated into the courteous algorithm, which is implemented and simulated to discuss the different results. Like the average waiting time, the average queue length is calculated using the mathematical model modeling in this paper and integrated into the courteous algorithm proposed in this study. Therefore, the analytical model was developed to measure the different delays and queues lengths to define the different thresholds. The outcomes show our projected work outperforms with 10% better results in comparison with the existing exertion. The projected work is implemented on NS3 and it gives a 50% improvement as compared to the obtainable method.

References

1. Srivastava A, Gupta MS, Kaur G (2020) Energy efficient transmission trends towards future green cognitive radio networks (5G): progress, taxonomy and open challenges. J Netw Comput Appl 102760
2. Solomon R (2020) Meshiness: mesh networks and the politics of connectivity. New York University, PhD diss.
3. Lodhi AK, Sattar SA (2019) Cluster head selection by optimized ability to restrict packet drop in wireless sensor networks. Soft Comput Data Anal 453–461.
4. Lodhi A, Rukmini MSS, Abdulsattar S, Tabassum SZ (2020) Performance improvement in wireless sensor networks by removing the packet drop from the node buffer. Mater Today: Proc
5. Arakadakis K, Pavlos C, Alexandros F (2020) Firmware over-the-air programming techniques for IoT networks--a survey. arXiv preprint arXiv:2009.02260
6. Al-Turjman F, Abujubbeh M, Malekloo A, Mostarda L (2020) UAVs assessment in software-defined IoT networks: an overview. Comput Commun 150:519–536
7. Mannweiler C, Sartori C, Wegmann B, Flinck H, Maeder A, Goerge J, Winkelmann R (2020) Evolution of mobile communication networks. In: Towards cognitive autonomous networks: network management automation for 5G and beyond 29–92
8. Ksouri C, Jemili I, Mosbah M, Belghith A (2020) Towards general Internet of Vehicles networking: routing protocols survey. Practice and Experience, Concurrency and Computation, p e5994
9. Loussaief F, Hend M, Hend K, Faouzi Z (2020) Radio resource management for vehicular communication via cellular device to device links: review and challenges. Telecommun Syst 1–29
10. Rukmini MSS, Amairullah KL (2020) Network lifetime enhancement in WSN using energy and buffer residual status with efficient mobile sink location placement. Solid State Technol 63(4):1329–1345
11. Babekier NEA (2018) Performance evaluation of enhanced inter-cell interference coordination with pico-cell adaptive antenna in long term evolution-advanced. PhD diss., Sudan University of Science and Technology
12. Bilbao MN (2019) Cristina perfecto del amo. PhD diss., University of the Basque Country

Chapter 8
Enhancing Software Quality Assurance by Using Knowledge Discovery and Bug Prediction Techniques

Alankrita Aggarwal, Kanwalvir Singh Dhindsa, and P. K. Suri

1 Introduction and Foreword

Software imperfection forecasting is standout among the most lively research areas in software engineering [1]. A rundown of defect inclined source code is gotten utilizing the defect prediction model. Identifying and finding the faults previous to the entrance of the piece or amid software development to build software reliable and imperfection forecast model is used [2]. As the span of software ventures get to be bigger, defect prediction will assume an essential part to bolster designer. The defect prediction process in which the prediction model is to produce examples is named with buggy/perfect or some bugs used to assess defect prediction execution. Distinctive sorts of measurements can be utilized as a part of the defect forecast model and broadly utilized measurements are code and procedure measurements [3].

Various models are worked to decide the defects. To develop the defect prediction model, machine learning methods are utilized. The two-level of machine inclining depend on preparing information (to prepare the information) and testing information (to break down model execution) [4].

Software quality assurance, source code consistency are key tasks in the corporate segment. There are two approaches to get the standard of source code and as per research analytics [5], technological and philosophical interpretation. The

A. Aggarwal (✉)
Department of Computer Science and Engineering, IKGPTU, Kapurthala, Jalandhar 144603, Punjab, India

K. S. Dhindsa
Department of Computer Science and Engineering, Baba Banda Singh Bahadur Engineering College, Fatehgarh Sahib 140406, Punjab, India

P. K. Suri
Department of Computer Science and Applications, Kurukshetra University, Kurukshetra 136038, Haryana, India

N. Marriwala et al. (eds.), *Soft Computing for Intelligent Systems*, Algorithms for Intelligent Systems, https://doi.org/10.1007/978-981-16-1048-6_8

professional concept is linked to other computer-style source code manuals, which emphasize readability, and certain language-specific rules are directed at retaining source code functionality, which includes testing and upgrading. The technical description then refers functional structure of code on usable parts in addition to certain consistency characteristics such as readability, security, checking, portability, sophistication, and others [6].

This research would assess the high-tech of source code concerning program metrics, a key purpose of which review current program metrics and checks the nature of the field and those metrics don't endure much. So it understood how the quality of the source code evaluates over years. This research does not, therefore, include other performance-related program metrics, productivity [7], among others [8–10].

Excellence is an observable fact involving a variety of variables that are dependent upon individual actions that can't be readily forbidden [11]. The methods of metrics may calculate and evaluate the variables of this type. You can find certain definitions for software metrics in the literature [12, 13]. The best motivation to measure is found in agreement with Daskalantonakis, which is to discover a mathematical value for any product or process attributes. These principles should contrast with one another and with the expectations that exist in an organization, such results, assumptions may be made about the product quality or the efficiency of the software method used.

Different frameworks and models have been proposed in the studies and there evaluation and analysis of the projects are also done by using different types of tools and algorithms [14–19].

As per the traditional implementations and study the greatest justification for measurements is that for certain software device attributes or software process attributes software calculation must have a numerical value [20]. Such values comparison with one another and from the standard that applies in a company. From results, assumptions may be made regarding product excellence or the nature of the software process used [21].

Somerville classified metrics into two groups- (a) control metrics that are linked with the software process (b) to forecast metrics that are related to software [22].

The emphasis in this research is to Predict Metrics, as they predict metrics compare to software's dynamic and static features [23]. It is possible to quantify the sophistication of the algorithm, volatility, bonding, continuity, inheritance, and other software attributes, according to certain characteristics. To analyze attributes can deduce excellence of software recommended points of enhancement about attempt, handling, testable and reusable [24].

The secondary indicators will provide an impression of the key machine module or table of databases [25]. The metric validation is deprived, as it ignores to employ modules in designing the system. Several studies have achieved a strong association between metric values and counts of faults, but only where there are a limited number of modules in the analyzed method. Here it is to be noted that it gave birth to the use of software tools used for designing that was reported. McClure has identified yet another difficulty metric [26]. This research focuses on the nature of the control variables and structures to guide the incantation of measures in the system [27].

Woodfield has released yet another difficulty method metric after some time. He observed that the metric is called by another component every time by a different context [10], a value used in another component, a given component must be understood. The provided variable must be checked in every new sense. Now it is to be noted that every revision is taking lesser effort compared to the previous [28], due to the learning from previous reviews. Consequently, the difficulty of increasing analysis is weighted by every function [29]. They have effectively implemented the metric in a review of student multi-procedure systems. We don't consider documentation on software applying this measure. We have discovered information regarding application of metric [30].

Henry and Kafura developed a further metric of machine sophistication. The metric of Henry and Kafura measures the difficulty of a process, which is based on two factors [31], the difficulty of code of process and complication of relations of the procedure to its context [32].

Methods of Henry and Kagura are more comprehensive with that of Yin and Winchester, metric, as follows the movement of knowledge rather than simply streaming over level boundaries. Some definitions, such as the definition of flows and definition of modules, are however puzzling. As a result, dissimilar researchers interpret the metric in different ways thereby troubling results of the assessment [33].

The measures of the binding between groups are the metrics [34]. Their empirical analysis suggests that coupling would very likely be a significant structural aspect to consider if one wants to construct quality models of OO architecture. For this, we could find an instrument for the extraction of metrics. Throughout the '90 s, work into machine measurements began strong. Some other OO metrics such as [35] have been developed; several works have been published evaluating and validating metrics. After 2000, the scenario about the software metrics is changed and is reported very less about new metrics [36].

Nowadays, most of the business as well as social media services are available online on different virtual platforms including mobile apps and web portals. These online platforms provide services with a higher degree of accuracy and performance without delay [35–37]. Even the small business ventures and entrepreneurs are now developing and launching their mobile apps so that their visibility and accessibility can be increased with the 24×7 availability as well as communication with the customers [38].

With the development and launching of mobile apps, there exist so many risk factors and vulnerabilities with each app because it is always required to test the mobile application on different parameters [39]. Simply deployment of a mobile app without proper and rigorous testing can be dangerous for the vendor. Many times it happens that the apps of e-wallets get exploited and the app users lose their money. Such situations hurt the overall reputation and reliability factors [40] of the seller or the company owning that app. It is always desired that the mobile app should be tested with different types of attacks so that the prior information on the behavior of the app can be obtained. Once the app is tested on different types of traffic, it becomes very useful to evaluate the performance of the app without actually launching the app on cyberspace. Also, we can predict the software using the soft computing methods like neural networks and classification [10, 41].

2 Testing Strategies

Several testing strategies are available which can be used for the evaluation of performance associated with the mobile app. This process of testing mobile apps on different dimensions is also known as a performance audit of the app [42]. In this process, the app is intentionally given malicious traffic so that the behavior in such an environment can be analyzed. There are many types of testing approaches that can be used for a performance audit of the mobile app (Fig. 1).

Following are the key testing strategies for mobile applications.

2.1 Functional Testing

This strategy ensures that all the functions, links, and buttons are working as per the desired operations in the mobile app [43]. Different options and further links are navigated in this process to confirm the overall flow of the application in this approach.

Fig. 1 Testing strategies for mobile applications

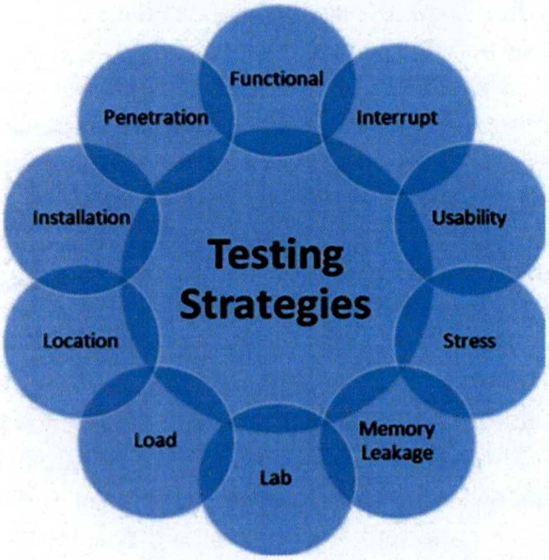

2.2 *Interrupt Testing*

As mobile apps work in the network environment, it is necessary to check the behavior of an application on assorted interrupts including.

- Network Coverage
- Incoming and Outgoing Calls
- Incoming and Outgoing SMS
- Battery Removal
- Media ON/OFF
- Power Cycle
- Data Transfer with Cable
- Roaming.

The app is considered effective on confirmation of all interrupt handling without any error, warnings, or system-generated exceptions [38].

2.3 *Usability Testing*

It ensures that the application is usable and working on different platforms and devices without compatibility issues. The ease in using the application with flexibility in handling different controls and options are considered in this testing strategy. It ensures the key dimensions and with the evaluations on assorted features and all units of the software.

2.4 *Stress Testing*

Stress testing aims to analyze the reliability and stability of the application on exposure to different types of abnormal traffic or inputs. Stress testing checks whether the application is getting any breakpoint on particular inputs. Also, the modes and points of failure in the application are identified in this testing [44]. In this approach, the elements with multiple perspectives have consistency in multiple features.

2.5 *Memory Leakage Testing*

As mobile devices are constrained in the memory (RAM, Internal Memory, SD Card Storage), it becomes necessary to check whether the application is consuming additional memory. An app is considered effective which works efficiently with fewer resources and higher performance [45].

2.6 Lab Testing

Laboratory or Lab testing provides the mobile app a simulated wireless environment without actual network coverage. It is implemented so that the coverage and behavior of Calls, Coverage, SMS, and other functions can be evaluated on different bandwidths.

2.7 Load Testing

The load testing includes the analysis of mobile app with exposure to enormous users at a time. Load testing checks [2] whether the app is getting crashed or hang if the heavy load of users is given to the app. For example, the mobile application is opened concurrently by one thousand users to check whether it can handle such load generated by a thousand users.

2.8 Location Testing

The performance of the mobile application in location testing includes whether the app can identify the network coverage and location correctly. The compatibility of the app in different networks is checked in this strategy. The assorted regions and places are tested with the presented outcomes to have consistent results with all-around consistency.

2.9 Installation Testing

This process ensures whether the application will be installed on all types of devices having different configurations. For example, the app can be tested to be installable on different versions of the mobile operating system and handsets of different memory constraints.

2.10 Penetration Testing

The process of penetration testing ensures whether the forms and online widgets of the app are fully secured. This type of testing applies to responsive and monetary applications like banking, online shopping-commerce, etc. The hackers should not be able to penetrate the payment gateway of the app with different tricks [46].

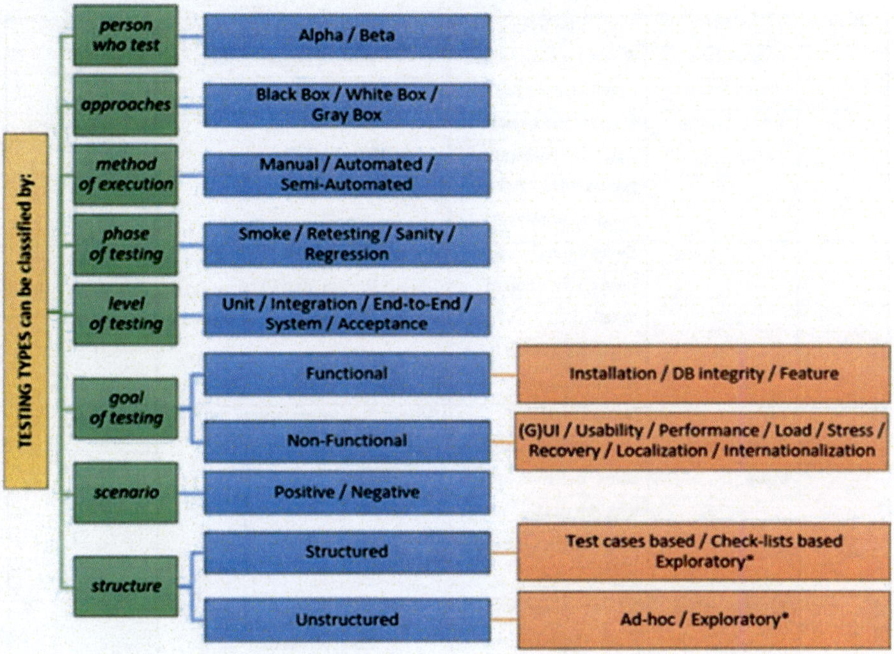

Fig. 2 Assortment of testing strategies

The figure below shows the different phases and their requirements at various levels and the associated deliverables (Figs. 2 and 3).

3 Metrics or Formulations

The number of metrics is used in SDP models and the past Software development process happening from the 1970s. Most metrics are listed as code metrics and process metrics. These are quite important to evaluate the overall criteria and are referred to as product metrics [47] (Fig. 4).

3.1 Code Metrics or Formulations

This Metric is used to calculate the density of the program and is more prone to errors as their source is very complex from the statistics composed from the present source code [34]. It is quite useful to have the analysis of source code with the multiple key features in the software product with prime features.

Sno.	Phase	Activity	Deliverables	Necessity
1.	Requirements/ Design Review	Review the software requirements/ design	• Review Defect Reports	Curiosity
2.	Test Planning	Plan for tests, once you have gathered a general idea of what needs to be tested.	• Test Plan • Test Estimation • Test Schedule	Farsightedness
3.	Test Designing	Design/ detail your tests on the basis of detailed requirements/design of the software.	• Test Cases / Test Scripts/Test Data • Requirements Traceability Matrix	Creativity
4.	Test Environment Setup	Setup the test environment (server/ client/ network, etc) with the goal to replicate the end-users environment.	• Test Environment	Rich company
5.	Test Execution	Execute your Test Cases/ Scripts in the Test Environment to check if they pass.	• Test Results (Incremental) • Defect Reports	Patience
6.	Test Reporting	Prepare various reports for your stakeholders.	• Test Results (Final) • Test/ Defect Metrics • Test Closure Report • Who Worked Late & on Weekends (WWLW) Report.	Diplomacy

Fig. 3 Phases and its requirements at various levels

Fig. 4 Types of metric

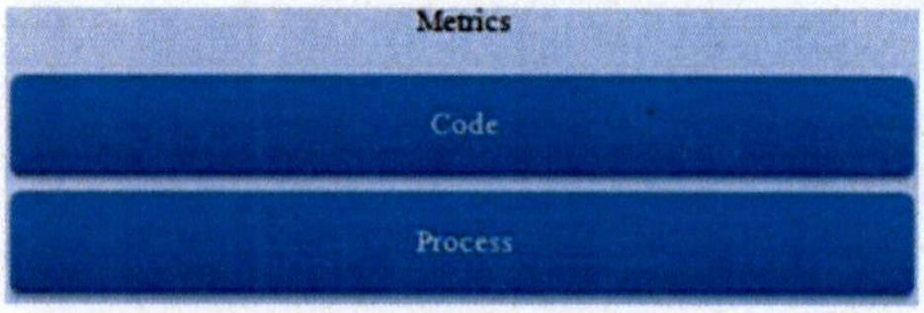

3.1.1 Size Metrics or Formulations

This metric calculates distance, weight, amount, and total significance of software items, and such metrics are called code lines (LOC). Line of Code's complexity is very basic and quick to comprehend [48].

3.1.2 Halstead Metrics or Formulations

This metrics deals with vocabulary, volume, effort, program length, difficulty, and time [49]. Most of the metrics are associated with magnitude or size and is used extensively.

x= Key dissimilar operators

y = Key dissimilar operands

a = Key Operators

b = Prime Operands.

Formulations Factors are

Program length: $c = a + b$

Volume: $V = c + logz$

Program vocabulary: $z = x + y$

Calculated program length: $L = x*\log(x) + y*\log(y)$

Difficulty: $D = x/2 + b/y$

Effort = $D*V$

Here the benefits are data flow is measured and demerits are it does not look upon floe of control.

3.1.3 McCabe Metric or Formulations

McCabe proposed cyclomatic metric and complexity of software products is represented by cyclomatic metric. Cyclomatic metric is represented by arcs, connected components, and nodes as control flow graphs of source code. Numbers of complex control path are measured by McCabe's cyclomatic metric. The cyclomatic complexity is diverse from Halstead and size metrics [50]. Demerits of data flow are not included.

3.1.4 Object-Oriented Metric or Formulations

OO metric is used to map to real-world entities and improves the development process.

3.1.5 CK Metrics or Formulations

CK Metrics is used to represent OO metric's characteristics like cohesion, inheritance, and coupling to design the CK metric by building a DPM.

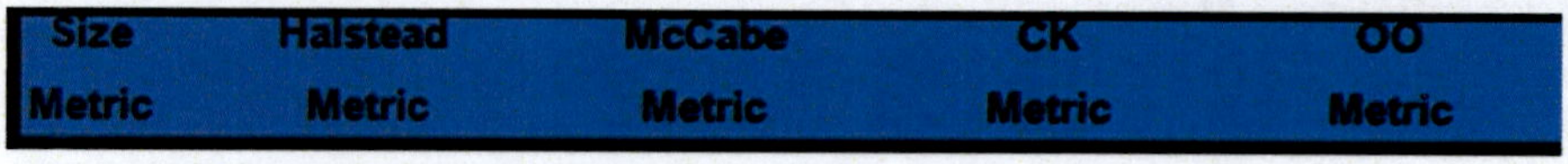

Fig. 5 Types of process metric

3.2 Process Metrics or Formulations

The process metric is acquired from software repository, i.e., issue and version control system having history of development [25] (Fig. 5).

3.2.1 Relative Code Change Churns

It is an excellent predictor to explain aspects like measuring total changes in code and. computed by number of code lines added and divided by total lines of code.

3.2.2 Change Metrics or Formulations

As discussed above in relative code change churn this metric includes deleted and added lines of code on the other hand file count and total LOC are not considered by considering maximum values and average values (Fig. 6).

4 Statistical Analysis Based Modules

4.1 Test Case Productivity (TCP)

The formulation offers test case writing productivity so that conclusive remark can be made.

Sno.	Process Metric	Metric Source
1.	Relative Code change churn	Version control
2.	Change Metric	Version control
3.	Change Entropy	Version control
4.	Code metric churn, Code Entropy	Version control
5.	Popularity	Version control
6.	Ownership	Version control
7.	Micro Instruction Metric	Mylyn

Fig. 6 Sources of process metric

Fig. 7 Example of the code of different projects

Test Case Name	Raw Steps
PROJECT_1	30
PROJECT_2	32
PROJECT_3	40
PROJECT_4	36
PROJECT_5	45
Total Raw Steps	183

$$\text{Test Case Productivity} = [[\text{Total Raw Test Steps}/\text{Efforts(Hours)}]^{*}\text{Step(s)}/\text{Hour}]]$$

Here we can see that effort if writing 183 steps is 9 h when TCP = 183/9 = 20.3

Therefore, Test case productivity = 21 steps/hour. Test case productivity can be compared to the value with the previous release so that effective conclusion can be made. Following Fig. 7 is the example of codes of different projects.

4.2 *Test Execution Summary*

This metric or formulation classifies test cases status along with reason to be accessible to a variety of test cases and provide cavernous view of release also trend for classification of reasons if not capable to fail test cases. Data can be collected for number of test case executed if it passes, fails for any reason, not able to run with reason, crunch of time, defect postponement, issue setup, out of scope (Fig. 8).

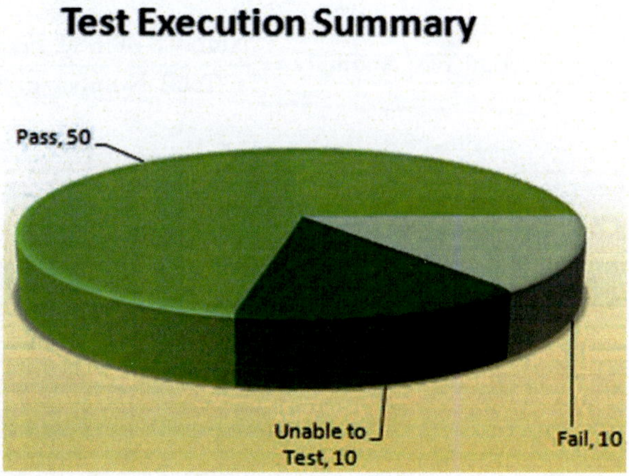

Fig. 8 Test execution summary

4.3 Anomaly Acceptance (AA)

This is to measure the valid number of anomalies identified during execution by the testing team and it can get better value if it is compared with the previous release of the software.

$$\text{Anomaly Acceptance} = \frac{\text{Number of Valid Defects}}{\text{Total Number of Defects}} * 100\%$$

4.4 Anomaly Refutation (AR)

A metric conclude amount of anomalies rejected during carrying out program and also gives the percentage of the invalid anomalies opened as well as controlled by the testing team.

$$\text{Anomaly Rejection} = \frac{\text{Number of Defects Rejected}}{\text{Total Number of Defects}} * 100\%$$

4.5 Bad Fix Anomaly (BFA)

This shows that anomaly resolution gives rise to new anomaly(s) are bad fix anomaly and determines effectiveness of anomaly resolution process and returns percentage of bad anomaly resolution which needs to be controlled

$$\text{Bad Fix Anomaly} = \frac{\text{Number of Bad Fix Defect(s)}}{\text{Total Number of Defects}} * 100\%$$

4.6 Variance of Effort

This metric gives the variance in the estimated effort. (Fig. 9)

$$\text{Effort Variance} = \frac{\text{Actual Effort} - \text{Estimated Effort}}{\text{Estimated Efforts}} * 100\%$$

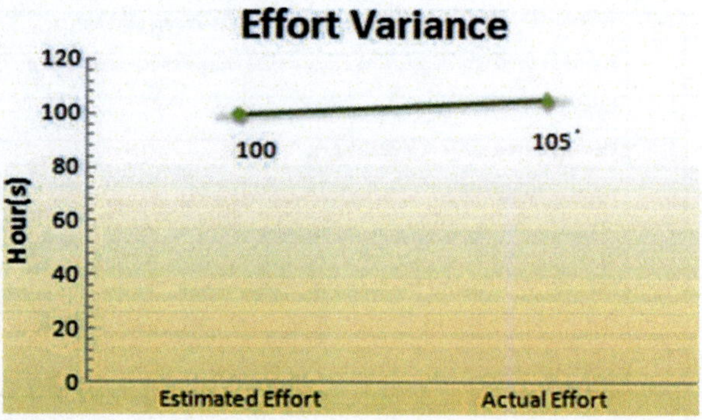

Fig. 9 Effort variance trend

4.7 Scope Change (SC)

To have the analysis on scope, there is a specific formulation on the metrics as mentioned and presented with the parameters

$$\text{Scope Change} = \frac{\text{Total Scope} - \text{Previous Scope}}{\text{Previous Scope}} * 100\%$$

where

Elevating Aspect and Scope = Traditional Scope + Novel Scope, if Scope Elevates

Elevating Aspect and Scope = Traditional Scope-Novel Scope, if Scope Reduces

Figure 10 shows the lines of code written and their detailed test case productivity.

For instance, most testing takes place in a phased process after framework requirements are defined and then put into place in testable initiatives. Conversely, specifications, planning, and training are always performed concurrently with an agile strategy.

5 Proposed Approach Using Training of Dataset

The proposed method is having the approach of statistical analysis and data science-based prediction using the integration of training of dataset. The dataset from the benchmark sources is fetched and trained with the model-based outcomes and the accuracy is found on the confusion matrix. Propagation function computes input *Pj* (*t*) from output so_i (*t*) as follows:.

$$p_j(t) = \sum_i o_i(t) w_{ij} \tag{1}$$

Sno.	Provides the information for the number of step(s) written per hour.	Test Case Productivity
1.	Provides statical view of execution for the release along with status and reason.	Test Execution Summary
2.	Indicates the stability and reliability of the application.	Defect Acceptance
3.	Provides the percentage of invalid defects.	Defect Rejection
4.	Indicates the effectiveness of the defect-resolution process	Bad Fix Defect
5.	Provides detail of the test case executed per day.	Test Execution Productivity
6.	Indicates the testing capability of the tester in identifying the defect.	Test Efficiency
7.	Provides indications about the quality of the product under test and at the time of release.	Defect Severity Index
8.	Provides scripting productivity for performance test flow.	Performance Scripting Productivity
9.	Provides classification with respect to number of test conducted along with status (Pass/Fail), for various types of performance testing.	Performance Execution Summary
10.	Gives the detail information of Client side data for execution	Performance Execution Data Client Side
11.	Gives the detail information of Server side data for execution	Performance Execution Data Server Side
12.	Indicates the quality of the Performance team in meeting the performance requirement(s).	Performance Test Efficiency
13.	Indicates quality of product under test with respect to performance criteria.	Performance Severity Index
14.	Indicates the scripting productivity for automation test script.	Automation Scripting Productivity
15.	Provides execution productivity per day.	Auto. Execution Productivity
16.	Gives percentage of manual TC automated	Automation Coverage
17.	Provides information on ROI	Cost Comparison
18.	Indicates schedule stability	Effort Variance
19.	Indicates requirement stability	Schedule Variance

Fig. 10 Case perspectives

Bias

The integration of bias is done to have the transparency and randomness with the dataset (Fig. 11).

$$p_j(t) = \sum_i o_i(t) w_{ij} + wo_j \tag{2}$$

5.1 Dataset in Analytics on Software Defects and Quality Assurance

The dataset of GitHub for Software Quality is taken as the benchmark dataset. The URL associated with the benchmark characteristics of software

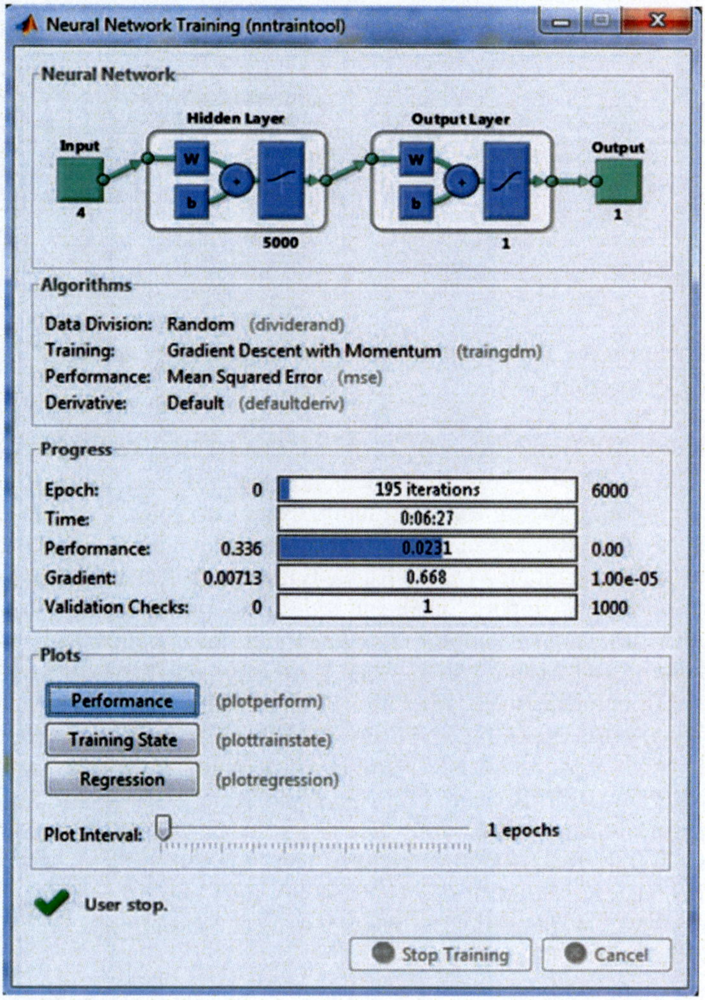

Fig. 11 Simulation process

quality is https://www.inf.u-szeged.hu/~ferenc/papers/GitHubBugDataSet/ from the master dataset https://www.inf.u-szeged.hu/~ferenc/papers/GitHubBugDataSet/GitHubBugDataSet-1.1.zip. After extracting dataset the training is done and analysis is performed on the dataset (Fig. 12).

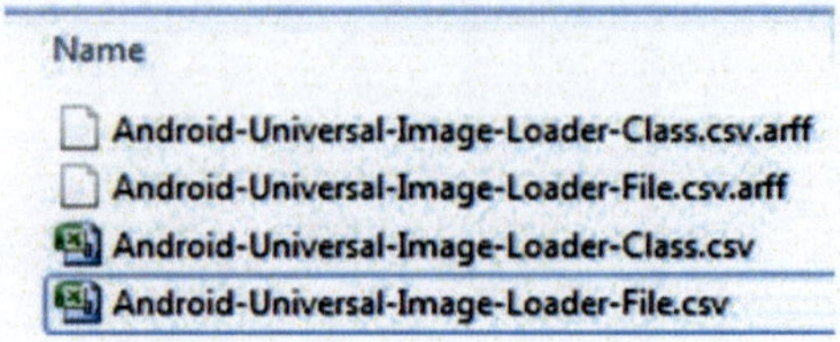

Fig. 12 Training of the dataset

5.2 *Results*

5.2.1 Attributes for Training the Model for Knowledge Discovery and Prediction

AD, Android Rules, Basic Rules, Brace Rules, CBO, CBOI, CC, CCL, CCO, CD, CI, CLC, CLLC, CLOC, Clone Implementation Rules, Clone Metric Rules, Code Size Rules, Cohesion Metric Rules, Column, Comment Rules, Complexity Metric Rules, Controversial Rules, Coupling Metric Rules, Coupling Rules, DIT, DLOC, Design Rules, Documentation Metric Rules, Empty Code Rules, End Column, End Line, Finalizer Rules, Import Statement Rules, Inheritance Metric Rules, J2EE Rules, JUnit Rules, Jakarta Commons Logging Rules, Java Logging Rules, JavaBean Rules, LCOM5, LDC, LLDC, LLOC, LOC, Line, MigratingToJUnit4 Rules, Migration Rules, Migration13 Rules, Migration14 Rules, Migration15 Rules, NA, NG, NII, NL, NLA, NLE, NLG, NLM, NLPA, NLPM, NLS, NM, NOA, NOC, NOD, NOI, NOP, NOS, NPA, NPM, NS, Naming Rules, Number of bugs, Optimization Rules, PDA, PUA, RFC, Security Code Guideline Rules, Size Metric Rules, Strict Exception Rules, String and StringBuffer Rules, TCD, TCLOC, TLLOC, TLOC, TNA, TNG, TNLA, TNLG, TNLM, TNLPA, TNLPM, TNLS, TNM, TNOS, TNPA, TNPM, TNS, Type Resolution Rules, Unnecessary and Unused Code Rules, Vulnerability Rules, WMC, WarningBlocker, WarningCritical, WarningInfo, WarningMajor, WarningMinor (Fig. 13, 14 and Table 1).

5.3 *Prediction Vector from Projected Approach*

[00101001000000011000100000000000000000001011110000]

The presented approach is quite effective in terms of higher accuracy and minimum error factor as per the confusion matrix. In the traditional approach of a greedy-based approach, the accuracy level is low and can be elevated with the statistical and data science-based integrations. With the implementation of the ensemble learning-based approach, the overall prediction and accuracy are elevated in the simulated attempts. By comparing both we obtained an 85.02% enhanced prediction result than recent approaches. This result is 11.03% improved than Bogner model. [11] (Figs. 15, 16 and Table 2).

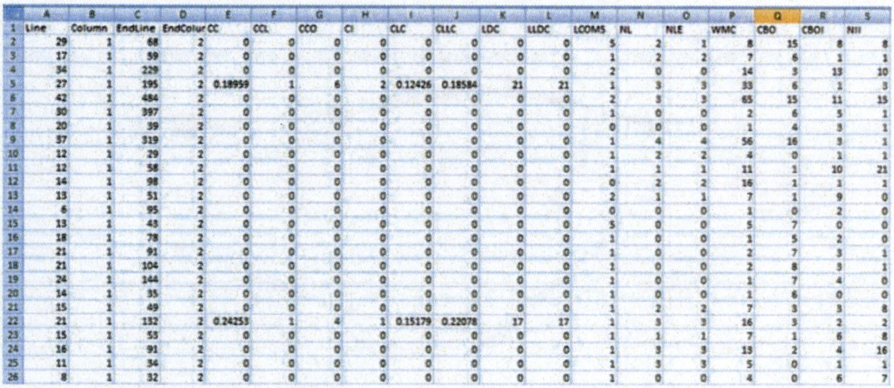

	Line	Column	EndLine	EndColur	CC	CCL	CCO	CI	CLC	CLLC	LDC	LLDC	LCOMS	NL	NLE	WMC	CBO	CBOI	NII
2	29	1	68	2	0	0	0	0	0	0	0	0	5	2	1	8	15	8	8
3	17	1	59	2	0	0	0	0	0	0	0	0	1	2	2	7	6	1	1
4	34	1	229	2	0	0	0	0	0	0	0	0	2	0	0	14	3	13	10
5	27	1	195	2	0.18959	1	6	2	0.12426	0.18584	21	21	1	3	3	33	6	1	3
6	42	1	484	2	0	0	0	0	0	0	0	0	2	3	3	65	15	11	15
7	30	1	397	2	0	0	0	0	0	0	0	0	1	0	0	2	6	5	1
8	20	1	39	2	0	0	0	0	0	0	0	0	0	0	0	1	4	3	1
9	37	1	319	2	0	0	0	0	0	0	0	0	1	4	4	56	16	3	1
10	12	1	29	2	0	0	0	0	0	0	0	0	1	2	2	4	0	1	1
11	12	1	58	2	0	0	0	0	0	0	0	0	1	1	1	11	1	10	21
12	14	1	98	2	0	0	0	0	0	0	0	0	0	2	2	16	1	1	1
13	13	1	51	2	0	0	0	0	0	0	0	0	2	1	1	7	1	9	0
14	6	1	95	2	0	0	0	0	0	0	0	0	0	0	0	1	0	2	0
15	13	1	43	2	0	0	0	0	0	0	0	0	5	0	0	5	7	0	0
16	18	1	78	2	0	0	0	0	0	0	0	0	1	0	0	1	5	2	0
17	21	1	91	2	0	0	0	0	0	0	0	0	1	0	0	2	7	3	1
18	21	1	104	2	0	0	0	0	0	0	0	0	1	0	0	2	8	3	1
19	24	1	144	2	0	0	0	0	0	0	0	0	1	0	0	1	7	4	0
20	14	1	35	2	0	0	0	0	0	0	0	0	1	0	0	1	6	0	0
21	15	1	49	2	0	0	0	0	0	0	0	0	1	2	2	7	3	3	8
22	21	1	132	2	0.24253	1	4	1	0.15179	0.22078	17	17	1	3	3	16	3	2	2
23	15	1	55	2	0	0	0	0	0	0	0	0	1	1	1	7	1	6	8
24	16	1	91	2	0	0	0	0	0	0	0	0	1	3	3	13	2	4	16
25	11	1	34	2	0	0	0	0	0	0	0	0	1	3	3	5	0	1	1
26	8	1	32	2	0	0	0	0	0	0	0	0	1	0	0	4	0	6	7

Fig. 13 Preprocessing of the dataset

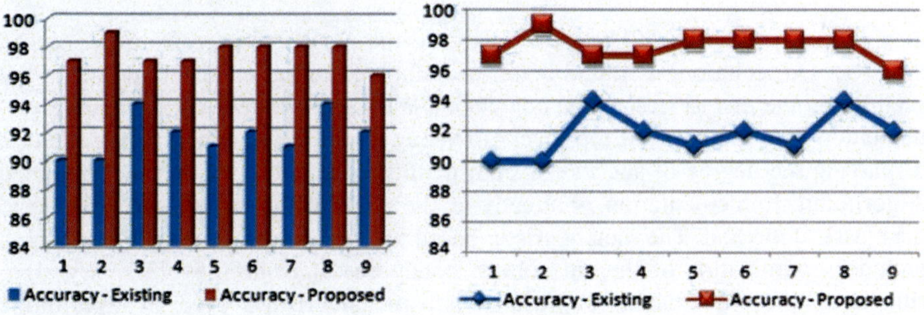

Fig. 14 Evaluation of outcomes

Table 1 Results analytics

Simulation attempt	Accuracy existing	Accuracy-proposed
1	90	97
2	90	99
3	94	97
4	92	97
5	91	98
6	92	98
7	91	98
8	94	98
9	92	96

6 Conclusion and Future Scope

After implementation and analyzing our proposed technique we obtained 85.02% enhancement in terms of prediction of software quality which is 11.03% higher

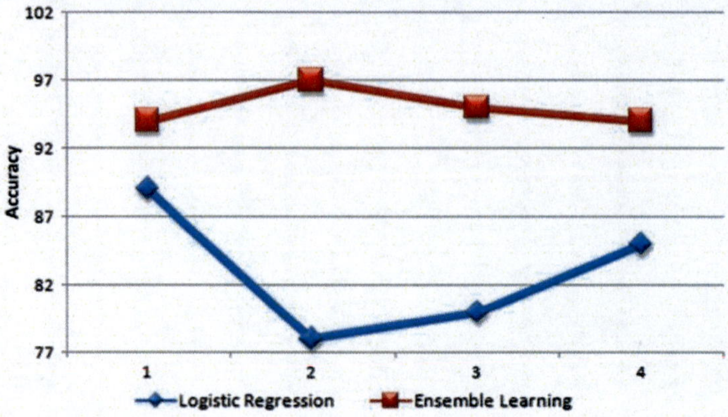

Fig. 15 Comparison of logistic regression and ensemble learning

than Bogner model [11]. The work analyzed the results fetched from the server, the proposed technique is found better as compared to the classical approach. We established the fact as well as the conclusion that the Projected Algorithmic Implementation is to the point of top-level efficiency. From the graphical representation displaying the degree of success of both methods, it is obvious that the Predicted Algorithmic Implementation is already at the peak of the output line relative to the classical method. The metaheuristic-based integration can be carried out, which includes optimization of the ant colony, honey bee algorithm, evolutionary algorithms, and several others. As we step into metaheuristics, even an algorithmic strategy will yield better outcomes. This theoretical work primarily addresses the performance metrics of Halstead applications for specific programming languages. The projected approach can be integrated with the advanced algorithms for optimizations and operations research whereby the elements of optimization and accuracy can be elevated. The scope of future work in this manuscript can be associated with global optimization-based evaluations with hybridization of soft computing and metaheuristics.

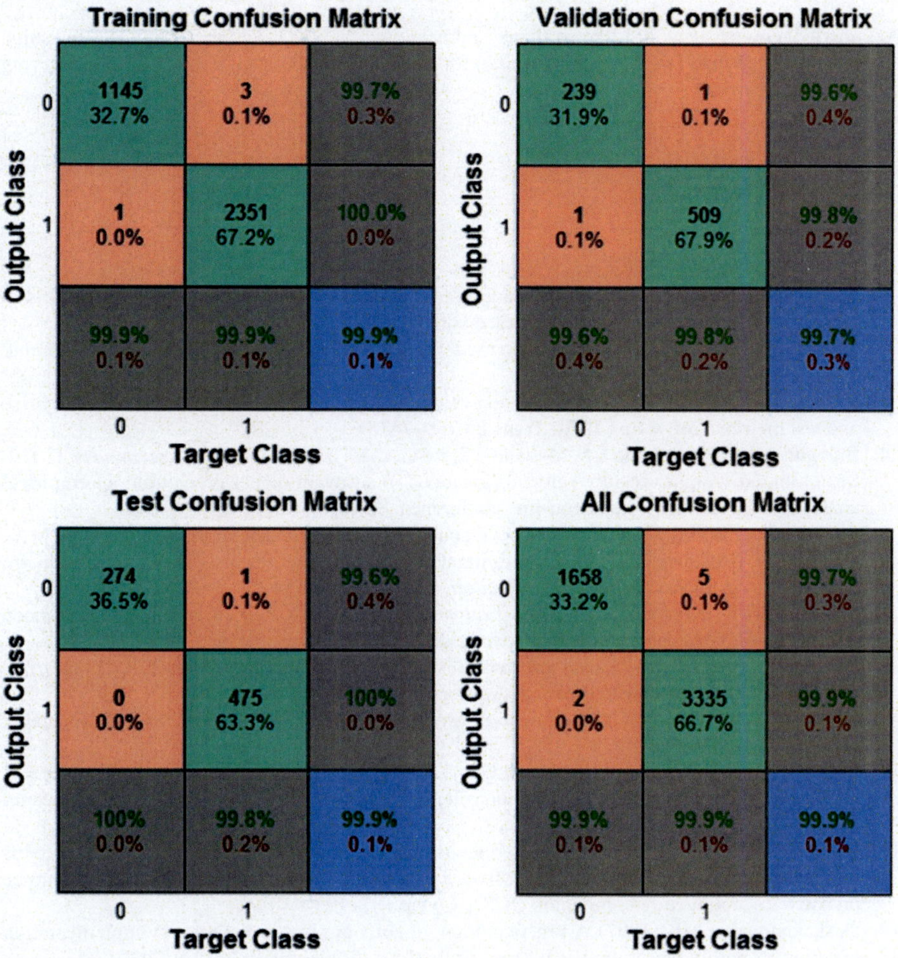

Fig. 16 Analysis of confusion matrix

Table 2 Evaluation of projected approach

Logistic regression	Ensemble learning
89	94
78	97
80	95
85	94

Acknowledgements I would like to thank and acknowledge IKG Punjab Technical University, Kapurthala, Jallandhar-144603, Punjab (India) for providing me the resources and help in carrying out the research work.

References

1. Yadav S, Kishan B (2020) Analysis and assessment of existing software quality models to predict the reliability of component-based software. Int J Emerg Trends Eng Res 8(6)
2. Jin C (2020) Software defect prediction model based on distance metric learning. Soft Comput 1–15
3. Schneidewind NF (1999) Measuring and evaluating maintenance process using reliability risk and test metrics. Softw Eng IEEE Trans 25:769–781
4. Balogun AO, Basri S, Jadid SA, Mahamad S, Al-momani MA, Bajeh AO, Alazzawi AK (2020) Search-based wrapper feature selection methods in software defect prediction: an empirical analysis. In: Computer science on-line conference. Springer, Cham, pp 492–503
5. Spinellis D (2003) Code quality: the open source perspective. Addison-Wesley, Boston, MA
6. Corwin BN, Braddock RL (1992) Operational performance metrics in a distributed system. In: Symposium on applied computing. Missouri, USA, pp 867–872
7. Subramanya R, Krishnan MS (2003) Empirical analysis of CK metrics for object-oriented design complexity: implication for software defects. IEEE Trans Softw Eng 29:297–310
8. Boehm B (1995) Cost models for future software life cycle processes: COCOMO 2.0, U.S.Center for Software Engineering, Amsterdam, pp 57–94
9. IFPUG (2008) International function point users group, online, last update: 03/2008. Available https://www.ifpug.org/
10. Mittal M, Kaur I, Pandey SC, Verma A, Goyal LM (2019) Opinion mining for the the tweets in healthcare sector using fuzzy association rule. EAI Endorsed Trans Pervasive Health Technol 4(16):1–10
11. Bogner J, Fritzsch J, Wagner S, Zimmermann A (2019) Microservices in the industry: insights into technologies characteristics and software quality. In: 2019 IEEE international conference on software architecture companion (ICSA-C). pp 187–195
12. Daskalantonakis MK (1992) A practical view of software measurement and implementation experiences within motorola. IEEE Trans Softw Eng 18:998–1010
13. Willis RR (1978) DAS-an automated system to support design analysis. In: 3rd International conference on software engineering. Georgia, USA, pp 109–115
14. Aggarwal A, Dhindsa KS, Suri PK (2019) Escalated methods for software defects audit in repercussion and effects construe to nature-inspired and behavior-driven mechanisms. IJEAT 8(6):1779–1783 ISSN: 22498958
15. Aggarwal A, Dhindsa KS, Suri PK (2019) Usage patterns and implementation of random forest approach for software risk and bugs predictions. IJITEE 8(9S):927–932 ISSN: 22783075
16. Aggarwal A, Dhindsa KS, Suri PK (2020) A pragmatic assessment of approaches and paradigms in software risk management frameworks. IJNCR 9(1):13–26 ISSN: 1947–928X, EISSN: 19479298
17. Aggarwal A, Dhindsa KS, Suri PK (2019) An empirical evaluation of assorted risk management models and frameworks in software. IJAEC 11(1):52–62 ISSN: 19423594, EISSN: 1942–3608
18. Aggarwal A, Dhindsa KS, Suri PK (2020) Design for software risk management using soft computing and simulated biological approach. IJAPUC 12(2):44–54 ISSN: 1937965X EISSN: 19379668
19. Alankrita A, Dhindsa KS, Suri PK (2020) Performance aware approach for software risk management using random forest algorithm. IJSI 9(1) ISSN: 2166–7160

20. Ince DC, Sheppard MJ (1988) System design metrics: a review and perspective. In: Second IEE/BCS conference, Liverpool, UK, pp 23–27
21. Briand LC, Morasca S, Basili VR (2002) An Operational process for goal-driven definition of measures. Softw Eng IEEE Trans 28:1106–1125
22. Numbers R (1999) Building productivity through measurement. Softw Testing Quality Eng Magaz 1:42–47
23. Bocco MG, Piattini M, Calero C (2005) A survey of metrics for UML class diagrams. J Object Technol 4:59–92
24. Metrics Eclipse Plugin (2005) https://sourceforge.net/projects/metrics
25. Emam KE, Benlarbi S, Goel N, Rai SN (2001) The Confounding effect of class size on the validity of object-oriented metrics. IEEE Trans Softw Eng 27:630–650
26. Li HF, Cheung WK (1987) An empirical study of software metrics. IEEE Trans Softw Eng 13:697-708
27. Zuse H (1995) In: History of software measurement. online Last Update 09 1995. Available https://irb.cs.tu-berlin.de/~zuse/metrics/History_00.html
28. Chidamber SR, Darcy DP, Kemerer CF (1998) Managerial use of metrics for object-oriented software: an exploratory analysis. Softw Eng IEEE Trans 24:629–639
29. Kafura D, Reddy G (1987) The use of software complexity metrics in software maintenance. IEEE Trans Softw Eng Arch New Jersey USA 13:335–343
30. Ramamurty B, Melton A (1988) A syntheses of software science measure and the cyclomatic number. IEEE Trans Softw Eng New Jersey-USA 1988(14):1116–1121
31. Navlakha JK (1987) A survey of system complexity metrics. Comput J 30:233–238, Oxford, UK
32. VanDoren E, Sciences K (2007) C. Springs: cyclomatic complexit. online Last Update 01 2007. Available https://www.sei.cmu.edu/str/descriptions/cyclomatic_body.html
33. Harrison R, Counsell S, Nithi R (1998) Coupling metrics for object-oriented design. In: 5th International software metrics symposium metrics, pp 150–156
34. Marchesi M (1998) OOA Metrics for the unified modeling language. In: Second euromicro conference, pp 67–73
35. Yin BH, Winchester JW (1978) The establishment and use of measures to evaluate the quality of software designs. In: Software quality assurance workshop on functional and performance. New York, USA, pp 45–52
36. Lind RK, Vairavan K (1989) An experimental investigation of software metrics and their relationship to software development effort. IEEE Trans Softw Eng New Jersey, USA, pp 649–653
37. McClure CL (1978) A model for program complexity analysis. In: 3rd International conference on software engineering, New Jersey, USA, pp 149–157
38. Morasca S, Briand LC, Basili VR, Weyuker EJ, Zelkowitz MV, Kitchenham B, Lawrence Pfleeger S, Fenton N (1995) Towards a framework for software measurement validation. Softw Eng IEEE Trans 23:187–189
39. Basili VR, Briand LC, Melo WL (1996) A validation of object-oriented design metrics as quality indicators. IEEE Trans Softw Eng New Jersey-USA 22:51–761
40. Alshayeb M, Li M (2003) An empirical validation of object-oriented metrics in two different iterative software processes. IEEE Trans Softw Eng Arch 29:1043–1049
41. Mittal M, Arora M, Pandey T (2019) Emoticon prediction on textual data using stacked LSTM model. In: International conference on communication and intelligent systems. Springer, Singapore, pp 259–269
42. Pressman RS (1997) Software engineering a practitioner's approach. vol 852. 4th edn. McGraw-Hill, New York, USA
43. Briand LC, Morasca S, Basili VR (1996) Property-based software engineering Measurement. Softw Eng IEEE Trans 22:68–86
44. Tang M-H, Kao M-H, Chen M-H (1999) An empirical study on object-oriented metrics. In: Software metrics symposium. pp 242–249

45. Lorenz M, Kidd J (1994) Object-oriented software metrics: a practical guide, englewood cliffs. New Jersey, USA
46. Briand L, Devanbu W, Melo W (1997) An investigation into coupling measures for C++. In: 19th International conference on software engineering, Boston, USA, pp 412-421
47. NezhadShokouhi MM, Majidi MA, Rasoolzadegan A (2020) Software defect prediction using over-sampling and feature extraction based on Mahalanobis distance. J Supercomput 76(1):602–635
48. Fenton NE, Neil M (2005) Software metrics: roadma. In: International conference on software engineering. Limerick, Ireland, pp 357–370
49. McCabe TJ (1976) A complexity measure. IEEE Trans Softw Eng 2:308–320
50. Briand LC, Morasca S, Basili VR (1999) Defining and validating measures for object-based high-level design. Softw Eng IEEE Trans 25:722–743
51. Chatzigeorgiou A (2003) Mathematical assessment of object-oriented design quality. IEEE Trans Software Eng 29:1050–1053
52. Brito AF, Carapuça R (1994) Object-oriented software engineering: measuring and controlling the development process. In: 4th International conference on software quality, USA
53. Sommerville I (2004) In: Engenharia de software , Addison-Wesley, 6º Edição, São Paulo, SP
54. Hawk J (2007) Eclipse Plugin. https://www.virtualmachinery.com/jhawkprod.htm
55. Halstead MH (1977) In: Elements of software science operating and programming systems. vol 7. New York, USA, pp 128
56. Refactorittool: https://www.aqris.com/display/ap/RefactorIt (2008)
57. Henry S, Kafura D (1981) Software structure metrics based on information flow. Softw Eng. IEEE Trans 510–518
58. Chidamber SR, Kemerer CF (1994) A metrics suite for object-oriented design. IEEE Trans Softw Eng Piscataway-USA 20:476–493
59. Zhang J, Wu J, Chen C, Zheng Z, Lyu M (2020) R: CDS: A cross-version software defect prediction model with data selection. IEEE Access 8:110059–110072

Chapter 9
Early Detection of Lung Cancer Using Convolutional Neural Network

Ritenderveer Kaur and Rajat Joshi

1 Introduction

Doctors can examine the inside of living beings using medical equipments like MRI scan, X-rays, CT scan and ultra-sound. Cancer is the main reason of death for both men as well as women as symptoms only appear during the peak stages, which leads to the highest death rate among all other diseases [1]. There are several key facts that show that the rate of death is minimized through early detection of lung cancer. To detect lung cancer the earlier stage is a difficult task. As per the survey conducted by Rani et al. (2016), at the earlier stage detection of lung cancer 20% of patients are diagnosed [2]. It is predictable that about 1.2 million people are detected with the lung cancer each year that is aroung12.3% of the number of cancers diagnosed. The death rate of people from lung cancer every year is estimated as 17.8% of all cancer deaths [3]. If the disease is not detected at earlier stage, it results in increase in the death rate. In modern era, the use of computer systems by the radiologists helps to diagnose abnormalities in the body tissues and hence detect the disease [4, 5]. Therefore, the detection of lung cancer at the earlier stage through improved image enhancement and classification approach is the main motive of this research [6, 7].

There are numerous morphological variants in these cancer cells. Therefore, to avoid classification errors, computer-aided diagnosis (CAD) is a useful tool. Among the four main types of cancer, large cell carcinoma is most easily found due to its severe atypical nature. Therefore, we focused our attention on the other three types of classification-adenocarcinoma, squamous cell carcinoma, and small cell carcinoma that are now and then confused with each other in cytological samples. CAD is the best option to provide better support to pathologists and helps clinical technicians to diagnosis any malignancies. In this research, our aim is to automatically detect cancer types using CNN as a machine learning approach [4]. Facts have proved that machine learning is very effective in discovering distinguishing features and improving the

R. Kaur · R. Joshi (✉)
Adesh Institute of Technology, Gharuan 140413, India

N. Marriwala et al. (eds.), *Soft Computing for Intelligent Systems*, Algorithms for Intelligent Systems, https://doi.org/10.1007/978-981-16-1048-6_9

accuracy of many visual recognition systems. In order to fully utilize the power of machine learning, one required very large training data, which might be not possible in the field of medical imaging in the near future. Using such features have presented a prior knowledge-based system that minimized the space problem [5].

Recently, machine learning has been very successful in solving various visual recognition issues and continuously breaking performance records in their respective challenges. CNN finds application in medical image analysis [6]. Although most medical images are 3D tensors, the traditional CNN method is based on a 2D kernel. Existing deep learning methods for medical image analysis usually convert 3D data into a 2D multi-channel representation, such as a three-plane representation, and then input them into a 2D CNN. Although 2D CNN has been shown to be useful in solving these problems, it essentially loses the 3D context of the original image, which limits the performance of the entire system [7].

2 Related Work

Senthil Kumar et al. [7] have designed lung cancer detection system using optimization approach and compared the performed with existing approaches. Adaptive median filter as pre-processing approach is best suitable in medical CT images. After that the brightness of an image has been enhanced through adaptive histogram equalization. Particularly, this experiment has been performed using MATLAB for 20 samples and highest accuracy of 95.89% has been attained using proposed guaranteed convergence particle swarm optimization approach [7]. Arulmurugan and Anandakumar (2018) have used wavelet features to trained ANN structure, which are used later to classify lung cancer scanned images. The authors have divided the data into 70:30 ratio, for training and validation respectively. The overall accuracy of 92.6% has been achieved [8]. Singh et al. (2018) have detected benign and malignant categories of lung cancer using multi-layer classifier. The training has been performed on the pre-processed images. Using K-nearest neighbors the features are extracted and then passed to the MLP classifiers. From the experiment, it has been determined that the obtained accuracy, i.e., 88.55% by applying MLP classifier is enhanced as compared to different classifiers [9]. Da et al. (2018) have utilized particle swarm optimization (PSO) in addition to convolutional neural network (CNN) to design a lung cancer detection system. The experiments have been performed on the CT images. The performance has been examined based on the different parameters named as accuracy sensitivity along with specificity of 97.62%, 92.20%, and 98.64%, respectively [10]. Chauhan and Jaiswal (2016) have used PCA; the canny edge detection technique is used for pre-processing as well as post-processing. In the very first step, edge detection is performed after that extraction of features is utilized for the purpose of obtaining the optimal number of features for identification among infected as well as non-infected disease. The developed method has introduced in MATLAB and evaluated with actual CT scan images. In this work, the dataset comprises five categories in which each class is composed of about 120 cancer and 200 kb non-cancerous images. Connection has been done with the SURF system and it was found that the PCA approach was better than the SURF method [11]. Kuruvilla and Gunavathi (2014) presented the computer-aided classification approach using CT images of lungs formed through the ANN approach. For classification purpose, neural network is utilized in both feedforward and feedforward backpropagation neural network. The results show that the proposed learning function 1 offers 93.3% precision, 100% specificity and sensitivity of 91.4%, and mean square error of 0.998 [12].

3 Methodology

The designed new model is shown in Fig. 1. The entire detection model is divided into four distinct segments namely; pre-processing, extraction of features, training as well as testing. Each stage is discussed one by one.

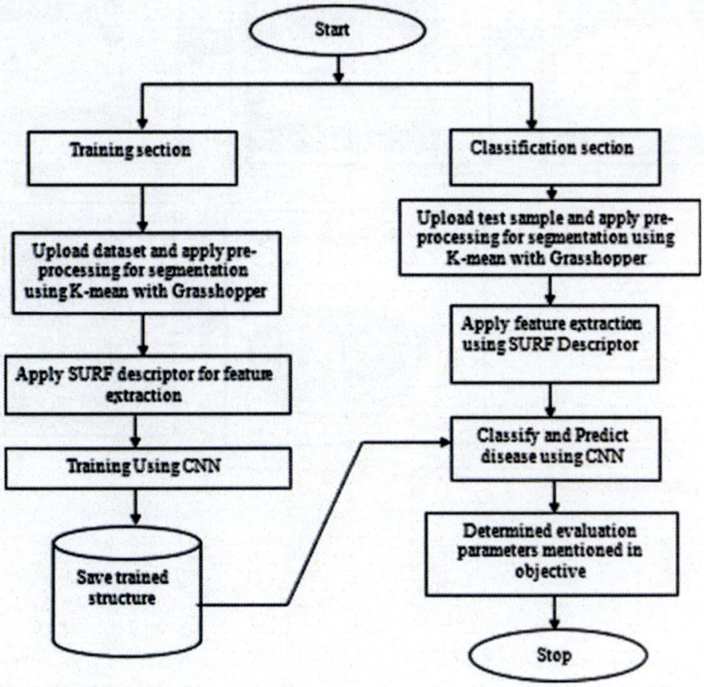

Fig. 1 Proposed work

3.1 Pre-processing

The main purpose of this step is to improve the image data that contains unnecessary distortions or enhances some of the required features of the image for further processing. It also eliminates the gap in a particular image without affecting the necessary details that play an important role in the designed model. Here, pre-processing composed of four different tasks: (i) select region, (ii) color labeled image, (iii) labeled image, and (iv) segmented image. The obtained image for each step is depicted in Fig. 2.

The segmentation of image has been performed using K-means with GoA. To compute the region of Interest (RoI), the segmentation scheme of image is utilized. The designed algorithm is written below.

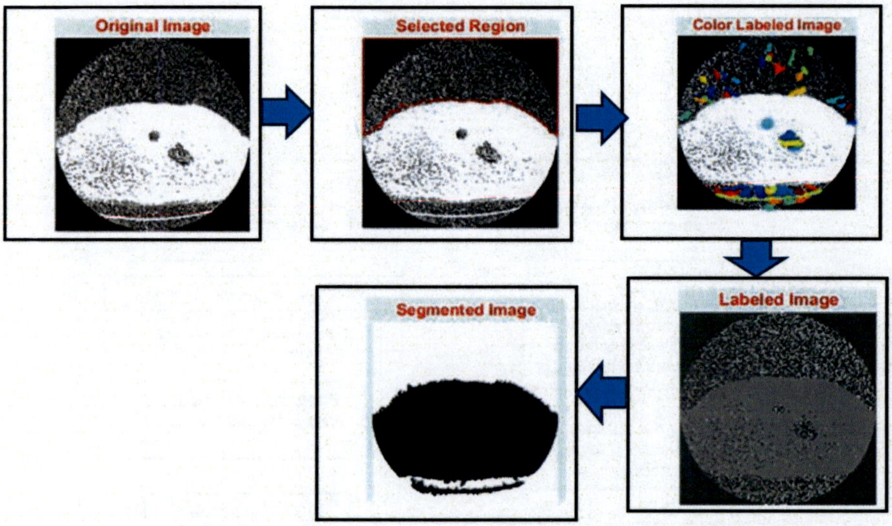

Fig. 2 Pre-processing steps

Algorithm 1: K-means with GOA

Input: Lung Image → Lung Image from the Dataset
Output: ROI → Lung ROI Image
Initialize an estimated group (BG-Background and FG-Foreground)
Initialize an estimated group
Calculate size of Lung Image, [R, C, P] =size (Lung Image)
For i→1 to all R
  For j→1 to all C
If MR Image pixel distance (i,j)==Foreground
    ROI 1=FG (i,j)
   Else
    ROI 2=BG (i,j)
   End End
  End
End
  ROI=min [ROI1 ROI2]
New Lung Image = double (ROI)
No. of part = 2
S_{img}Index = Seperate (New Lung_Image, No. of part)
Seg_Label_Img=reshape (S_{img}Index, R, C)
Data_Pos=find (Seg_Col_Label_Img>0)
Data=Seg_Col_Label_Img(Data_Pos)
Initialize GOA parameter – No. of Iterations (T)
– Size of population for Grasshopper (S)
– Lower_Bound (LB)
– Upper_Bound (UB)
– Fitness function
– Number of selection (N)
Determine T = Size (New Lung Image)
Computation of fitness function is done as;

$$f\,(fit) = \begin{cases} 1 & if\ pixel\ is\ less \\ 0 & otherwise \end{cases}$$

For I = 1 → T

$$fs = \sum_{i=1}^{P} Data\,(I)$$

$$ft = \frac{\sum_{i=1}^{P} Data(I)}{Length\ of\ feature}$$

$f(fit)$ denotes the fitness function that described by above expression

$$Threshold_{value} = GOA(P, T, LB, UB, N, f(fit))$$

End
While T ~= Maximum
$Threshold = Threshold_{value}$
Mask_Img=Morphological (S_{img} Index, Threshold)
Boundaries = bw_boundaries (Mask_Img)
Segmented_Region = Boundaries
For i→1: P
Segmented_Image = Lung_Image * Segmented_Region
End
Return: Segmented_Image as ROI of Lung_Image
END

3.2 Feature Extraction

The removal of unwanted feature is the process to improve the low-level image quality, which is the interesting part of the image. Feature extraction is the essential part of the model to determine normal and abnormal cells in CT images. Here, SURF is used as a feature extraction approach. This is used to return image pattern. It is an un-named feature extraction approach, which represents the pixel pattern of an image. The extracted pixels using SURF are depicted in Fig. 3.

In this research threshold-based segmentation method has been used to select the appropriate features that represent lung cancer properties. After the extracted features of image, we have to apply the CNN as training algorithm.

Algorithm 2: SURF Descriptor

```
Input: ROI →ROI of lung image
Output: SURF-ROI→ SURF feature set of ROI image
Calculate size of ROI, [Row, Col] = size (ROI)
For X=1→Row
    For Y =1→ Col
    Scaled Image = ROI (X, Y, 8) // Scaling of ROI into 8 X 8
    Extrema Point = Extrema (Scaled Image (X, Y))
    Key-point= Localization (Extrema Point (X, Y))
    Check: Variation occurs after Key-point orientation
        If Variation occurs
            Discard those key points
        Else
            SURF-ROI = Key-point (X, Y)
        End
    End
End
Return:SURF-ROI as a SURF feature set of ROI image
End
```

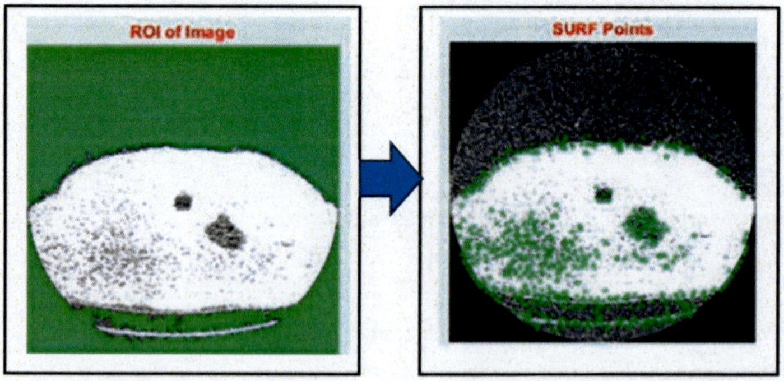

Fig. 3 Feature extraction process

3.3 Feature Selection and Feature Classification

Feature selection is the process used to select a required feature subset to minimize the dimensionality of features space. This approach helps to increase the accuracy and reduced the computation time taken by the designed model. Feature selection has been performed using Grasshopper approach. After obtained the best pixels, the model is trained using CCN method. The trained structure of pattern net CNN is shown in Fig. 4.

Algorithm 3: Pattern Net CNN

Input: SURF-ROI and T→ SURF feature of ROI image as training data and their types as a target or category

Output: Pattern-Net-Structure→ Trained structure with Classified Results

Initialize the Pattern net based CNN with ROI: – Epoch Amount (E) // Iterations utilized through CNN

– Total amount of neurons (N) // Consumed as a carrier

– Performance: MSE, Gradient, Error Histogram and Validation

–Division of Data: Randomly

For i = 1 → SURF-ROI
If SURF-ROI belongs to Type 1
Group (1) = Feature from the ROI 1 // Benign
Else
Group (2) = Feature from the ROI 2 // Malignant
End
End
Initialized the pattern net using ROI and T
Pattern-Net-Structure = Pattern-net (N)
Training parameter are defined and set as per the requirements after that train the system
Pattern-Net-Structure = Train (Pattern-Net-Structure, ROI, Group)
Test Data Group = Sim (Pattern-Net-Structure, Current Lung Image)
If Test Data Group = 1
Classified Results = Benign
Else
Classified Results = Malignant
Return: Pattern-Net-Structure as a trained structure with Classified Results
End

After following the above defined steps, at the end, performance has been analyzed in terms of different parameters as discussed in Result and Discussion section.

4 Result and Discussions

The designed model's performance has been evaluated in terms of True Positive Rate (TPR), False Positive Rate (FPR), True Negative Rate (TNR), False Negative Rate (FNR), Classification Time (s), classification Accuracy (%), and classification Error.

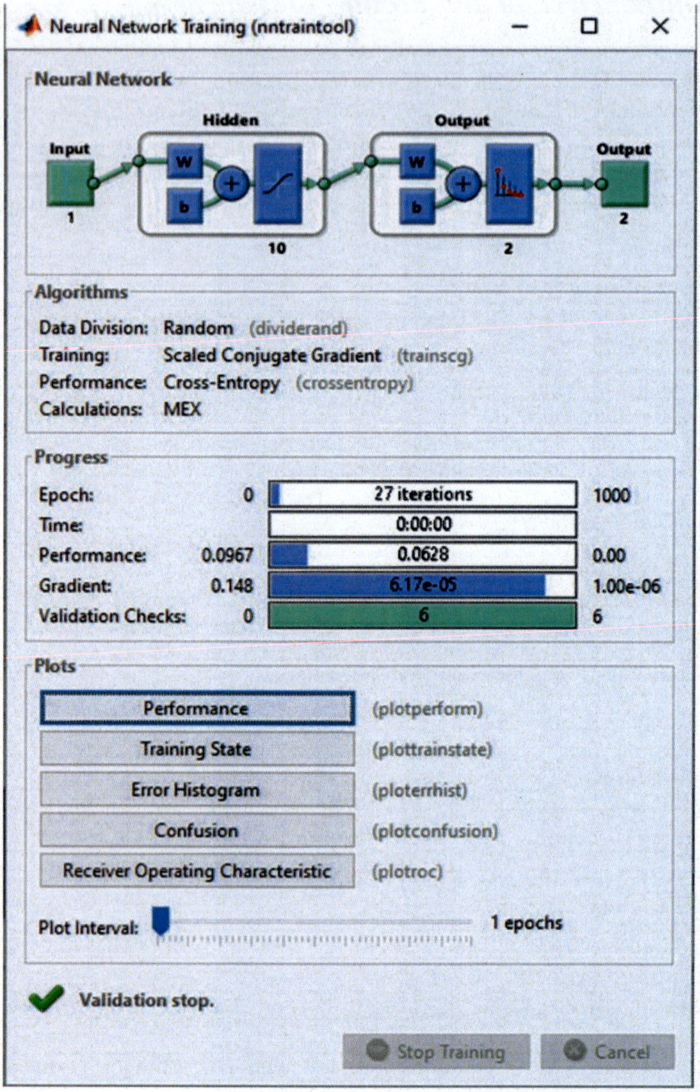

Fig. 4 Trained CNN

The lung sample image is uploaded seven times and obtained values are presented in Table 1.

The graphical representation of these parameters is given below one after another.

Figure 5 represents the true positive rate of the proposed work. From the given figure the no. of sample is depicted through x-axis and y-axis depicts the rate of true positive. The rate of true positive measures the amount of actual positive which is accurately determined. The value of true positive rate is 99.11 and 99.008 at first

Table 1 Classification parameters

No. of samples	TPR	TNR	FPR	FNR	Classification Time (s)	Classification accuracy	Classification error
1	99.1114	96.6780	1.3098	1.8845	0.0989	99.6842	0.7158
2	98.6789	96.450	1.312	1.8758	0.1374	98.8765	0.7567
3	98.3456	96.345	1.4678	1.845	0.1768	99.7654	0.6885
4	99.0089	95.678	1.345	1.8678	0.1561	97.9889	0.7064
5	99.10	95.456	1.425	1.8345	0.1981	98.5898	0.7516
6	98.2345	96.45	1.389	1.8623	0.1657	99.567	0.7741
7	98.567	96.008	1.452	1.8734	0.1419	98.9816	0.7085

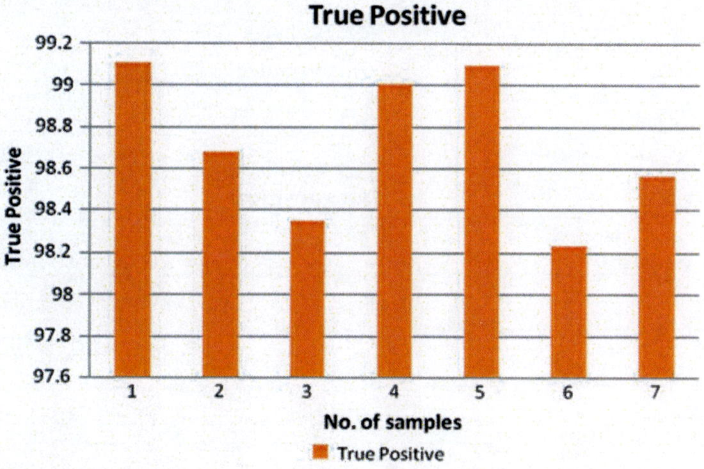

Fig. 5 True positive

sample and at sample fourth correspondingly. The obtained value of true positive rate in average is 98.7.

Figure 6 represents the true negative of the proposed work. From the depicted figure the no. of samples is plotted along x-axis and y-axis depicts the rate of true negative. The amount of actual negatives which is accurately identified is analyzed by the true negative rate. At sample second the value of true negative rate is 96.450 and at sample fifth the value of true negative rate is 95.456. The average value of true negative rate is 96.1.

Figure 7 represents the false positive of the proposed work. From the given figure, the no. of samples is depicted through x-axis and y-axis depicts the rate of false positive. The possibility of falsely rejecting null hypothesis corresponds to particular analysis which denotes the rate of false positive. The value of thirdsample is 1.467 and the value of seventh sample is 1.452. The average value of false positive rate is 1.38.

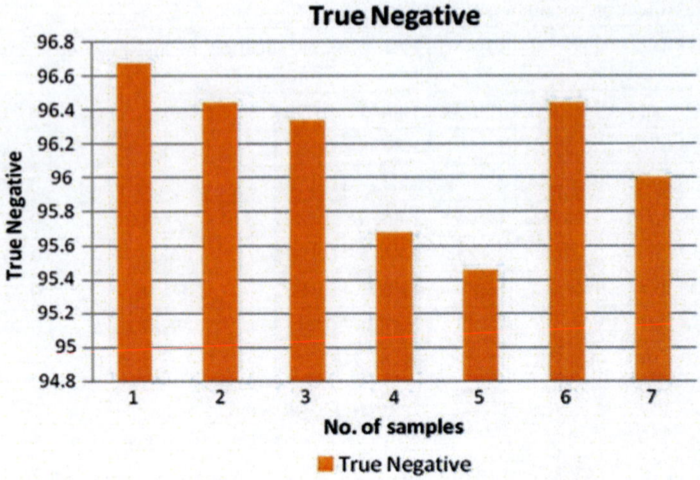

Fig. 6 True negative

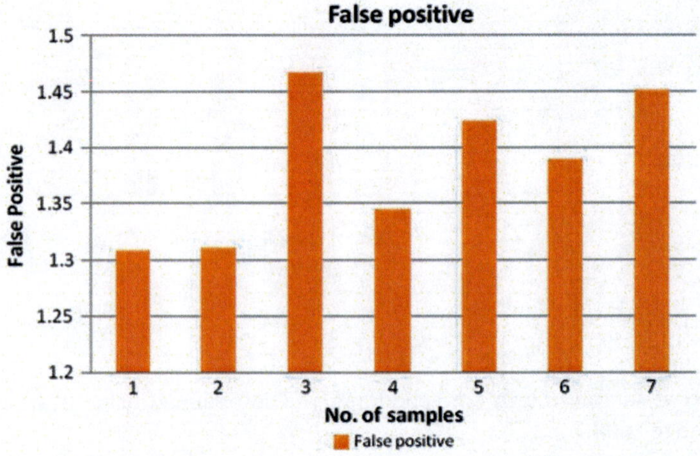

Fig. 7 False positive

Figure 8 represents the false negative of the proposed work. The rate of false negative is corresponding to the amount of individuals including known positive circumstances for which the result of test is negative. The average value of false negative rate is 1.86.

The total time consumed by the system for performing some task is defined by execution, also included the time spent performing run-time as well as services of system on its behalf. Defined implementation is the system utilized to examine the time taken to execute. Figure 9 shows the classification time against the number of

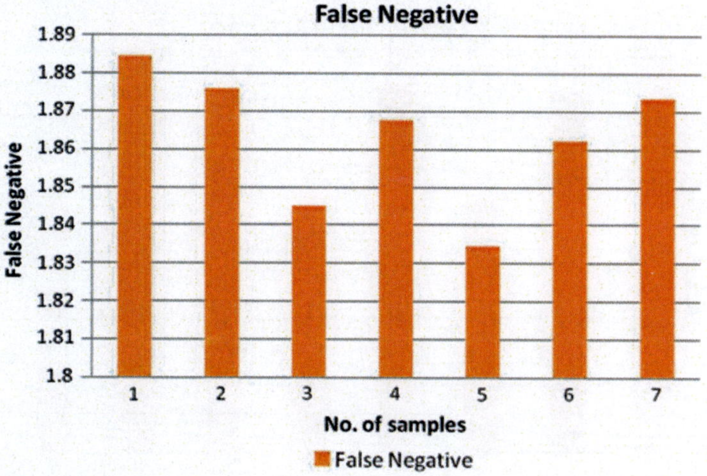

Fig. 8 False negative

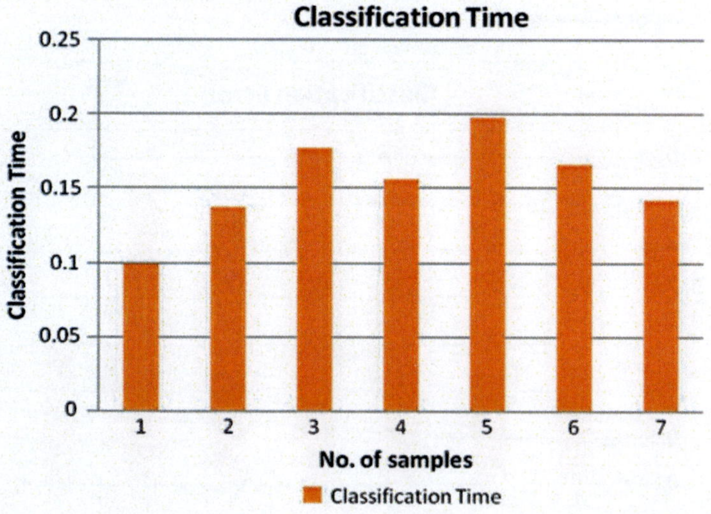

Fig. 9 Classification time

text samples maximum and minimum obtained execution time to perform the task have 0.0989 and 0.1374 in seconds.

Figure 10 represents the accuracy of the proposed work is the capacity of machine to measure the precise value. The average value of accuracy is 99%.

In Fig. 11, the error in percentage has depicted against the number of samples, the error doesn't need to be increased according to the increasing number of samples. In which maximum and minimum error occurred 0.7741 and 0.6885 correspondingly.

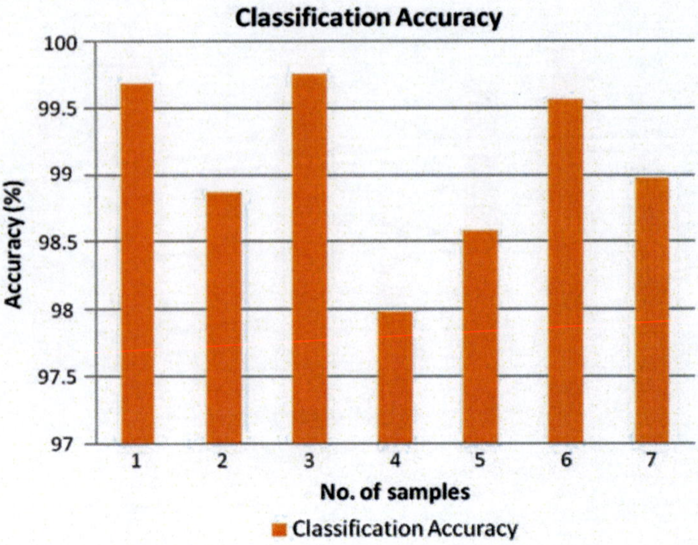

Fig. 10 Classification accuracy

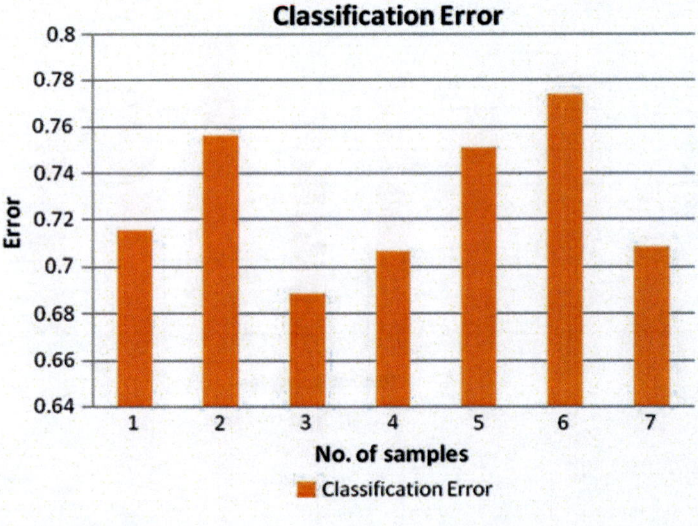

Fig. 11 Classification error

In the form of column graph the error is plotted along vertically against number of samples horizontally.

The contrast of proposed work is given in Table 2. In which shows the value of true positive of proposed work is 98.7 and corresponds to Kumar et al. [7] is 95.28 correspondingly.

Table 2 Comparison with proposed work

Parameters	Proposed	Kumar et al. [7]
True positive	98.7	95.28

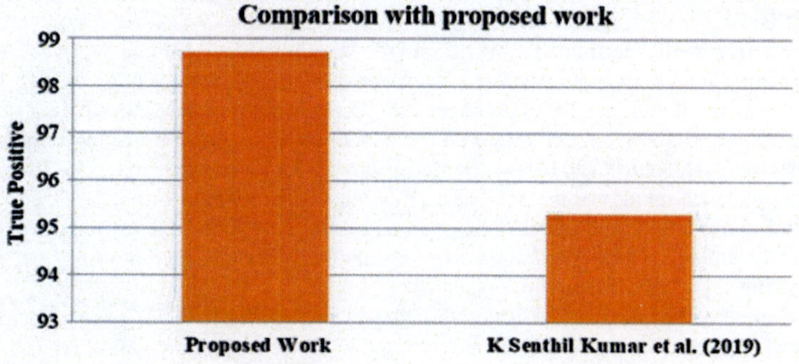

Fig. 12 Comparison of proposed and Kumar et al. [7]

Figure 12 represents the comparison of proposed work and Kumar et al. [7] on the basis of parameter named as true positive. The comparison shows that our proposed work has high positive rate as compared to Kumar et al. [7] which means that our work shows good results.

5 Conclusion

In the proposed work, early detection of lung cancer classification system uses SURF descriptor along with CNN as classifier have been proposed. The evaluation-based optimization algorithm named as Genetic Algorithm is used to optimize the data value on the basis of evaluation criterion of fitness functions. To extract features from CT images the neural network-based classification mechanism, i.e., convolution neural network has been utilized. This classifier is mainly three distinct layers, namely, input layer, pooling layer along with convolution layer. Previously various works have been presented particularly for CT images to examine the performance of this presented work on the basis of few parameters named as error (%), classification time, true positive, true negative, false positive, false negative, and accuracy (%). The value of true positive of proposed work is 98.7 and Kumar et al. [7] are 95.28. The average value of accuracy is 99%.

Acknowledgements We are thankful to Inder Kumar Gujral, Punjab Technical University, Jalandhar, Punjab for providing us the opportunity to carry research work.

References

1. Shimizu S, Shirato H, Ogura S, Akita-Dosaka H, Kitamura K, Nishioka T, Miyasaka K (2001) Detection of lung tumor movement in real-time tumor-tracking radiotherapy. Int. J. Radiation Oncol* Biol* Phys 51(2):304–310
2. Swensen SJ, Jett JR, Hartman TE, Midthun DE, Mandrekar SJ, Hillman SL, Allen KL (2005) CT screening for lung cancer: five-year prospective experience. Radiology 235(1):259–265
3. Kavitha P, Prabakaran S (2019) A novel hybrid segmentation method with particle swarm optimization and fuzzy c-mean based on partitioning the image for detecting lung cancer
4. Rani B, Goel AK, Kaur R (2016) A modified approach for lung cancer detection using bacterial forging optimization algorithm. Int. J. Sci. Res. Eng. Technol 5(1)
5. Joon P, Bajaj SB, Jatain A (2019) Segmentation and detection of lung cancer using image processing and clustering techniques. In: Progress in advanced computing and intelligent engineering. Springer, Singapore, pp 13–23
6. Thabsheera AA, Thasleema TM, Rajesh R (2019) Lung cancer detection using CT scan images: a review on various image processing techniques. In: Data analytics and learning. Springer, Singapore, pp 413–419
7. Senthil Kumar K, Venkatalakshmi K, Karthikeyan K (2019) Lung cancer detection using image segmentation by means of various evolutionary algorithms. In: Computational and mathematical methods in medicine
8. Arulmurugan R, Anandakumar H (2018) Early detection of lung cancer using wavelet feature descriptor and feed forward back propagation neural networks classifier. In: Computational vision and bio inspired computing, Springer, Cham, pp 103–110
9. Makaju S, Prasad PWC, Alsadoon A, Singh AK, Elchouemi A (2018) Lung cancer detection using CT scan images. Proc Comput Sci 125:107–114
10. Da Silva GLF, Valente TLA, Silva AC, de Paiva AC, Gattass M (2018) Convolutional neural network-based PSO for lung nodule false positive reduction on CT images. Comput Methods Programs Biomed 162:109–118
11. Chauhan D, Jaiswal V (2016) An efficient data mining classification approach for detecting lung cancer disease. In: 2016 International conference on communication and electronics systems (ICCES) October , IEEE, pp 1–8
12. Kuruvilla J, Gunavathi K (2014) Lung cancer classification using neural networks for CT images. Comput Methods Programs Biomed 113 (1):202–209

Chapter 10
Design and Analysis of Thin Micro-Mechanical Suspended Dielectric RF-MEMS Switch for 5G and IoT Applications

Bikramjit Sharma, Manvinder Sharma, Bhim Sain Singla, and Sumeet Goyal

1 Introduction

In 1959, the possibilities of manufacturing ultraminiaturize systems for a variety of applications that may involve multi-scale formulation in terms of manipulation of atoms and molecules were proposed by Feynman. MEMS (Micro Electro Mechanical Systems) are micron-sized devices that are made by combining mechanical and electrical components [1–4]. The devices can act as either a sensor, wherein an electrical signal is created by the change in physical property, or as an actuator wherein by applying electrical signal, the physical effect is seen. MEMS devices have shown many potential applications in the field of communication, industrial automation, biotechnology, military, motorized engineering, user electronics, etc. This is because MEMS have many extra advantages like low price, low power ingestion, relaxed integration with electronics, high conflict to vibration, shock and radiation, high dependability, high accuracy and low-cost bunch production [5]. One of the fields in which the MEMS devices have shown high possible presentations is the ground of RF communication. The MEMS resonators have become the most preferred choices for detecting and high-frequency requests like oscillators, filters and mixers because of the inherent advantages associated with MEMS [6, 7]. MEMS resonators have almost replaced quartz crystal technology whose main drawbacks are large size,

B. Sharma (✉)
Thapar Institute of Engineering and Technology, Patiala, India
e-mail: bikramjit@thapar.edu

M. Sharma · S. Goyal
Chandigarh Group of Colleges, Landran, India

B. S. Singla
College of Engineering and Management, Punjabi University, Patiala, India

N. Marriwala et al. (eds.), *Soft Computing for Intelligent Systems*, Algorithms for Intelligent Systems, https://doi.org/10.1007/978-981-16-1048-6_10

which ultimately disturbs the miniaturization, high price and non-compatibility with integrated circuit technology [8, 9].

In arrears to temperature and high Q stable properties of quartz crystal oscillators, they have been important watch foundations in consumer, marketable, industrial and armed products for many years. The frequency control components of off-chip piezoelectric are becoming very small through years of efforts and they provide excellent frequency generating and sifting functions to components the demanding necessities from the designers of handset [10–12]. The unique encapsulation and fabrication necessities of the devices even though shorten them close to unbearable to assimilate onto the mature silicon-based IC stages [13]. As the handset/mobile market is continuing to grow and is continuously developing MEMS-based RF mechanisms like filters, switches, oscillators, resonators, VCO, etc. within the RF unit of mobile handsets opening the logical route towards integration [14–17].

There has been a great and significant evolution in the services and devices of mobile communication systems across the last few decades. Starting from the first generation (1G) that included cellular network to the fourth generation (4G-LTE) that includes long-term evolution has spread significantly. The upcoming 5G will prove to be visionary in the communication field. The specifications of 5G are of high level, which include data increase in manifolds, connected devices are increased, increase in the user data rate, increased battery life of the devices and reduced latency. A common platform is realized to satisfy the IoT and 5G specifications such as antenna transmission (multi-node), radio links, usage of spectrum and network dimension [18]. The requirements and demands of the emerging 5G world appear to be very diverse and challenging as no research industry has witnessed before. In the last two decades, the mobile handsets have been used massively and this leads to the trend of integrating wireless services supported by wireless mobile handsets. This trend has been increasing exponentially rather than linearly. 5G systems provide large mobile network capacity such as video resolution will be of very high quality, high data rates as compared to 4G wireless network systems offered [19]. Figure 1 shows the front-end structure of RF-MEMS passive components employed in IoT and 5G communication network.

The emerging Internet of Things (IoT) connects daily objects with the environment in our daily lives. With the significant development of IoT, the mobile network capacity is increased to 1000 times with the emerging 5G and IoT through internet technology. The implementation and employment of this new technology are employed by various components like wideband switches, attenuators, filters, impedance tuners, hybrid devices and antenna arrays. The 5G and IoT network performance is supported by upgrading all these devices [20]. The existing radio technologies such as Global System for Mobile communication (GSM), Wireless Fidelity (WiFi) and LTE are integrated with the 5G and IoT technology. The demand for frequency range for 5G system would be higher and wider than the existing technologies so that different services can be covered while reducing the power consumption and hardware redundancy. 5G systems use millimetre wave technology so that a large spectrum is available while reducing the size of the antenna. Smart antennas are used because of their capability of phase array and beam forming so that high

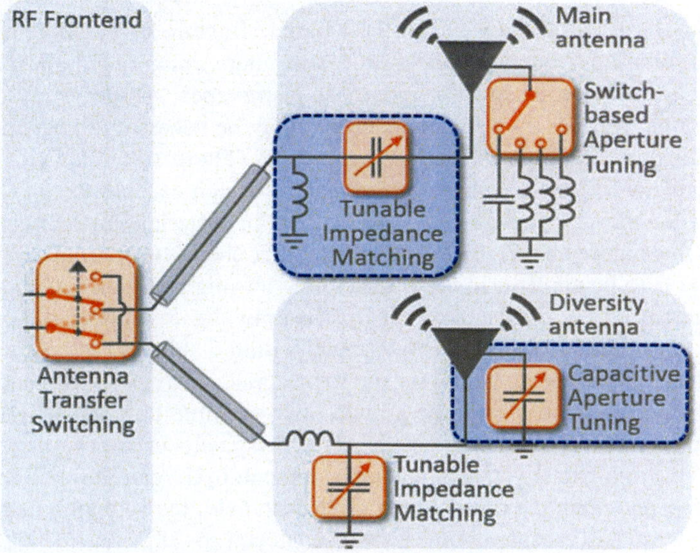

Fig. 1 Front-end structure of RF-MEMS passive components employed in IoT and 5G communication network

precision is obtained when the antenna beam is pointed to a desired location. The transmission lines and RF-MEMS switches of high quality are employed to realize the smart antennas in emerging IoT and 5G communication systems [21]. The paradigm of Internet of things (IoT) displays the development of ongoing technology through which the daily life environment and objects experience their individual identity in the modern digital world through the internet. A large variety of wireless connections can be accommodated by the 5G mobile networks provided the frame of reference of IoT. IoT supports M2M (Machine-to-Machine) applications by fulfilling the requirements in terms of QoS (Quality of Service), reliability and energy as well as spectral efficiency. IoT technology enables the addressing of all challenging requirements of 5G applications such as smart home, smart buildings, smart agriculture and so on. It is necessary to innovate and re-engineer the algorithms and network architecture. This leads to the demand for software solutions and novel hardware. The RF-MEMS solutions have proved to be quite successful in fulfilling the demand of protocols of 5G communication such as higher frequencies of operation, reduction in the redundancy of hardware, coverage of various services by reconfiguring and consumption of power. The technology of RF-MEMS has proved to be the most promising in the concerns of 5G smartphones and base stations. The RF-MEMS devices working at millimetre wave frequency tend to address the challenge of high-frequency operation, which leads to the high data rated to the individual users in 5G and IoT technology [22, 23].

The basic MEMS employ a cantilever-based structure or diaphragm-based or micro bridge-based structure. Special processing steps commonly known as micromachining are required to manufacture these membranes, cantilever rays, resonant configurations, etc. For a given application, it may be necessary to have integrated MEMS using one or more of the basic structures. These structures provide some feasible designs for micro sensors and actuators, which execute the preferred task in most of the clever structures [24]. The choice of materials for fabricating these devices is the major and affecting issue and another one is the micromachining technology that is to be applied. In these structures, sensing and then actuation happen as a consequence of moving a piezo-electric layer by the use of potential or voltage. This type of excitation led to the development within diaphragm or another method to make free-standing beam within the micro bridge structure, or in the cantilever ray [25]. In the previous two cases, the growth interprets into upward curvature in the diaphragm or in the free-standing ray, hence, which results in a net vertical displacement from the non-excited equilibrium development. In the cantilever circumstance, nevertheless, and upon the presentation of electric field, the actuation had occurred by a vertical upward measure of the cantilever slope [26–28].

It is very much evident that in the active parts of the system structure such as free-standing ray, diaphragm and in the cantilever ray, the design in the micro-actuator consists of the piezoelectric layer and the leading electrodes for calculating the results of the electric field in the particular layer.

Micromachining is being employed to make the membranes, cantilever rays and resonant structures. Silicon micromachining is an important issue for the massive advancement of MEMS in the previous era. This has referred to the shaping of minute mechanical portions, which are made of silicon substrates and more lately other resources [29]. It is being used to construct such features, which can be categorized as channels, clamped rays, cantilevers, membranes, gears, etc. and these all features can be gathered to make a variety of sensors. They have proven to be very efficient systems when we compare them with the conventional systems because the conventional systems lack in their mass production, which is completely compensated by the MEMS system because of their several advantages including miniaturization [30]. There are many inherent disadvantages of the devices, which are used currently or we can say the conventional devices that have provoked the need for MEMS or the micromaching grounded systems for microwave and RF applications. There are broadly main three categorizations of the motivations for comprising the fabrication technologies used in millimetre wave systems, microwave and RF systems which are completely based on MEMS technology [31]. Talking about the first one, the size of the microwave devices is reduced when the frequency is increased. Thus, it has become very important and crucial to look upon the dimensions of the components being employed in the millimetre wave systems [32, 33].

These claim for great accuracy manufacture technologies through which micromachining has offered a feasible direction. This method has been providing us with the integration capabilities of the system besides the other mentioned before [34]. Many efforts are made to employ the micromachining structures in such a way that at the lower frequencies of 1–2 cm, there is a coordination to decrease the effective dielectric constant of the substrate used at the microwave level and millimetre systems. When we micromachine these substrates, it improves the radiation pattern or the capability of the antennas and also increases the bandwidth [35].

Most of the devices that are completely based on MEMS system always improve for the reduction in insertion losses and thereby increase in the bandwidth. For many micromachined devices such as tunable capacitors, micro shifts inductors and RF shifts, this characteristic is completely effective. The PIN diodes are incompetent at higher frequencies in case of RF switching systems [36]. The RF switches that are based on MEMS have actuation voltages, which are very low. The distributed components are being replaced by the lumped components and devices, which are micromachined because they are very flexible in integration and have proved to be very developed in case of bandwidth. Likewise, many currently employed configurations are being replaced by the MEMS-based devices or the micromachined components because they tend to have very large insertion losses at higher frequencies. These MEMS technology-based devices have proven to achieve higher Q-factor for even very large frequencies up to 10 MHz and even up to 2 GHz for surface acoustic wave filters [37]. The wireless communication marketplaces have expanded so intensely that it has become very important to emphasize our study on these micromachined and MEMS-based technology because the market has been shifted to the consumer applications from the traditional defence-based applications [38]. The power handling capability of these systems, which is required, has been decreased due to the manifold increase in the production volume. Because of these developments, there is an increase in the performance of the devices and components based on MEMS technology, which have their applications employed in RF systems, microwave systems and the sub-millimetre systems [39, 40]. Figure 2 shows a micromachined RF-MEMS device.

The handling techniques and methods for MEMS systems are increasing considerably because of their significant applications employed in millimetre and microwave-based systems and methods [41].

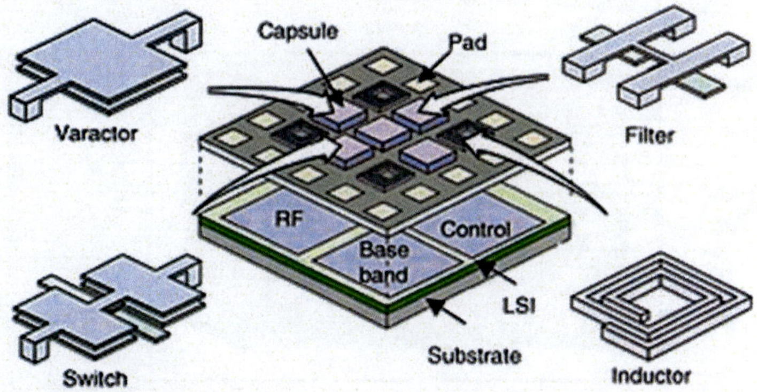

Fig. 2 RF-MEMS Device

RF-MEMS switches are the micromachined components, which employ any program in mechanical form to attain an open or short circuit (close) circuit in the radiofrequency transmission line. Radiofrequency MEMS switches are categorized by axis of deflection (lateral, vertical), actuation method (magnetic, electrostatic, electrothermal, piezoelectric), clamp configuration (fixed–fixed beam, cantilever), circuit configuration (series, parallel) or contact interface (ohmic, capacitive). The two foremost kinds of forces that are mostly employed for the actuation of RF switches are electrostatic as well as electromagnetic. The electromagnetic force has high current consumption but a low actuation voltage [42–44], while the electrostatic force has a great actuation voltage and nil current ingestion. The RF switches based on electrostatic switches are most common. These find their application in mm-wave regions and in the region of the microwave. Microwave circuits mostly employ electrostatic switches. Electrostatic-based RF MEMS devices offer high isolation and low insertion loss, linearity, Better Q factor and high power handling [45]. There are two categories of electrostatically actuated RF switch, one of which is series and the second is shunt. The capacitive and ohmic coupling switches both can be employed whichever as a parallel or a serial configuration switch, commonly ohmic switches have been used in serial method. For parallel or shunt switches, mostly capacitive coupling is preferred [46].

The series switch is predominantly open-circuited (open) and it gets short-circuited (close) when the switch is actuated. On the other hand, shunt switch is initially short-circuited (close) and it gets open-circuited (open) when a mandatory voltage is being applied [47]. Figure 2 describes the series contact switch in which the transmission line is being usually OFF state, which signifies that the required signal will only permit through the line once the membrane of switch is hauled down and the conductor traces together each side of gap. These types of switches have good isolation characteristic but the reliability and control over insertion loss are the chief matters need to be distinguished. This is due to self-welding might take place at the metal to metal exchange area amid transmission line as well as the membrane [48].

A capacitive parallel or shunt switch is shown in Fig. 3. In this configuration,

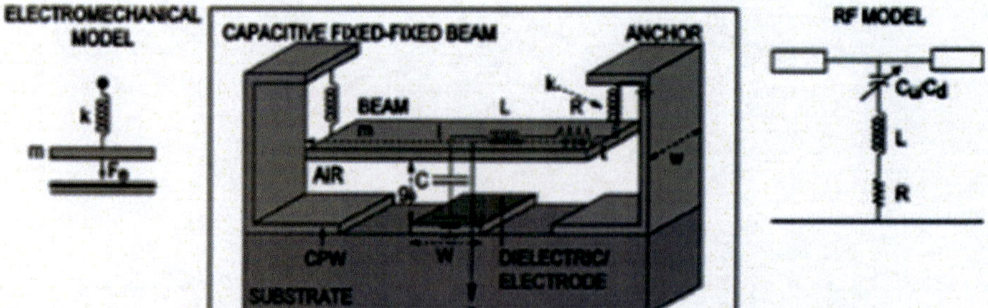

Fig. 3 Series contact RF MEMS switch

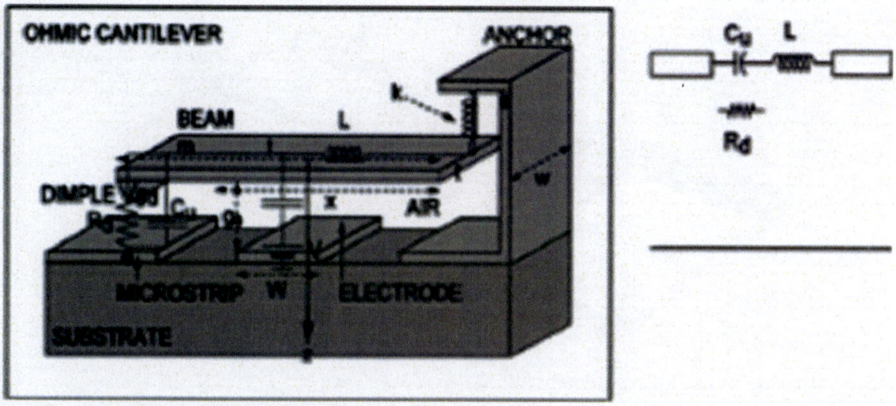

Fig. 4 Parallel/Shunt contact RF-MEMS switch

the transmission line is normally short-circuited (closed switch) or in ON condition and there is a laid a dielectric layer amid transmission line as well in the membrane of switch. A parallel plate capacitor structure is formed when we pull down the membrane of the switch. The input signal is conducted towards ground via the parallel plates of capacitor [49, 50]. These switches are mostly used in RF ranges. The RF features of parallel-type switches get exaggerated by capacitance ratio amongst open-circuited and short-circuited (ON/OFF) state. Due to this principle, at higher frequencies, the isolation characteristic of this type is better (Fig. 4).

When a DC voltage is applied in such a manner that each of the ends of the switch gets differently charged and thus DC voltage creates an electrostatic force. This electrostatic force creates magnetism amid the two parts. This force causes the switch to be short-circuited or in closed position and permits the RF power to stream over the switch. However, the dielectric layer between the parallel plates does not permit the electricity to drift over it [51].

Now when we turn off the voltage, the switch gets open-circuited or goes in open condition back because of the dielectric layer. In the absence of the layer of dielectric, the DC current can be passed/flowed through the switch producing it to switch short-circuited or in closed condition permanently [52].

RF-MEMS switches are manufactured with air gaps, therefore, they offer low off state capacitance hence very high resolution.

2 Design and Modelling

The design of capacitive parallel switch of RF MEMS switch consists of square polycrystalline silicon with density 2320 kg/m^3 is modelled, which is suspended 0.9 μm above a 0.1 μm thin silicon nitride film. The relative dielectric constant of silicon nitride is taken as 7.5. The spring stiffness is taken as 1×10^{15} Pa/m. The

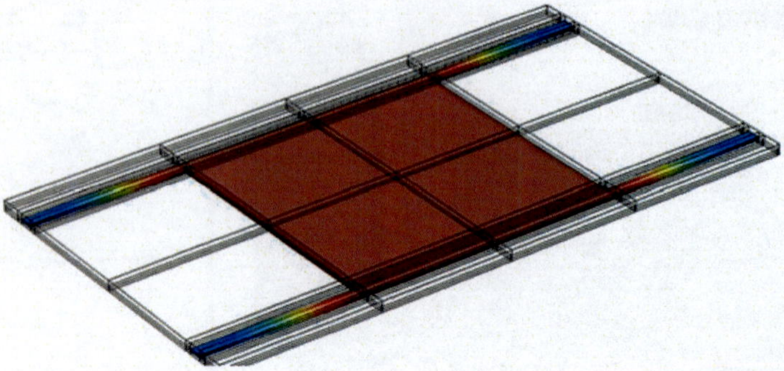

Fig. 5 Structure of RF-MEMS switch

width of the design is 111 μm and length is 61 milometres. The silicon counter electrode is placed under the substrate and is grounded. The plate is connected with the substrate with four rectangular shaped flexures at corners but it is electrically isolated. Figure 5 shows the geometry of the designed RF switch.

DC voltage (5 V in step) is supplied to the structure. The interface of electromechanics meshes the gap between nitride and polysilicon. The smooth step function represents the dielectric constant in the gap. The gap takes value of nitride dielectric constant when polysilicon is in contact with nitride. The penalty method or barrier method is used to describe the contact between nitride and polysilicon. To represent the surface of nitride, stiff and non-linear springs are used. These springs have low stiffness when the polysilicon is away from the nitride and have negligible influence on deformation. The distance of spring becomes much stiffer when the gap is reduced and it resists further closure.

Contact forces are given by

$$F_c = t_n - \mathrm{e}_n . g \quad g < 0 \tag{1}$$

$$F_c = t_n + \exp\left(-\frac{\mathrm{e}_n}{t_n} . g\right) \quad g \geq 0 \tag{2}$$

where e_n is penalty stiffness, t_n is input estimate of contact force, g is gap between polysilicon and nitride.

To compute electrical characteristics, the following equations are used in model.

$$\nabla_m . \left(\in_{0,\mathrm{vac}} (\in_r - I) E_m\right) = \rho_q \tag{3}$$

$$\nabla_s . \left(\in_{0,\mathrm{vac}} E\right) = 0 \tag{4}$$

$$E_m = -\nabla_m V \tag{5}$$

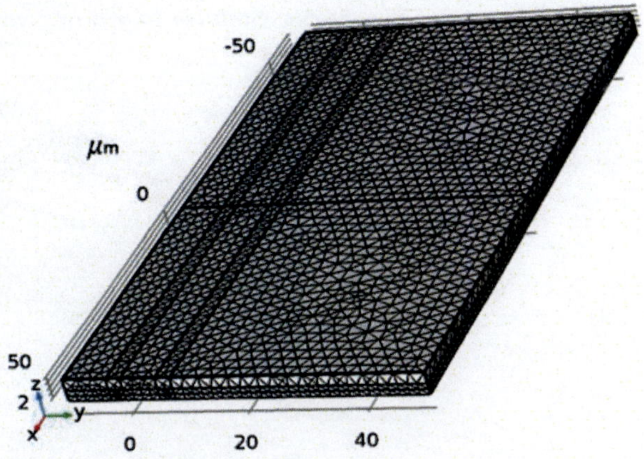

Fig. 6 Meshing of design

$$E = -\nabla_s V \tag{6}$$

Fine meshing is conducted on modelled RF switch structure than extra fine meshing so to reduce computational load. The meshing is done by tetrahedral mesh with minimum element quality taken 0.2041. In meshing, 54,098 tetrahedron values, 23,812 triangles, 1948 edge elements and 72 vertex elements are used with curvature factor of 0.5 and maximum element growth rate of 1.45. Figure 6 shows the meshing of the RF-MEMS switch model.

3 Results, Analysis and Discussion

First, a small voltage of 1 mV is applied to the structure of polysilicon. To measure the DC capacitance of the device, this applied voltage is sufficient. After a time period of 25 μs, the applied voltage is increased to 5 V, which is greater than pull in voltage using a step function which has risen time of 10 μs. This voltage pulls down onto nitride and results in a significant and sudden change of capacitance of structure.

Device pulled in the spatial dependence of total displacement is shown in Fig. 7. The bending occurs primarily in the vicinity of their attachment points and the flexures. Most of the structure is in contact with nitride.

Figure 8 shows the contact forces acting on polysilicon, which gives important observation that the largest forces occurred in vicinity of flexures of structure. Figure 9 shows the electric potential over the structure.

Figure 10 shows the displacement of RF switch as function of time. Due to the inertia, with the timescale of voltage change, the switch takes some longer time to change.

Figure 11 shows the capacitance of device as a function of time. It has been observed that due to pull in the capacitance of structure changes from 0.1 to 1.5 pF.

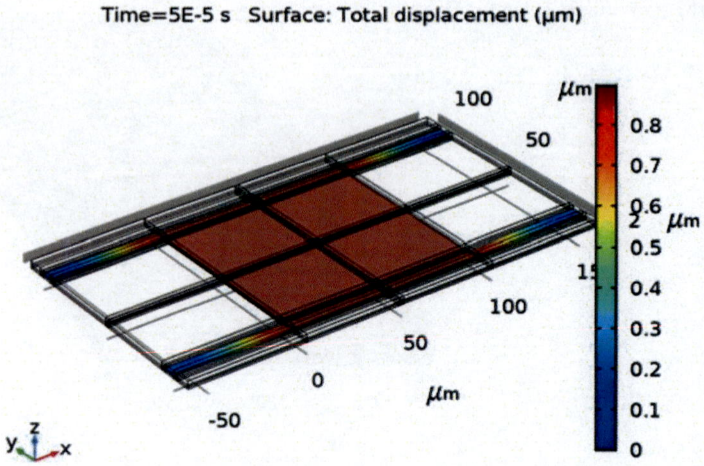

Fig. 7 Total displacement of polysilicon when pulled in

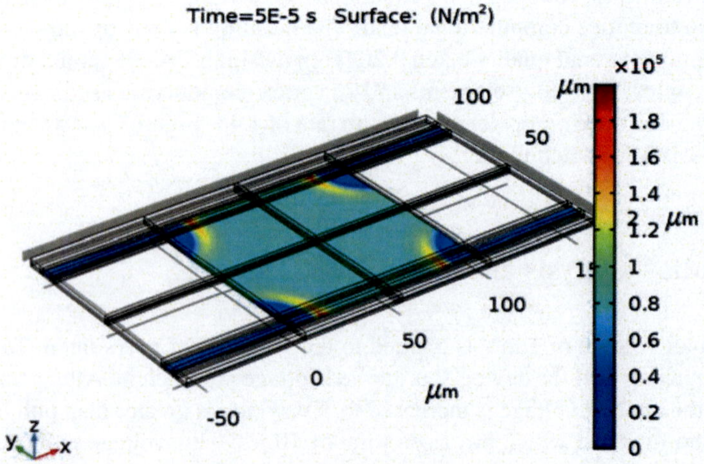

Fig. 8 Contact forces acting on polysilicon when pulled in

The capacitance increases by factor of 55 approximately and it changes on significant shorter timescale than displacement.

4 Conclusion

Due to less costing, smaller size and less power consumption, the RF-MEMS switches are used in 5G and IoT systems. In this paper, RF MEMS switch is modelled and

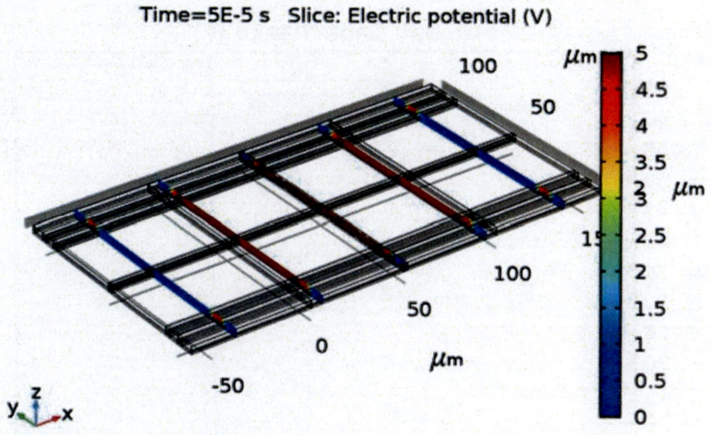

Fig. 9 Electric potential

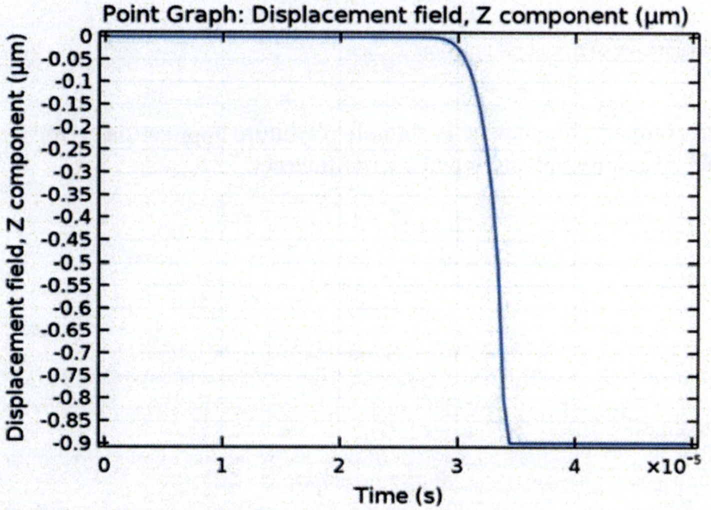

Fig. 10 Displacement field of the centre of device as function of time

analysed, which consists of thin micromechanical bridge, which is suspended over a layer of dielectric. A higher value of DC voltage, which is greater than device's pull in voltage, is applied to the structure, due to this pull in the capacitance of structure changes from 0.1 to 1.5 pF. The capacitance increases by factor of 55 approximately and it changes on significant shorter timescale than displacement. However, due to the inertia, with the timescale of voltage change, the switch takes some longer time to change. The proposed structure can be used for RF-MEMS circuits and finds application in satellite communication, wireless communication and radar systems, etc.; however, it is crucial to tune the stiffness and contact force. The design is small

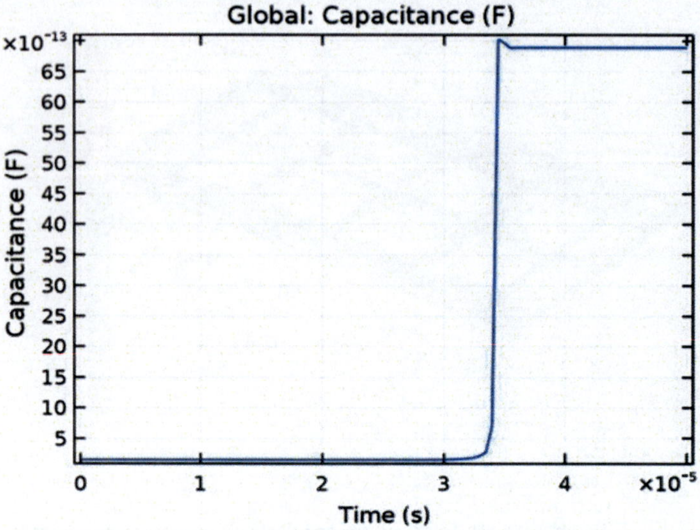

Fig. 11 Capacitance of device as function of time

in size and compact. The device is suitable for future applications lying in the field of IoT, IoE, 5G communication and Tactile Internet.

References

1. Rebeiz GM (2003) RF MEMS-Theory, design and technology. Wiley Inc.
2. Cho I-J, Song T, Baek S-H, Yoon S-H (2005) A low-voltage and low-power RF MEMS series and shunt switches actuated by combination of electromagnetic and electrostatic forces. IEEE Trans Microwave Theory Techniques 53(7)
3. Lin TH et al (2009) A study on the performance and reliability of magnetostatic actuated RF MEMS switches. MicroelectronicsReliability 49:59–65
4. Han W, Pryputniewicz RJ (2007) Design fabrication, andcharacterization of surface micromachined MEMS cantilever components. In: Sem-proceedings. pp 1162–1170
5. Yanjue G et al (2009) Design optimization of cantilever beam MEMS switch. In: Industrial mechatronics and automation. ICIMA
6. Rebeiz GM (2003) RF MEMS Theory: Theory design and technology. Wiley, Colorado
7. Wang Z et al (2005) Contact physics modeling and optimization design of RF-MEMS cantilever switches. In: Antennas and propagation society international symposium, vol 1A. IEEE, pp 81–84
8. Spasos M et al (2011) RF-MEMS switch actuation pulse optimization using Taguchi's method. Microsyst Technol Micro Nanosyst-Inf Storage Process Syst 17:1351–1359
9. Varadan VK, Vinoy KJ, Jose KA (2003) RF MEMS and their applications. New York, Wiley
10. Sharma M, Singh H (2018) SIW based leaky wave antenna with semi C-shaped slots and its modeling, design and parametric considerations for different materials of dielectric. In: 2018 Fifth international conference on parallel, distributed and grid computing (PDGC), IEEE, pp 252–258
11. Peroulis D, Pacheco SP (2003) Electromechanical considerations in developing low voltage RF MEMS switches. IEEE Trans Microwave Theory Techniques pp 259–270

12. AbbaspourSani E, Afrang S (2006) A low voltage MEMS structure for RF capacitor switches. Progress in Electromag Res 157–167
13. Wang Z et al (2005) Contact physics modelling and optimization design of RF MEMS cantilever switches. IEEE
14. Bakri-Kassam M, Mansour RR (2013) High power latching RF MEMS switches. IEEE Trans Microwave Theory Techniques 63
15. Vakilian M et al (2012) Optimization of cantilever–based MEMS switch used in reconfigurable antennas. In: IEEEICSE2012 Proceedings 2012. Kuala Lumpur, Malaysia
16. Goldsmith C, Ehmke J, Malczewski A, Pillans B, Eshelman S, Yao Z, Brank , Eberly M (2001) Lifetime characterization of capacitive RF MEMS switches. In: IEEE International microwave symposium diagnosis. pp 227–230
17. Bozler C, Drangmeister R, Duffy S, Gouker M, Knecht J, Kushner L, Parr R, Rabe S, Travis L (2000) MEMS microswitch arrays for reconfigurable distributed microwave components. In: IEEE International Microwave Symposium Diagnosis. pp 153–156
18. Iannacci J (2017) RF-MEMS for high-performance and widely reconfigurable passive components–A review with focus on future telecommunications internet of things (IoT) and 5G applications. J King Saud University-Sci 29(4):436–443
19. Iannacci J (2018) RF-MEMS technology as an enabler of 5G: low-loss ohmic switch tested up to 110 GHz. Sens Actuators A 279:624–629
20. Iannacci J (2019) Compact modelling-based coupled simulation of RF-MEMS networks for 5G and internet of things (IoT) applications. Microsyst Technol 25(1):329–338
21. Dubuc D, Grenier K, Iannacci J (2018) RF-MEMS for smart communication systems and future 5G applications. In: Smart Sensors and MEMs Woodhead Publishing, pp 499–539
22. Ma L-Y, Soin N, Daut MHM, Hatta SFWM (2019) Comprehensive study on RF-MEMS switches used for 5G scenario. IEEE Access 7:107506–107522
23. Fitzgerald P, Berney H, Lakshmanan R, Coburn N, Geary S, Mulvey B (2018) Devices and sensors applicable to 5G system implementations. In: 2018 IEEE MTT-S International Microwave Workshop Series on 5G Hardware and System Technologies (IMWS-5G), IEEE, pp 1–3
24. Chan KY, Daneshmand M, Mansour RR, Ramer R (2009) Scalable RF MEMS switch matrices: methodology and design. IEEE Trans Microwave Theory Techniq 57(6)
25. Sinha S, Bansal D, Rangra KJ (2012) RF MEMS T-type switch for redundancy switch matrix applications in space telecommunication. ISSS-2012
26. Sharma M, Singh S, Khosla D, Goyal S, Gupta A (2018) Waveguide diplexer: design and analysis for 5G communication. In: 2018 Fifth international conference on parallel, distributed and grid computing (PDGC), IEEE, pp 586–590
27. Peroulis D, Pacheco SP, Sarabandi K, Katehi LPB (2001) Alleviating the adverse effects of residual stress in RF MEMS switches. In: 31st European microwave conference October. pp 1–4
28. Dahleh RA, Mansour RR (2007) A novel wraped beam design that enhances RF performance of capacitive MEMS switches. pp 1813–1816
29. Chan K, Daneshmand M, Mansour RR, Ramer R (2009) Scalable RF MEMS switch matrices: methodology and design. IEEE Trans MTT 57(6):1612–1621
30. Hijazi YS, Vlasov YA, Larkins GL (2003) Design of a superconducting MEMS shunt switch for RF applications. IEEE Trans Appl Supercond. 13(2):696–699
31. Sutagundar M, Sheepramati BG, Jangamshetti DS (2014) Research issues in MEMS resonator. Int Journal Eng Sci 4(8):29–39
32. Hector J (2007) In: Introduction to microelectromechanical microwave system
33. Sharma M, Singh H, Singh S, Gupta A, Goyal S, Kakkar R (2020) A novel approach of object detection using point feature matching technique for colored images. In: Proceedings of ICRIC 2019. Springer, Cham, pp 561–576
34. Varadan VK, Vinoy KA, Jode KA (2003) RF MEMS and their applications
35. Tillmans H, Raedt W, Beyne E (2003) MEMS for Wireless communication. J Micromech Microeng

36. Van Caekenberghe K (2009) RF MEMS technology for radar sensors. IEEE Trans
37. Mafinejad, Kouzani AZ, Mafinezhad K, Nabovatti H (2009) Design and simulation of a low voltage wide band RF MEMS switch. In: The proceedings of the IEEE international conference on systems, man, and cybernetics October 2009. San Antonio, TX, USA
38. Manvinder S, Gupta AK (2020) An algorithm for target detection, identification, tracking and estimation of motion for passive homing missile autopilot guidance. In: Mobile radio communications and 5G networks, Springer, Singapore, pp 57–71
39. Zhang Y, Onodera K, Maeda R (2006) RF micro-electro-mechanical-systems capacitive switches using ultra thin hafnium oxide dielectric. In: Japanese J Appl Phy 45(1A):300–304
40. Fruehling A, Khater MA, Lung B, Peroulis D (2010) CMOS-based monitoring of contact events up to 4 MHz in Ohmic RF MEMS switches. IEEE Trans 978–1–4244–7732
41. Sharma* M, Singh H (2019) Substrate integrated waveguide based leaky wave antenna for high frequency applications and IoT. Int J Sens Wireless Commun Control 9:1. https://doi.org/10.2174/2210327909666190401210659
42. Grant PD, Denhoff MW (2004) A comparison between RF MEMS switches and semiconductor switches. In: Proceedings of international conference on MEMS, NANO and smart systems (ICMENS'04) held on July 2004
43. Choi J-Y, Lee J-S (2006) Development of RF MEMS switches for high power applications. IEEE Trans Paper-31449713
44. Lee Y-S, Jang Y–H (2010) Design a 50–110 GHz ohmic contact RF MEMS silicon switch with high isolation. IEEE Trans
45. Llamas MA, Girbau D, Pausas E, Pradell L, Aouba S, Villeneuve C, Puyal V, Pons P, Plana R, Colpo S, Giacomozzi F (2009) Capacitive and resistive RF-MEMS Switches 2.5D and 3D electromagnetic and circuit modelling. In: The proceeding of Spain conference on electron device, February
46. Lahiri SK, Saha H, Kundu A (2009) RF MEMS SWITCH: an overview at-a glance. In: International conference on computers and devices for communication
47. Maninder K, Rangra KJ, Kumar D, Kumar S, Girdhar N, Boora P, Maurya S, Singh S (2008) Effect of multilayer structure on electromechanical characteristics of RF MEMS capacitive symmetric toggle switch for X-band applications. In: Proceeding of ISSS-08 held on July 24–26
48. Rebeiz GM (2003) RF MEMS switches: status of the technology. In: 12th International conference on solid state sensors, actuators and microsystems, June 8–12, Boston
49. Gaddi R, Bellei M, Gnudi A, Margesin B, Giacomozzi F (2004) Interdigitated low-loss ohmic RF-MEMS switches. NSTI-Nanotech 2004, vol 2. ISBN 0–9728422–8–4
50. Tilmanns HAC (2002) MEMS Components for wireless communications in: euro sensors. In: 16th European conference on solid state transducers, Prague, Czech Republic, September 15–18, pp 1–34
51. Varadan VK, Vinoy KJ, Jose KA (2003) Pennsylvania State University, USA. "RF MEMS and Their Applications. Wiley Inc.
52. Van Caekenberghe K (2009) RF MEMS on the Radar. In: The IEEE microwave magazine

Chapter 11
Designing and Development of Stemmer of Dogri Using Unsupervised Learning

Parul Gupta and Shubhnandan S. Jamwal

1 Introduction

The development of morphological analyzers with high accuracy for Indian languages, which are highly inflected, remains a challenge. There are many Indian languages that are under-resourced, like Dogri language, which is spoken by 5.5 crore people. Dogri language has various dialects such as Poonchi, Mirpuri, Pothwari, Siraji, Bhaderwahi, Kishtwari, Gadi, Pogli etc., which are spoken in various districts of the Jammu province.

There are some peculiar features of the Dogri language like orthography, which makes the development of the NLP applications very difficult. There are some consonants such as घ (gh), झ (jh), ढ (dh), ध (dh), भ (bh) in Devanāgarī script, which exist only in Dogri orthography. They are phonetically used for representing tonal क (k), च (c), ट(ṭ), त (t), प (p) at the initial stage of a word, e.g., घर (ghar/house), झड़ना (jhaṛanā/shed), ढक्कन (ḍhakkan/lid), धमकी (dhamakī/threat), भारी (bharī/heavy) etc. Moreover, Dogri language is sometimes also called as tonal language and we have three types of tones, i.e., level tone, low or low-rising tone, high or high-falling tone. The different tone that appears in Dogri language conveys many sentiments. The non availability of other simple tools for developing any NLP applications for Dogri puts major obstacle for every researcher to the work in this direction for the language in question. This research work focuses on the development of one such tool, namely, Stemmer for Dogri language.

For understanding the natural language, word-level analysis, sentence-level analysis, context-level analysis and discourse-level analysis are required. Morphological analysis is the first step to understand a given sentence and stemming is the main task done at the morphological level. Stemming is not just related to natural language processing domain, but it is equally important in Information

P. Gupta (✉) · S. S. Jamwal
PGDCSIT, University of Jammu, Jammu, J&K, India

N. Marriwala et al. (eds.), *Soft Computing for Intelligent Systems*, Algorithms for Intelligent Systems, https://doi.org/10.1007/978-981-16-1048-6_11

Retrieval [1]. Some probabilistic approaches are also used for the development of the stemmer, but it requires a vast amount of monolingual corpus.

The basic approach in developing any search and information retrieval system uses stemmer in order to reduce morphological variants to their stem. For example, the following inflected words अक्‍ख, अक्‍खै, अक्‍खी,अक्‍खा, अक्‍खे, अक्‍खां,अक्‍खें, अक्‍खीं, अक्‍खो are reduced to same stem अक्‍ख(eye). In this research paper, we are presenting the development of stemmer for the Dogri language using the unsupervised machine learning approaches.

2 Literature Review

Sheth and Patel [1] suggested DHIYA stemmer for Gujarati language, which is based on the morphology of Gujarati language. The inflections which appeared in Gujarati text were identified, and based on it, the ruleset was created. EMILLE corpus is used for training and evaluation of the performance of the stemmer and accuracy of the stemmer was reported as 92.41%. Kasthuri et al. [2] proposed an unsupervised stemming hybridized with partial lemmatization for four morphologically different languages such as English, French, Tamil and Hindi. An innovative attempt has been made to develop a stemming algorithm for a novel conflation method that exploits the quality of words and uses some standard Natural Language Processing tools like Levenshtein Distance and Longest Common Subsequence for the stemming process. Even though it is recommended for other Indian languages, no significant development is seen in the implementation of other Indian languages. Patel and Patel [3] have also developed a Gujarati Stemmer and named it as GUJSTER, based on an algorithm for reducing affixes from the Gujarati words and for getting an optimal output as compared with the previous hybrid algorithm.

The accuracy of the morph analyzer can be achieved if the stemmer works with more accuracy, therefore, Sunitha and Kalyani [4] presented an unsupervised stemmer for improving the performance of Telugu rule-based morph analyzer and observed an increase in performance of rule based, that is, from 77 to 84.2% for words which are in the hundreds. But, the accuracy can still be increased if the corpus is increased. Attempts have been made to improve the refinement of the stemmer, therefore, Narayan et al. [5] proposed a novel corpus-based method for stemmer refinement, which can provide improvement in both classification and retrieval. The proposed method models the given words as generated from a multinomial distribution over the topics available in the corpus and includes a procedure like sequential hypothesis testing that enables grouping together distributionally similar words. It has also been claimed that the system can refine any stemmer. Melucci and Orio [6] presented a method based on Hidden Markov Models to generate statistical stemmers. A list of words are used as a training set and HMM parameters are estimated and are then used to calculate the most probable stem. Dangui and Naik [7] have also developed common stemmer for the languages using the Devanagari script by using a supervised approach and to evaluate stemmer to measure the performance of stemmer. Al-Shammari and Lin [8] presented Educated Text Stemmer for Arabic (ETS), which is a dictionary free, simple and highly effective Arabic stemming algorithm. Frakes et al. [9] analyzed four affix removal stemming algorithms and strength and similarity based on the Hamming distance measure were evaluated in different ways.

The use and development of the stemmer are very primitive approach that has been used over the years for the development of morphological analyzers. Lovins [10] wrote the first stemmer in1968. Snover and Brent [11], Brent et al. [12] and Goldsmith and John [13] have also carried out a lot of work on morphology by using unsupervised learning approaches.

3 Proposed Architecture

In Indian languages, Dogri is an Indo-Aryan language with rich morphology and high inflection. In this language, several words have a common root, which requires stemming. For example, words रानी, रानयै, रानयि, रानयिां, रानयिं and रानयिो will be reduced to रानी (queen). In Natural Language Processing, it is the initial step to identify the root or base of the word. Stemmer is the basic tool used to reduce a word to its root without semantics. By reducing the index size of the word, the performance of Information Retrieval systems can be increased. A word is usually of two types, simple and compound words. A simple word cannot be broken or decomposed further, whereas a compound word is composed of a root and suffix. The रानी is a simple word and रानयै, रानयि, रानयिां, रानयिं, रानयिो are compound words.

In the proposed algorithm, we have an input text file, databases of root words and possible suffixes, a Dogri corpus file that contain the Dogri text. The steps of creating the new database of the root words and checking the new corpus, which is used as test data are explained in the following steps. In the proposed algorithm, we have

TXFL = Input Text file
SNT_k = *K*th Sentence from M Sentences in TXFL, where *M* is the number of sentences in TXFL.
WRD_i = Number of words in a current sentence, where $i = 1, 2, 3 \ldots N$.
RT_i = Ith root of WRD_i.
SFX_i = Ith suffix of WRD_i.
MINEDST = Minimum Edit distance.
DVCB = Database containing set of RT_i.
PB_m = Mth probablity of split WRD_i, where '*m*' defines the number of splits in a WRD_i.
DSFX = Database containing set of SFX_i.
DGRICRPS = Dogri Corpus File.

1. Repeat for SNT_k from TXFL.
2. Split SNT_k into WRD_i (i = 1, 2, 3…N), where WRD_i= RT_i+ SFX_i.
3. Repeat for WRD_i in SNT_k.
4. If $WRD_i \in$DVCB, then Display WRD_i as RT_i and Stop.
5. Calculate the PB_m of different splits using Split—methods from DGRICRPS. Get the value of RT_i and SFX_i.
6. IF PB_m = 0, then If $SFX_i \in$ DSFX: stem the SFX_i from the WRD_i, update DVCB with the new entry of RT_i and display RT_i. Otherwise, If ($SFX_i \notin$ DSFX), Update DVCB with the new entry of RT_i and display RT_i and Stop.
7. IF $PB_m \neq 0$, then perform MINEDST on the MAX(PB_m).
8. IF MINEDST = 0, then If $SFX_i \in$ DSFX: stem the SFX_i from the WRD_i, update DVCB with the new entry of RT_i and display RT_i. Otherwise ($SFX_i \notin$ DSFX) Update DVCB with the new entry of RT_i and display RT_i and Stop.
9. IF MINEDST $\neq 0$, then perform RT_i matching with DVCB using Brute force approach. Get the most appropriate match and display it. Update DSFX accordingly.

3.1 Word Segmentation

Our methodology is incomplete in accordance with Goldsmith (2001) approach. It depends on split-all strategy. For unsupervised learning, training data words of Dogri are taken from CALTS, University of Hyderabad. These words have been split to give n-gram (n = 1, 2, 3 … l) suffix, where l is the length of the word. Then we compute suffix and stem probability. These probabilities are multiplied to give a split probability. The optimal segment corresponds to maximum split probability. Some postprocessing steps have been taken to refine the learned suffixes. Figure 1 outlines the steps involved in our algorithm. The details of these steps are presented below:

The words were segmented by performing splits after each character of the word as:

STM = Stem
SFX = Suffix

$$W_i = \{STM_{i1,}SFX_{i1};\ STM_{i2,}SFX_{i2};\ STM_{i3,}SFX_{i3};\ldots STM_{ij,}SFX_{ij};\}$$

where W_i corresponds to current word and j denotes total number of splits in W_i.

For example, the word 'चलदी'[Walks(Female)] is split into the following stem and suffix: {च लदी,चल दी, चलद ी, चलदी NULL}. The suffixes correspond to n = 1, 2, 3,…, l, where l is length of the word. Next, we have made three separate

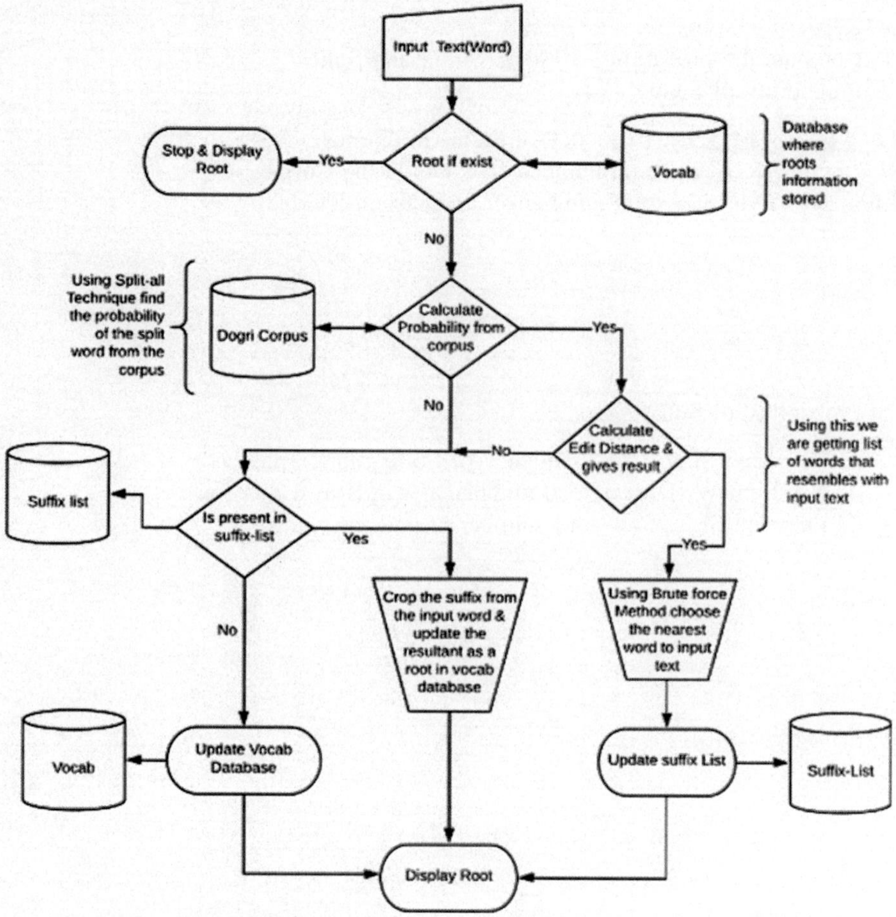

Fig. 1 Flowchart of proposed algorithm

arrangements of stem, suffix and split. Stem and suffix probability is computed based on their frequency in the corpus as:

1. Read list of words, i.e. WRD_i.
2. Repeat for WRD_i in list of words.
3. Perform splitting or segment words as:

$$\text{WRD}_i = \text{Split}_k\{\text{STM}_{i1}, \text{SFX}_{i1};\ \text{STM}_{i2,}\text{SFX}_{i2};\ \text{STM}_{i3,}\text{SFX}_{i3}; \ldots \text{STM}_{ij,}\text{SFX}_{ij};\}$$

4. Repeat for $Split_k$
5. Compute the probability of Stem, Suffix and Split.
6. Probability of Stem:

Fq = freq(STM_{ij}), i.e. frequency of $Stem_{ij}$ in Corpus.
N = Sum(WRD), i.e. total number of words in the Corpus.
Ln = len(STM_{ij}), i.e. total number of character in STM_{ij}.

$$P(STM_{ij}) = \frac{Fq}{N * Ln}$$

7. Probability of Suffix:

 Fq = freq(SFX_{ij}), i.e. frequency of $Suffix_{ij}$ in Corpus.
 N = Sum(WRD), i.e. total number of words in the Corpus.
 Ln = len(SFX_{ij}), i.e. total number of character in SFX_{ij}.

$$\mathrm{P}(SFX_{ij}) = \frac{Fq}{N * Ln}$$

8. Probability of Split:

$$P(\mathrm{Split}_k) = P(\mathrm{STM}_{ij}) \times P(\mathrm{SFX}_{ij})$$

9. Choose best split, i.e. the split with the highest probability value.

3.2 *Stemming of New Words*

The output of the proposed unsupervised stemming algorithm is the root. But the error occurs when we get zero probability for all the splits of a particular word. It happens when such a word is not present in our corpus. In that case, we use a list of all the suffixes, we search for the existence of any of the suffixes in the word. It may be the case that either one or multiple suffixes exist in the word. It may also be the case that none of the suffixes exist in the word, in that case, word itself is taken as stem. If only one suffix matches, we split the word into two, considering the other part as a stem. In case multiple suffixes exist, we have used the longest suffix matching during stemming. This means that the longest possible suffix of the target word matching with some suffix in the list is dropped. The suffix list is arranged in

Table 1 Suffixes exist in word 'चलदी'

…	दी	ी	…

decreasing order of length to facilitate it. We start matching from first suffix to last suffix, and when a match is found, we segment the word in its stem and suffix. Eg: Probability(चलदी[Walks(Female)]) = 0, i.e. the word is not present in Corpus and vocab list. The list of suffixes that exists in word चलदी[Walks(Female)] are (Table 1):

As 'दी' is the longest match and is, therefore, dropped. 'चल' is taken as root word and added to the vocab list.

3.3 Root Refinement Using Levenshtein Distance

In computational linguistics and computer science, edit distance is a way of quantifying how dissimilar two strings (e.g., words) are (दक्खिन, दक्खि) to one another by counting the minimum number of operations required to transform one string into the other. Once we get the root word, it is possible that the stem may not be the meaningful root. In order to get the meaningful root, we use Levenshtein distance for finding words from the vocabulary that is very close to the target word, i.e. words that require fewer modifications to make them similar to the target word. For example (Table 2),

After performing root refinement if only one root is returned from the vocabulary, we take it as the final stem. In case multiple root words are returned, we apply the brute force approach to get the final stem based on their frequency in the corpus.

4 Experimentation and Results

The proposed algorithm has been implemented in Python 3.8.4 and to assess the proposed stemmer, we evaluate the percentage of accuracy of the stemmed data. The training data set has been constructed by extracting 17,643 words from 56 documents taken from random Dogri documents. Six different test sets have been created by randomly extracting words from Dogri corpus. For our experimentation, we are converting Dogri text into its WX notation, which is a transliteration scheme for representing Indian languages in ASCII. This scheme originated at IIT Kanpur for computational processing of Indian languages and is widely used among the natural language processing (NLP) community in India. Based on the split

Table 2 List of words similar to the calculated root

Target word	Result of root refinement
दक्खिन	दक्खि
खाद्ध	खा, खाद्

Table 3 Test result using our proposed model

Test Run	Total words	Accurately stemmed	Inaccurately stemmed	% age of accuracy	Error rate
Test run-1	50	32	18	64	0.36
Test run-2	155	106	49	68	0.32
Test run-3	356	250	106	70	0.30
Test run-4	658	440	218	67	0.33
Test run-5	932	671	261	72	0.28
Test run-6	1256	904	352	71	0.29
Average	567	400	167	69	0.31

probability, we are finalizing the stem for a particular word. The split with maximum probability is considered to be optimal segments of the word. The optimal breakpoint largely depends upon the probability distribution of split of the word, language and corpus. Some heuristics have been considered while taking an optimal split. First, we restrict the minimum length of stem to three. This heuristic avoids reducing words like cala, capa, cama etc. to same stem ‘ca’. This is due to the increase in stem probability due to the reduction of otherwise dissimilar words into the same stem. This is actually an indication of over-stemming.

We use random Dogri data to test the algorithm. Six different runs have been performed, where each run contains a varying number of test words. As the number of runs increases, the training data, suffix list and the vocabulary size also get increased. During the first test run, we have comparatively small training data, therefore, it is more likely that new words which are not present in training data will encounter. Consequently, it leads to less percentage of data being stemmed accurately. Simultaneously, new encountering words are added to our training data, as a result of which the probability of more words being stemmed increases. The results of six test runs are shown in Table and Figure (Table 3).

It is observed that errors are due to the words that have their stems less than three characters, or these are new words present in test data. When these words are included in successive test runs, positive stems are attained, which increases the percentage of successive test runs. However, we have fallen in the percentage of accuracy of stemmed words in Test Run-4. The cause for this fall in percentage is due to the occurrence of new words, i.e. words that are not present in our training data. When all such new words are added to the training data, the percentage of accuracy stemmed words increases in Test Run-5.

5 Conclusion

In this paper, first-ever attempt has been made for the construction of stemmer for Dogri language, and quite significant amount of accuracy has been achieved. Pandey and Siddiqui [14] also used the approach of unsupervised learning and it was also presumed that this can be easily applied to a new language because it is language independent and also because it does not require any language-specific rules as input. We have also applied similar techniques for the Dogri language, and percentage of accuracy achieved is 69%. The approach used is to learn suffixes and does not require any linguistic input. It is also observed from the test data that if some specific rules are embedded the accuracy can be drastically increased. The performance of our algorithm enhances as the training data size increases and it can be further enhanced by handling word having stem length less than three. As not much work has been done for the automation of handling Dogri language, our approach can be used for the rapid development of morphological analyzers, sentiment analysis, information retrieval and machine translation in Dogri language.

References

1. Sheth J, Patel B (2014) Dhiya: a stemmer for morphological level analysis of Gujarati language. In: International conference on issues and challenges in intelligent computing techniques (ICICT), April, IEEE
2. Kasthuri M, Bishop S, Khaddaj S (2017) PLIS: proposed language independent stemmer for information retrieval systems using dynamic programming. In: World congress on computing and communication technologies (WCCCT) February, IEEE
3. Patel CD, Patel JM (2017) GUJSTER: a rule based stemmer using dictionary approach. In: International conference on inventive communication and computational technologies (ICICCT) July, IEEE
4. Sunitha KVN, Kalyani N (2010) A novel approach to improve rule based Telugu morphological analyzer. In: World congress on nature and biologically inspired computing (NaBIC) January, IEEE
5. Bhamidipati NL, Pal SK (2007) Stemming via distribution-based word segregation for classification and retrieval. IEEE Trans Syst Man Cybernet 37(2)
6. Melucci M, Orio N (2003) A novel method for stemmer generation based on hidden markov models. In: Proceedings of the twelfth international conference on Information and knowledge management, CIKM, pp 131–138
7. Dangui SR, Naik N (2015) A lightweight stemmer for devanagari script. In: Compute '15: proceedings of the 8th annual ACM India conference October, pp 55–62
8. Al-Shammari ET, Lin J (2008) Towards an error-free Arabic stemming. In: Publication: iNEWS '08: proceedings of the 2nd ACM workshop on improving non english web searching October, pp 9–16
9. Frakes WB, Fox CJ (2003) Strength and similarity of affix removal stemming algorithms. Publication, ACM SIGIR Forum
10. Lovins JB (1968) Development of a stemming algorithm. Mech Transl Comput Linguistics 11 (1):22–31
11. Snover MG, Brent MR (2001) A Bayesian model for morpheme and paradigm identification. In: Proceedings of the 39th annual meeting of the ACL, pp 482–490
12. Brent M, Murthy R, Lundberg A (1995) Discovering morphemic suffixes: a case study in minimum description length induction. In: Proceedings of the fifth international workshop on artificial intelligence and statistics

13. John G (2001) Unsupervised learning of the morphology of a natural language. Comput Linguistics 27(2):153–198
14. Pandey AK, Siddiqui TJ (2008) An unsupervised Hindi stemmer with heuristic improvements. In: ACM Proceedings of the second workshop on analytics for noisy unstructured text data July, Singapore, pp 99–105

Chapter 12
Credit Card Fraud Detection Techniques: A Review

Ankit Mohari, Joyeeta Dowerah, Kashyavee Das, Faiyaz Koucher, and Dibya Jyoti Bora

1 Introduction

The most prominent form of payment is a credit card. Since the variety of credit card users is growing worldwide, fraud is improved, and fraud is increasing in the purchasing of virtual cards, where only card information such as card number, expiry date, security pin, etc. is required. These transactions are often made on the internet or via the phone. If anyone gets access to the specifics of the card, so fraud can be easily committed. Nowadays, many online transactions are made using credit cards. It is inevitable for the user to keep their credit card details confidential and the service provider must take the initiative to protect the users' details; credit card fraud can be characterized, as there might be some cases where the owner and the card-issuing authorities are unaware of a situation where a random person uses their card for their own personal purposes. Many frauds, which are detected worldwide, include monitoring the behaviors of user numbers in order to interpret, estimate, or prevent offensive activity that consists of fraud default and interference. Methods of fraud detection are increasingly designed to prevent the actions of offenders from adjusting to their fraudulent techniques, such frauds are known as:

- Online and offline credit card frauds
- Card pilferage
- Account insolvency
- Interference of device
- Fraud for applications

A. Mohari (✉) · J. Dowerah · K. Das · F. Koucher · D. J. Bora
Department of Information Technology, School of Computing Science, Kaziranga University, Assam, India
e-mail: cs16msiit067@kazirangauniversity.in

D. J. Bora
Assistant Professor, Department Of Information Technology, School of Computing Sciences, Kaziranga University, Assam, India

N. Marriwala et al. (eds.), *Soft Computing for Intelligent Systems*, Algorithms for Intelligent Systems, https://doi.org/10.1007/978-981-16-1048-6_12

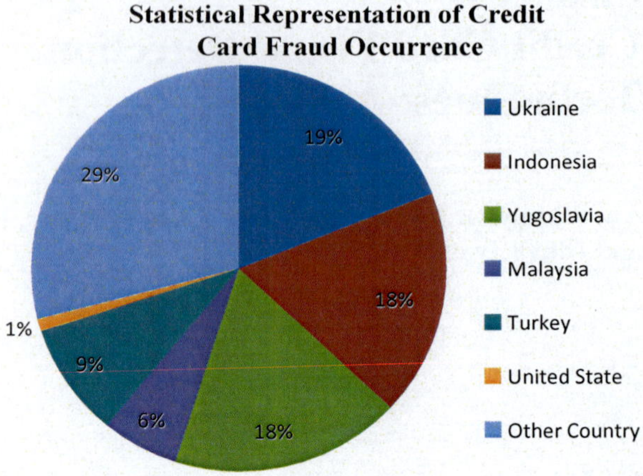

Fig. 1 Outlining the countries where credit card fraud remains prominent

- Forge card
- Communication through media/telephone fraud

Some risky countries are outlined in Fig. 1, on the basis of knowledge expressed in [1] 2012. The United States has introduced a minimum level of fraud rate, but credit purchases are the highest. With a staggering 19%, Ukraine has the highest amount of fraud rate, along with Indonesia at an 18.3% fraud rate. The riskiest nation after these two is Yugoslavia with a rate of 17.8. Malaysia having 5.9%, Turkey having 9% and finally the US account for less amount of fraud rate. Figure 1 does not clearly depict the other country's risk for credit card fraud at a rate of 29%.

2 Technical Terminologies

2.1 Some Techniques to Detect Credit Card Fraud Are Explained Below

i. *Naive Bayes:* In this classification method, the probability of an object associated with a particular category or class with a certain feature is learned. Naive Bayes algorithm can fit model fast and provide high accuracy when applied to big data and need less training.

$$P(A|B) = \frac{P(B|A).P(A)}{P(B)} \quad (1)$$

where $P(A)$ and $P(B)$ are just the probability of A and B occurring, $P(B|A)$ is the probability of event B occurring, ensuring that event A is true, $P(A|B)$ is the conditional probability of event A occurring, event B is being true.

ii. *Logistic Regression:* Logistic regression is less inclined to the overfitting, but it can overfit in high-dimensional datasets. We can consider regularization techniques to avoid overfitting. Any big outliers will be transformed into the range of 0 and 1. It helps mainly to solve classification problem and supply us with the knowledge of weather the event is happening or not.

iii. *Random Forest:* First, we begin by choosing random samples from the dataset that is provided. Then, this algorithm will be used to create a decision tree for every sample that is generated. Then, the prediction result is found out for each decision tree. For each expected outcome, a voting mechanism is carried out. Therefore, ultimately, as the final prediction outcome, the most voted prediction outcome is selected.

iv. *AdaBoost:* AdaBoost or Adaptive Boosting improves the performance of the weak classifier; here each instance in the training dataset is weighted.

$$\text{Initial Weight} : \text{Weight}(xi) = \frac{1}{n} \tag{2}$$

where xi is the ith training instance.
n is the number of training instance.

v. *Artificial Neural Network:* A collection of nodes is interlinked, which is designed to depict the functioning similar to the human brain, which is known as Artificial Neural Network. All the other nodes that are present in the adjacent layers of the node are being assigned the weighted connection.

vi. *Genetic Algorithm:* This algorithm is being inspired by natural evolution. So, from here the algorithm has been introduced. The binary strings used in chromosomes describe the population of candidate solutions. It assumes that the chances of survival and the amount of reproduction are higher with the higher quality chromosomes, which means it increases the health value.

vii. *Hidden Markov Model (HMM):* A Hidden Markov model is also described as double embedded stochastic process using which highly complex stochastic processes can be generated. Within the underlying framework, a Markov process that has an unnoticed stage is presumed to be available. The definite transformation of the state's present inside the simpler Markov models is the only unknown parameters that are present.

viii. *KNN Classifier:* In the case of classification and regression, KNN is generally the non-parametric algorithm that is used. The input of this algorithm consists of K-nearest training examples in the feature space for classification and regression, whereas on the other hand, the output generally depends on whether the KNN belongs to the classification category or regression category.

ix. *Decision tree:* Decision tree is a tree-shaped structure that expresses mainly independent attributes and dependent attributes and is a data mining technique. Classification rules that are derived from the decision trees are generally expressions of IF–THEN and each rule is needed to be produced, all the tests must succeed. Decision tree is generally a non-parametric algorithm.

3 Literature Review

A way of using machine learning for the detection of credit card fraud was suggested by Ratna Sree Vall et al. [2]. Supervised learning provided the unexpected input example of the associate degree and is designed to perform predictions. The supervised methods used in this paper are Random Forest, Logistic Regression, Naive Bayes, and a boosting technique (AdaBoost) to enhance the classification algorithm. AdaBoost or Adaptive Boosting is a boosting method, which provides us with a single "strong classifier" with the combination of multiple "weak classifiers". It was. Therefore. concluded here that compared to logistic regression and even the Naïve Bayes techniques with a boosting technique, the random forest classifier is stronger.

Mehndiratta et al. [3] In this paper, various credit card fraud detection techniques are reviewed and supported on some parameters. Here, through collecting historical data, predictive analysis methods can be used to notice fraud. Different methods, such as Artificial Neural Network, Hidden Markov Model, Genetic Algorithm, Naive Bayes, KNN classifier, are used here. In this paper, mainly fraud prediction is done using two phases that are feature extraction and classification, and it decides to use a hybrid approach in the future for credit card fraud detection.

Zarrabi et al. [4] Deep Autoencoder is proposed by the author and that serves as the best extraction of the details of the features from the fraud transaction that occurred in the credit card. For the classmark problems, softmax tools are used here by the author. For classifying a form of fraud, an AutoEncoder is used here, which maps the data into a high-dimensional space. Deep learning can be said as one of the most effective methods for detecting the credit card fraud. The dynamic distribution of the data in the network types becomes difficult to understand. For extraction, the best features of data with a high amount of precision and low variance the networks via Deep Autoencoder were used.

Al-Khatib [5] Fraud detection is often a drawback of the classification of legit transactions from deceitful transactions. Fraud detection includes watching the defrayal behavior of users/customers so as to work out, detect, or turning away from undesirable behavior. The utilization of credit cards is common in contemporary society. Several issues are also being faced by the developer to relate to credit card fraud detection. A number of trendy fraud detection techniques are employed by the research worker: Neural Networks, Genetic Algorithms (GAs), Rule induction, Expert systems, Case-based reasoning(CBR), Inductive logic programming (ILP), Regression, Artificial intelligence, etc. have been used for detecting deceitful transactions. Comparative study on data mining techniques is used to detect deceitful coverage, to obtain low false alarm rate. Multiple algorithms procedure can be used in this analysis to achieve greater cost savings.

Patidar et al. [6] used a dataset in combination with genetic algorithms (GA), for credit card fraud detection to train three layers of backpropagation ANN. Genetic algorithms were responsible for determining the network layout in this analysis, dealing with the network topology, the number of hidden layers, and nodes in each layer.

Sisodia et al. [7] the study of the output of several classifier-sampling methods when applied to the credit card fraud dataset in which classes are imbalanced was presented. Principal component analysis (PCA) for real data and the variables time, quantity, and class obtained 28 main components in the data. We have found 10,000; 15,000, and 20,000 instances available from the three datasets we have implemented. The authors analyzed five over-sampling and four methods to under-sampling, respectively, and few cost-sensitive and ensemble classifiers are applied.

Liu et al. [8] Here the author used two kinds of random forests to coach the behavior options of traditional and deceitful transactions, so compare this each and differentiated on the premise of a classifier, performance on the detection of credit card fraud. The data used are associated with an e-commerce company in China that is used to research the performance of those two kinds of random forest models. In this paper, to identify deceitful, B2C dataset is used by the authors. Random Forest classifier is used and which obtained better result only on the small dataset and cannot handle any imbalanced dataset, which makes it inefficient, as fraud datasets are mostly imbalanced.

Roy et al. [9] proposed a deep learning method for detecting fraud transactions. 80 million transactions have been detected as fraud. They have used cloud-based environment to obtained high performance. The researchers have concluded a deep learning method with tuning parameters for deceitful transaction detection, which helps the financial institution for the prevention of illegal practices.

Pojee et al. [10] proposed a modern process that includes payment of invoices or bills. This technique is referred to as the "No Cash" smartphone program and is primarily used by retailers through whom consumer payment services can be eased. In this situation, there is no need for the NFC-Enabled Point of Sales (PoS) Machines Approach and only cell phones are required. This system is designed and which minimized the problem of the customers bringing the card and provide them a easy payment workflow. The customer's shopping experience is enhanced as the program NoCash, which has several features, and is used based on growth in the number of NFC-based mobile devices. Application clients can refer to the history of the cost and the costs will be reduced, which are not necessary.

Estevez et al. [11] Through this research paper, we have got some ideas that every year tens of billion dollar losses are estimated in global telecommunications fraud. Many researchers have implemented many techniques. The system consists of both classification and prediction modules, to develop a system for the detection of subscription fraud for the fixed telephone lines. The classification module classifies subscribers according to their past historical conduct into four distinct categories: Subscription is fraudulent, otherwise fraudulent, insolvent, and regular. Some of the methods, which are implemented in the research paper, are creating a dataset, categories of subscriptions, system architecture, classification, and prediction module. Information over ten-thousand real subscribers of a major telecom company in Chile was on the database and the classification module was implemented using fuzzy rules. 2.2% subscription fraud prevalence was found in this database. A multilayer perceptron neural network was implemented for the prediction model. True fraudsters identified were 56.2%, screening only 3.5% of all the customers in the test dataset. This research was carried out on fixed telecommunications, but the methods suggested here may be applied to fraud in mobile communications and also other markets. It also demonstrates the possibility of substantially avoiding telephone service fraud by examining the application details and consumer data at the time of application.

Choi et al. [12] in this paper, from 2016 to 2018, have surveyed deceitful transaction techniques using machine learning and deep learning algorithms and analyzed the fraud detection techniques' limitations and advantages. Feature selection, sampling, and applying some supervised and unsupervised algorithms to detect deceitful transactions are the techniques included in the paper on the financial dataset. In 2015, the ultimate model was validated from the data occurring in the actual financial transaction in Korea. In conclusion, machine-learning techniques performed better and detect most deceitful transactions than the artificial neural network. In the machine learning method, the maximum detection rate and the lowest detection rate were 1 and 0.736, respectively, and when all of the algorithms were analyzed, and then 0.98618 was the average detection rate. To maximum detection rate in all ratios, the lowest detection rate and the lowest detection rate of the artificial neural network were 0.914, 0.651, and 0.77228 respectively.

Feedzai et al. [13] In this paper, the open-source tool, IBM Proactive Technology Online (PROTON), was presented to deal with the uncertainty. All levels of the architecture and logic of an event-processing engine were strongly affected by the inclusion of uncertainty. In the complex event processing, programmatic language new capabilities as building blocks and basic primitives were implemented to proceed with the implementation of event-driven applications. At first, the application mechanism is in the domain of credit card fraud detection. To support all the varieties of uncertainty, a few Complex Event Processing (CEP) engines are there. Some limitations are also identified from the recent probabilistic engines, and there is the absence of supporting uncertainty, for specifying the complex events.

Phua et al. [14] In online transaction, there can be numerous amount of frauds that can occur, which is analyzed by viewing the behavior of the user and if there is seen any deviation in the spending behavior then it may be a fraudulent transaction, so far that there are many data mining methods are used by bank and credit card companies like Decision Tree, Rule-based mining, ANN and fuzzy clustering, hidden Markov model, or a hybrid approach of these methods, so any one of these methods will be used to detect the behavioral activities of the customers based on the past experience, this paper mainly will compare the different techniques that would detect fraud more precisely. After comparing the algorithms, it was found out that every algorithm has drawbacks as well as some advantages as ANN and Hidden Markov Model can handle large data, but the imitation is that the process becomes slow and it is also expensive, it is also seen that decision tree may be easy to understand but has a limitation as it cannot handle complex data. So, to increase its accuracy, Precision and Recall is used, so that the amount of false positive and false negative can be reduced to a certain extent, and finally, F1 score is found out, which is the harmonic mean of precision and recall, which depicts that if the F1 score is higher, it indicates it as a good model. Therefore, as we can see, a combination of those methods mentioned above can be used to detect deceitful transactions, which were adding new features, and sampling methods could be performed to train the model accurately.

Saini et al. [15] In this paper, anomaly detection algorithms are used, which include Isolation Forest and Local Outlier Factor. They have analyzed the dataset with various statistical methods. Only 10% sample was used for training the model with the outlier detection algorithms. Comparative analysis with two algorithms was done and the outcome was not that desirable one. As Isolation forest only can give 28% precision rate on 10% sample data and 33% precision rate on 100% dataset, and Local Outlier Factor precision rate was only 2% (Table 1).

4 Open Issues

Based on the observation in [16], 2019, some open issues faced by the researchers have been discussed. Like no paper benchmark and dataset without outliers are available, and if analysis with the standard dataset is not done, then the proper outcome will not be obtained.

Because there's no literature on credit card fraud algorithms, benefits and drawbacks will be present, and their limitations will be hidden. Algorithms will be combined to overcome their disadvantages.

Limitations of good metrics to evaluate the result causes incapability for researchers for comparing different techniques and approaches and determining the importance of the most efficient system for credit card fraud detection.

5 Conclusion

The credit card fraud detection methods have gained popularity in the past decade with the evolution of statistical model, machine-learning algorithms, data mining techniques. The fraud transactions prediction has feature extraction and classification that are two phases. Within the first phase, the feature extraction technique is applied and within the second phase, classification is applied for fraud transaction detection. Fraud transaction detection is that the major issue of prediction because of a frequent and enormous number of transactions.

During this research study, various techniques of credit card fraud detection are reviewed. In the future, hybrid approaches are going to be designed for credit card fraud detection.

Table 1 Table of comparison

Author's_Name	Technique	Features	Results
Ratna Sree Valli, Jyothi, Varun Sai, Rohith Sai Subash	To detect fraud machine learning method was used	Naïve Bayes, Logistic Regression, Random Forest, and a boosting technique (AdaBoost)	Applying boosting technique with random forest performs better
Sonal Mehndiratta and Gupta	Predictive analysis methods	Genetic Algorithm, Artificial Neural Network, Hidden Markov Model, KNN Classifier, Naive Bayes	For fraud detection hybrid approach was made
Kazimi, Zarrabi	Deep autoencoder	Softmax software, Autoencoder	Higher accuracy and low variance are achieved within these networks
Adnan M. Al-Khatib	Data mining	Neural Networks, Rule Induction, Expert System, Case-Based Reasoning(CBR), Genetic Algorithms(GAs), Inductive logic programming(ILP), Regression, Artificial intelligence etc	Multiple algorithms are analyzed based on their performance with noisy and missing data, scalability, accuracy, etc
Patidar, Sharma	Neural networks	With genetic algorithms, backpropagation neural network is combined	Dealing with the number of nodes in each layer, the number of hidden layer and the network topology
Sisodia, Reddy, Bhandari	Proposed a novel approach and applied PCA on real dataset	Certain performance metrics have been used to measure the quality of prediction of various methods, which showed the proposed approach's efficiency against others	The precision value and execution time are not as per the demand
Modi and Dayma	To provide a secure mechanism several methods are integrated	Applying any of these methods the normal usage pattern of clients identified by their past activities	The proposed algorithm achieves high performance in terms of execution time by accuracy factor get compromised

(continued)

Table 1 (continued)

Author's_Name	Technique	Features	Results
Changjun Jiang Shiyang Xuan, Zhenchuan Li, Lutao Zheng, Shuo Wang, and Guanjun Liu	For the identification and detection of fraud from credit cards the B2C dataset is used	Analyze that random forests provide good results if the dataset is small	Imbalanced data make it less effective than any other dataset
Sun, Roy, Mahoney, Alonzi, Adams and Beling	For online transaction, deep learning approach was made for fraud detection	The proposed model outperformed and prevented frauds in any online transaction through credit cards	There is a need to improve the accuracy of the proposed algorithm
Dastgir Pojee and Sajjad Zulphekari and Fahim Rarh and Varsha Shah	NoCash mobile application was proposed	Proposed application help to minimize the fraudulent activities	The expense history and any unwanted costs need to be minimized
Estevez, Held and Perez	Prevent subscription fraud	Classification module is implemented using fuzzy rules	By analyzing the customer antecedents, the application information and subscription fraud in telecommunications can be prevented at the time of application
Dahee Choi and Kyungho Lee	Machine learning and deep learning methodology	Applying supervised and unsupervised algorithms, feature selection, and sampling	ANN detection rate is better
Ivo Correia Feedzai, Fabiana Foumier, Inna Skarbovsky	IBM Proactive Technology Online (PROTON) open-source tool to cope with uncertainty	CEP engines	There is an uncertainty while defining complex events in the rules
Gayler, Phua, Lee, Smith,	Data mining	Decision tree, rule-based mining, ANN and fuzzy clustering, hidden Markov model, or hybrid approach	The methods were combined to detect deceitful transactions and to train the model accuracy sampling methods
Aditya Saini, Swarna Deep Sarkar, Shadab Ahmed	Anomaly detection algorithms	Isolation Forest, Local Outlier Factor	The precision rate is not as per demand

6 Future Research Work

An experimental study and analysis of methods are discussed here to analyze their efficiency. A comparative analysis will be carried out on various algorithms to find out the most efficient one.

Acknowledgements We would like to take the opportunity to thank Dr. Dibya Jyoti Bora, Assistant Professor, Kaziranga University to provide the necessary support and suggestions for our research work.

References

1. Chaudhary K, Yadav J, Mallick B (2012) A review of fraud detection techniques: credit card. Int J Comput Appl 45(1)
2. Ratna Sree Valli K, Jyothi P,Varun Sai G, Rohith Sai Subash R (2020) Credit card fraud detection using machine learning algorithms. Quest J Res Humanities Social Sci 8(2): 04–11 ISSN(Online): 2321–9467
3. Mehndiratta S, Gupta K (2018) Credit card fraud detection techniques: a review. IJCSMC 8(8)
4. Kazemi ZH (2017) Using deep networks for fraud detection in the credit card transactions. In: IEEE 4th International conference in knowledge-based engineering and innovation (KBEI). pp 0630–0633
5. Al-Khatib AM (2012) Electronic payment fraud detection techniques. World Comput Sci Info Technol J (WCSIT) 2(4):137–141 ISSN: 2221–0741
6. Patidar R, Sharma L (2011) Credit card fraud detection using neural network. Int J Soft Comput Eng (IJSCE)
7. Sisodia DS, Reddy NK, Bhandari S (2017) Performance evaluation of class balancing techniques for credit card fraud detection. In: 2017 IEEE International conference on power, control, signals
8. Liu G, Li Z, Zheng L, Wand S, Xuan CJ (2011) Random forest for credit card fraud detection. In: IEEE 15th International conference on networking, sensing and control (ICNSC)
9. Roy A, Sun J, Mohoney R, Alonzi, Adams S, Beling P (2006) Deep learning detecting fraud in credit card transactions. Syst Appl 31(2):337–344
10. Pojee D, Zulphekari S, Rarh F, Shah V (2017) Secure and quick NFC payment with data mining and intelligent fraud detection. In: 2017 2nd International conference on communication and electronics systems (ICCES)
11. Estevez PA, Held CM, Perez CA (2006) Subscription fraud prevention in telecommunications using fuzzy rules and neural networks. Expert Syst Appl 31(2):337–334
12. Choi D, Lee K (2018) An artificial intelligence approach to financial fraud detection under IoT environment: a survey and implementation
13. Feedzai IC, Foumier F, Skarbovsky I (2015) The uncertain case of credit card fraud detection. In: The 9th ACM international conference on distributed event based systems(DEBS15)
14. Phua C, Lee V, Smith, Gayler KR (2010) A comprehensive survey of data mining-based fraud detection research. arXiv preprint arXiv:1009.6119
15. Saini A, Sarkar SD, Ahmed S, Maniraj SP (2019) Credit card fraud detection using machine learning and data science. Int J Eng Res Technol 8(09) ISSN: 2278–0181
16. Sorournejad S, Zojaji Z, Atani R, Monadjemi AH (2016) A survey of credit card fraud detection techniques: data and technique oriented perspective. 22:46:13 UTC

Chapter 13
Extracting Knowledge in Large Synthetic Datasets Using Educational Data Mining and Machine Learning Models

Jaikumar M. Patil and Sunil R. Gupta

1 Introduction

EDM is a recent phenomenon in the field of data mining and Knowledge Discovery in Databases (KDD), concentrating on the mining of user trends and the development of helpful information of educational knowledge practices such as selection systems, enrolment systems, course administration systems and any other systems agreement with students at various rates. Researchers in this area are based on attempting estimable knowledge to unleash informative help to organizations to better control their learners or help students better manage their education and achievements for improving their prosperity. Observing the data and experience of students to create determination trees or association rules, to make better judgments or enhance student performance, is an essential area of research that focuses mainly on analyzing and understanding the educational data of students designating their scholastic performance and producing unique rules and classifications; and prediction classification is the most common and useful data mining methodology used for the classification and appraisal of values. Educational Data Mining the exception to the statement, so in this academic research, it was used to analyze the alternatives collected by the students through a survey and also provide classifications based on the data collected to predict and identify the output of the students in their next semester. The purpose of this analysis is to define relationships and academic success between the personal and social factors of students. The newly learned information will allow both students and teachers to achieve better improved educational results, by understanding inherent underperformers at the commencement of the term year and pay more innumerable awareness to them to aid them in their institutional period and get sufficient consequences. Indeed, not only exclusively underperformers can assist

J. M. Patil (✉)
Department of Computer Science and Engineering, SSGMCE, Sheagon, India

S. R. Gupta
Department of Computer Science and Engineering, PRMIT&R, Amravati, India

N. Marriwala et al. (eds.), *Soft Computing for Intelligent Systems*, Algorithms for Intelligent Systems, https://doi.org/10.1007/978-981-16-1048-6_13

from this investigation but also expected well-performers could avail from this study by producing more meaningful endeavours to persevere better schemes and analysis by more substantial compensation and deliberation from their supervisors. Since getting educational data mining as a new field of study, it already has made contributions to the theory of learning and its teaching practice. The aim of this work is to offer a way to increase student performance and inspire educational institutions to encourage students in different aspects of their learning by analyzing data and evaluating it. With this static machine learning method, we have illustrated the overall performance of various learning techniques with synthetic datasets.

This paper describes the background information of research insection 1 and a brief state of art in Sect. 2, which is done by various existing authors. Section 3 includes the proposed system methodology including numerous EDM techniques and flow of execution and tools and techniques. The experimental analysis has done with Weka 3.8 environment that demonstrates in Sect. 4 and finally conclusion describes in Sect. 5.

2 Literature Review

EDM generates techniques for investigating data of character types that appear from instructional settings. Also during these methods, students are presented with a more reliable understating and knowledge manner. Detailed studies have been carried out in this region. It distinguishes the weak performers and interpretations of the factors that affect the students' educational enforcement at schools, colleges and even multiversities. In the subordinate section, we explain some extant investigation done by different contributors.

There are several problems that have been discussed in classroom contexts by Data Mining. Baker [1–3] identifies important points for EDM to encourage positive models, develop domain models, review pedagogical support given by link adaptation, scientific studies regarding learning and five methods such as prediction, convergence, association mining, sublimation of human judgement data and template discovery. Castro et al. [4] recommend EDM implementations addressing the measurement of the learning output of the student, applications offering course adaptation and learning suggestions related to the learning actions of the learner, methods dealing with the assessment of learning content and web-based educational courses, applications requiring feedback to both teachers and students. We think that even more educational applications are possible that have employed Data Mining techniques in educational environments.

Rastrollo-Guerrero et al. [5] proposed to evaluate and predict Students success using machine learning methods, based on the data gathered in this study, supervised learning was the most commonly used technique for predicting students' behaviour, producing accurate and reliable results. Specifically, the authors used the Support Vector Machine algorithm most and made the most reliable predictions [6, 7]. Besides

Support Vector Machine, Decision Tree, Naive Bayes and Random Forest were also well studied algorithmic proposals, which produced good results.

Baker [8] describes the challenges of educational data mining; basically, this author has done a lot of research in EDM since 1995 to till date. This article eliminates the bugs of the traditional data mining approach. The six major problems have defined in the entire research and how to overcome those using Baker Learning Analytics Prizes (BLAP).

Tsiakmaki et al. [9] proposed to examine the learning of student's success prediction using deep neural networks in higher education. Since predictive models have so far been poorly studied in the field of Educational Data Mining through transfer learning methods, we find this research to be an important step in that direction.

Arora et al. [10] proposed estimation of the educational output of students using a linear regression model; this study performed two distinct investigations for assessment, such as linear and logistic regression. They were collected from the Delhi university college dataset in real-time and predict the performance of the individual students [11, 12]. Eventually, the system acquires 71% and 85% classification accuracies for all methods.

Bindhia et al. [13] Proposed for students' academic success, students' success by their higher secondary value, IQ and EQ values. This is calculated only by certain modern algorithms. This is specified based on training, learning, analysis and predictive level. First of all, students or teachers must enter their marks in the interface given and students must attend an intellectual quotient and emotional quotient questions set with certain psychologist's guidance. In the prediction stage, the students will either pass or fail.

Daud et al. [14] proposed an approach Predicting Student Performance using Advanced Learning Analytics to achieve better accuracy levels, more than one data mining techniques are combined to test output using ensemble learning approach. Here, we have added Boosting (Ensemble) Stacking functions, Random Forest and Generalized Boosted System based on the Decision Tree, Support Vector Machine and Naïve Bayes base learners. Such ensemble classifiers are implemented for the assessment of Faculty Results on Faculty dataset and student outcomes. Ensemble classifier efficiency is evaluated according to accuracy, sensitivity and specificity [15, 16].

Krishna et al. [17] predicted and classified by way of tree logistic regression. He extracted the data from a Moodle-based integrated learning course and developed a student model. Identification and Regression Trees (CART) decision tree algorithm was used to identify and forecast at-risk students, based on the influence of four online activities: message sharing, community wiki content development, opening course files and taking online quiz.

Cen et al. [18], Romero and Ventura [19] improved reality learning efficiency; this work introduces an award-winning AIR-based mobile education program, codenamed AIR-EDUTECH, which was created to assist high school students in learning chemistry. The AIR-EDUTECH introduced new AIR technologies to help the students understand and learn basic molecular chemistry concepts better. This offers interactive 3D simulation and visual contact with the structures studied, offering a broader and more retentive awareness and increasing intuition around basic chemical reactions.

Tapingkae et al. [20] proposed the performance improvement strategy of teachers through the study of teaching evaluation data often cantered on the recognition of the teacher's teaching attitude and skills based on student input that can be applied to improved decision-making and quality improvement. Based on this research project, we had provided the automated decision-making framework and achieved it.

Chandra et al. [21], Yousafzai BK et al. [22] proposed graduation eligibility of students for analysis using data warehouse, the purpose of this research is to build a framework using data warehouse capable of seeing success student progress and failure of the course in each term. This program is supposed to be capable of predicting eligibility for graduation from graduates. This research approach consists of some steps to implement data warehouse and is then accompanied by the generation of the review report and online analytical processing (OLAP).

3 Methodology

This proposed EDM frame the framework for our work while obtaining information on our dataset. The approach provides the flow of the work used in this paper. Here the approach begins with the problem statement, then proceeds to data collection, then the dataset has been preprocessed, then we move to the Data Mining Classification, followed by the evaluation of the results, and finally, we begin the manner of information description. A wide range of Educational Data Mining methods have emerged over the last several years. Some are roughly similar to those seen in the use of data mining in other domains, whereas others are unique to educational data mining. The four major methods that are frequently used by the EDM community are prediction models, structure discovery, relationship mining and discovery with models. Data mining is a process of applying computer-based methodology, including new techniques for knowledge discovery, to data. In education to extract the information from the large datasets generated by modern experimental and observational methods. Data mining seems to be a very optimistic and productive approach to decision-making processes. The categorization is a fundamental method of data processing and is often used. Awareness of training examples is required to understand the identification. Figure 1 demonstrates numerous classification techniques that are used for effective mining on a large synthetic dataset.

3.1 Experimental Design

Before applying data mining methods in synthetic dataset, we should frame an experimental setup for our work. The methodology starts from the problem statement, then proceeded to data collection, preprocessing the dataset has been done, then moved

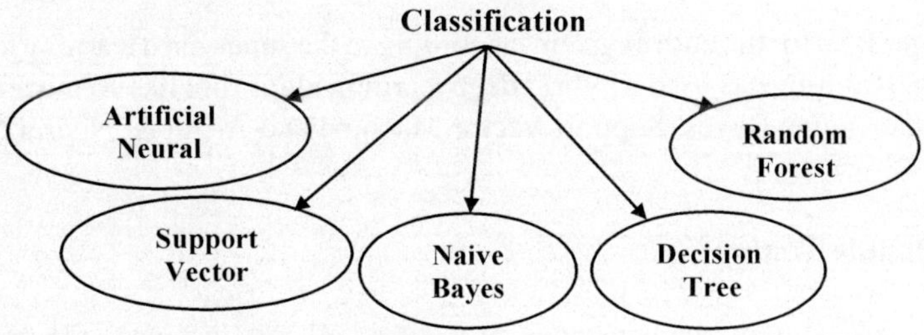

Fig. 1 Machine learning classification for EDM

to feature extraction and selection, later Data Mining Classification followed by the evaluation of module training, testing and finally entered into the process of prediction analysis.

3.1.1 Data Collection

We collected data from various sources like Kaggle, UCI Machine learning repository, as well as some real-time data sources.

3.1.2 Pre-Processing and Normalization

This phase introduces data reduction techniques like data balancing, data sampling, data normalization etc. After this, the processed data will be balanced with the respective class and will be normalized. Using data pre-processing and data normalization, we create data sampling using a systematic sampling technique and normalization for data reduction or attribute reduction from the input dataset.

3.1.3 Feature Extraction and Selection

This process extracts the various features from input data, the extracted feature should be normalized using feature selection threshold, which will remove redundant as well as worst features for training. It extracts various hybrid features from the normalized dataset with relational features and training has been done using a particular optimization algorithm selection method.

3.1.4 Module Training

The selected features are forwarded to the training module using machine learning classifier, after successful execution of module, it generates complete Background

Knowledge (BK) for the entire system, according to the supervised learning approach. The module training has used a hybrid deep learning algorithm like Recurrent Neural Network like Naive Bayes, Support Vector Machine and Artificial Neural Network.

3.1.5 Module Testing

Here we used an individual instance process or all test instances for getting the prediction accuracy of the system, this phase evaluates the effectiveness of the system with the various datasets. The module testing uses hybrid deep learning algorithms like Naïve Bayes, Support Vector Machine and Artificial Neural Network.

3.1.6 Prediction Analysis

Finally, it demonstrates class prediction of each test instance and evaluates the classification accuracy of the entire system using confusion metrics. The dataset has used for evaluating the proposed classification system. The dataset has been taken UCI machine learning repository and some real-time dataset, which are collected from colleges, educational institutes etc. The attributes have been categorized based on importance like basic, intermediate and important etc.

4 Results and Discussion

In past times, mining techniques have also been used to derive trends through raw statistics to get meaningful information. Furthermore, artificial intelligence has been used to perform many tasks expertly, involving recognition, classification, analysis, forecasting and identification. Similar application techniques were also quite helpful in finding associations not traditionally considered, making automatic choices and identifying promotional advantage to analyse vast quantities of knowledge. In the below Table, we demonstrate some classification results (Table 1).

Table 1 Classification accuracy of various machine learning algorithms using EDM in Weka environment

Algorithm	Accuracy	Precision	Recall	F-score
SVM	0.93	0.94	0.95	0.95
ANN	0.92	0.91	0.89	0.90
Naïve Bayes	0.89	0.88	0.93	0.91
Random forest	0.87	0.86	0.92	0.89
J48	0.90	0.92	0.85	0.88

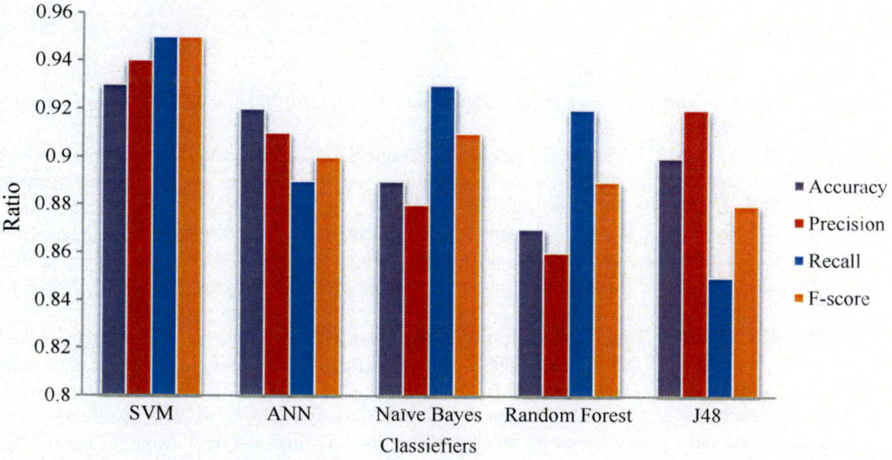

Fig. 2 Data classification results using various machine learning algorithms on large student data

This experiment basically demonstrates the various machine learning algorithms performance on a large student's synthetic dataset Fig. 2.

5 Conclusion

The EDM basically used to evaluate the performance of individual candidates on large historical data, this work similarly demonstrates on factors effects on students' performance. In the intellectual sense, computer vision and machine intelligence have recently been approved, enabling institutions to efficiently distribute human and physical capital, manage student success and increase performance effectiveness across the education curriculum. By implementing several of the most commonly used statistical and machine learning strategies, including such artificial neural networks, Bayesian modelling and random forest, the methods mentioned in this paper have shown various solutions to handle numerous significant learning difficulties. Future studies might include the development of computer algorithms to classify the ability of the students, the improvement of decision-making, assigning of tutors using specific parameters, the suggestion of lessons and materials, the improvement of the recruitment process and the prediction of student success. It is also possible to apply so many different machine learning techniques, including such working with imbalanced data, choosing the best characteristics, using method sets and implementing methods for data reductions. In the experimental analysis, we achieved an average accuracy of around 90% over the various supervised learning classifiers. Proposed method focus to extract some abstract features from high-dimensional dataset for interesting work in future directions.

References

1. Baker, Yacef K (2009) The state of educational data mining in 2009: a review and future visions. J Educ Data Mining 1(1): 3–17
2. Babu I, Siva Balan RV, Mathai PP (2019) Envisaging the academic performance of student. In: 2019 International conference on computing, communication, and intelligent systems (ICCCIS), IEEE, pp 90–94. https://doi.org/10.1109/ICCCIS48478.2019.8974531
3. Chaki PK, Sazal MMH, Barua B, Hossain MS, Mohammad KS (2019) An approach of teachers' quality improvement by analyzing teaching evaluations data. In: 2019 Second international conference on advanced computational and communication paradigms (ICACCP), IEEE, pp 1–5
4. Vellido CA, Nebot A, Mugica F (2007) Applying data mining techniques to e-learning problems. In: Evolution of teaching and learning paradigms in intelligent environment. Springer, Berlin, Heidelberg, pp 183–221
5. RastrolloGuerrero JL, Gomez-Pulido JA, Duran-Dominguez A (2020) Analyzing and predicting student's performance by means of machine learning: a review. Appl Sci 10(3):1042
6. Kumar M, Singh AJ (2019) Performance analysis of students using machine learning and data mining approach. Int J Eng Adv Technol 8.3:75-79
7. Alaa H, Hashim AS, Awadh WA (2018) Predicting student performance in higher education institutions using decision tree analysis. Int J Interactive Multimedia Artif Intell 5:26–31. https://doi.org/10.9781/ijimai.2018.02.004
8. Baker RS (2019) Challenges for the future of educational data mining: the Baker learning analytics Prizes. JEDM J Educ Data Mining 11(1):1–17
9. Tsiakmaki M, Kotsiopoulos GK, Kotsiantis S (2002) Transfer learning from deep neural networks for predicting student performance. Appl Sci 10.6:2145. https://doi.org/10.3390/app10062145
10. Arora S, Agarwal M, Kawatra R (2020) Prediction of educationist's performance using regression model. In: 2020 7th International conference on computing for sustainable global development (INDIACom), IEEE, pp 88–93. https://doi.org/10.23919/INDIACom49435.2020.9083708
11. Yassein NA, Helali R, Mohomad SB (2017) Predicting student academic performance in KSA using data mining techniques. J Inf Technol Softw Eng 7(5): 1–5.https://doi.org/10.4172/2165-7866.1000213
12. Binh HT, Duy BT (2017) Predicting students' performance based on learning style by using artificial neural networks. In: 9th International conference on knowledge and systems engineering, Vietnam, Corpus ID: 19600065. https://doi.org/10.1109/KSE.2017.8119433
13. Francis BK, Babu SS (2019) Predicting academic performance of students using a hybrid data mining approach. J Med Syst 43.6:162, Springer. https://doi.org/10.1007/s10916-019-1295-4
14. Daud A, Aljohani NR, Abbasi RA, Lytras MD, Abbas F, Alowibdi JS (2017) Predicting student performance using advanced learning analytics.In: Proceedings of the 26th international conference on world wide web companion. pp 415–421. https://doi.org/10.1145/3041021.3054164
15. Zaldivar-Colado A, Aguilar-Calderon JA, Garcia-Sanchez OV, Zurita-Cruz CE, Moncada-Estrada M, Bernal-Guadiana R (2014) Artificial neural networks for the prediction of students academic performance. In: INTED2014 Proceedings, IATED. pp 4584–4589
16. Kaviyarasi R, Balasubramanian T (2020) Predicting the academic performance of college students through machine learning techniques. Res J Comput Inf Technol Sci 6(6):1–10 ISSN 2320–6527
17. Krishna M, Rani BSBP, Kalyan Chakravarthi G, Madhavrao B, Chowdary B (2002) Predicting student performance using classification and regression trees algorithm. Int J Innov Technol Explor Eng (IJITEE) 9(3) ISSN: 2278–3075
18. Cen L, Ruta D, Said Al Qassem LMM, Ng J (2020) Augmented immersive reality (AIR) for improved learning performance: a quantitative evaluation. IEEE Trans Learn Technol.https://doi.org/10.1109/TLT.2019.2937525

19. Romero C, Ventura S (2020) Educational data mining and learning analytics: an updated survey. Wiley Interdisciplinary Rev: Data Mining Knowled Discovery 10(3):e1355. https://doi.org/10.1002/widm.1355
20. Tapingkae P, Panjaburee P, Hwang GJ, Srisawasdi N (2020) Effects of a formative assessment-based contextual gaming approach on students digital citizenship behaviours, learning motivations, and perceptions. Comput Educ 159:103998
21. Chandra E, Girsang AS, Hadinata R, Isa SM (2018) Analysis students' graduation eligibility using data warehouse. In: 2018 International conference on information management and technology (ICIMTech), IEEE, pp 61–64
22. Yousafzai BK, Hayat M, Afzal S (2020) Application of machine learning and data mining in predicting the performance of intermediate and secondary education level student. Educ Info Technol 1–21 https://doi.org/10.1007/s10639-020-10189-1
23. Mallikarjun RB, Ramana Murthy BV (2020) Prediction of student's educational performance using machine learning techniques. In: Data engineering and communication technology. Springer, Singapore, pp 429–440. https://doi.org/10.1007/978-981-15-1097-7_36

Chapter 14
An Integrated Approach of Conventional and Deep Learning Method for Underwater Image Enhancement

Rashmi S. Nair and Rohit Agrawal

1 Introduction

Underwater imaging involves many applications such as port security, hull inspection, offshore oilfield inspection, real-time marine life and environment monitoring [1, 2]. Deep sea is hazardous for human divers as it is thousands of feet deep and existence of unknown underwater conditions. Underwater sensor networks are employed to gather information in an efficient and automated manner.

Complex and poor visibility of underwater images is attributed by water turbidity, light scattering, light absorption and light reflection, thus, affecting the image pixel distribution in Red, Green, Blue (R, G, B) channels and low histogram range. Artificial lights are used for underwater image acquisition to compensate light absorption.

Underwater image communication involves processing and transmission of a large amount of irrelevant and repetitive information. There are various conventional ways to enhance and restore these images. One such method is underwater image restoration and enhancement algorithms based on light propagation physics [3–5]. But the disadvantage of this method is that they are affected by long computation time and high computation resources.

A second approach that is more elementary and productive is based on the characteristics of images. Image-based methods such as Integrated Color Model (ICM) and Unsupervised Color Correction Method (UCCM) [6, 7] apply color equalization on either of the Red, Green, Blue channels to overcome color contrast problem, whereas ICM with Rayleigh distribution approach [8] is an image-based method

R. S. Nair (✉)
Department of Computer Applications, National Institute of Technology, Tiruchirappalli, Tamil Nadu 620015, India

R. Agrawal
Department of Production Engineering, National Institute of Technology, Tiruchirappalli, Tamil Nadu 620015, India

N. Marriwala et al. (eds.), *Soft Computing for Intelligent Systems*, Algorithms for Intelligent Systems, https://doi.org/10.1007/978-981-16-1048-6_14

applies Rayleigh distribution on each image channel irrespective of the different characteristics of the three channels. Enhanced image of ICM, UCCM and ICM with Rayleigh Distribution approach [6–8] is still subjected to blue–green illumination and noise. Prevalence of bluish tone in underwater image is due to the attenuation of green and blue channels as they have high frequency and shorter wavelength.

Conventional image compression techniques such as JPEG and JPEG2000 [9, 10] produce image artifacts on high compression rates. Deep learning-based underwater image compression and restoration networks are trained on large image datasets to achieve excellent performance. Training on large image datasets requires high computational resources. In this viewpoint, this article integrates both traditional and deep learning technique for pre-processing and transmitting data using a smaller number of bits, respectively. For successful implementation of a technique that provides both image enhancement and good compression, this work develops the following research questions:

- How to minimize the prevalence of greenish blue tone in underwater image.
- How less computational resources and removal of image artifacts with high compression rate can be achieved.
- Development of an integrated method by combining both image compression and image enhancement.

To solve these research questions, this work integrates both conventional and deep learning-based methods on image compression and restoration, by using a two-step process:

- Pre-processing of the input image to improve its color and contrast.
- A learning-free image encoder–decoder convolutional neural network.

The paper consists of a review of related works in Sect. 2. Methodology and architecture of the proposed model are presented in Sect. 3. Case study is presented in Sect. 4. Results, discussion and performance comparison with standard techniques are presented in Sect. 5. Implications of the proposed work are presented in Sect. 6, followed by conclusion, limitations and upcoming possible research work are presented in Sect. 7.

2 Literature Review

A systematic literature review is presented in this section. This section is further divided into three sections. First sub-section shows studies on image compression by considering conventional as well as deep learning technique. Second section includes studies on image enhancement based on considering conventional as well deep learning technique. Third section includes research gaps and problem formulation.

2.1 Image Compression Based on Conventional and Deep Learning

Conventional image compression techniques such as JPEG [9] suffered from 'blocking' and 'contouring' artifacts, DWT [11] exhibited blurred edges for high-frequency sub-band coefficients, EZW [12] and SPIHT [13] suffered from large requirement of memory, complex list manipulation, repeated search operations, WDR [14] is limited by prescribed unitary scanning order and CWDR [15] suffers from much longer time than SPIHT.

Vector quantization (VQ) is a much-preferred method for image compression as it is simple to decode [16]. Algorithms such as Linde–Buzo–Gray (LBG) [17], Fuzzy VQ [18], Enhanced Linde–Buzo–Gray (ELBG) [19] make use of VQ technique to optimize the size of codebooks. Optimization algorithms based on VQ such as Particle Swarm optimization (PSO) [20], Evolutionary fuzzy PSO (EFPSO) [21], Quantum PSO [22], Honey bee mating optimization (HBMO) [23] provide methods to optimize the size of codebooks to provide high compression and enhanced reconstructed images. But they all have been found unsatisfactory for underwater environment.

Toderici et al. [24] proposed an LSTM-based variable rate compression framework. Theis et al. [25] proposed a compression framework that used a smooth estimation of discrete rounding function. Ballé et al. [26] made use of Generalized Divisive Normalization (GDN), which made use of additive noise to replace rounding quantization. Li et al. [27] showed the significance of importance map of an image in a content-weighted compression method. But the disadvantage of all these learning methods was it needs large image datasets to achieve excellent performance.

Jiang et al. [28] proposed an asymmetric image compression model that consists of two Convolutional Neural Networks (CNNs) named Compact CNN (ComCNN) and Reconstruction CNN (RecCNN). At the sender side, it consists of compact representation ComCNN followed by a traditional encoder (such as JPEG, JPEG2000). At the receiver side, it consists of traditional decoder followed by RecCNN, which provides high-quality decoded image. This method overcomes the gradient descent problem faced by deep symmetric auto-encoders. It also provides artifact removal and maintains image edge and texture information. But the compression performance of the end to end framework is not as significant as BPG (Better Portable Graphics).

Choi et al. [29] proposed a variable-rate, conditional image auto-encoder. Conditional auto-encoder makes use of the Lagrange multiplier and the quantization bin size for controlling the image quality. It overcomes the need for training separate networks by deploying a single variable-rate trained network whose performance is conditionally controlled. It outperformed traditional image codecs such as JPEG2000 and BPG and achieved similar results as learned image compression models. Zhou et al. [30] proposed a variational image auto-encoder consisting of a pyramidal encoder, a uniform quantizer modeled by Laplacian distribution, a non-pyramidal decoder and a post-processing element. Pyramidal encoder helps to learn the features of image at each scale. Uniform quantizer helps improve the compression rate. Post-processing helps to enhance the reconstructed image quality at low bit rates. The efficiency of this method is affected by the computational cost associated with the usage of a large number of convolutional layers.

Liu et al. [31] proposed an image compression prototype based on Picture Wise Just Noticeable Difference (PW-JND). The prototype consists of two phases known as training and predicting phase. In the training phase, a CNN-based predictor is trained. In the predicting phase, the predictor from the trained phase is used to classify the distorted image to be lossy or lossless image when compared with the test image. The results are further subjected to search strategy for accurate PW-JND prediction. The drawback of this methodology is the requirement of a large number of comparisons and a large set of distorted images to make the accurate prediction. Distortions caused by learned compression models are not captured by perceptual metrics such as PSNR and MS-SSIM. So, Patel et al. [32] proposed a Deep Perceptual Compression (DPC) framework that used Learned Perceptual Image Patch Similarity (LPIPS) as loss function in concoction with MS-SSIM. DPC consisted of a symmetric encoder and decoder, which were jointly optimized using bitrates, LPIPS and MS-SSIM loss. At several bit rates, DPC provides better compression than the conventional compression methods.

Liu et al. [33] proposed a DeepN-JPEG-based solution for low data traffic cost and end devices storage space. DeepN-JPEG redesigned the quantization table of JPEG to reduce important feature loss at high compression rates. The most significant features are found by computing the standard deviation of JPEG DCT coefficient over 8×8 image blocks. The quantization step for these significant features is derived using a piece-wise linear mapping function (PLM) to raise the compression rate with minimal drop in accuracy.

Hussain and Jeong [34] proposed a real-time deep learning model using ReLUs for image compression, which supports the training of encoder and decoder in reduced time and provides better rationalization. This method provides faster and better learning due to sparsity induced by ReLUs. To overcome the drawbacks of Carrato and Marsi [35] parallel-structure Neural Network (CMZ), Benbenisti et al. [36] proposed a simple monochromatic image compression model. The parallel compression model worked at fixed bitrates providing high compression efficiency. There was no entropy encoding and could not perform at par with JPEG algorithm. To preserve texture and global structure, Agustsson et al. [37] proposed two GAN compression models optimized using adversarial losses to get a better quality of reconstructed image. Generative compression (GC) was used in bandwidth-constrained scenarios. Selective generative compression (SC) used semantic label map to recover unimportant regions in an image. Human face could not be accurately recovered by both GC and SC models. SC model performance was constrained by the availability of a semantic label map.

2.2 *Image Enhancement Based on Conventional and Deep Learning*

Conventional image enhancement techniques include histogram stretching [38], which could result in blurring, Dark Channel Prior (DCP) [4] suffers from high computational complexity, thus making the entire process time-consuming, and optimized contrast enhancement [39] suffers from loss of details and edges in the dehazed images. Other conventional image enhancement methods based on filters such as swift visibility restoration from an 8-bit image [40], trilateral filter for high contrast images and meshes [41], a general histogram modification framework for efficient contrast enhancement [42], unsupervised prediction of perceptual image fog density and image defogging [43] and a robust framework consisting of both trilateral filter and S-shape map functions [44] are used to improve the image contrast and visibility. But these methods were found unsuitable for underwater images due to the different characteristics of underwater environment.

Conventional methods based on histogram equalization [45] such as histogram splitting and clipping for enhancement of low radiance retinal images [46], Brightness Preserving Bi-histogram Equalization (BBHE) [47], Minimum Mean Brightness Error Bi-histogram Equalization (MMBEBHE) [48] provide uniform radiance at cost of loss in information content.

Convolutional Neural Network (CNN)-based methods such as Underwater Image Enhancement-net (UIE-net) [49], Underwater CNN (UWCNN) [50], Generative Adversarial Networks (GANs) [51], WaterGAN [52], CycleGAN [53], depth map using CNN [54], multiscale architecture consisting of a combination of refined and un-refined CNN network [55], Perception-inspired single image dehazing network with refinement (PDR-net) [56] were designed to correct color and remove haze. Disadvantage of all these methods was that they all needed large image datasets to achieve excellent performance.

Research Gaps
From the literature survey, it has been found that many recent studies available to enhance the quality of underwater images and compression techniques suffer from long computation time, high computation resources [3–5] and loss of information. Very few studies have been reported on integrated approach of conventional and learning technique. This study provides a method that integrates features of both conventional and learning techniques for image compression and enhancement.

3 Methodology

For image enhancement, contrast correction and color correction methods are used to neutralize the effects of light scattering and absorption. To reduce the computational time and complexity of deep learning technique, an hourglass shape encoder–decoder

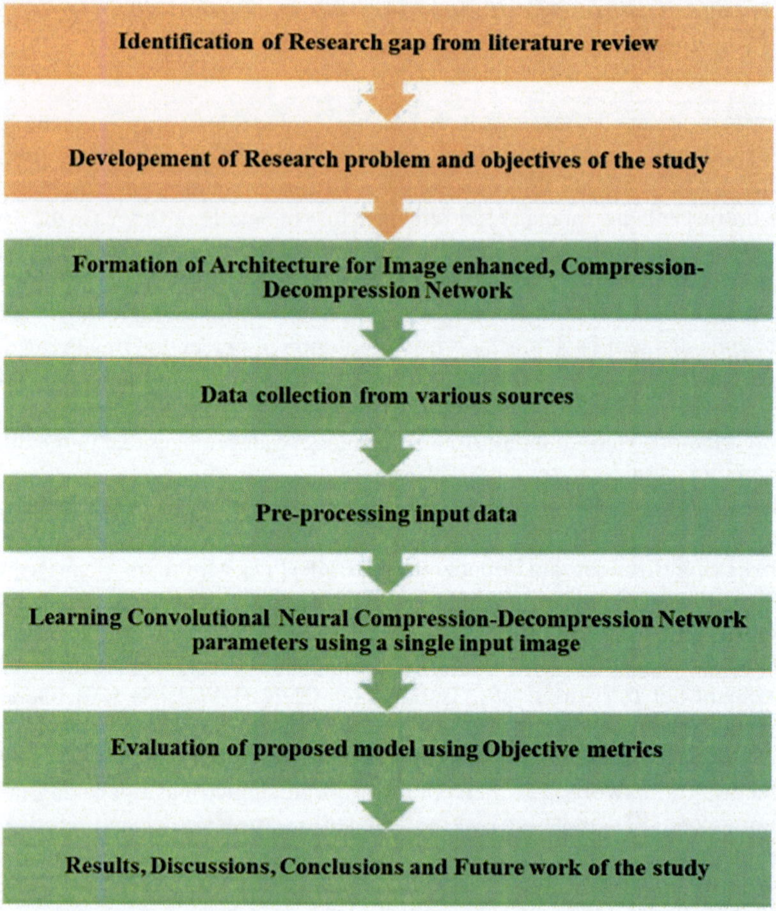

Fig. 1 Framework for the study

neural network is used whose features are learnt using a single input image. In this study, pre-processing-based conventional technique is integrated with learning-free image encoder–decoder deep learning technique. First, the architecture of proposed enhancement-compression model is presented followed by detailed description of each component of the model. The flowchart of this work is presented in Fig. 1.

3.1 Architecture of the Proposed Enhancement-Compression Model

Input underwater image is resized to $128 \times 128 \times 3$ for ease of computation. Resized input image is subjected to pre-processing, where it is first subjected to contrast correction followed by color correction. The output of pre-processing module is subjected to encoder for compression. The compressed stream at receiver side is subjected to decoder to retrieve the decompressed image.

3.2 *Pre-Processing*

Pre-processing step is as shown in Fig. 2. It involves contrast correction and color correction to neutralize the effects of light scattering and absorption.

Contrast correction consists of basically three steps:

Step 1: Color balancing.
Step 2: Histogram Stretching (HS).
Step 3: Bilateral filter on RGB channels to remove noise incurred due to Step 1 and Step 2.

To provide color balancing, the G and B channels of an image are multiplied by constant δ_g and δ_b, respectively. Gray World assumption [57] governs the constant value. The average of R, B, G channels of an image is equivalent to its gray using Eqs. (1) and (2).

$$\left(\mathrm{NR_{avg}} + \mathrm{NG_{avg}} + \mathrm{NB_{avg}}\right) = 0.5, \tag{1}$$

where $\mathrm{NR_{avg}}$, $\mathrm{NG_{avg}}$, $\mathrm{NB_{avg}}$ represent normalized average of R, G and B channels, respectively.

$$\delta_g = 0.5/\mathrm{NG_{avg}} \quad \text{and} \quad \delta_b = 0.5/\mathrm{NB_{avg}} \tag{2}$$

Color balancing is not applied for Red channel as it may result in oversaturation. HS can be computed using Eq. 3.

$$\varphi_{\mathrm{out}} = (\varphi_{\mathrm{in}} - \alpha_{\mathrm{min}})((\beta_{\mathrm{max}} - \beta_{\mathrm{min}})/(\alpha_{\mathrm{max}} - \alpha_{\mathrm{min}})) + \beta_{\mathrm{min}}, \tag{3}$$

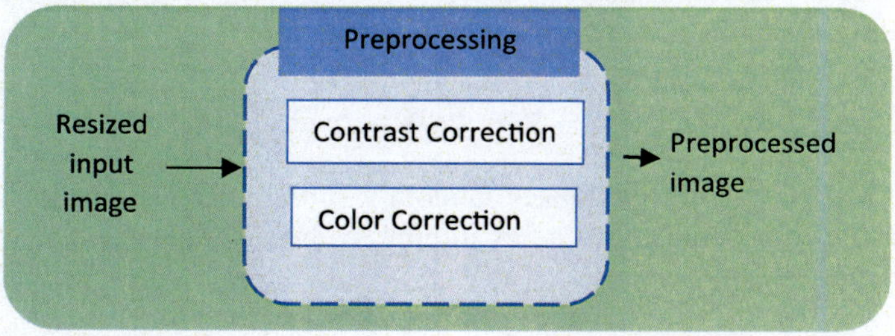

Fig. 2 Pre-processing module

where input and output pixels are represented by φ_{in} and φ_{out}, $[\alpha_{\min}, \alpha_{\max}]$ represents the actual stretching range and $[\beta_{\min}, \beta_{\max}]$ represents the desired stretching range.

$[\alpha_{\min}, \alpha_{\max}]$ range value depends on whether the histogram is normally distributed or not. If histogram is normally distributed, 0.1% and 0.99% are the stretching range of the histogram. Else higher and lower proportion of each Red-Blue-Green channel intensity values are split to compute $[\alpha_{\min}, \alpha_{\max}]$.

The desired stretching range $[\beta_{\min}, \beta_{\max}]$ can be defined using Eq. (4).

$$\beta_{\lambda\,\min} = x_\lambda - y_\lambda, \tag{4}$$

where $\lambda \in$ {red, green, blue} channels, x_λ is the channel mode, y_λ represents the standard deviation values of Rayleigh Distribution.

$\beta_{\lambda\,\max}$ depends on x_λ, y_λ and mean of channel (u_λ) values. If u_λ has several solutions, then $\beta_{\lambda\,\max}$ value is taken as the average of these solutions, else $\beta_{\lambda\,\max} =$ 255.

After the input image is contrast corrected, it is subjected to Color correction. RGB image is converted to La^*b^* color model. L component value ranges between [0–100] and a^* and b^* component value range between [−128,128]. Brightness of the image (L component) is adjusted to stretching histogram lower and upper bound between [0–100]. a^* and b^* components are modified using Eq. (5).

$$\varphi_{\Delta\text{out}} = \varphi_{\Delta\text{in}} = *1.3^{1-|\varphi_{\text{in}}/128|}, \tag{5}$$

where $\Delta \in \{a^*, b^*\}$ and $\varphi_{\Delta\text{in}}$ represent input and $\varphi_{\Delta out}$ represents output pixel values.

On experimental basis, 1.3 has been found as the optimal value for this method. After the above-mentioned modification, image is converted from La^*b^* color model to RGB color model.

Learning-free Image Encoder—Decoder

An encoder–decoder network is used, which has an hourglass shape. Encoder network consists of average pooling layers to reduce the size. Decoder network consists of up-sampling layers for size restoration. It makes use of Mean Square Error (MSE) [58] as the loss function and Adam optimizer [59] with 0.0001 as the learning rate.

Random code vector ϕ is mapped into preprocessed image ψ of size $3 \times H \times W$ using function F with Δ as network weights as shown in Eq. (6).

$$\psi = F_\Delta(\phi), \tag{6}$$

here Δ acts as a network parameter randomly initialized. The encoder–decoder network can be considered as an energy minimization of the type as shown in Eq. 7.

$$\psi^* = \min_\psi E(\psi, \psi_0) + R(\psi), \tag{7}$$

where $R(\psi)$ represents the Regularizer. Here $R(\psi)$ is replaced with the prior captured by encoder–decoder network as shown in Eq. (8).

$$\Delta^* = \arg\min_{\Delta} E(F_{\Delta}(\phi), \psi_0) \tag{8}$$

Using gradient descent as an optimizer, local minimizer Δ^* is computed to obtain Eq. (9).

$$\psi^* = F_{\Delta^*}(\phi) \tag{9}$$

4 Case Study

In this study, three different underwater image datasets have been considered to evaluate the proposed model. The considered datasets are Fish dataset, USR-248 dataset and UIEB dataset. The reason for selecting above datasets is that they provide images taken at various depths of deep sea, affected by numerous environmental conditions. The dataset contains a combination of images with varied color range, contrast and resolution.

4.1 Description of Dataset

Fish image dataset [60] used for the experiment is taken from Fish4Knowledge, which is funded by EUSFP (European Union Seventh Framework Programme). Fish dataset consists of images of Amphiprion Clarkia, Canthigaster Valentine, ChromisChrysura, DascyllusReticulatus, MyripristisKuntee, Neoniphon Samara, PlectroglyphidodonDickii. Some of the images are crowded and some are blurred due to underwater lighting effects. Underwater Image Enhancement Benchmark Dataset (UIEB) [61] consists of 890 raw underwater images. Underwater Super Resolution (USR-248) [62] consists of both 'high' (640 × 480) and 'low' (80 × 60, 160 × 120 and 320 × 240) resolution underwater images. The images in the dataset have been collected during oceanic exploration, field experiments and from online resources such as FlickerTM images and YouTubeTM videos.

The efficiency of the proposed model can be evaluated using objective metrics such as PSNR [58] and Structural Similarity Index Measure (SSIM) [58]. PSNR computes the peak signal-to-noise ratio. Higher value of PSNR indicates better quality of the reconstructed image. SSIM measures the perceptual difference between original and reconstructed images.

Consider A represents the input image and B represents the rebuilt image of size $p \times q$. Then PSNR and SSIM can be defined using the formula given in Eqs. (10) and (11), respectively.

$$\begin{aligned} &\text{PSNR(A B)} = 10\log_{10}\big(255^2/MSE(\text{A}, \text{B})\big) \\ &\text{where,} \\ &\text{MSE(A, B)} = \tfrac{1}{pq}\sum_{i=1}^{p}\sum_{j=1}^{q}\big(\text{A}_{ij} - \text{B}_{ij}\big)^2 \end{aligned} \tag{10}$$

$$\text{SSIM(A, B)} = l(\text{A, B})c(\text{A, B})\text{s(A, B)}, \tag{11}$$

where Luminance comparison is made using $l(\text{A, B})$ function, contrast comparison is made using $c(\text{A, B})$ function and structure comparison is done using s(A, B) function.

5 Results and Discussions

This study proposed an energy efficient, free learning convolutional neural compression-decompression network with a pre-processing step. The architecture of the method is presented in Sect. 3. Experiments on three different datasets having different image characteristics accomplish the motive of the study. The proposed study has been compared with SRCNN and Bi-Cubic methods. PSNR value and SSIM value have been analyzed for comparison. Tables 1, 2 and 3 show the comparison of PSNR and SSIM value for the proposed method with SRCNN and Bi-Cubic by considering Fish dataset, USR-248 dataset and UIEB dataset, respectively. The resulted images achieved from proposed method have been compared with SRCNN and Bi-Cubic methods and are presented in Appendix Tables 1, 2 and 3. Experimental results on three databases Fish-dataset, USR-248 dataset and UIEB dataset

Table 1 Comparison of results for fish dataset

Dataset	PSNR value			SSIM value		
	Proposed method	SRCNN	Bi-cubic	Proposed method	SRCNN	Bi-cubic
Fish0_RGHS	31.889	28.250	27.866	0.809	0.860	0.865
Fish1_RGHS	28.874	22.279	21.993	0.846	0.887	0.886
Fish2_RGHS	29.518	23.025	23.000	0.860	0.867	0.867
Fish3_RGHS	29.913	29.306	29.677	0.930	0.960	0.968
Fish4_RGHS	29.904	23.810	23.698	0.837	0.887	0.897
Fish5_RGHS	31.346	30.128	29.113	0.909	0.947	0.937
Fish6_RGHS	32.370	31.262	30.925	0.949	0.963	0.967
Fish7_RGHS	32.052	31.782	29.384	0.929	0.960	0.946
Fish8_RGHS	30.684	27.759	27.372	0.877	0.941	0.935
Fish9_RGHS	31.070	27.612	25.785	0.898	0.935	0.926
Fish10_RGHS	30.508	26.875	26.305	0.865	0.934	0.928

Table 2 Comparison of results for USR-248 dataset

Dataset	PSNR value			SSIM value		
	Proposed method	SRCNN	Bi-cubic	Proposed method	SRCNN	Bi-cubic
SFish0_RGHS	29.487	26.272	25.499	0.914	0.896	0.873
SFish1_RGHS	31.569	30.419	30.082	0.835	0.909	0.903
SFish2_RGHS	29.834	25.910	25.770	0.825	0.884	0.889
SFish3_RGHS	29.684	25.955	25.580	0.776	0.869	0.861
SFish4_RGHS	30.257	24.126	24.065	0.838	0.876	0.877
SFish5_RGHS	28.830	25.646	25.641	0.822	0.888	0.889
SFish6_RGHS	29.178	24.317	24.156	0.758	0.825	0.817
SFish7_RGHS	30.251	28.049	28.448	0.798	0.885	0.900
SFish8_RGHS	29.231	26.387	25.915	0.853	0.895	0.882
SFish9_RGHS	29.590	25.508	24.706	0.820	0.881	0.852
SFish10_RGHS	30.819	30.306	30.432	0.921	0.953	0.948

Table 3 Comparison of results for UIEB dataset

Dataset	PSNR value			SSIM value		
	Proposed method	SRCNN	Bi-cubic	Proposed method	SRCNN	Bi-cubic
UFish0_RGHS	30.382	35.014	34.860	0.798	0.916	0.923
UFish1_RGHS	31.733	32.435	32.521	0.863	0.919	0.925
UFish2_RGHS	31.348	37.057	36.139	0.857	0.928	0.936
UFish3_RGHS	29.359	29.072	28.854	0.840	0.873	0.872
UFish4_RGHS	28.650	25.285	24.931	0.794	0.777	0.773
UFish5_RGHS	29.411	28.220	27.367	0.835	0.838	0.822
UFish6_RGHS	30.694	29.980	30.073	0.876	0.908	0.907
UFish7_RGHS	33.173	39.635	39.965	0.906	0.963	0.969
UFish8_RGHS	30.721	42.971	43.502	0.887	0.971	0.975
UFish9_RGHS	30.661	40.153	40.045	0.781	0.947	0.948
UFish10_RGHS	29.616	37.975	38.547	0.792	0.919	0.929

are shown in Tables 1, 2 and 3, respectively. For visualization of the achieved result, a comparison of images has been presented in Appendix Tables 4, 5 and 6.

Both visual and objective metrics results are shown for evaluation. PSNR values of the proposed study are much better when compared with SRCNN and Bi-cubic techniques for Fish and USR-128 datasets. Higher PSNR values suggest a better quality of image. But at the same time, SSIM values of underwater images for the proposed method are lesser than SRCNN and Bi-cubic methods for the same dataset. In the case of UIEB dataset, the PSNR and SSIM values of the proposed method are lesser in most of the images when compared with SRCNN and Bi-cubic methods. As per human visual analysis, the proposed method output images shown in Appendix Table 6 show better visual quality than SRCNN and Bi-cubic methods. For images in Appendix Table 6, Green channel is more predominant than other colors and thus the proposed color correction could bring enough, significant changes than SRCNN and Bi-cubic methods.

6 Implications

Underwater imaging has various applications and one of the major applications is in fish aquaculture. The monitoring of fish behavior is of prime importance for fish breeding and efficient usage of fish feed to reduce the overall cost and sea waste. Transmission of data from one node to another node in the deep-sea network or on terrestrial node needs large bandwidth and consumes a lot of energy in nodes. Due to deep sea environmental factors, the quality of the acquired images is affected and does not serve the required purpose. The proposed method not only reduces the requirement of bandwidth needed for data transmission but also provides enhanced quality to the reconstructed image at the receiver end. This work helps in proper behavior monitoring of underwater images. Intrusion detections can be checked, and proper action can be taken to save the fishes from harmful predators.

6.1 Unique Contribution

The important contributions of this work are:

- The proposed methodology is a combination of both conventional and deep learning techniques.
- The method requires a single image to train the network parameters instead of large datasets of images. Thus, reducing the complexity of the training model. Enhancement of underwater image by reducing the prevalence of bluish green tone in it.
- This work considered datasets of different characteristics for comparison to validate the proposed integrated method.

7 Conclusions

Underwater imaging has numerous applications such as monitoring of marine habitat, ocean engineering, photography, archaeology, etc. Employing human beings to collect data from deep sea is hazardous in nature due to underwater environment. So, images are collected from cameras fitted in Remotely Operated Vehicles (ROVs) and Autonomous Underwater Vehicles (AUVs). These images are then transmitted through communication channels to the terrestrial nodes and other underwater nodes for monitoring purpose. They have more visual and spatial redundancy than terrestrial images. Underwater imaging is greatly affected by the underwater environment. Due to varied artificial lighting conditions at different depths and scattering of light by water particles, underwater images have low contrast, low detail and are blurry in nature. The basic requirement for underwater imaging is to provide a lightweight method to transfer data, which consumes minimal transmission bandwidth producing good-quality reconstructed images.

A pre-processing step for color and contrast correction is applied to enhance the quality of the image. The processed image is then given as input to the encoder module at the sender side. Encoder–decoder module is solely dependent on untrained convolutional neural network structure and the pre-processed image. The proposed encoder–decoder network uses randomly initialized weights. Advantage of the proposed framework over the convolutional compression–decompression networks is that the encoder–decoder network parameters (i.e., network weights) are not learned from large image datasets. It can be learnt from a single input image. It provides better PSNR values for Fish and USR-248 dataset. Even though PSNR values are less for images in UIEB dataset, visual quality is far better than SRCNN and Bi-cubic methods. The present study helps in providing an energy efficient, image quality enhancement and free learning convolutional neural compression—decompression network.

7.1 Limitations and Future Work

Present study compares three datasets, namely, Fish dataset, USR-248 dataset and UIEB dataset, which can be further expanded by including more dataset. Also, the proposed method has been compared with SRCNN and Bi-Cubic methods. In future, it can be compared with several existing methods to validate its advantages over other methods. The proposed method tries to provide a lightweight technique, which does not need large dataset for its training and combines the properties of both conventional and deep learning techniques. Even then there is a limitation faced by the proposed method. Higher perceptual difference between original and reconstructed images is considered as one of the drawbacks of the proposed method even though they have high significance as per human visual analysis. The reconstructed images do not have the exact features of the original image as they lose information content during the transmission process. The proposed method can be extended in future for video transmission, where the individual frames can be considered as input images. To further improve on the compression, both inter and intraframe dependency can be considered.

Appendix

See Tables 4, 5 and 6.

Table 4 Comparison of result for Fish-dataset with SRCNN and Bi-cubic methods

S. no.	Original image	Proposed model	SRCNN	Bi-cubic
Fish_0				
Fish_1				
Fish_2				
Fish_3				
Fish_4				

Table 5 Comparison of result for USR-248 dataset with SRCNN and Bi-cubic methods

S. no.	Original image	Proposed model	SRCNN	Bi-cubic
SFish_0				
SFish_1				
SFish_2				
SFish_3				
SFish_4				

Table 6 Comparison of results for UIEB dataset with SRCNN and Bi-cubic methods

S. no.	Original image	Proposed model	SRCNN	Bi-cubic
UFish_0				
UFish_5				
UFish_8				
UFish_9				
UFish_10				

References

1. Magadza TBHT, Ramanaiah OBV (2015) VSP: Video streaming protocol for underwater sensor network. In: 2015 4th international conference on reliability, infocom technologies and optimization (ICRITO) (Trends and future directions). IEEE, pp 1–6
2. Han S, Chen R, Noh Y, Gerla M (2014) Real-time video streaming from mobile underwater sensors In: Proceedings of the international conference on underwater networks & systems, pp 1–8
3. Zhao X, Jin T, Qu S (2015) Deriving inherent optical properties from background color and underwater image enhancement. Ocean Eng 94:163–172
4. He K, Sun J, Tang X (2010) Single image haze removal using dark channel prior. IEEE Trans Pattern Anal Mach Intell 33(12):2341–2353
5. Chiang JY, Chen YC (2011) Underwater image enhancement by wavelength compensation and dehazing. IEEE Trans Image Process 21(4):1756–1769
6. Iqbal K, Salam RA, Osman A, Talib AZ (2007) Underwater image enhancement using an integrated colour model. IAENG Int J Comput Sci 34(2)
7. Iqbal K, Odetayo M, James A, Salam RA, Talib AZH (2010) Enhancing the low quality images using unsupervised colour correction method. In: 2010 ieee international conference on systems, man and cybernetics. IEEE, pp 1703–1709
8. Ghani ASA, Isa NAM (2015) Underwater image quality enhancement through integrated color model with Rayleigh distribution. Appl Soft Comput 27:219–230
9. Pennebaker WB, Mitchell JL (1992) JPEG: still image data compression standard. Springer Science & Business Media
10. Rabbani M (2002) JPEG2000: Image compression fundamentals, standards and practice. J Electron Imaging 11(2):286
11. Al Muhit A, Islam MS, Othman M (2004) VLSI implementation of discrete wavelet transform (DWT) for image compression. In: 2nd international conference on autonomous robots and agents, vol 4(4), pp 421–433

12. Shapiro JM (1993) Embedded image coding using zerotrees of wavelet coefficients. IEEE Trans Signal Process 41(12):3445–3462
13. Said A, Pearlman WA (1996) A new, fast, and efficient image codec based on set partitioning in hierarchical trees. IEEE Trans Circuits Syst Video Technol 6(3):243–250
14. Tian J, Wells Jr RO (1996) A lossy image codec based on index coding. In: Data compression conference, p 456
15. Yuan Y, Mandal MK (2004) Embedded color image coding using context-modeled wavelet difference reduction. In: 2004 IEEE international conference on acoustics, speech, and signal processing, vol 3. IEEE, pp iii–61. https://doi.org/10.1109/ICASSP.2004.1326481
16. Karri C, Jena U (2016) Fast vector quantization using a Bat algorithm for image compression. Eng Sci Technol Int J 19(2):769–781
17. Linde Y, Buzo A, Gray R (1980) An algorithm for vector quantizer design. IEEE Trans Commun 28(1):84–95
18. Tsekouras GE, Tsolakis DM (2013) Fuzzy clustering-based vector quantization for image compression. In: Computational intelligence in image processing). Springer, Berlin, Heidelberg, pp 93–105
19. Patané G, Russo M (2001) The enhanced LBG algorithm. Neural Netw 14(9):1219–1237
20. Kumar M, Kapoor R, Goel T (2010) Vector quantization based on self-adaptive particle swarm optimization. Int J Nonlinear Sci 9(3):311–319
21. Feng HM, Chen CY, Ye F (2007) Evolutionary fuzzy particle swarm optimization vector quantization learning scheme in image compression. Expert Syst Appl 32(1):213–222
22. Wang Y, Feng XY, Huang YX, Pu DB, Zhou WG, Liang YC, Zhou CG (2007) A novel quantum swarm evolutionary algorithm and its applications. Neurocomputing 70(4–6):633–640
23. Horng MH, Jiang TW (2011) Image vector quantization algorithm via honey bee mating optimization. Expert Syst Appl 38(3):1382–1392
24. Toderici G, O'Malley SM, Hwang SJ, Vincent D, Minnen D et al. (2015). Variable rate image compression with recurrent neural networks. arXiv preprint arXiv:1511.06085
25. Theis L, Shi W, Cunningham A, Huszár F (2017) Lossy image compression with compressive autoencoders. arXiv preprint arXiv:1703.00395
26. Ballé J, Laparra V, Simoncelli EP (2016) End-to-end optimized image compression. arXiv preprint arXiv:1611.01704
27. Li M, Zuo W, Gu S, Zhao D, Zhang D (2018) Learning convolutional networks for content-weighted image compression. In: Proceedings of the IEEE conference on computer vision and pattern recognition, pp 3214–3223
28. Jiang F, Tao W, Liu S, Ren J, Guo X, Zhao D (2017) An end-to-end compression framework based on convolutional neural networks. IEEE Trans Circuits Syst Video Technol 28(10):3007–3018
29. Choi Y, El-Khamy M, Lee J (2019) Variable rate deep image compression with a conditional autoencoder. In: Proceedings of the IEEE international conference on computer vision, pp 3146–3154
30. Zhou L, Cai C, Gao Y, Su S, Wu J (2018) Variational autoencoder for low bit-rate image compression. In: CVPR Workshops, pp 2617–2620
31. Liu H, Zhang Y, Zhang H, Fan C, Kwong S, Kuo CCJ, Fan X (2019) Deep learning-based picture-wise just noticeable distortion prediction model for image compression. IEEE Trans Image Process 29:641–656
32. Patel Y, Appalaraju S, Manmatha R (2019) Deep perceptual compression. arXiv preprint arXiv:1907.08310
33. Liu Z, Liu T, Wen W, Jiang L, Xu J, Wang Y, Quan G (2018). DeepN-JPEG: a deep neural network favorable JPEG-based image compression framework. In: Proceedings of the 55th annual design automation conference, pp 1–6
34. Hussain F, Jeong J (2016) Efficient deep neural network for digital image compression employing rectified linear neurons. J Sensors
35. Carrato SERGIO, Marsi STEFANO (1992) Parallel structure based on neural networks for image compression. Electron Lett 28(12):1152–1153

36. Benbenisti Y, Kornreich D, Mitchell HB, Schaefer PA (1999) Fixed bit-rate image compression using a parallel-structure multilayer neural network. IEEE Trans Neural Networks 10(5):1166–1172
37. Agustsson E, Tschannen M, Mentzer F, Timofte R, Gool LV (2019) Generative adversarial networks for extreme learned image compression. In: Proceedings of the IEEE international conference on computer vision, pp 221–231
38. Alparslan E, Fuatince M (1981) Image enhancement by local histogram stretching. ITSMC 11:376–385
39. Kim JH, Jang WD, Sim JY, Kim CS (2013) Optimized contrast enhancement for real-time image and video dehazing. J Vis Commun Image Represent 24(3):410–425
40. Tarel JP, Hautiere N (2009) Fast visibility restoration from a single color or gray level image. In: 2009 IEEE 12th international conference on computer vision). (IEEE, pp 2201–2208)
41. Choudhury P, Tumblin J (2005) The trilateral filter for high contrast images and meshes. In: ACM SIGGRAPH 2005 Courses, pp 5-es
42. Gu K, Zhai G, Wang S, Liu M, Zhoi J, Lin W (2015) A general histogram modification framework for efficient contrast enhancement. In: 2015 IEEE international symposium on circuits and systems (ISCAS). IEEE, pp 2816–2819
43. Choi LK, You J, Bovik AC (2015) Referenceless prediction of perceptual fog density and perceptual image defogging. IEEE Trans Image Process 24(11):3888–3901
44. Pal NS, Lal S, Shinghal K (2019) A robust framework for visibility enhancement of foggy images. Eng Sci Technol Int J 22(1):22–32
45. Kim YT, Bi CEUBP (1997) Histogram equalization . IEEE Trans Consum Electron 43(1):1–8
46. Singh N, Kaur L, Singh K (2019) Histogram equalization techniques for enhancement of low radiance retinal images for early detection of diabetic retinopathy. Eng Sci Technol Int J 22(3):736–745
47. Kim YT (1997) Contrast enhancement using brightness preserving bi-histogram equalization. IEEE Trans Consum Electron 43(1):1–8
48. Chen SD, Ramli AR (2003) Minimum mean brightness error bi-histogram equalization in contrast enhancement. IEEE Trans Consum Electron 49(4):1310–1319
49. Wang Y, Zhang J, Cao Y, Wang Z (2017). A deep CNN method for underwater image enhancement. In: 2017 IEEE international conference on image processing (ICIP). IEEE, pp 1382–1386
50. Anwar S, Li C, Porikli F (2018) Deep underwater image enhancement. arXiv preprint arXiv: 1807.03528
51. Goodfellow I, Pouget-Abadie J, Mirza M, Xu B, Warde-Farley D, Ozair S et al (2014) Generative adversarial nets. In: Advances in neural information processing systems, pp 2672–2680
52. Li J, Skinner KA, Eustice RM, Johnson-Roberson M (2017) WaterGAN: Unsupervised generative network to enable real-time color correction of monocular underwater images. IEEE Rob Automat Lett 3(1):387–394
53. Fabbri C, Islam MJ, Sattar J (2018) Enhancing underwater imagery using generative adversarial networks. In: 2018 IEEE international conference on robotics and automation (ICRA)). IEEE, pp 7159–7165
54. Ding X, Wang Y, Zhang J, Fu X (2017) Underwater image dehaze using scene depth estimation with adaptive color correction. In: OCEANS 2017-Aberdeen. IEEE, pp 1–5
55. Cao K, Peng YT, Cosman PC (2018) Underwater image restoration using deep networks to estimate background light and scene depth. In: 2018 IEEE southwest symposium on image analysis and interpretation (SSIAI). IEEE, pp 1–4
56. Li C, Guo C, Guo J, Han P, Fu H, Cong R (2019) PDR-Net: Perception-inspired single image dehazing network with refinement. IEEE Trans Multimedia 22(3):704–716
57. Chambah M, Semani D, Renouf A, Courtellemont P, Rizzi A (2003) Underwater color constancy: enhancement of automatic live fish recognition. In: Color Imaging IX: processing, hardcopy, and applications, vol. 5293. International Society for Optics and Photonics, pp 157–168

58. Hore A, Ziou D (2010) Image quality metrics: PSNR vs. SSIM. In: 2010 20th international conference on pattern recognition. IEEE, pp 2366–2369
59. Kingma DP, Ba J (2014) Adam: A method for stochastic optimization. arXiv preprint arXiv:1412.6980
60. Fisher RB, Chen-Burger YH, Giordano D, Hardman L, Lin FP (eds) (2016) Fish4Knowledge: collecting and analyzing massive coral reef fish video data, vol 104. Springer
61. Li C, Guo C, Ren W, Cong R, Hou J, Kwong S, Tao D (2019) An underwater image enhancement benchmark dataset and beyond. IEEE Trans Image Process 29:4376–4389
62. Islam MJ, Enan SS, Luo P, Sattar J (2019) Underwater image super-resolution using deep residual multipliers. arXiv preprint arXiv:1909.09437

Chapter 15
A Survey on Plant Dısease Detectıon Methods for Buıldıng a Robust Plant Dısease Detectıon System

A. Firos, Seema Khanum, and M. Gunasekaran

1 Introduction

Farming has become significantly more vital compared to the past few decades where vegetation was merely used to cheer populace and creatures. This is since the method that vegetation is currently used to make an influence and diverse wellsprings of energy to improve the livelihood states of mankind. With the use of modern techniques, the tainted region of the harvest is used to recognize the illness of the harvest automatically. The discovery and recognition of harvest infection were normally done by use of bare eyes, by the specialist in olden times. But this needs incessant observation by the specialist and it is excessively costly in big farming. So cultivator needs to do a lot of hard work in several underdeveloped countries. The recognition and the categorization of the crop will be done more rapidly at all stage, if machine vision can be used. Usually, the machine vision system comprises of a digital camera and computer application software. There is a range of applications for the application software. With feature extraction, the input data are transforming into a set of features. For the identification of a variety of features of the harvest, image processing is used, which gives new method to investigate the issues of the

A. Firos (✉)
Department of Computer Science and Engineering, Rajiv Gandhi University, Rono Hills, Doimukh 791112, India
e-mail: firos.a@rgu.ac.in

S. Khanum
Department of Computer Science, Government Arts College, Salem, Tamil Nadu 636007, India
e-mail: seem.kh@gmail.com

M. Gunasekaran
Department of Computer Science, Government Arts College, Dharmapuri, Tamil Nadu 636705, India

N. Marriwala et al. (eds.), *Soft Computing for Intelligent Systems*, Algorithms for Intelligent Systems, https://doi.org/10.1007/978-981-16-1048-6_15

crop. The image processing is employed as a division of the agrarian applications for the associated reasons:

- To recognize the sick side of plant, organic product
- To recognize subjective region by sickness.
- To recognize the state of sickness area.
- To determine the dimension and state of organic crop.

There are a huge number of illnesses that control vegetation and can create overwhelming misfortune in diverse economies and societies. This may also create an unbelievable ecological disaster. For this reason, it is wise to do disease identification early so that it would be handy to keep a deliberate detachment from this kind of misfortune. Disease identification can be done through few methods including physical- and computer-based techniques. It is not practical to observe each crop and handpick the disease if vegetation is spread in a huge area. In these situations, we may employ a camera that will grab the stills from crop and communicate to the designated server room, which helps to maintain the plants, and craft it solid by giving automated meditation or advice to the owner. This is really necessary to differentiate the unkind bug in time by keeping in mind that there are a number of diseases that are not visible on the leaves and others would be visible in the last phase when they have just done collateral damage on the plants. In such situations, mechanized frameworks would be essential to identify the reason for the diseases for making it sure such disease would not happen in the future. Such mechanized frameworks employing a set of composite computations and logical tools [1]. So the problems influence the growth and the yield of crops. The illnesses in crops can reduce the production and fall apart the variety of such crops and its extinction from expansion. Plant diseases particularly leaf infections are usually checked with the help of fungicides, pesticides and bug sprays. However, immoderate use of such synthetic mixtures for the handling of crop diseases may harm their natural production and may lead to diverse damages to inhabitants and creatures. The risk of toxic build-up on plants because of the employment of insect repellents on crops that have been inclined by diverse types of diseases has been documented as a notable sponsor of groundwater contamination and pollution. Extreme quantity of use of insect repellent by planters raises the cost of generation, which can ampere more remarkable adversity [2]. Also, note that there is a call from nature lovers to minimize their use because of the above facts. One notable way for achieving this is by appraising the gravity of the inclined area of the crop sickness concentrating on the ailing region, with an appropriate quantity and grouping of insect repellent [3]. The exploitation of personal investigation and physical methods are for the most part used to choose problem significance in the generation practice though that may create a few faults and erroneous results. Diverse tactics can be employed to improve the exact level be that as it may. Grid tallying is one such tactic that may be used. This method is tough to use and also uses a much time.

Incorporation of picture-preparing procedures as a foremost examination in the agriculture area has influenced a range of methods to improve the progression of the horticultural field. A range of methods is used by a few experts to recognize, determine and illustrate leaf diseases on crops. A set of such methods incorporate AI, Support Vector Machine, Neural Networks, Bounding Box, Moment Analysis, Image processing, etc.

In this paper, Sect. 2 presents the existing pest detection techniques. Section 3 introduces the automated pest detection system. Section 4 introduces the proposed pest detection system, which is employing a KNN classifier. Section 5 gives the conclusion and future scope of this study.

2 Existing Pest Detection Techniques

The inspiration after this survey is to weigh the gravity of annoyance, which grounds diseases on leaves employing the K-Nearest Nabors (KNN) characterization computation and limit strategies. This part covers the survey of obtainable researches and implementations in connection to the investigation. Literature and techniques composed by people or inventors for weighing the best factors for crop protection were taken into consideration. Sub-sections including the description basis and incorporation of picture handling, picture preparing methods in relation to contamination gravity and influence of bug on crops are discussed.

Gupta et al. [4] displayed an automated segmentation of powdery mildew sickness from cherry leaves utilizing image processing. Their method uses a mechanized key ejection of base from the image and later removes the coveted tainted part. A mixture of morphological methods and power-based thresholding are employed, which makes an approach computationally fruitful and less complicated. Incorporation of open arXiv e-prints data [5, 6] is used to test their estimated computation. The process proposed for the detection of fine mold is programmed and it has much quality making the process suitable for modern applications. The undesired distortion, mismatching light and low differentiation in the middle of disease are the major difficulties viewed in part of fine buildup; all of them are considered in their work. The sued morphological administrators with power alteration gave them a greater result and also algorithmically better.

Raichaudhuri et al. [7] introduced a model to identify wheat leaf infections. They used robotized method and image processing techniques to identify the wheat leaf infections. The *k* mean estimation and vigilant channel are employed for the image grabbed and division it was considering. The arrangement confirmation is done with an expert through PCA or GLCM. SVM or ANN was used for course of action. It has been identified that approaches like center channel, histogram adjust, picture smoothening, picture sharpening, etc. can be employed for doing picture transformation. The estimation methods like Gabor channel, shading co-occasion methods, wavelet transformation, etc. can be used for surface factor elicitation.

Marfo [8] invented the triangular thresholding method with a collected computation to differentiate and assess the gravity of escalation caused by diseases in the beginning stage. This method may be used to distinguish trouble caused by bacterial infections on plant cuttings. The computation is measured to give a performance approximately 97% accurate results. In his paper, he portrayed picture securing as the key method. Here computerized camera was utilized to grab pictures in controlled circumstances with the dull foundation at the beginning. The subject (leaf) was reasonably zoomed with the help of the computerized camera to assure that the image grabbed has just the leaf and dark base with the correct picture values. Also, the image portion was used to disseminate the image into a range of locales with respect to analogous factors in the image. For suitable dissemination, the image is transformed to gray scale from the RGB pose. This may be done by identifying the normal of the three shading portions in the actual nature. Here, the process happens in three stages. First, get the assumed picture. Then we extract the red, green and blue (RGB) factors of pixel [9] with the help of their relating numbers. Finally, displace the initial RGB factors with the new values. The transformation is done by computing the normal of the three factors.

If we have a true-color image picture like the image shown in Fig. 1 and we need to disseminate it into gray scale with the help of usual methods, the result would be as shown in Fig. 2.

This is because this method figures the average of the three colors. Since the three colors have different wavelengths and give in the order of the image, they bring incorrect results. This can be changed by figuring the usual vision of the obligation made by every shade in the image.

Fig. 1 Sample true-color image

Fig. 2 Gray scale of Fig. 1

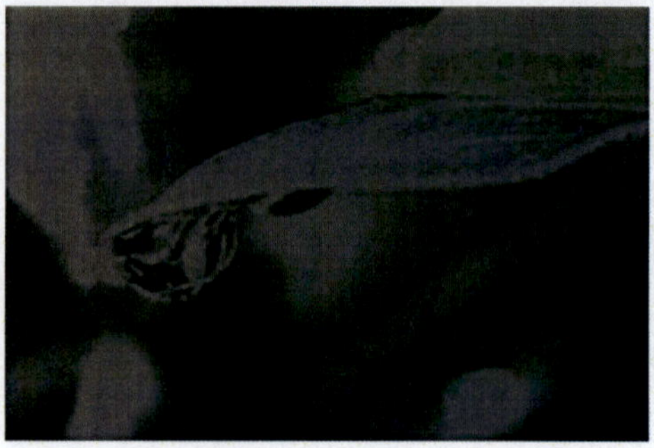

These factors can be defined as:

- $G_1 = (R + G_1 + B)/3$
  where G_1 = dark, R = red, G_2 = green and B= blue
- Then, $Gray = (Red + Blue + Green)/3$

Computation to achieve this is defined as:

- For all pixel in image
- *Red = pixel.Red*
- *Green = pixel.Green*
- *Blue = pixel.blue*
- $Dim = (Red + Green + Blue)/3$
- *Pixel.Red = Gray*
- *Pixel.Green = Gray*
- *Pixel.Blue = Gray*

The third step is the Leaf district separation which is a partition procedure, the image was initially transformed into gray scale from the original form. This caused partitioning because of the diversity in the gray evaluation of the two (base and real image). The base shading of the picture has dark factors while the genuine picture was spoken to as white. After picture partitioning, the paired image including leaf area is acquired by locale filling and expelling each gap in the white portion. The picture is then checked from left to right using the MATLAB programming to confirm the aggregate (quantity of pixels in the leaf). Eventually, impure locale portion would center around division, to get just the ill part. The motive for this inspection is to measure the significance of development caused by disease on leaves with the help of triangle and simple limit methods. These ideas would help prospect investigators with enhancing methods for examination and even go into points of interest of the research.

Barnes [10] introduces the issue of automatic discovery and recognition of infections in digital pictures of potatoes. The presented clarification contains categorizing different pixels. Human intervention is needed to boost regions of infections and non-infections in a pact of training pictures. For fault detection, each pixel gives either positive or false positive. For false positive distinguishing proof, every pixel is prearranged by a variety of pre-decided defect categories. For training, the structure must have the ability to distinguish singular pixels in new images potatoes with high precision. After partitioning the potato from the picture background, a very big set of candidate features, based on statistical data connecting to the color and texture of the area close to a given pixel. Here instigator uses picture acquisition, image pre-processing, image segmentation, etc. in feature extraction and feature selection. It makes use of an adaptive boosting algorithm called AdaBoost [11, 12] to routinely choose fine features for a particular pattern categorization job. An irrelevant pact of features is selected from a vast collection of competitor places of interest, which compute factual factors of the shading and exterior circulation of the image area around a given pixel. When the framework training stage is finished, just these chosen factors are extracted for the new images introduced to the framework, and are then used to portray query pixels comparing to potatoes.

Shaikh et al. [13] explain about the realization of image processing to find out anomalies and infections in citrus leaves. Their study model comprises four parts: image pre-processing including normalization and contrast adjustment first. It improves certain significant aspects, which are useful for advance computing. Image segmentation is done next. This will do a portrayal of an image into something that is more vital protest of interest from basis. Likelihood management is done to sense the pull out and hardness area of cover leaves that figure accepting darken mutate YCbCr (A DigitalMPEG image-standard value) and bi-level thresholding. To the end, categorization is done with the help of Hidden Markov Model (HMM).

Dhaware et al. [14] introduced the issue of plant leaf disease classification, which depends on leaf image processing. Plant leaf infection recognition and classification include the phases like picture grabbing, picture pre-processing, picture segmentation, feature extraction and classification. They discussed the methods for image pre-processing, image segmentation algorithm used for automatic recognition and the study on a variety of plant leaf infection segmentation techniques that may be used for leaves disease segmentation. They used Support Vector Machine (SVM) method for image classification. SVM is a supervised learning technique, which is typically useful for pattern recognition and classification [15, 16]. This technique creates the hyper-planes in high-dimensional space for groups, the data spread into different classes. SVM does the classification by spotting the perfect hyper-plane, which differentiates the information of diverse groups. The hyper-plane having the largest gap among two groups is the best hyper-plane for SVM.

Jumb et al. [17] introduced the color image segmentation process. In this technique, foreground substances are notable clearly from the background. Since HSV color space is analogous to the way human eyes recognize color RGB image is transformed to HSV (Hue, Saturation, Value) color [18, 19] model and *V* (Value) is taken out. V corresponds straight to the idea of strength/brightness in the color basis. Next, an Otsu's multi-thresholding is applied on V channel to get the best thresholds from the image. The result of Otsu's multi-thresholding may comprise excess segmented portions. So, K-means clustering is done to merge the excess segmented portions. Background subtraction is carried out along with morphological processing at the end. The trial outcomes are acquired using metrics like PSNR and MSE, which establish the proposed method and give superior outcome as compared with other methods.

Ustad et al. [20] introduced a method intended to build up an infection recognition and categorization system for grapes plant with the help of image processing. The captured grape leaf image is done with complex background segmentation of K-means clustering algorithm which discovers the infection part of image. Color, shape and other factors were elicited from the portions of image based on the most matched features. Then classifications are done on the basis of diseases with SVM classifier to diverse infection groups. Generally, observed infections in grape plants are Black rot and Downey mildew [21, 22].

Gondal et al. [23] introduced an automatic method for early on pest discovery. They analyze the harvests to differentiate bug invasion and arrange the sort of bugs on vegetation. SVM is utilized for the arrangement of images. This method is easier when compared with the other automatic methods and shows good results. The images of leaves were taken and diverse preprocessing procedures were done on them. For bug position detection, 'limit technique' was done to isolate base from the insect on images. This method is very simple and correct in the detection of bugs from the images. Characteristic factors of the images are elicited, which can be used as given to SVM to sort images with factors and without bugs. For training and classification, a collection of 100 images were put for training with SVM. The primary segment of the record stores the characterization estimate for each image. 0 is denoted for leaf without any bugs and 1 denoted for leaf with white flies. The main section is utilized for gathering or arrangement and the remaining is utilized as information and SVM is utilizing these two data sources. Subsequent to training any doubtful image with predefined property is given to SVM for characterization.

Gavhale et al. [24] study the image processing techniques and incorporated it as a part of doing early detection of crop infections with leaf features examination. The goal of this work is to implement a real image analysis and arrangement techniques for extraction and grouping of leaf ailments. Leaf image is captured, and the image is processed to decide the status of each plant. They have done image preprocessing including RGB to various color space change, image improvement, fragment the locale image factors and K-mean clustering to decide the deformity and seriousness portions of plant leaves for extraction and grouping. Feature extraction utilizing factual GLCM and color highlight by means of mean qualities is done. At last, characterization is accomplished utilizing SVM. This technique will guarantee that when plant leaves infected it would be recognized correctly.

Shaaban et al. [25] introduced Evaluation K-mean and Fuzzy c-mean image segmentation-based clustering classifier. It was followed by thresholding and level set segmentation parts to provide accurate infection fragments. Their proposal took advantage of the K-means clustering. The execution and evaluation of the given image partitioning methods were evaluated with K-mean and Fuzzy c-mean calculations for precision, execution time, clustering classifier accuracy and features. Their database included 40 images executed by K-mean and Fuzzy c-mean image segmentation-based clustering classifier. Their test confirmed the viability of the proposed Fuzzy c-mean image segmentation-based clustering classifier. The measures of mean estimations of Peak Signal-To-Noise Ratio (PSNR) [26] and Mean Square Error (MSE) and inconsistency are utilized for performance evaluation of K-mean and Fuzzy c-mean image segmentation.

Anjna et al. [27] talk about Image segmentation as the most critical procedure of image preparation. It is extremely helpful to use image processing and reporting applications in light of the fact that with these kinds of applications, we may avoid wasteful processing of the entire image. There are various image segmentation techniques that help in image segmentation with certain image factors like pixel power esteems, color, textures, etc. The different image segmentation techniques are studied

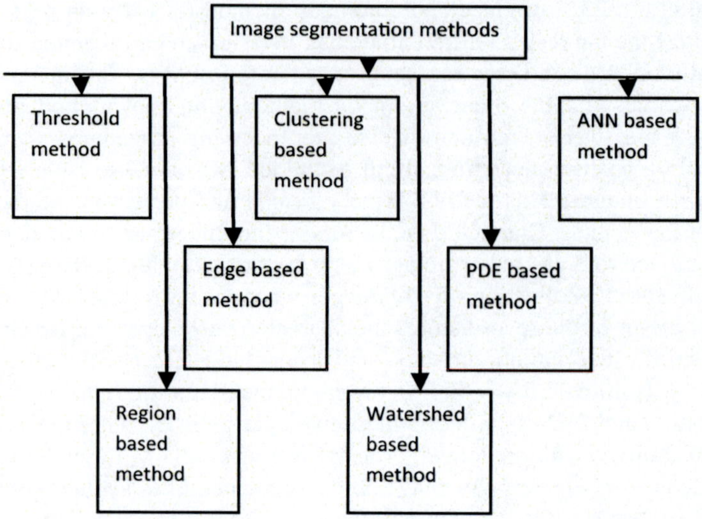

Fig. 3 Image segmentation techniques

and inspected in this paper. Various image segmentation techniques discussed in the paper are depicted in Fig. 3.

Rao KMM [28] discusses the ways toward enhancing and improving the raw images that are taken through advanced cameras, sensors, and through numerous other refined means like satellites, space shuttles and IoT-based flying gadgets for different applications. It is worth noting here that someone rightly pointed that 'a photo speaks a thousand words'. The above saying has significance to processing and investigating with images. Most experts in computer vision and image processing use powerful and better devices and in addition to the legitimate methodologies that give distinctive thoughts on a similar image by giving means to understand the substance of the image as well as give meaning and importance of the image [16]. These new and current techniques for handling images show the signs of improvement in meaning detection and comprehension of images with connection between its segments, its specific situation and its history in the event that it is a piece of a grouping and from the earlier learning picked up from a scope of images.

Gui et al. [29] introduced a novel methodology to focus on trim ailment portions with respect to the normal area. This is definitely not hard to execute and error free. This paper differentiated between Itti procedure and other novel techniques to choose momentous portions in the image. They use low-level parts of luminance and shading in CIELab shading space. This is added with multi-scale test by changing the degree of each size of the channel and makes the outstanding quality layout. This methodology used low-level components of luminance and shading and it merged with multi-scale test to choose saliency maps in images, and subsequently Fuzzy K-means (USFCM) congregation estimation was used.

Kuhkan [30] confirms that the most important challenges to the application of the classification algorithm is the same impact of all characteristics is perceived in classification, while some of the characteristics are less important for classification. This idea may influence the process of classification and reduce the accuracy of the KNN algorithm. Using a new method in this study, a definite weight is allocated to a variety of features based on their significance, so that the similar effect of all features is avoided in doing the categorization and divergence of classification process, thereby increasing the accuracy of KNN classification. They claimed considerable improvement of the classification by the proposed algorithm.

Deya et al. [31] incorporate Otsu thresholding system. This technique relies upon image dealing with figuring for the division of leaf disease situations. The proposed technique is successful in perceiving and watching the outside disease features. This approach has three stages. The primary stage was the photo acquiring stage. At first, the honest to integrity word test is recorded in its automatic edge using a level bed propelled scanner. In the next stage, the assessment picture was subjected to a preprocessing stage. Measure and resourceful value is completely reduced in this stage. Finally, the image getting ready and count process is done to find the infected region.

Zhou et al. [32] say leaves are the most vital piece of any plant, which causes plant to make likely life to develop. It changes over daylight into sustenance by a procedure of photosynthesis. Green color factor, chlorophyll helps in photosynthesis. Many leaf diseases, for example, fine buildup, wool mold, leaf sear can be found in plants. The fine mold on leaf surface covers the area of leaf, which fills in as an obstruction in process of photosynthesis. There may be a hindrance and unsuitable finding in the problem identification of plant leaves, which may prompt lower profitable yield quality and produce distress. In this way, an automatic framework is necessary to differentiate the illness with the aim that suitable actions could be taken in advance.

Camargo et al. [33] discuss the issue of creating image handling algorithms that perceive variations from the norm in the yields. Maize plant images are considered for the investigation by the authors. The authors have considered different diseases like Brown stripe fleece mold, Stem borer, etc. for classification. The accuracy of the classification has been estimated up to 95%. They were preparing leaf images in light of texture, shape and color. Two algorithms are utilized at the preparing side. To extract HSI estimates, color change is utilized. Then feed forward neural network is utilized for accurate estimation of diseases.

Revathi et al. [34] proposed a new approach of distinguishing evidence of cotton trim diseases from RGB images. The authors proposed an enhanced PSO feature choice strategy, which embraces client features like differences, texture, color and edge to remove the unessential factors. Utilizing back propagation neural network (BPN), Fuzzy and SVM classifiers obtained features are removed with the assistance of genetic algorithm (GA), feature determination and edge CYMK color feature. Comparative study of three classification models was surveyed to test this proposal. Six sorts of diseases, for example, Root spoil, Leaf Blight, Micro Nurient, Verticilium Wilt, Bacterial Blight, Fusarium wither have been precisely arranged to assess its efficiency.

3 Automated Pest Detection System

3.1 Statement of Problem

Pest detection on plant leaves is the most vital thing for making our yield strong. This if not done at the right time may direct to ruthless failure. Countries like India, other parts of the world, farmers suffer massive losses due to infections that hit their farming. According to a survey done by Indian Council of Agricultural Research (ICAR), Pests eat away 35% of the total crop production in India. Plant diseases such as Indian cassava mosaic virus (ICMV) and Anthracnose Disease that affect cassava plants affect planters in their yield. Others like MUT disease of the Jo war crop caused by the fungus Sphacelotheca sorghi, Sugar cane mosaic caused by a virus and wilt of the pea crop caused by the fungus Fusarium Vasinfectum also anguish crop production in India [35]. Apart from the loss to planters, tremendous use of fungicides and pesticides for pest discovery add to the cost of making and may also contaminate our natural surroundings. So, we should reduce their use. To attain this, the contaminated region must be targeted by applying the correct amount and concentration of insect repellent by assessing the ruthlessness of the infection using image processing techniques.

3.2 Research Objectives

Basically, a pest detection system study assesses the level to which infected pest distress vegetation most particularly, plant leaves. Purposely, the investigation looks for:

1. Sense and recognize pest on plant leaf through feature extraction methods.
2. A flexible database is used for training. In this, we may put in new disease type and its solution
3. Propose the suitable remedy for the pest/disease and advises in which group it belongs based on the disease ruthlessness. Here, the KNN classifier and thresholding are intended to be used to recognize which pest belongs to which group.

3.3 Research Question

1. How best can image processing techniques be employed to identify pest which causes infection on plant leaf?
2. Which techniques may be employed to assess and measure the ruthlessness infection on leaves?
3. Is this the right time to identify the infection?

4. What quantity of insect repellent could be used on plant leaves?

3.4 Scope of the Study

The investigation mainly focused on identifying and assessing the ruthlessness of pest-caused infections on plant leaves. Image processing techniques and methods are also taken into consideration. This study looked into some infections that affect plant leaves through pest and the potential of assessing the infections on leaves with the help of image processing techniques particularly with KNN and unique thresholding techniques.

3.5 Significance of the Research

Innovation is a continuous process and it is highly unlikely that we can keep a deliberate distance from its use. So it is left to the clients to use it sensibly to improve their living circumstances. In any case, this investigation would empower the people in the agriculture field to find the importance of image handling in their field of work. It will also help to get high profit because of the way that plant diseases can be predicted and reinstated making use of suitable disease detection. This study also would help to reduce the wastage of water bodies. The study will likewise help reduce the cost of generation that conveys huge loss to planters due to excessive use of pesticide on their plants. The results of this study can give very important suggestions and input to the experts allied fields so as to keep suitable actions for better farming, especially in situations where there is an outbreak of plant diseases. This study would fill in as a type of viewpoint matter for different analysts who might want to lead an additional study into the concerns recognized with image handling and plant diseases.

4 Proposed Model

The proposed pest detection system is employing a KNN classifier apathetic algorithm. There is no express training stage or it is extremely negligible. This likewise means the training stage is entirely quick. The absence of speculation means that KNN keeps all the training information.

If some diseases are found on a crop, the system will take the snapshot of the crop and will send it for analysis and then feature extraction is done. This is the important module as to identify the disease we require the features of crop as well as the features of the image to make the comparison. In this module, the features such as area, eccentricity, etc. are extracted and stored for further usage and then,

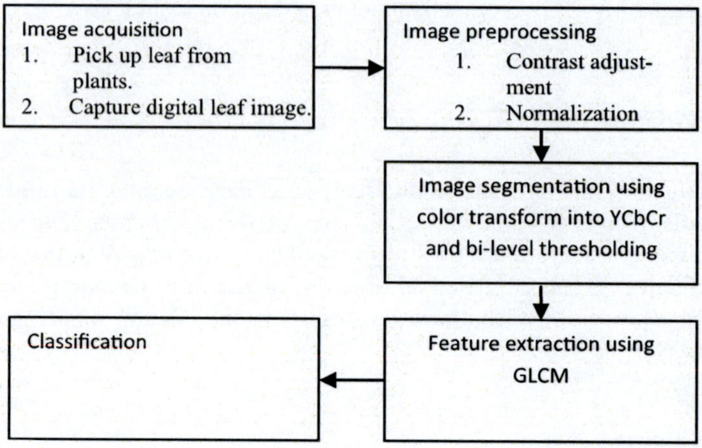

Fig. 4 Block diagram of the proposed methodology

segmentation is done. In this stage, the features of the images are segmented to create a differentiator between them. Here thresholding will apply, which is carried out to segment the image properties. Classification is the next step. Once the extraction and segmentation are done, there is a need for classify to find the disease. This classification is fulfilled by the hybrid of KNN and thresholding techniques. With the help of the extracted and segmented features, the KNN-thresholding classified the data and gives the result. Output module focuses on the final output if the crop has some kind of bug. It will tell the bug name and the effects made by them, also remedy will be suggested. The process discussed here is depicted in Fig. 4 and Algorithm 1.

Algorithm 1: Segmentation by K-means clustering operation

Input: Grape leaf image.

Output: Segment groups of grape leaf image.

Start

1. Read input image
2. images are converted to gray scale
3. Apply enrichment
4. Resize the image
5. Apply k-means clustering process.
6. Discover the centroid of the pixels.
7. Divide the pixels into cluster.
8. Represent the clustered image.
9. Segmented output.

Stop

5 Conclusion

An automated and robust pest detection system design for efficient plantation is a very appealing topic. Due to its applicability in a variety of areas in agriculture and environmental science, it has attained the interest of many researchers recently. In this paper, we present a variety of automated and pest detection methods and explain each of them by literature survey. Different datasets are used for each of the techniques wherein the data are grabbed by diverse means such as sensors and images by assigning these gadgets at different places. Machine learning methods like decision trees, support vector machines, Hidden Markov Models are assessed for automated pest detection system and afterward the robust and automated pest detection system with the help of KNN classifier is presented. This is a very simple KNN classifier-based pest detection model. This work opens up avenues for researchers to identify and work on areas of semi-heightened crop-based image processing, which enhances the quality of farming by detecting and sometimes suggesting the areas of improvement via classification of crops in the agricultural land.

References

1. Kim J, Kim B, Savarese S Comparing image classification methods: k-nearest-neighbor and support-vector-machines. In: Applied mathematics in electrical and computer engineering. Ann Arbor, MI 48109–2122, ISBN: 978–1–61804–064–0
2. Bania RK (2018) Handwritten Assamese character recognition using texture and diagonal orientation features with artificial neural network. Int J Appl Eng Res 13(10):7797–7805. ISSN 0973–4562, Research India Publications. https://www.ripublication.com
3. Jawarah CV, Biswas PK, (1977) Investigation on Fuzzy thresholding based on fuzzy clustring. Pattern Recog 30(10):1605–1613. Pattern Recognition Society, Published by Elsevier Science Ltd Printed in Great Britain
4. Gupta V, Sengar N, Dutta MK, Travieso CM, Alonso JB (2017) Automated segmentation of powdery mildew disease from cherry leaves using image processing. In: 2017 international conference and workshop on bioinspired intelligence (IWOBI), Funchal 2017, 1–4. https://doi.org/10.1109/IWOBI.2017.8006454
5. Kaganami HG, Beij Z (2009) Region based detection versus edge detection. In: IEEE Transactions on Intelligent information hiding and multimedia signal processing, pp 1217–1221
6. Dobrescu A, Giuffrida MV, Tsaftaris SA (2010) Leveraging multiple datasets for deep leaf counting. Retrieved from https://arxiv.org/abs/1709.01472 on 01 October 2020
7. Raichaudhuri R, Sharma R (2016) On analysis of wheat leaf infection by using image processing. In: Proceedings of the international conference on data engineering and communication technology, advances in intelligent systems and computing (1), 978–981
8. Marfo R (2016) Measuring the severity of fungi caused disease on leaf using triangular thresholding method. M.Sc Thesis, Kwame Nkrumah University of Science and Technology, Kumasi, Ashanti, Ghana, june 2016. Retrieved from https://ir.knust.edu.gh/xmlui/handle/123456789/9922 on 01 October 2020
9. Kang WX, Yang QQ, Liang RR (2009) The comparative research on image segmentation algorithms. In: IEEE Conference on ETCS, pp 703–707
10. Barnes M (2012) Computer vision based detection and identification of potato blemishes. PhD thesis, University of Lincoln. May 2012, retrieved from https://eprints.lincoln.ac.uk/id/eprint/14568/ on 01 October 2020

11. Robert YF, Schapire E (1999) A short introduction to boosting. J Japanese Soc Artif Intell 14(5):771–780
12. Dehariya VK, Shrivastava SK, Jain RC (2010) Clustering of image data set using k-means and fuzzy k-means algorithms. In: International conference on CICN, pp 386–391
13. Shaikh RP, Dhole SA (2017) Citrus leaf unhealthy region detection by using image processing technique. In: 2017 international conference of electronics, communication and aerospace technology (ICECA), Coimbatore, 420–423. https://doi.org/10.1109/ICECA.2017.8203719
14. Dhaware CG, Wanjale KH (2017) A modern approach for plant leaf disease classification which depends on leaf image processing. 2017 international conference on computer communication and informatics (ICCCI). Coimbatore, pp 1–4. https://doi.org/10.1109/ICCCI.2017.8117733
15. Zhu C, Ni J, Li Y, Gu G (2009) General tendencies in segmentation of medical ultrasound images. IN: International conference on ICICSE, pp 113–117
16. Rahman MH, Islam MR (2013) Segmentation of color image using adaptive thresholding and masking with watershed algorithm. IEEE
17. Jumb V, Sohani M, Shrivas A, Color image segmentation using K-means clustering and otsu's adaptive thresholding. Int J Innov Technol Expl Eng (IJITEE) 3(9). ISSN: 2278–3075
18. Otsu N (1979) A threshold selection from gray level histograms. IEEE Trans Syst Man Cybernet 9:62–66
19. Saikumar T, Yugander P, Murthy P, Smitha B (2011) Colour based image segmentation using fuzzy c-means clustering. In: International conference on computer and software modeling IPCSIT, vol 14. IACSIT Press
20. Ustad MS, Korke AG, Bhaldar HK (2017) Novel algorithm for detection and classification of grape leaf disease using k-mean clustering. Int J Innov Res Comput Commun Eng 5(4)
21. Kumar A, Kumar P (2008) A new framework for color image segmentation using watershed algorithm. Comput En Intell Syst 2(3):41–46
22. Vijayanandh R, Balakrishnan G (2012) Hillclimbing segmentation with fuzzy C-means based human skin region detection using bayes rule. European J Sci Res (EJSR) 76(1):95–107
23. Gondal MD, Khan YN (2015) Early pest detection from crop using image processing and computational intelligence. FAST-NU Res J (FRJ) 1(1)
24. Gavhale KR, Gawande U (2014) Unhealthy region of citrus leaf detection using image processing techniques. In: International conference for convergence of technology—2014. IEEE. 978–1–4799–3759–2/14/$31.00©2014
25. Shaaban HRM, Habib AA (2015) Performance evaluation of k-mean and fuzzy c-mean image segmentation based clustering classifier. (IJACSA) Int J Adv Comput Sci Appl 6(12)
26. Mythili C, Kavitha V (2012) Color image segmentation using ERKFCM. Int J Comput Appl (IJCA) 41(20):21–28
27. Anjna, Kaur R (2017) Review of image segmentation technique. Int J Adv Res Comput Sci 8(4). ISSN NO: 0976–5697
28. Rao KMM (1996) Image processing for medical applications. In: Proceedings of 14th World Conference on NDF, 8th–13th Dec 1996
29. Gui J, Hao L, Zhang Q, Bao X (2015) New method for soybean leaf disease detection based on modified salient regions. Int J Multimed Ubiquit Eng 10(6):45–52
30. Kuhkan M (2016) A method to improve the accuracy of k-nearest neighbor algorithm. Int J Comput Eng Inform Technol 8(6):90–95. Available online at: www.ijceit.org E-ISSN 2412–8856
31. Deya AK, Sharma, M, Meshramb MR (2016) Image processing based leaf rot disease, detection of betel vine. International conference on computational modeling & security (CMS 2016). Procedia Comput Sci 85:748–754
32. Zhou B, Xu J, Zhao J, Li A, Xia Q (2015) Research on cucumber Downy Mildew detection system based on SVM classification algorithm. In: 3rd international conference on material, mechanical and manufacturing engineering (IC3ME 2015)
33. Camargo A, Smith JS (2009) An Image-Processing based algorithm to automatically identify plant disease visual symptoms. ELSEVIER 17(1):9–21

34. Revathi P, Hemalatha M (2014) Classification of cotton diseases using cross information gain minimal resource allocation network classifier with enhanced particle swarm optimization. J Theoret Appl Inform Technol 60(1)
35. Mitra A Plant diseases and pests of India. Nature, SP–272, EP–272, VL–140, IS–3537, AB 2, Nos. 8 and 9, Feb. and March 1937, 1476–4687, https://doi.org/10.1038/140272a0

Chapter 16
Troubleshooting Fluctuations in Power System and Network Harmonic Analysis

Harwinder Karwal, Umesh Sehgal, and Twinkle Bedi

1 Introduction

For keeping our expensive electronic devices safe, a good power source is an essentiality today. Simply a quality electrical power refers to the problem-free working of our gadgets while consuming it. The current and voltage fluctuations and erratic sine wave damage our gadgets. The sinusoidal waveforms of AC are distorted by harmonic frequencies spoiling the power quality, resulting in the heating of gadgets, catching fire, fluctuations in motor rotation, fusing conductors and transistors, fire catching and fatal shocking hazards in neutral wiring, which are also caused by overloading due to harmonics. The performance of electrical and electronic systems keeps on decreasing due to hazards of harmonics. We can decrease harmonics by innovative methodologies. The harmonic distortion phenomenon has raised a concern for the engineers to limit it to acceptable proportions. Earlier, harmonic problems could often be tolerated because of the conservative electronic architecture.

But today's highly sophisticated, sensitive and costly electronic gadgets need to be taken care of by controlling the harmonics currents to keep the sine wave pure, thus well in shape. Our team has worked objectively and invented certain methodologies to troubleshoot the fluctuations in network harmonic analysis.

H. Karwal (✉) · U. Sehgal · T. Bedi
GNA University, Sri Hargobindgarh, Phagwara 144401, Punjab, India

N. Marriwala et al. (eds.), *Soft Computing for Intelligent Systems*, Algorithms for Intelligent Systems, https://doi.org/10.1007/978-981-16-1048-6_16

2 Related Work

2.1 *Ideal Power System*

The ideal electrical power is represented by perfect sine wave of current and voltage [1] as shown in Fig. 1. Practically, however, the conditions are never ideal, so these sine waves are often distorted to large and harmful extents. This deviation from perfect sinusoidal waveforms is usually expressed in terms of harmonic distortion of the voltage and current waveforms. The concern of harmonics has grown serious because of their numerous hazards, for example, overheating, erratic functioning and decrease in the life of modern electronic gadgets based on microprocessors silicon chips and complicated circuit architecture.

A quality power supply has constant and uniform current and voltage maintaining the overall health of our equipment's, increasing their life with optimum performance, etc.

2.2 *Types of AC Loads*

(a) **Linear load**: It involves direct proportionality between electrical current (I) and voltage (V) at any time, making a sine wave. Therefore, the linear load follows Ohm's law, i.e. V = IR. For example, it happens in capacitors, transformers and electrical motors.

(b) **Non-linear Load**: Unlike the linear load, the non-linear load does not follow the Ohm's law, i.e. V ≠ IR and does not make sinusoidal waves between current and voltage. For example, it happens in TVs, computers, printers, rectifiers SMPS electronic instrumentation, etc.

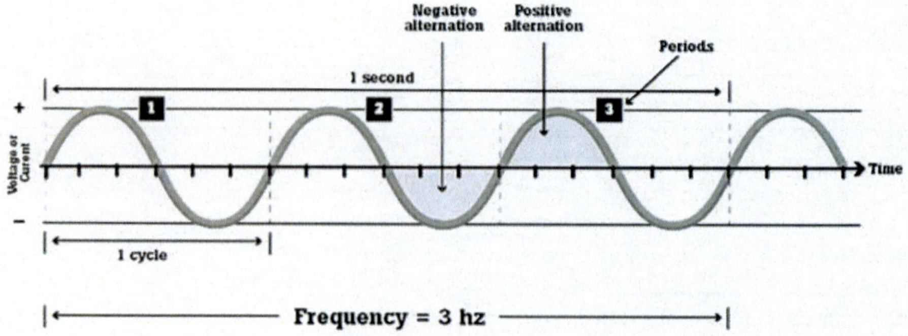

Fig. 1 The sinusoidal AC power with a frequency of 3 Hz

The non-linear loads absorb the sinusoidal electrical power in fast short pulses, producing distorted sine waves, which result in unnecessary tripping of magnetic circuit breakers, overheating, fire, destroying smooth operation of motors leading to mechanical faults soon.

2.3 Harmonics

The quality and performance of electrical power is highly destroyed by harmonics penetrating from the non-linear loads. The technology needs to find the source of harmonics and take user-friendly corrective measurements.

The non-linear AC load sends back waves with the same frequency ω as that of the source sinusoidal waves, at whole number multiples $n\omega$ into the source wave, creating harmonic distortion. The additional waves ($n\omega$) added to the source wave (ω) are known as THD (Total Harmonic Distortion).

Customarily, the ratio of root mean square (RMS) of higher harmonic frequencies to that of first harmonic frequency is the main criterion to measure the quality of the source electrical power. As per IEE-519 standards, THD can be calculated by the formula as follows (Fig. 2):

$$\mathrm{THD}_F = \frac{\sqrt{V_2^2 + V_3^2 + V_3^2 + \cdots}}{V_1}$$

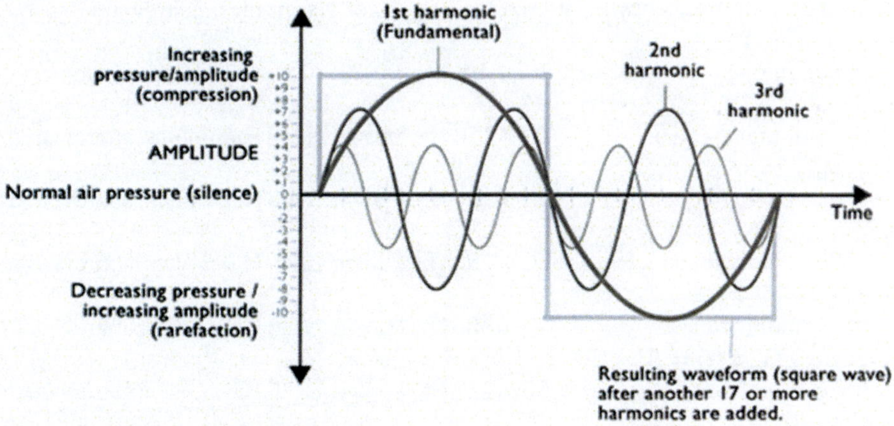

Fig. 2 Distortion of sine wave by harmonics

2.4 Classification of Harmonics

There are two categories of harmonics as follows:

(a) **current harmonics**
(b) **voltage harmonics**

The current harmonics are produced when current is passed through a non-linear AC load like inverters, computers, smartphones, variable speed drives, etc. While the voltage harmonics do not originate directly from the non-linear AC loads but in fact produced by current harmonics. The voltage harmonics are produced by high impedance transformers, long electrical cables, etc. [2]. The above image shows the distorted current waveform (original shape-pure sine wave) across the non-linear AC load representing the current harmonics in the power system. The simplest harmonic current formulation is as follows:

$$I_h = I_f / h,$$

where I_h = harmonic current; I_f = fundamental current and h = harmonic order.

2.5 Harmonic Sequencing

The harmonics are voltages or currents that operate at a frequency that is an integer multiple ($n\omega$) of the fundamental frequency (ω) [3]. The harmonics ($n\omega$, n is an integer) is the distortion in the source-electrical current (ω, fundamental frequency). For example, if we have $\omega = 50$ Hz for the source electrical current wave, then.

1st harmonic frequency = 50 Hz, known as 1st Harmonics, abbreviated as 1_f, $\nu_{\max}(2\pi ft) = \mathbf{E_1}$.

2nd harmonic frequency = 2 × 50 = 100 Hz, known as 2nd Harmonics, abbreviated as $2f$, $\nu_{2\max}(4\pi ft) = \mathbf{E_2}$.

3rd harmonic frequency = 3 × 50 = 150 Hz, known as 3rd Harmonics, abbreviated as $3f$, $\nu_{3\max}(6\pi ft) = \mathbf{E_3}$.

4th harmonic frequency = 4 × 50 = 200 Hz, known as 4th Harmonics, abbreviated as $4f$, $\nu_{4\max}(8\pi ft) = \mathbf{E_4}$.

5th harmonic frequency = 5 × 50 = 250 Hz, known as 5th Harmonics, abbreviated as $5f$, $\nu_{5\max}(10\pi ft) = \mathbf{E_5}$.

6th harmonic frequency = 6 × 50 = 300 Hz, known as 6th Harmonics, abbreviated as $6f$, $\nu_{5\max}(12\pi ft) = \mathbf{E_6}$.

7th harmonic frequency = 7 × 50 = 350 Hz, known as 7th Harmonics, abbreviated as $7f$, $\nu_{7\max}(14\pi f\mathrm{t}) = \mathbf{E_7}$.

nth harmonic frequency = n × 50 = n50 Hz, known as nth Harmonics, abbreviated as nf, $\nu_{7\max}(2n\pi ft) = \mathbf{E_n}$.

Therefore, we have

$$\mathbf{E_T = E_1 + E_2 + E_3 + \cdots E_n}$$

The distorted waves produced by harmonics are shown at the r.h.s. column. The resultant wave is produced by both harmonics and phase difference [3].

The positive sequence harmonics like $4f$, $7f$, $10f$, etc. rotate in the same direction as the fundamental frequency does (forward direction), while the negative sequence harmonics like $2f$, $5f$, $8f$ rotate opposite to the fundamental frequency (reverse direction).

The positive sequence harmonics add up to the fundamental wave frequencies and cause overheating of electrical appliances. On the other hand, the negative sequence harmonics are present between the phases causing non-uniform rotation of electrical motors (Fig. 3).

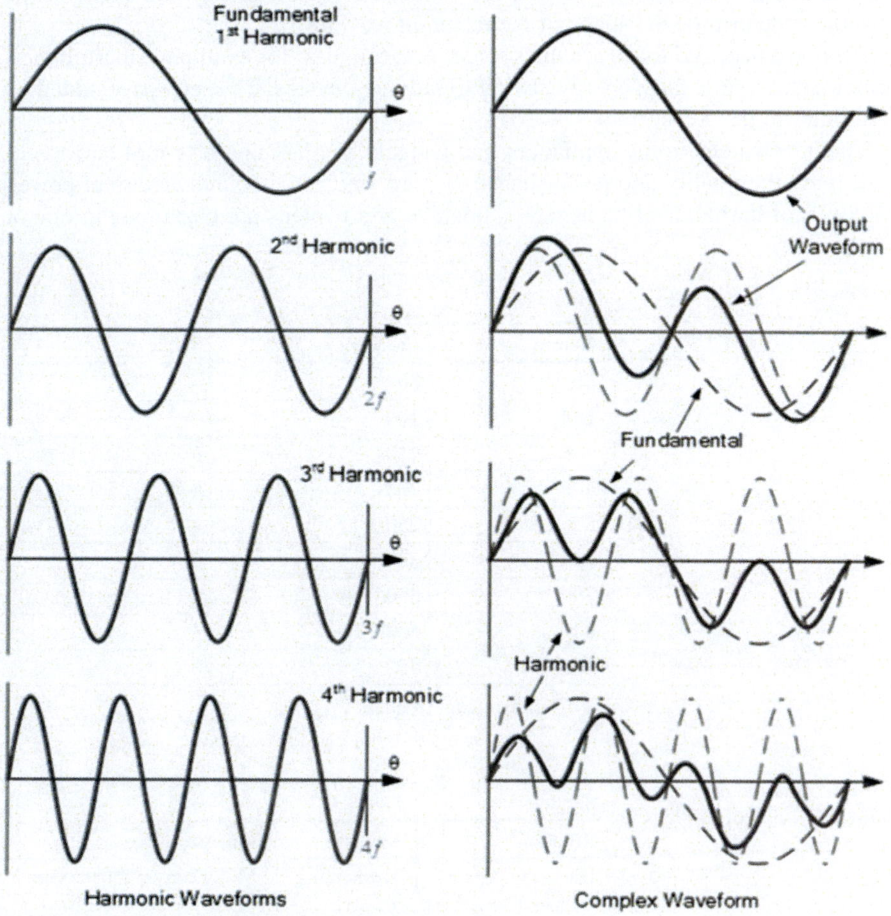

Fig. 3 Complex waveforms due to harmonics

2.6 *Triplens*

The harmonics that are multiples of 3rd harmonics are known as triplens e.g. ($3f$, $6f$, $9f$ ...). The triplens are not displaced because of their zero rotational sequence. They are always present between the phases, neutral or ground [4].

The triplens do not cancel each other like positive and negative sequences harmonics, but they add up arithmetically in the common neutral wire, resulting in overheating and destroying its efficiency. It has been elaborated in the following table with 50 Hz, (Tables 1 and 2).

Harmonics Summary

Harmonics deteriorate the quality of source electrical power by adding up frequencies integral multiple ($n\omega$) of the fundamental one (ω). The frequency and shape of the harmonics determine the shape of the resultant waveform.

All non-linear AC loads are the source of harmonics, for example, smartphones, computers, servers, fans, energy-saving lightening devices, TVs, etc. have added up harmonics to the AC current.

The modern electronic appliances and gadgets are the main origin of harmonics destroying the quality and performance of pure sinusoidal source electrical power. All kinds of harmonics like negative, positive and triplens are hazardous in one or another way.

Table 1 Elaboration of Triplens

Name	Frequency	Sequence
FF*	50	+
2nd	100	−
3rd	150	0
4th	200	+
5th	250	−
6th	300	0
7th	350	+
8th	400	−
9th	450	0

**FF* Fundamental frequency

Table 2 Rotation & Harmonic Effect of Triplens

Sequence	+	-
Rotation	Forward	Reverse
Harmonic effect	Overheating	Motor torque fluctuation

2.7 Fourier Analysis of Electric Waves

Prior to any implementation of the FFT algorithm, a computerized Fourier Transform Analysis of the harmonic waveform may be of great use. The FTS for a periodic function can be written as follows:

$$x(t) = a_0 + \sum_{n=1}^{\infty}\left(a_n \cos\frac{2\pi nt}{T} + b_n \sin\frac{2\pi nt}{T}\right) \quad E_1,$$

where.

$a_0 =$ Average value.
$a_n =$ Coefficient of series component of nth harmonic.
$b_n =$ Coefficient of rectangular component of nth harmonic.
The mathematical expression for the nth harmonic vector is as follows:

$$\mathrm{A_n Ø_n} = \mathrm{an} + \mathrm{jbm} \ldots \mathrm{E_2}$$

With the angle of phase and magnitude, it takes the following form:

$$A_n = \sqrt{\mathrm{a}^2\mathrm{n} + \mathrm{b}^2 n}, \quad \mathrm{Ø_n} = \tan^{-1}\mathrm{b_n} \ldots \mathrm{E_3}$$

2.8 Wavelets—A Short Note

Unlike Fourier Transform (FT), the Wavelet Transform (WT) can explain any type of signal both in time and frequency domain. Therefore, WT can be of tremendous help for customized accurate measurements. The Heisenberg Principle applies to both STFT and MRA [5] For a function for a signal $x(t)$, the Continuous Wave Transform (CWT) has the following expression

$$\mathrm{CWT}_{(x,a,b)} = \frac{1}{\sqrt{|a|}}\int_{-\infty}^{\infty} x(t)^{\psi^*_{a,b}}\left(\frac{t-b}{a}\right)\mathrm{d}t,$$

where.

$a =$ scale parameter.
$b =$ translation parameter.
$\psi(t) =$ mother wavelet.
In little bit more details, the expression can be written as:
$\psi_{a,b}(t) = 1/\sqrt{|a|}\psi(t - b/a)$.

For a limited length, we have its integral form as follows:

$$\int_{-\infty}^{\infty} \psi(t)\,\mathrm{d}t = 0$$

2.9 *THD and Power Factor*

For higher efficiency of power systems, the Total Harmonic Distortion (THD) should tend to zero, higher value of power factor. For this, International standards, e.g. IEC 61,000-3-2, have set limits on different kinds of harmonics for particular devices in use [6]. For a perfectly sinusoidal waveform, we have Eq. (1) for power factor calculation:

$$\text{Power Factor} = \cos\left(\theta_v - \theta_i\right), \tag{1}$$

here θ_v and θ_i are voltage and current phases, respectively. However, for a wave with some kind of THD, the Eq. (2) is valid

$$\text{Power Factor} = \frac{P_{\text{avg}}}{(V_{\text{rms}})(I_{\text{rms}})} \tag{2}$$

Power Factor without THD

$$V_{\text{rms}} = \frac{V_{pk}}{\sqrt{2}} \quad \text{and} \quad I_{\text{rms}} = \frac{I_{pk}}{\sqrt{2}}$$

THD in Power Factor

$$I_{\text{rms}} = \sqrt{I_{\text{dc}}^2 \sum_{k=1}^{\infty} I_{k\_\text{rms}}^2} \tag{3}$$

For THD = 0, only the power source will have only fundamental frequency,

$$P_{\text{avg}} = V_{1\_\text{rms}} \times I_{1\_\text{rms}} \times (\text{displacement factor}) \tag{4}$$

On the other hand, in case of apparent power, $(V_{\text{rms}})(I_{\text{rms}})$ will have harmonics, so the term in the denominator of Eq. (2) will be higher than the expected for the fundamental frequency.

Using Eq. (3) and (4) into Eq. (2),

$$\text{Power Factor} = \frac{V_{1\_\text{rms}} \times I_{1\_\text{rms}} \times (\text{Displacement Factor})}{V_{1\_\text{rms}} \times \sqrt{I_{dc}^2 + \sum_{k=1}^{\infty} I_{k\_\text{rms}}^2}}$$

$$= \frac{I_{1\_\text{rms}} \times (\text{Displacement Factor})}{\sqrt{I_{dc}^2 + \sum_{k=1}^{\infty} I_{k\_\text{rms}}^2}} \quad (5)$$

2.10 Distortion Factor and THD

The phase difference between voltage and current (θ_v–θ_i) gives rise to displacement factor. And the harmonic distortion gives rise to distortion factor. We have

$$\text{Distortion factor} = \frac{I_{1\_\text{rms}}}{\sqrt{I_{\text{dc}}^2 + \sum_{k=1}^{\infty} I_{k\_\text{rms}}^2}} = \frac{I_{1\_\text{rms}}}{I_{\text{rms}}}$$

For better analysis, we can discuss the relationship b/w distortion factor and THD [7], where

$$THD = \frac{\sqrt{\sum_{k \neq 1} I_{k\_rms}^2}}{I_{1\_rms}}$$

We can calculate the distortion factor in terms of THD by the following simple equation:

$$\text{Distortion factor} = \sqrt{\frac{1}{1 + \text{THD}^2}}$$

Therefore, the power factor can be calculated in terms of displacement factor and THD as follows:

$$\text{Power factor} = \text{displacement factor} \times \text{distortion factor}$$

$$\text{Power factor} = \cos(\theta_v - \theta_i) \times \sqrt{\frac{1}{1 + \text{THD}^2}}$$

For a complete understanding and correction of power factor, the distortion factor must be used for all calculations [8].

2.11 *ITIC (and CBEMA) Curves 2000*

The Information Technology Industry Council (ITIC) standard curve (ITIC Curve) [9] that explains the voltage fluctuations should be tolerated by most of Information Technology Equipments (ITE).

It is not meant for design base for the manufacture of devices and AC distribution system. Rather the ITIC curve explains steady-state tolerance of increase or decrease up to 10%, voltage increases in RMS amplitude of up to 120% of the RMS nominal voltage, for 0.5 s duration, high or low frequency, voltage sags, etc. (Fig. 4).

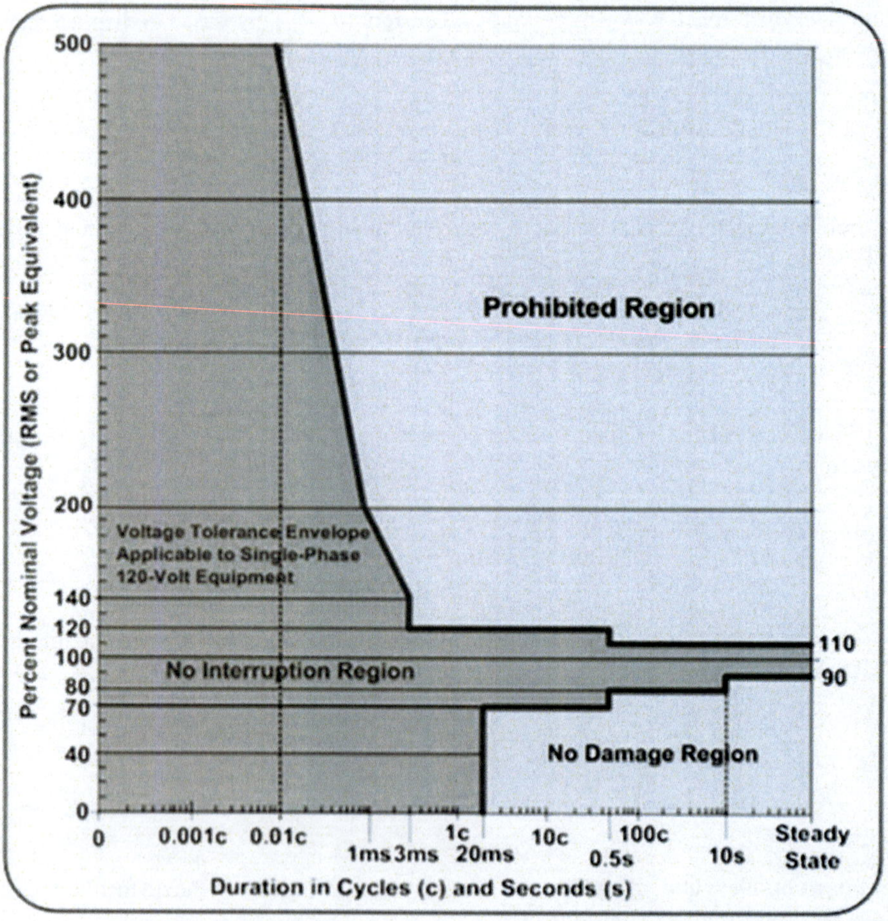

Fig. 4 ITIC standard curve 2000

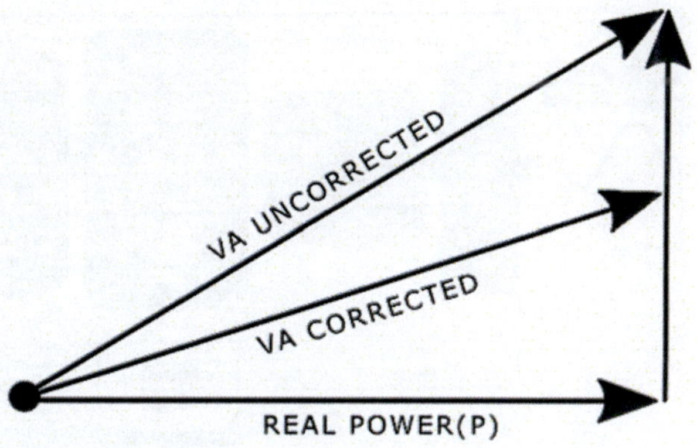

Fig. 5 Power factor correction by capacitors

2.12 *Harmonic Effects on Equipments [10]*

i. **Transformers**: The harmonics in transformers result into loss of iron and copper by putting extra stress on insulators and hysteresis along with thermal fatigue by overheating, problem of resonance (between transformer inductance and the system capacitance), core vibrations
ii. **Motors and Generators**: The overheating of rotor caused by Harmonics dumps the efficiency of machines and elevates their torque values, killing smooth operation of the mechanical parts resulting into short life of various components of the machines.
iii. **Transformers**: Harmonics pollute the pure sinusoidal waveform causing over-stressing, eddy currents and loss of iron and copper, core vibration and thermal fatigue by overheating.
iv. **Communication network**: High noise is caused by the AC power lines containing harmonics running close to telephone lines.

Capacitors: The capacitors are quite effective in keeping the power factor close to 1, saving power and money of the users. Actually, the capacitors are enemy of inductance. For this purpose, we can use capacitors in series and parallel manner. However, the over-harmonics may lead to capacitor bank failures and dielectric breakdown of insulated cables, etc. (Fig. 5).

3 Proposed Modeling

After an extensive analysis of old techniques and their limitations to analyze and to filter out THD in the source load, [11–13] thus, for getting sinusoidal waveform output, we have devised a probable model by combining the following in one device [14],

i. Using Wavelet Transform in place of FFT analysis [15–17]

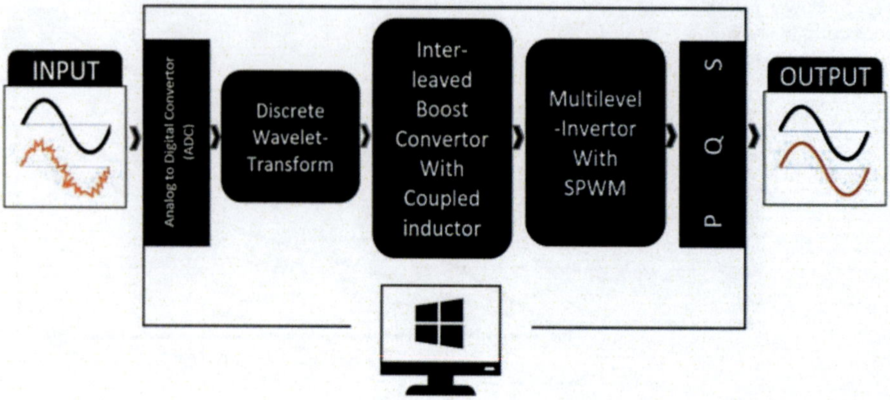

Fig. 6 Proposed model for troubleshooting fluctuations in power system [22]

ii. Boosting Solar output by interleaved converters [18–20]
iii. Improving output of NCC by multilevel inverters [21] (Fig. 6).

The proposed model has scanning, analyzing and corrective capabilities to produce pure sinusoidal waveform from the polluted source current and voltage with an overall computer control.

4 Results and Discussions

The in-lab results drawn with oscilloscope and Metlab have confirmed that

i. DWT is found to a better choice for measuring PQ indices [23, 24].
ii. While low to high DC/DC application, the Boost converters with coupled inductors are excellently able to reduce output voltage ripple.
iii. The multilevel inverter topologies, while keeping the modularity preserved, guarantee boosting of voltage.

The modern appliances and gadgets have polluted the quality of power system by producing harmonics beyond acceptable extent. The power quality has emerged as a serious concern for more technical research with the introduction of sophisticated electronic equipment's like computers and peripherals, smartphones, washing machines, electrical motors, varying speed drives, etc. in the last decade.

The results drawn from the old experiments in this paper may benefit a lot in reducing harmonics with better analysis bringing improvement in overall quality of the electrical powers being used by the people for their expensive gadgets [25–28].

5 Conclusion

The proposed model is found to be useful as follows:

i. Protection against voltage sags and interruptions;
ii. Maintaining power quality by refining sinusoidal wave;
iii. Benchmarking electrical power and quality;
iv. Enhancing the life and working of sensitive gadgets;
v. Avoiding overheating of equipments.

References

1. Beaty HW, Santoso S, Dugan RC, McGranaghan MF (2002) Electrical power systems quality. McGraw-Hill Education, United Kingdom
2. van Niekerk CR, Rens APJ, Hoffman AJ (2002) Identification of types of distortion sources in power systems by applying neural networks. IEEE AFRICON Mbabane, Swaziland 2:829–834
3. El-Bayeh C (2016) Simple method for determining harmonic sequences in a machine, transformer or network. Int J Digital Informat Wireless Commun 6:112–121. https://doi.org/10.17781/P002027
4. Chicco G, Postolache P, Cornel T (2011) Triplen harmonics: myths and reality. Lancet 81:1541–1549. https://doi.org/10.1016/j.epsr.2011.03.007
5. Hosseini SA, Amjady N, Velayati MH (2015) A fourier based wavelet approach using Heisenberg's uncertainty principle and Shannon's entropy criterion to monitor power system small signal oscillations. IEEE Trans Power Syst 30(6):3314–3326. https://doi.org/10.1109/TPWRS.2014.2377180
6. Kawasaki S, Ogasawara G (2017) Influence analyses of harmonics on distribution system in consideration of non-linear loads and estimation of harmonic source. J Int Council Electri Eng 7(1):76–82. https://doi.org/10.1080/22348972.2017.1324267
7. Rohollah A Comparison of power quality indices and apparent power (kVA) ratings in different autotransformer-based 30-pulse AC–DC converters. Electrical Engineering Department, Technical and Vocational University, Qom Boys Technical
8. Power Systems Engineering Research Center A National Science Foundation Industry/University Cooperative Research Center since 1996; Cornell, Arizona State, Berkeley, Carnegie Mellon, Colorado School of Mines Georgia Tech, Illinois, Iowa State, Texas A&M, Washington State, Wisconsin
9. Power Quality Standards: CBEMA, ITIC, SEMI F47, IEC 61000–4–11/34. Mark Stephens, PE Manager Industrial Studies Electric Power Research Institute 942 Corridor Park Blvd Knoxville, Tennessee 37932
10. IEE colloquium on sources and effects of harmonic distortion in power systems (Digest No.1997/096). IEE colloquium on sources and effects of harmonic distortion in power systems (digest No: 1997/096), London, UK, 1997, p. 05
11. Apetrei V, Filote C, Graur A (2014) Harmonic Analysis Based on Discrete Wavelet Transform in Electric Power Systems. 2014 9th International Conference on Ecological Vehicles and Renewable Energies, EVER 2014. https://doi.org/10.1109/EVER.2014.6844070
12. Rabi BJ, Sekar TC (2013) A review and study of harmonic mitigation techniques. In: Proceedings—ICETEEEM 2012, international conference on emerging trends in electrical engineering and energy management. https://doi.org/10.1109/ICETEEEM.2012.6494450
13. Rao RK, Ram SST, Kumar G (2011) Effective harmonic mitigation techniques using wavelets based analysis. Int J Eng Sci Technol

14. College), Qom, Iran; https://www.elsevier.es/en-revista-journal-applied-research-technology-jart-81-articulo-comparison-power-quality-indices-apparent-S1665642317300408
15. Ashok Kumar L, Indragandhi V, Pius S (2020) Design and implementation of advanced harmonic filter for variable frequency automotive drives. EPE J. https://doi.org/10.1080/09398368.2020.1750866
16. Lee SS, Yang Y, Yam S, Lee K (2020) A novel boost cascaded multilevel inverter. IEEE Trans Industri Electron 1–1. https://doi.org/10.1109/TIE.2020.3016252
17. Montaño J, Borras D, Bravo J, Castilla M, Lopez A, Gutiérrez J (2010) Wavelet-fourier analysis of electric signal disturbances. 34-954552831
18. Schwanz D, Bollen M, Larsson A (2016) A review of solutions for harmonic mitigation. In: 2016 17th international conference on harmonics and quality of power (ICHQP), Belo, Horizonte, pp 30–35. https://doi.org/10.1109/ICHQP.2016.7783422
19. Nandakumar S, Prabhuraj S, Sakda S (2017) Two-phase interleaved boost converter using coupled inductor for fuel cell applications. Energy Procedia1 38:199–204. https://doi.org/10.1016/j.egypro.2017.10.150. ISSN 1876 6102
20. Mohammad RM, Mohammad AB (2019) Switched linear control of interleaved boost converters. Int J Electri Power Energy Syst 109:526–534. https://doi.org/10.1016/j.ijepes.2019.02.030. ISSN 0142-0615
21. Chen C, Wang C, Feng H (2009) Research of an interleaved boost converter with four interleaved boost convert cells. 396–399. https://doi.org/10.1109/PRIMEASIA.2009.5397361
22. Kazem HA (2013) Harmonic mitigation techniques applied to power distribution networks. Adv Power Electro https://doi.org/10.1155/2013/591680
23. Pothisarn C, Klomjit J, Ngaopitakkul A, Jettanasen C, Asfani DA, Yulistya, Negara I (2020) Comparison of various mother wavelets for fault classification in electrical systems Appl Sci 10:1203 https://doi.org/10.3390/app10041203
24. Avdaković S, Čišija N (2015) Wavelets as a tool for power system dynamic events analysis—state of-the-art and future applications. J Electri Syst Informat Technol 2(1):47–57. ISSN 2314-7172
25. Asad, Engr, Hassan R, Sherwani F (2014) An analytical comparison between open loop, PID and fuzzy logic based DC-DC boost convertor. Int J Electri Ro Electron Commun Eng 8
26. Rao RK, Ravi GK (2011) Effective harmonic mitigation techniques using wavelets based analysis
27. Sekar T, Justus Rabi B (2012) A review and study of harmonic mitigation techniques. In: 2012 international conference on emerging trends in electrical engineering and energy management (ICETEEEM), 93–97
28. Fleet P (2019) Discrete wavelet transformations. https://doi.org/10.1002/9781119555414

Prof. Harwinder Karwal, M.Sc. (Chemistry), M.Sc. (Computer Science), M.B.A. (HR & Marketing) Multi-disciplinary Research Scholar having a wid-range experience in Science & Technology with keen interest inGreen Computing, Cybersecurity, AI, ME and Neural Networks to solve multi-disciplinary problems and an established programmer. GNA University, Pb. INDIA.

Dr. Umesh Sehgal Ph.D. M.Phil., M.Sc. (Computer Science), M.C.A. Assistant Professor, Research Scholar having a wide-range experience in Science & Technology with keen interest inGreen Computing, Cybersecurity, AI, ME and Neural Networks to solve multidisciplinary problems, having more than 200 publications in national & International journals GNA University, Pb. INDIA.

Prof. Twinkle Bedi, M.Tech. (Computer Science Engineering) MBA & specialization in Data Analytics from Punjab University Chandigarh; Multidisciplinary Research Scholar having a wide-range experience in Science & Technology with keen interest in Green Computing, Cybersecurity, AI, ME and Neural Networks to solve multidisciplinary problems in fashion technology & patterns, having more than 10years of teaching to UG and PG students of Computer science Engineering at Punjab University Chandigarh, INDIA.

Chapter 17
Classification of Herbal Plant and Comparative Analysis of SVM and KNN Classifier Models on the Leaf Features Using Machine Learning

Priya Pinder Kaur and Sukhdev Singh

1 Introduction

In today's era of the modern and technological world in which technology is used everywhere for ease of complex problem solution, herbal plants are available in all the regions of the country not in one piece of world but plants are obtained everywhere where there is life. Herbal plants are mainly used to cure diseases from ancient times to fight various diseases, which were not cured by other medicinal ailments. Herbal plants are considered as most useful plants for medicinal purposes and also to clean air, which disinfect germs and other harmful air pollutants for living species. These plants not only provide food, fruits, shelter but also used for treatment purposes. Shape is considered as an important feature for the identification and classification of plant species. Most researchers used this feature in their research as it is available all the time. In this paper, mainly features are extracted from herbal leaves and then machine is trained. Afterward, dataset is tested and classifies those leaves to obtain optimum results using SVM and KNN classifier.

2 Literature Review

Today researchers carried out their research by using various techniques of machine learning and in the deep learning field. Some of the studies already studied are discussed below related to machine learning techniques to classify and recognition herbal plants through various features of plant leaves.

P. P. Kaur (✉)
Punjabi University Patiala, Punjab, India

S. Singh
M. M. Modi College, Patiala, Punjab, India

N. Marriwala et al. (eds.), *Soft Computing for Intelligent Systems*, Algorithms for Intelligent Systems, https://doi.org/10.1007/978-981-16-1048-6_17

Kaur et al. [1] It purposed that different machine learning techniques used for the recognition of herbal plants and various methods were discussed that are suitable identification of herbal plant species and which physical features are considered as the most important for the plant to be recognized in image processing, machine learning and in deep learning.

Singh et al. [2] The paper gives a comparative study of features were discussed, which gives better result and used most widely in the field of computer vision and image processing for recognition and identification of herbal plants. Shape, color, texture and vein features were discussed in detail.

Singh et al. [3] In this review on plant recognition system on the basis of leaf done in which different methods for classification of leaf on the basis of their shape, color, texture and vein were discussed with their problem addressed and how they solve using best method/technique for recognition, classification and identification of plant using machine learning.

Ramesh et al. [4] Proposed a system for identifying healthy and diseased leaf through random forest algorithm. 160 leaves of papaya are contained in the dataset and texture, color and shape features were extracted. For object detection, the histogram of an oriented gradient method is used. Accuracy attained through random forest was 70%, it improved by combining local features with global features such as SURF (speed up robust features) and SIFT (scale invariant feature transform), DENSE along with BOVW (Bag Of Visual Word).

Vijayshree et al. [5] Purposed automatic system identifies and classifies herbal medicinal plants. 500 leaves along with 21 features containing 50 different species were extracted using color, texture and shape. The experimental output shows that using neural network for identification and classification of texture feature resulted in 93.3% accuracy, while using all features, it gives 99.2% accuracy.

Dahigaonkar et al. [6] Identified ayurvedic medicinal plants used leaves on the basis of their shape, color and texture feature. A total of 32 different plants are classified using features such as solidity, eccentricity, entropy, extent, standard deviation, contrast, equivalent diameter, mean and class. They attained through support vector machine classifier as 96.66% of accuracy

Kan et al. [7] Purposed classification of herbal plants using automated system, which used leaves. Shape-5 and texture-10 features were extracted and implemented using SVM classifier for classification of the plant. From medicinal plant specimen library of Anhui University of Traditional Chinese Medicine were taken and its dataset contained 240 leaves. The experiment resulted into 93.3% recognition rate.

De Luna et al. [8] Developed the system for the identification of Philippine herbal medicine plant through leaf features by ANN. Dataset contains 600 images from 12 different herbal plants having 50 from each. Basic features, length, width, circularity, convexity, aspect ratio, area, perimeter of convex hull, solidity, vein density, are used that defined the venation, shape and structure of leaf. The classification and analysis of data are done through Python and Matlab. The experiments conducted for the identification of herbal plants through various algorithms mentioned as—Naïve Bayes, logistic regression, KNN, classification and regression trees, linear discriminant analysis, SVM, Neural network and resulted in 98.6% recognition.

Isnanto et al. [9] Proposed a system for recognition of herbal plants using their leaf patterns. Extracted features of the image used Hu's seven moments invariant, image segmentation-Otsu method and for recognition—Euclidean and Canberra Distance were implemented. The experimental results gave 86.67% accuracy in the identification system through Euclidean Distance while using Canberra distance 72% accuracy was achieved. 100% accuracy attained for nine types of leaves using Euclidean distance and five types of leaves using Canberra distance.

3 Proposed Methodology

In this proposed methodology, the colored image of leaf from herbal plant is captured through a mobile camera, the leaf is placed on white plain paper as a background. Then the image is properly oriented in same direction so that when length and width, area and pixels of different quadrants of leaf were calculated, it does not carry any problem. The extra white space cropped and resized the original colored image to 20%, then applying preprocessing convert that colored image to HSV, gray scale and binary image. Then features are extracted from those leaves, after collecting features dataset is used for training and testing purposes. For training, 80% of data are used and 20% of data are used for testing. Classifier, namely KNN and SVM, has been used to evaluate the accuracy and classification report, which gives precision, support, F1-value and recall. The experiment was conducted using Python (Fig. 1).

3.1 Image Acquisition

The sample of leaves contained in the dataset obtained from herbal plants collected from herbal and botanical gardens of Chandigarh and Punjab. Different samples of leaf images are collected by using a mobile camera in natural daylight. The leaves of various herbal plants are collected for the experimental purpose for training and testing to classify plant through their leaves. Below is the table represented the sample of leaves from different plants, namely, haar shinghaar, periwinkle, currey patta, rose and tulsi (Table 1).

3.2 Pre-Processing

This step is done after collected images of leaves, which involves image enhancement, quality improvement and segmented the desired area of leaf to find out the features. The original image of herbal plant leaf has been taken on the plain white paper background; the extra white space has been cropped and resized the image for faster, takes less memory space and for easy evaluation purpose (Fig. 2).

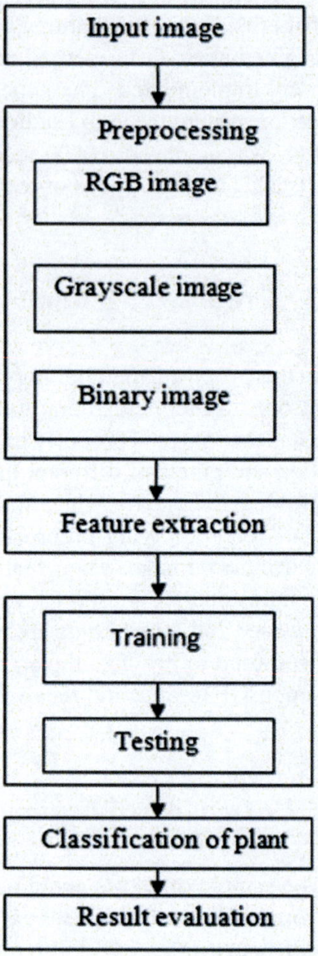

Fig. 1 Proposed methodology follows

3.3 *Feature Extraction*

Feature extraction is the main part to recognize the plant by extracting various features of plant leaf. Different shapes and sizes of specific type of plant leaf feature have been used for classification purpose. Various morphological features have been calculated such as height and width of leaf, total area of rectangle of enclosing leaf, total area of leaf, perimeter of leaf in pixels, four quadrants of rectangle in pixels, four quadrants of leaf in pixels and the remaining extra pixels in each quadrant. In Fig. 3, sample of leaf shows their description.

Figure 4 shows the different quadrants 1, 2, 3 and 4 in which rectangle and leaf pixels are calculated, which comes under different quadrants. In Fig. 5, the screenshot of features extraction of single leaf is represented.

Table 1 Sample of leaves from five different plants

3.4 Classification

For the classification purpose, SVM and KNN classifier is used in which features extracted from leaf are stored in the dataset and then that dataset is used for training and testing purposes. 80% data are used for training and 20% data are used for testing, then using SVM and KNN classifier model for classification of herbal plant. The machine is trained to obtain more accuracy using various features.

3.5 Performance Evaluation

After the classification of herbal plant to which category plant belongs, the performance has been evaluated based on the results given by different classifier models, SVM and KNN. It shows that how many plants belong to specific category and generates confusion matrix and classification report that contains precision, recall, F1-score and support values. The below table shows the number of leaves taken for training and testing and the category to which they belong, number of features used for classifying them using SVC model and KNN model (Table 2).

Fig. 2 Images of haar shinghar, periwinkle, currey patta and tulsi depicting RGB image, Gray scale image and binary image of different leaves of herbal plants

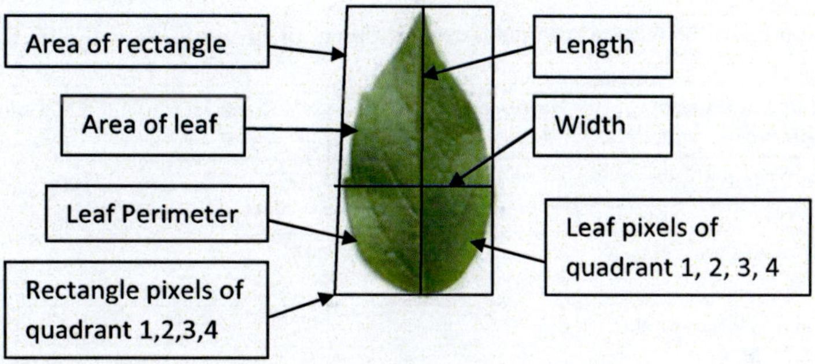

Fig. 3 Sample of leaf shows different features that are calculated

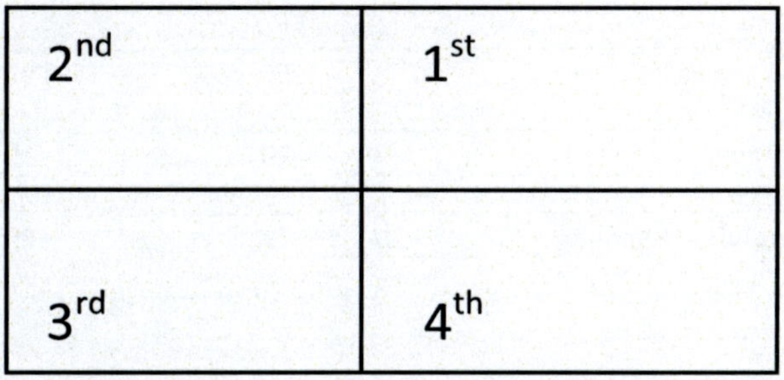

Fig. 4 Shows different quadrants

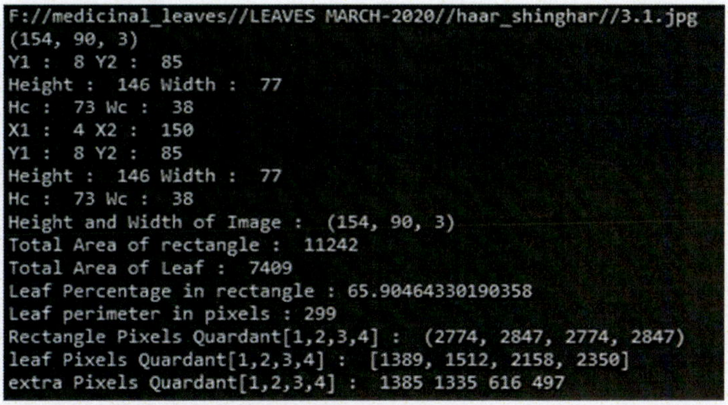

```
F://medicinal_leaves//LEAVES MARCH-2020//haar_shinghar//3.1.jpg
(154, 90, 3)
Y1 :  8 Y2 :  85
Height :  146 Width :  77
Hc :  73 Wc :  38
X1 :  4 X2 :  150
Y1 :  8 Y2 :  85
Height :  146 Width :  77
Hc :  73 Wc :  38
Height and Width of Image :  (154, 90, 3)
Total Area of rectangle :  11242
Total Area of Leaf :  7409
Leaf Percentage in rectangle : 65.90464330190358
Leaf perimeter in pixels : 299
Rectangle Pixels Quardant[1,2,3,4] :  (2774, 2847, 2774, 2847)
leaf Pixels Quardant[1,2,3,4] :  [1389, 1512, 2158, 2350]
extra Pixels Quardant[1,2,3,4] :  1385 1335 616 497
```

Fig. 5 Represent the sample output of leaf feature extraction

4 Simulation Results and Discussion

Results while using SVM model and KNN model on a different combination of plants using four features are stated below (Fig. 6):

Table 2 Represent the accuracy attained using different plants by using two different classifier models

Name of plant and their number of leaves	Number of features	SVM model (%)	KNN model (%)
Periwinkle-31 leaves and tulsi-25 leaves	4	100	100
Haar shinghar-44 and tulsi-25 leaves	4	93	86
Haar shinghar-79 leaves and periwinkle-75 leaves	4	97	90
Haar shinghar-30, tulsi-25 and periwinkle-30	4	83	89
Periwinkle-31 leaves and tulsi-25 leaves	6	100	100
Haar shinghar-44 and tulsi-25 leaves	6	93	93
Haar shinghar-79 and periwinkle-75 leaves	6	90	90
Haar shinghar-34, tulsi-25 and Periwinkle-35	6	79	89

Result while using SVM model and KNN model on a different combination of plants using six features mentioned below (Fig. 7):

Discussion—1: (for SVM Model)

Case 1. Periwinkle 31 leaves and tulsi 25 leaves:

Four features (height, width, area of leaf, perimeter of leaf in pixels)—resulted into 100% accuracy and six features (height of leaf, width of leaf, perimeter of leaf in pixels, area of leaf, area of rectangle enclosing leaf and the percentage of leaf in the rectangle)—resulted into 100% accuracy. Even with the increase in the number of features, the accuracy remains same.

Case 2. Haar shinghar 44 leaves and tulsi 25 leaves:

Four features—resulted into 93% accuracy and six features—resulted into 93% accuracy. Even with the increase in the number of features, the accuracy remains same.

Case 3. Haar shinghar 79 leaves and periwinkle 75 leaves:

Four features—resulted into 97% accuracy and six features—resulted into 90% accuracy. With the increase in the number of features, the accuracy was decreased by 7% analyzed that when the features are increased it becomes more complex for system to detect six features accurately.

a

	precision	recall	f1-score	support
haar shinghar	1.00	0.75	0.86	8
periwinkle	0.90	1.00	0.95	9
tulsi	0.00	0.00	0.00	1
accuracy			0.83	18
macro avg	0.63	0.58	0.60	18
weighted avg	0.89	0.83	0.85	18

b

	precision	recall	f1-score	support
haar shinghar	1.00	0.75	0.86	8
periwinkle	1.00	1.00	1.00	10
tulsi	0.00	0.00	0.00	0
accuracy			0.89	18
macro avg	0.67	0.58	0.62	18
weighted avg	1.00	0.89	0.94	18

c

	precision	recall	f1-score	support
haar shinghar	1.00	0.94	0.97	17
periwinkle	0.93	1.00	0.97	14
accuracy			0.97	31
macro avg	0.97	0.97	0.97	31
weighted avg	0.97	0.97	0.97	31

d

	precision	recall	f1-score	support
haar shinghar	0.88	0.93	0.90	15
periwinkle	0.93	0.88	0.90	16
accuracy			0.90	31
macro avg	0.90	0.90	0.90	31
weighted avg	0.91	0.90	0.90	31

e

	precision	recall	f1-score	support
haar shinghar	1.00	0.91	0.95	11
tulsi	0.75	1.00	0.86	3
accuracy			0.93	14
macro avg	0.88	0.95	0.90	14
weighted avg	0.95	0.93	0.93	14

f

	precision	recall	f1-score	support
haar shinghar	0.90	0.90	0.90	10
tulsi	0.75	0.75	0.75	4
accuracy			0.86	14
macro avg	0.82	0.82	0.82	14
weighted avg	0.86	0.86	0.86	14

g

	precision	recall	f1-score	support
periwinkle	1.00	1.00	1.00	8
tulsi	1.00	1.00	1.00	4
accuracy			1.00	12
macro avg	1.00	1.00	1.00	12
weighted avg	1.00	1.00	1.00	12

h

	precision	recall	f1-score	support
periwinkle	1.00	1.00	1.00	8
tulsi	1.00	1.00	1.00	4
accuracy			1.00	12
macro avg	1.00	1.00	1.00	12
weighted avg	1.00	1.00	1.00	12

Fig. 6 **a** Results of haar shinghar, periwinkle and tulsi for SVM model, **b** Results of haar shinghar, periwinkle and tulsi for KNN model, **c** Results of haar shinghar and periwinkle for SVM model, **d** Results of haar shinghar and periwinkle for KNN model, **e** Results of haar shinghar and tulsi for SVM model, **f** Results of haar shinghar and tulsi for KNN model, **g** Results of periwinkle and tulsi for SVM model, **h** Results of periwinkle and tulsi for KNN model

Case 4. Haar shinghar 30 leaves, periwinkle 30 leaves and tulsi 25 leaves:

Four features—resulted into 83% accuracy due to introduce of one extra plant and six features—resulted into 79% accuracy. With the increase in the number of

a

	precision	recall	f1-score	support
haar shinghar	1.00	0.91	0.95	11
tulsi	0.75	1.00	0.86	3
accuracy			0.93	14
macro avg	0.88	0.95	0.90	14
weighted avg	0.95	0.93	0.93	14

b

	precision	recall	f1-score	support
haar shinghar	1.00	0.91	0.95	11
tulsi	0.75	1.00	0.86	3
accuracy			0.93	14
macro avg	0.88	0.95	0.90	14
weighted avg	0.95	0.93	0.93	14

c

	precision	recall	f1-score	support
haar shinghar	1.00	0.84	0.91	19
periwinkle	0.80	1.00	0.89	12
accuracy			0.90	31
macro avg	0.90	0.92	0.90	31
weighted avg	0.92	0.90	0.90	31

d

	precision	recall	f1-score	support
haar shinghar	1.00	0.84	0.91	19
periwinkle	0.80	1.00	0.89	12
accuracy			0.90	31
macro avg	0.90	0.92	0.90	31
weighted avg	0.92	0.90	0.90	31

e

	precision	recall	f1-score	support
haar shinghar	1.00	0.67	0.80	12
periwinkle	1.00	1.00	1.00	7
tulsi	0.00	0.00	0.00	0
accuracy			0.79	19
macro avg	0.67	0.56	0.60	19
weighted avg	1.00	0.79	0.87	19

f

	precision	recall	f1-score	support
haar shinghar	1.00	0.80	0.89	10
periwinkle	1.00	1.00	1.00	7
tulsi	0.50	1.00	0.67	2
accuracy			0.89	19
macro avg	0.83	0.93	0.85	19
weighted avg	0.95	0.89	0.91	19

g

	precision	recall	f1-score	support
periwinkle	1.00	1.00	1.00	8
tulsi	1.00	1.00	1.00	4
accuracy			1.00	12
macro avg	1.00	1.00	1.00	12
weighted avg	1.00	1.00	1.00	12

h

	precision	recall	f1-score	support
periwinkle	1.00	1.00	1.00	8
tulsi	1.00	1.00	1.00	4
accuracy			1.00	12
macro avg	1.00	1.00	1.00	12
weighted avg	1.00	1.00	1.00	12

Fig. 7 **a** Results of haar shinghar and tulsi for SVM model, **b** Results of haar shinghar and tulsi for KNN model, **c** Results of haar shinghar and Periwinkle for SVM model, **d** Results of haar shinghar and Periwinkle for KNN model, **e** Results of haar shinghar, Periwinkle and tulsi for SVM model, **f** Results of haar shinghar, Periwinkle and tulsi for KNN model, **g** Results of periwinkle and tulsi for SVM model, **h** Results of periwinkle and tulsi for KNN model

features, the accuracy was decreased by 4.3% and analyzed that when the features are increased, it becomes complex for system to classify them.

Discussion—2: (for KNN Model)

Case 1. Periwinkle 31 leaves and tulsi 25 leaves:

Four features—resulted into 100% accuracy and six features—resulted into 100% accuracy. Even with the increase in the number of features, the accuracy remains same.

Case 2. Haar shinghar 44 leaves and tulsi 25 leaves:

Four features—resulted into 86% accuracy and six features—resulted into 93% accuracy. Even with the increase in the number of features, the accuracy increased by 7%.

Case 3. Haar shinghar 79 leaves and periwinkle 75 leaves:

Four features—resulted into 90% accuracy and six features—resulted into 90% accuracy. With the increase in the number of features, the accuracy remains same.

Case 4. Haar shinghar 30 leaves, periwinkle 30 leaves and tulsi 25 leaves:

Four features—resulted into 89% accuracy and six features—resulted into 89% accuracy. With the increase in the number of plants and number of features, there is no change in accuracy, it remains same.

Discussion—3

We are considering two plants, i.e. Case—I, II, III and comparing both SVM and KNN models. Analysis shows that SVM model is more effective considering four features, while both are equally effective in accuracy considering six features.

When we consider three plants in this case, i.e. Case—IV, KNN model is more effective over SVM model considering both four features and six features.

5 Conclusion

In this paper, the main focus is on comparative results of both the classifiers SVM and KNN classifier models, both the models give good results in their different respective cases mentioned above. Features of herbal plants were extracted from their leaf, shape is considered as the main feature on the basis of that other morphological features such as height of leaf, width of leaf, area of leaf, perimeter of leaf in pixels, area of rectangle enclosing leaf and the percentage of leaf in the rectangle are calculated. On the basis of these features of leaf dataset, two classifier models are used for the classification of different types of plants. Then both the models were provided 80% of dataset for training and 20% of dataset used for testing purpose. After training and testing, the SVM (support vector machine) model works well on two different combinations of plants using four features (height, width, area of leaf and perimeter of leaf in pixels) resulted into Case I—100%, Case II—93%, Case III—97%. Accuracy using for six features (height of leaf, width of leaf, area of leaf, perimeter of leaf in pixels, area of rectangle enclosing leaf and the percentage of leaf in the rectangle): Case I—100%, Case II—93%, Case III—90% . While using another model KNN (k-nearest neighbor) model, it gives a better result than SVM model for two different combination of plants, there is improvement in accuracy rate from four features

Table 3 Comparison of accuracy of both the models (SVM and KNN)

	SVM		KNN	
	4 Features (%)	6 Features (%)	4 Features (%)	6 Features (%)
Case I	100	100	100	100
Case II	93	93	86	93
Case III	97	90	90	90

Table 4 Comparison of result for another case for both the models

	SVM		KNN	
	4 Features (%)	6 Features (%)	4 Features (%)	6 Features (%)
Case IV	83%	79%	89%	89%
Remarks	Figure 6a	Figure 6b	Figure 7e	Figure 7f

(Case I 100%, Case II 86%, Case III 90%) to six features (Case I 100%, Case II 93%, Case III 90%), results refer to Table 3. SVM classifier is more effective than KNN classifier model for two plants having four features and accuracy remains same in the case when features are increased by 2 (i.e. six features). In Table 4, it shows when the number of plants increased then the SVM model gives less accuracy than KNN model, in this case, four features give 83% accuracy using SVM and 79% accuracy using six features. In the case of KNN classifier model, four give 89% same as six features resulted.

References

1. Kaur PP, Singh S, Pathak M (2020) Review of machine learning herbal plant recognition system. In: 2nd proceedings of international conference on intelligent communication and computational research ICICCR-India
2. Singh S, Kaur PP (2019) A study of geometric features extraction from plant leafs. J Informat Computat Sci 9(7):101–109
3. Singh S, Kaur PP (2019) Review on plant recognition system based on leaf features. J Emerging Technol Innov Res (JETIR), 6(3):352–361
4. Ramesh S, Hebbar R, Niveditha M, Pooja R, Prasad Bhat N, Shashank N, Vinod PV (2018) Plant disease detection using machine learning. In: IEEE international conference on design innovations for 3Cs compute communicate control, pp 41–45
5. Vijayshree T, Gopal A (2018) Identification of herbal plant leaves using image processing algorithm: review. Res J Pharm Biol Chem Sci 9(4):1221–1228
6. Dahigaonkar TD, Kalyana RT (2018) Identification of ayurvedic medicinal plants by image processing of leaf samples. Int Res J Eng Technol (IRJET) 5(5):351–355
7. Kan HX, Jin L, Zhou FL (2017) Classification of Medicinal Plant leaf image based on multi-feature extraction. Pattern Recognit Image Anal 27(3):581–587

8. De Luna RG, Baldovino RG, Cotoco EA, De Ocampo ALP, Valenzuela IC, Culaba AB, Gokongwei EPD (2017) Identification of Philippine herbal medicine plant leaf using artificial neural network. In: IEEE 9th international conference on humanoid, nanotechnology, information technology, communication and control, environment and management (HNICEM)
9. Isnanto RR, Zahra AA, Julietta P (2016) Pattern recognition on herbs leaves using region-based invariants feature extraction. In: 3rd international conference on information technology and computer, electrical engineering (ICITACEE), (1), pp 455–459

Chapter 18
Face Recognition in Unconstrained Environment Using Deep Learning

Rajeshwar Moghekar and Sachin Ahuja

1 Introduction

Face recognition is one of the interesting and challenging problems in the field of computer vision. Among the various biometric features used to identify an individual like iris, fingerprint, palm print, face etc., face has an advantage over other biometric techniques in that it does not require active cooperation of the person [1]. This makes it best suitable for identifying person from the images captured from surveillance cameras. Surveillance cameras installed in both public and private places are being monitored by human operator to recognize or identify the persons of interest. Monitoring by human operator has many drawbacks in terms of scalability and reliability. Face recognition techniques have achieved remarkable results in constrained environment. Face recognition in unconstrained environment has many challenges in the form of variations in pose, resolution, illumination and occlusion, etc. [2, 3]. The availability of large face datasets in the public domain and its wide applications in security encouraged the researchers to focus on this area. CelebFaces + database introduced in 2014 consists of 202,599 images of 10,177 subjects [4]. Labelled face in the wild dataset contains 13,233 images of 5739 subjects and was created in 2007 [5]. CAS-PEAL is a database of face, which was created by placing nine cameras and consists of 99,594 face images with 1040 subjects [6]. Indian Movies Face database (IMFDB) is created using manual cropping of face images from the videos/movies, which makes it one of the most challenging face datasets [7].

R. Moghekar (✉)
Hyderabad Institute of Technology and Management, Hyderabad, Telangana, India
e-mail: rajeshwarm.cse@hitam.org

S. Ahuja
Chitkara University Institute of Engineering & Technology, Chitkara University, Rajpura, Punjab, India

N. Marriwala et al. (eds.), *Soft Computing for Intelligent Systems*, Algorithms for Intelligent Systems, https://doi.org/10.1007/978-981-16-1048-6_18

The present study has two goals. The first goal is to create the dataset with a combination of face images from the internet and face images captured from real-time video, which is discussed in Sect. 3. This face dataset is unique in its nature as the image quality of the face images downloaded from web and those images extracted from the video have a lot of variations. The second goal is to develop a deep learning model for face recognition and train on the dataset created, which is robust to the challenges faced in unconstrained environment, i.e. pose, illumination, resolution, etc. as discussed in Sect. 4. Convolution neural network was considered for study as the results achieved by it on Imagenet Large Scale Visual Recognition Challenge (ILSVRC) clearly indicate that they are the best choice for computer vision applications like Image classification and object detection [8]. Section 5 discusses the results achieved using the proposed model, which is followed by conclusion in Sect. 6.

2 Related Work

There are many face recognition methods proposed in the literature, which can be classified based on whether the features of the face are first extracted and then fed to the model or the model extracts the features on its own. The facial descriptors like SIFT [9], HOG [10], SURF [11] and LBP [12] fall under the category where the features are extracted and fed to the model. But, the approach to perform face recognition has changed after AlexNet, a deep neural network won the ImageNet competition by a large margin [13]. These deep learning methods based on convolutional neural networks have multiple layers, which extract the features and learn representations that have different levels of abstraction. DeepFace [14] and DeepId series [15, 16] are deep learning models that extract the features on their own. They provide the benchmark results on the labelled face in the wild.

Karaahan et al. [17] analyzed the impact of low-quality images due to noise, distortions, blur and occlusion on the performance of the deep learning models. They evaluated using three deep neural networks on LFW dataset. The results show that blur, occlusion and noise reduce the performance of deep models but colour distortion has no effect on them. They even carried out by occluding different parts of the face like sunglasses, nose, mouth and other parts. They observed VGGFace was affected more by nose occlusion when compared with eyes covered with sunglasses. Liu et al. [18] addressed the challenge of face attributes prediction from the wild by using LeNet to detect the face and Alexnet to retrieve the attributes/representation of face. LeNet as a face detector performed better than other face detectors like SURF Cascade and Face++ . Fontaine et al. [19] developed face recognition using very few images, which differ in expressions, orientation and lighting conditions. They used modified Robust Sparse Coding after aligning the faces mesh wrap, which deforms the input image such that its features match with that of the reference image. Wang et al. [20] collected face images from surveillance video and applied fine-tuning to VGGFace and achieved face recognition accuracy of 92.1%.

Fig. 1 Images downloaded from Internet

3 Creating Dataset

To create a unique dataset other than the existing datasets, the images were collected from internet and from the video captured from the camera placed on the university campus.

3.1 Dataset from Internet

We collected the images of 40 celebrities randomly from different categories including politicians, sports person and actors. We selected only those celebrities who had more than 400 images in the Google search. Later we used HAAR cascade face detector available in OpenCV to detect the face images from the downloaded images. It was found that the detector could not detect faces from the images as compared to deep neural network model available in OpenCV. Figure 1 shows a few images downloaded from internet. The face images in Fig. 2 were not detected by HAAR cascade but detected by deep model, which made us confine to deep model for face detection. Face images were manually checked to remove the images other than the subject under consideration as the image downloaded may have face images of others.

3.2 Dataset from Camera

We placed the camera in a fixed position in the university and recorded the video. Later we detected the faces in the video using face detector and manually segregated the face images. Subjects having more than 200 face images were only considered for inclusion in the dataset. Face images of 40 different subjects were collected. Figure 3 shows sample face images collected from video with variations in resolution, Fig. 4 shows face images with variation in lighting conditions and Fig. 5 shows face images with variations in pose.

Fig. 2 Faces in the images not detected by HAAR cascade face detector

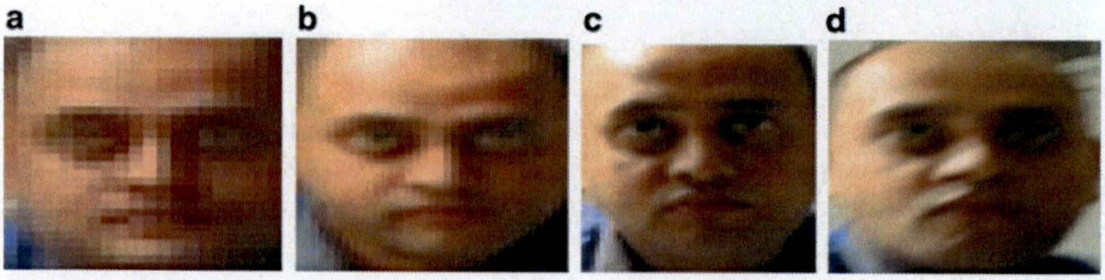

Fig. 3 **a** 20X30 Pixels, **b** 35X56 Pixels, **c** 51X72 Pixels, **d** 85X124 Pixels

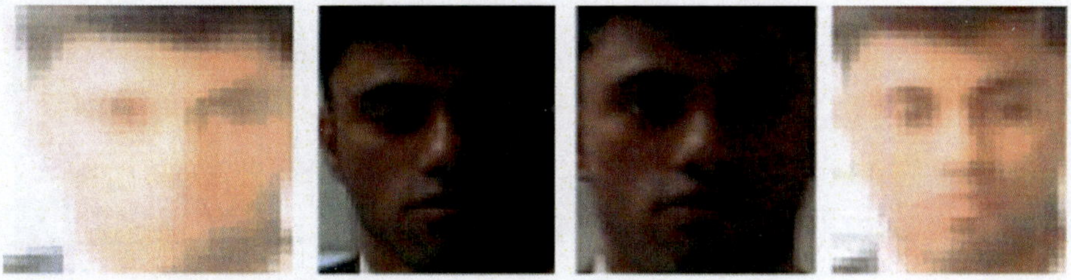

Fig. 4 Face images with variation in light conditions

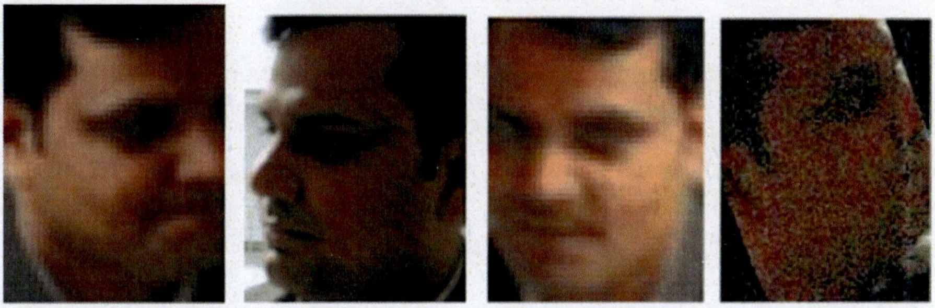

Fig. 5 Face images with variation in pose

3.3 Data Augmentation

Machine learning algorithms would be more effective when they have access to more data. In specific applications like medical-related, we get insufficient data for image classifications tasks. The models will not generalize well over the validation and test datasets if the small datasets on which the model is trained, which leads to overfitting. Overfitting can be reduced by applying regularization, dropouts and data augmentation. Perez and Wang [21] have studied the effectiveness of different data augmentation techniques on image classification tasks. The results show that augmentation of data using simple transformations like rotating; flipping and cropping improve the performance of the model. The number of face images for each subject was fixed at approximately 330. The dataset was split in the ratio of 70:15:15 with 70% for training and 15% each for validation and test. Subjects that have less number of images have been increased by applying translation, flip, scale and noise. Figure 6 shows the face images generated after applying transformation.

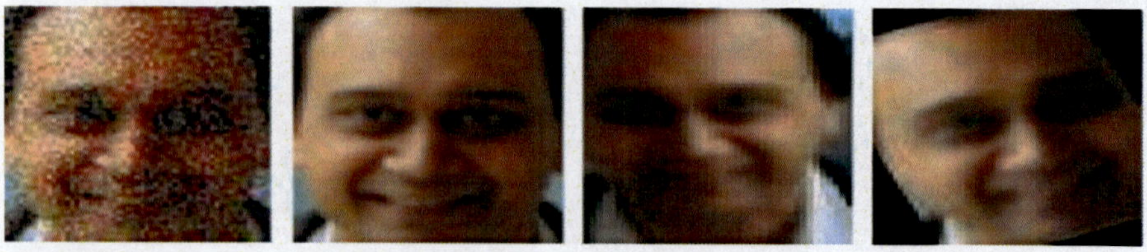

Fig. 6 Face Images after applying translation, flip, scale and noise

4 Deep Neural Network Model

Our deep neural network model 1 as shown in Fig. 7 has four convolution layers with filter size 5 × 5. Max pooling layer of size 2 × 2 after each convolution layer followed by a fully connected layer. Figure 8a and b shows that the Model 1 achieved a training accuracy of 62.7% and training loss of 1.45 and validation accuracy of 71.3% and validation loss of 1.12.

Model 2 was created as shown in Fig. 9 has eight convolution layers with filter

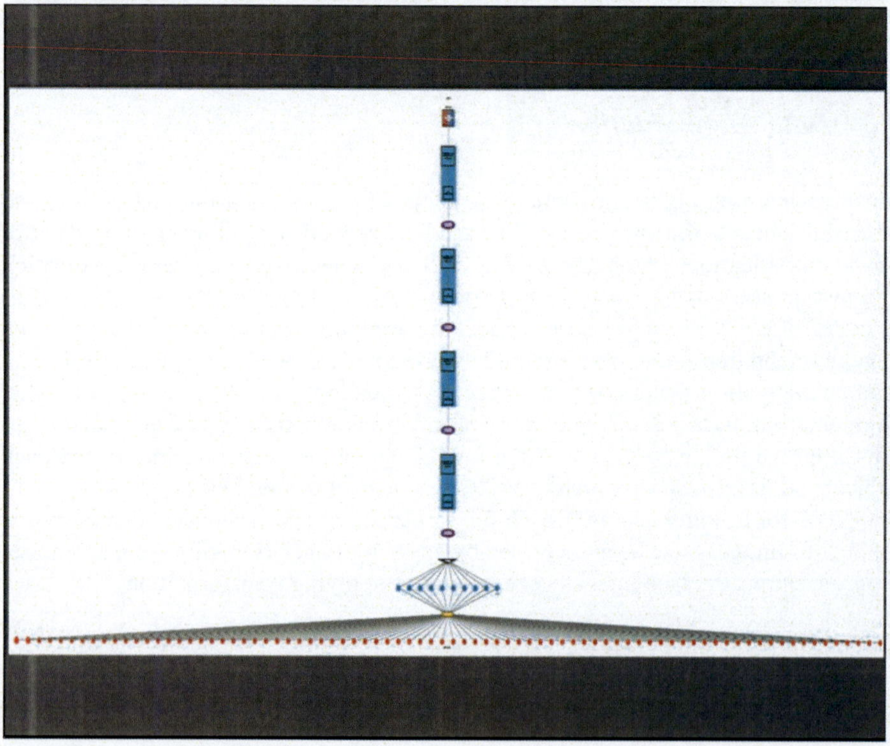

Fig. 7 Model 1 with four convolution layers

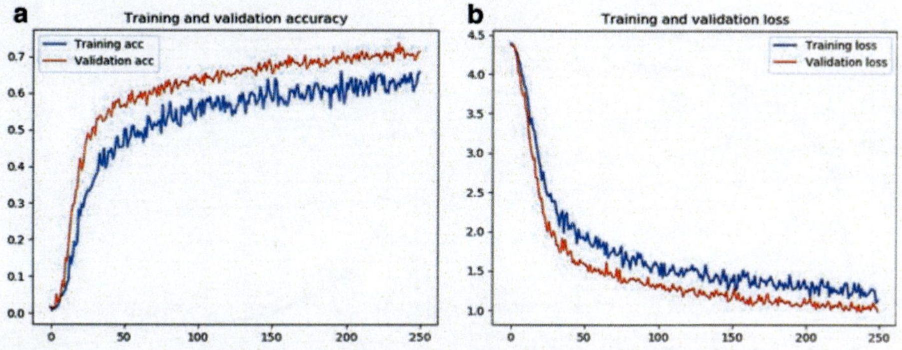

Fig. 8 **a** Training and validation accuracy, **b** Training and validation loss

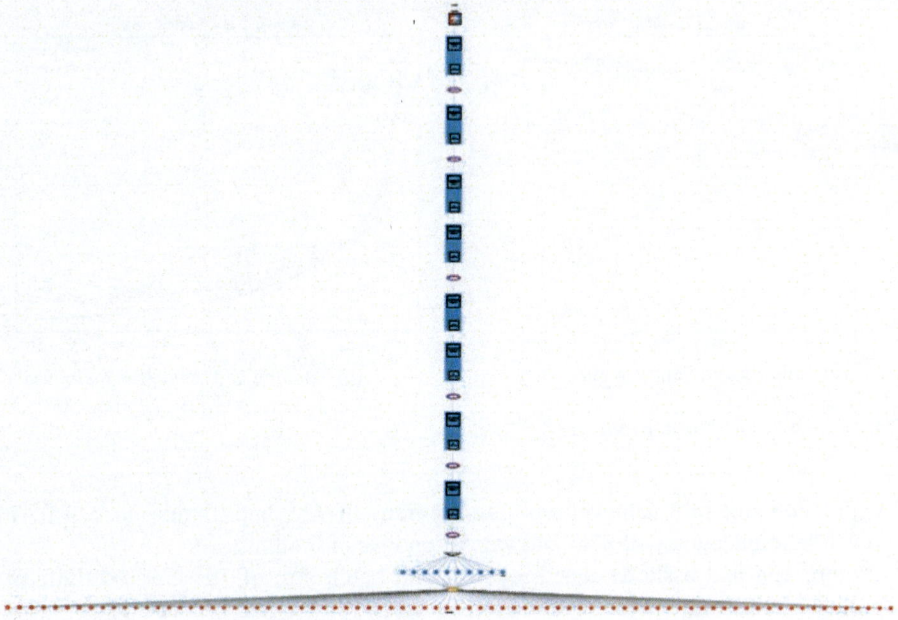

Fig. 9 Model 2 with eight convolutional layers

size 5 × 5. Max pooling layers of size 2 × 2 were used followed by a fully connected layer.

The model accepts input size 224 × 224 with RMSProp optimizer and achieved training accuracy of 90.7 and training loss of 0.38 and validation accuracy of 88.3 and validation loss of 0.67 as shown in Fig. 10a, b.

Impact of batch size on the performance of the Model 2 has been studied with batch size of 8, 16 and 32 with RMSProp optimizer. Figure 11a, b shows that Model

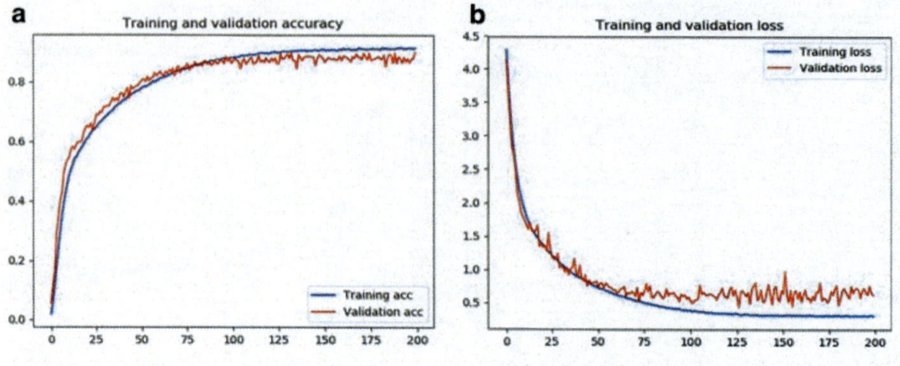

Fig. 10 **a** Training and validation accuracy, **b** Training and validation loss

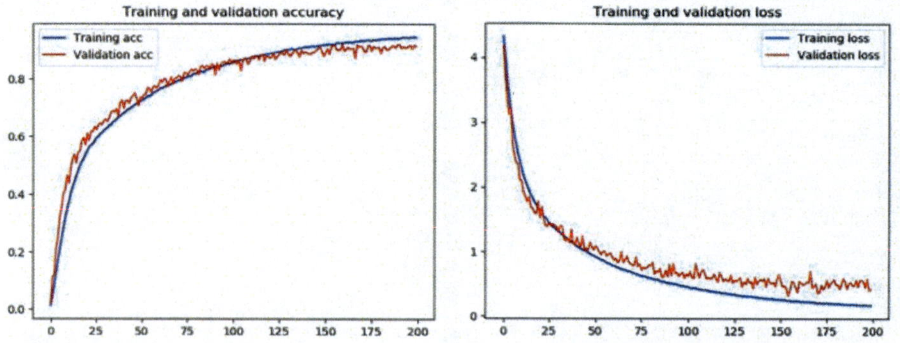

(a) Training and validation accuracy (b) Training and validation loss

Fig. 11 **a**, **b** Model 2 with Batch size 8

2 with batch size of 8 achieved training accuracy of 92.9 and training loss of 0.41 and validation accuracy of 87.3 and validation loss of 0.68.

Figure 12a and b shows that Model 2 with batch size of 16 achieved training accuracy of 93.7 and 0.34 and validation accuracy of 89.3 and loss of 0.57.

Figure 13a and b shows that Model 2 with batch size of 32 achieved training accuracy of 95.2 and loss of 0.35 and validation accuracy of 93.3 and loss of 0.69.

Figure 14a and b shows that Model 2 with batch size of 32 and Adam optimizer has achieved training accuracy of 99.94 and training loss 0.09 and validation accuracy of 99.61 and loss of 0.21 (Fig. 15).

VGGFace dataset contains 2.6 M face images with 2622 different subjects [22]. VGGFace model was fine-tuned and trained on huge dataset by keeping the weights of the initial layers, which extract the general features. Figure 16 shows the VGGFace fine-tuned model, which achieved training accuracy of 99.82 and training loss of 0.23 and validation accuracy of 99.54 and validation loss of 0.32 as depicted in Fig. 17a and b.

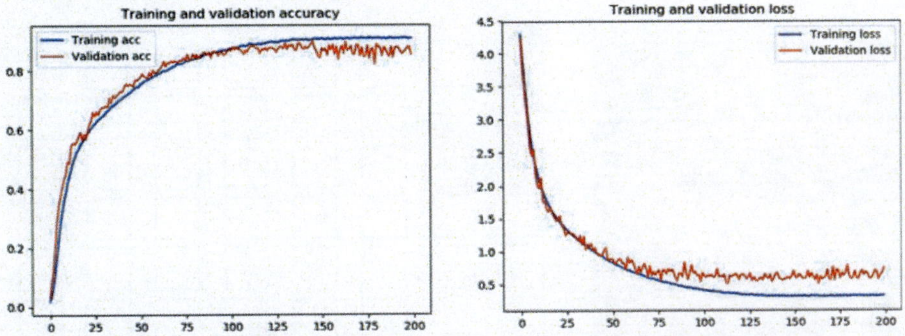

(a) Training and validation accuracy (b) Training and validation loss

Fig. 12 **a, b** Model 2 with Batch size 16

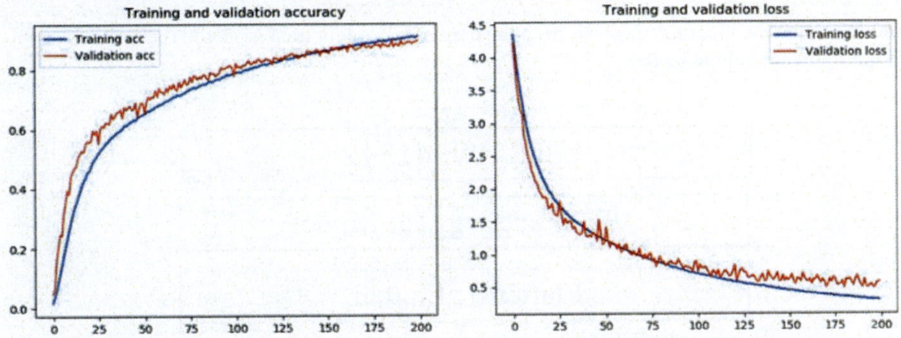

(a) Training and validation accuracy (b) Training and validation loss

Fig. 13 **a, b** Model 2 with Batch size 32

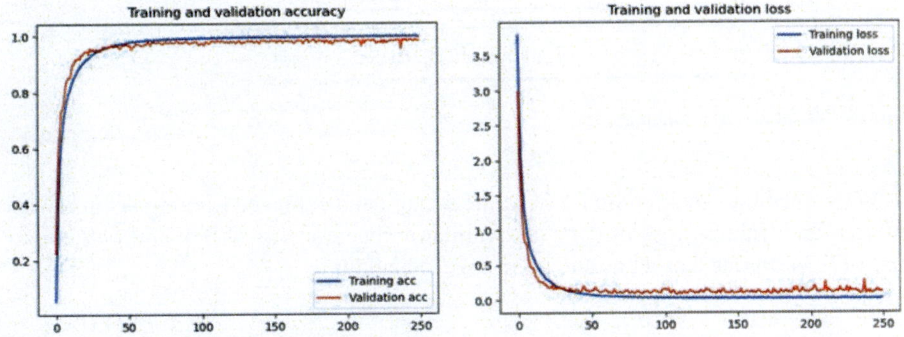

(a) Training and validation accuracy (b) Training and validation loss

Fig. 14 **a, b** Model 2 with batch size of 32 and Adam optimizer

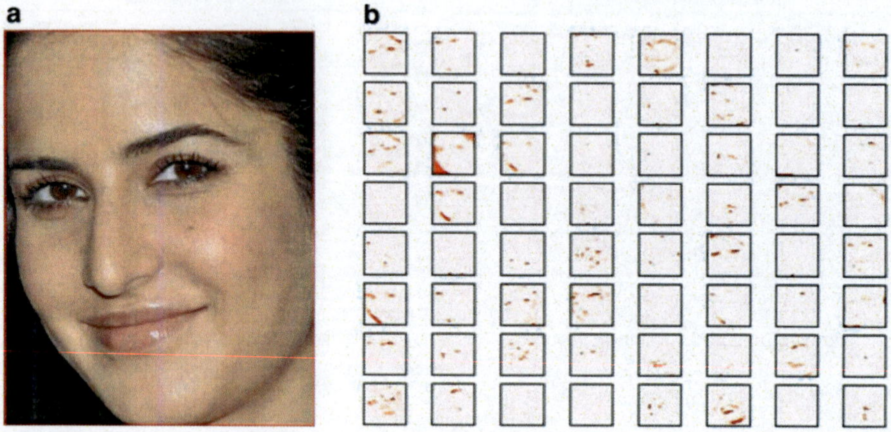

Fig. 15 **a** Sample input image given from the dataset to the proposed model **b** The features learnt by the sixth layer of the model

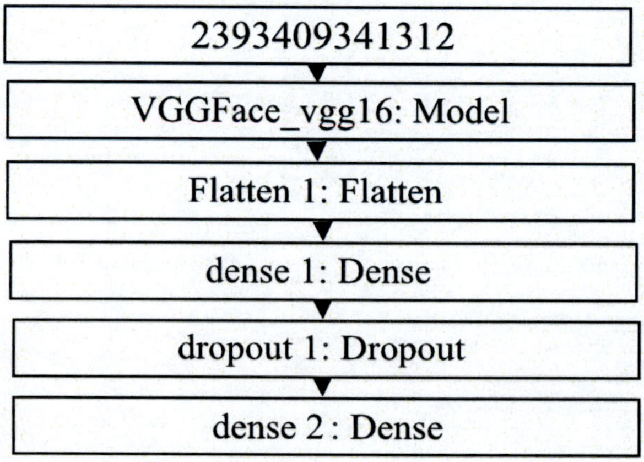

Fig. 16 VGGFace fine-tuned model

We trained the Alexnet model from the scratch and achieved training accuracy of 99.02% and training loss of 0.19 and validation accuracy of 98.1% and validation loss of 0.34 on our dataset as shown in Fig. 18a and b.

5 Results

As shown in Table 1, our proposed Model 2 which has 5,671,824 with small model size gives good results on our dataset as compared to VGGFace fine-tuned model and Alexnet. The impact of hyperparameter batch size on our model was studied and

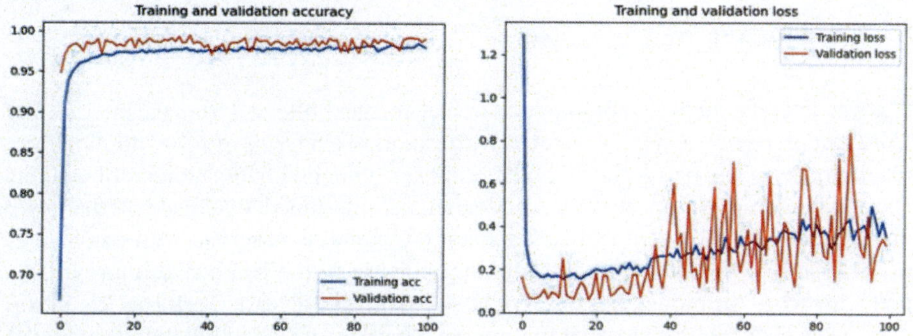

(a) Training and validation accuracy (b) Training and validation loss

Fig. 17 Results obtained using VGGFace fine-tuned model

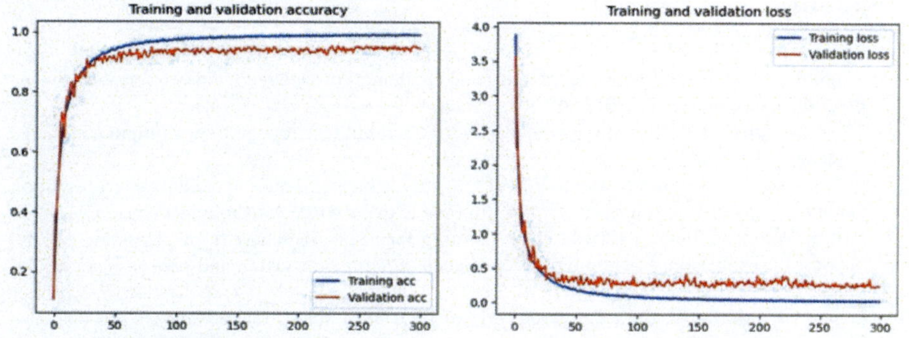

(a) Training and validation accuracy (b) Training and validation loss of Alexnet

Fig. 18 Alexnet model

Table 1 Results obtained our dataset

Model	Training accuracy	Testing accuracy	Number of parameters	Model size in MB
Alexnet	99.02	98.65	28,159,832	338
Fine-tuned VGGFace	99.82	99.68	40,487,824	293
Our model 2	99.94	99.75	5,671,824	68

results indicate as the batch size increases the performance of the model increases. When Adam optimizer is used in place of RMSProp optimizer with batch size of 32, the model training accuracy increases from 95.2 to 99.94%.

6 Conclusion

A novel dataset, which is a combination of face images collected from the internet and from that of a fixed camera, is created with variations in terms of pose, illumination, resolution, etc. The performance of the proposed model with less number of weights and model size performs better than Alexnet and fine-tuned VGGFace on the novel dataset created. The effect of batch size on the model is observed with batch sizes of 8, 16 and 32. Results demonstrate that increase in the batch size improves the performance of the model. It is observed that Adam optimizer outperforms RMSProp optimizer. There is further scope to improve the performance of the proposed model with less parameter on the testing data.

References

1. Singh S, Prasad SVAV (2018) Techniques and challenges of face recognition: a critical review. Procedia Comput Sci 143:536–543
2. Guo G, Zhang N (2019) A survey on deep learning based face recognition. Comput Vis Image Underst 189:102805
3. Lahasan B, Lutfi SL, San-Segundo R (2019) A survey on techniques to handle face recognition challenges: occlusion, single sample per subject and expression. Artif Intell Rev 52(2):949–979
4. Sun Y, Wang X, Tang X (2014) Deep learning face representation from predicting 10,000 classes. In: Proceedings of the IEEE conference on computer vision and pattern recognition, pp 1891–1898
5. Huang GB, Mattar M, Berg T, Learned-Miller E (2008) Labeled faces in the wild: a database for studying face recognition in unconstrained environments
6. Gao W, Cao B, Shan S, Chen X, Zhou D, Zhang X, Zhao D (2007) The CAS-PEAL large-scale Chinese face database and baseline evaluations. IEEE Trans Syst Man Cybernet Part A Syst Hum 38(1):149–161
7. Setty S, Husain M, Beham P, Gudavalli J, Kandasamy M, Vaddi R, Hemadri V, Karure JC, Raju R, Rajan B, Kumar V (2013) Indian movie face database: a benchmark for face recognition under wide variations. In: 2013 fourth national conference on computer vision, pattern recognition, image processing and graphics (NCVPRIPG). IEEE, pp 1–5
8. Russakovsky O, Deng J, Su H, Krause J, Satheesh S, Ma S, Huang Z, Karpathy A, Khosla A, Bernstein M, Berg AC (2015) Imagenet large scale visual recognition challenge. Int J Comput Vision 115(3):211–252
9. Geng C, Jiang X (2009) Face recognition using sift features. In: 2009 16th IEEE international conference on image processing (ICIP). IEEE, pp 3313–3316
10. Dadi HS, Pillutla GKM (2016) Improved face recognition rate using HOG features and SVM classifier. IOSR J Electron Commun Eng 11(4):34–44
11. Dreuw P, Steingrube P, Hanselmann H, Ney H, Aachen G (2009) SURF-Face: face recognition under viewpoint consistency constraints. In BMVC, pp 1–11
12. Ahonen T, Hadid A, Pietikainen M (2006) Face description with local binary patterns: application to face recognition. IEEE Trans Pattern Anal Mach Intell 28(12):2037–2041
13. Krizhevsky A, Sutskever I, Hinton GE (2012) Imagenet classification with deep convolutional neural networks. In: Advances in neural information processing systems, pp 1097–1105
14. Taigman Y, Yang M, Ranzato MA, Wolf L (2014) Deepface: Closing the gap to human-level performance in face verification. In: Proceedings of the IEEE conference on computer vision and pattern recognition, pp 1701–1708

15. Sun Y, Chen Y, Wang X, Tang X (2014) Deep learning face representation by joint identification-verification. In: Advances in neural information processing systems, pp 1988–1996
16. Sun Y, Liang D, Wang X, Tang X (2015) Deepid3: face recognition with very deep neural networks. arXiv preprint arXiv:1502.00873
17. Karahan S, Yildirum MK, Kirtac K, Rende FS, Butun G, Ekenel HK (2016) How image degradations affect deep CNN-based face recognition? In: 2016 international conference of the biometrics special interest group (BIOSIG). IEEE, pp 1–5
18. Liu Z, Luo P, Wang X, Tang X (2015) Deep learning face attributes in the wild. In: Proceedings of the IEEE international conference on computer vision, pp 3730–3738
19. Fontaine X, Achanta R, Süsstrunk S (2017) Face recognition in real-world images. In: 2017 IEEE international conference on acoustics, speech and signal processing (ICASSP), IEEE, pp 1482–1486
20. Wang Y, Bao T, Ding C, Zhu M (2017) Face recognition in real-world surveillance videos with deep learning method. In: 2017 2nd international conference on image, vision and computing (ICIVC). IEEE, pp 239–243
21. Perez L, Wang J (2017) The effectiveness of data augmentation in image classification using deep learning. arXiv preprint arXiv:1712.04621
22. Parkhi OM, Vedaldi A, Zisserman A (2015) Deep face recognition

Chapter 19 Randomized Neighbour Grey Wolf Optimizer

Shahnawaz Ali, Swati Jadon, and Ankush Sharma

1 Introduction

Nature-inspired algorithms (NIAs) are a powerful origin for motive in the field of engineering computations. It provides accuracy in solving the optimization issues [1]. NIAs are classified into two broad classes: evolutionary algorithms (EA) and SI-based algorithms. The essence of the solution gets intricate as the level of complexity is boosted [3, 4]. The GWO imposes the hunting behaviour and communal hierarchy of grey wolves. To integrate the behaviour of GWO, Mirjalili et al. [2] describe the position factor approach of wolf by taking average of positions of three best wolves.

In this proposed work, to increase the exploration skill, an alternate of GWO is imported. In the proposed method, the exploration work is guided by a randomly selected solution in explore area combined with positions of first three best solutions. This work is named as randomized neighbour grey wolf optimizer (RNGWO). In RNGWO a new randomly selected agent is added and randomly multiplied by coefficients A and C for enhancing the exploration skill of wolves. A group of 12 benchmark functions is calculated for experiments. The gained outcomes are validated for the correctness of the planned method.

Further partition of the paper is organized as follows: Sect. 2 briefly introduces the GWO algorithm. The proposed RNIGWO is described during the Sect. 3. Section 4 describes the achievement of planned innovation over 12 benchmark functions. Finally, the conclusion is explained in Sect. 5.

S. Ali (✉) · S. Jadon · A. Sharma
Gurukul Institute of Engineering and Technology, Kota, Rajasthan 325003, India

N. Marriwala et al. (eds.), *Soft Computing for Intelligent Systems*, Algorithms for Intelligent Systems, https://doi.org/10.1007/978-981-16-1048-6_19

2 GWO Algorithm

The GWO algorithm specifies the communal style of grey wolves [2]. This communal style involves approaching, tracking, quest and enclosing the prey. Wolves have some specific characteristics like sharp teeth, fluffy tassel, and they perform exploitation in a pack of 5–12 wolves. The ranking of wolves is as follows:

Rank 1: Alpha wolf is the commander of the community. It is responsible for any judgement related to how to execute the tracking, hunting, pursuit, relaxing place, exploration etc.

Rank 2: Beta wolf behaves like assistant of alpha wolf. This wolf assists the commander in any type of decisions, judgement and leadership. If in some cases the alpha wolf is accidentally injured or die, then this wolf takes the responsibility of alpha wolf. It acts as a counsellor to the commander wolf [3].

Rank 3: Delta wolf dispatches all the information to the commander wolf (alpha) and assistant wolf (beta). The delta wolf is further classified into different types according to their work style, and is as follows:

Sentinels-wolf: act safeguard of the group.
Scout-wolf: watch the boundaries.
Senior-wolf: erstwhile alpha or beta wolf.
Hunter-wolf: helps in exploitation.
Attendant-wolf: attend the injured wolf.

Figure 1 specifies the social behaviour of grey wolves.

2.1 Computation Model

The computation model describes various phases which are as follows:

1. Communal rank:

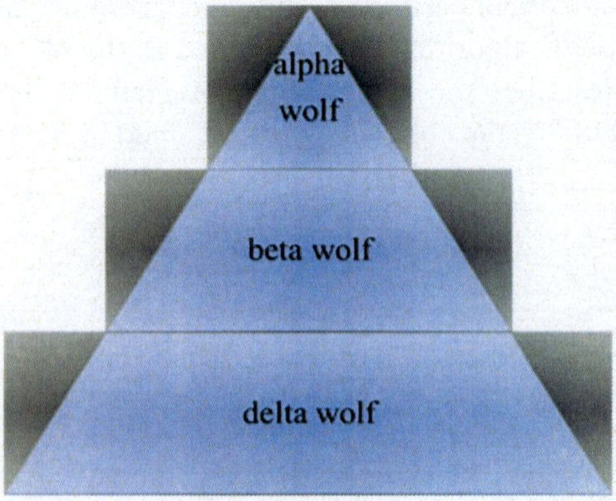

Fig. 1 Communal style of wolves

For explaining the computation model, the communal rank of wolves is marked as per their fitness value. First, find out the three best solutions like $\text{alpha}_{\text{wol}}$, beta_{wol} and $\text{delta}_{\text{wol}}$ marked as first, second and third best solutions, respectively.

2. Enclose the prey:
Wolves enclose the prey when they perform hunting procedure [2]. The computation equations for enclose action are interpreted and described as follows:

$$\vec{D} = |\vec{C}_{\text{co}} * \vec{z}_{\text{pr}}(\text{it}) - \vec{z}_{\text{gwol}}(\text{it})| \tag{1}$$

$$\vec{z}(\text{it}+1) = \vec{z}_{\text{pr}}(it) - \vec{A}_{\text{co}} * \vec{D} \tag{2}$$

Here, the current iteration and the position vector are indicated by it and z_{gwol} (it), respectively. The position of the prey is indicated by z_{pr} (it); D represents the distance from the position of prey. The distance factor depends on A and D. When the iteration number of the algorithm is increased, the wolves come near to the prey for hunting action. As the initial locations are found randomly the wolves confine the prey [4].
The coefficients A_{co} and C_{co} are evaluated by the following formula:

$$\vec{A}_{\text{co}} = 2\vec{a} * \vec{r}_{A_1} - \vec{a} \tag{3}$$

$$\vec{C}_{\text{co}} = 2 * \vec{r}_{C_2} \tag{4}$$

Here, r_1 and r_2 are random vectors in the range [0,1] with component |a| stable between 2 and 0.
Again in the next iteration, the wolf explores the prey to find the next fittest solution between the grey wolf populace. This process is repeated until the terminus criterion is accomplished [5].

3. Hunting:
Grey wolf acts on the hunting process in a group in which first three best wolves, the $\text{alpha}_{\text{wol}}$, the beta_{wol} and the $\text{delta}_{\text{wol}}$, have the leadership capacity in the group of wolves [6]. The group of wolves will update their positions in various iteration. The following equations are for the position refurbishing:

$$\vec{Z}_{(\text{it}+1)} = \frac{\left[\vec{Z}_1 + \vec{Z}_2 + \vec{Z}_3\right]}{3} \tag{5}$$

$$\vec{Z}_1 = \vec{Z}_{\text{alpha,wol}} - \vec{A}_{\text{co}_1} * \vec{D}_{\text{alpha,wol}} \tag{6}$$

$$\vec{Z}_2 = \vec{Z}_{\text{beta,wol}} - \vec{A}\text{co}_2 * \vec{D}_{\text{beta,Wol}} \tag{7}$$

$$\vec{Z}_3 = \vec{Z}_{\text{delta,wol}} - \vec{A}_{\text{co}_3} * \vec{D}_{\text{delta,wol}} \tag{8}$$

Here, the position vectors of first three best solutions are represented by $z_{\text{alpha,wol}}$, $z_{\text{beta,wol}}$ and $z_{\text{delta,wol}}$. The coefficient vectors are represented by A_{co1}, A_{co2}, A_{co3} and A_{co} in Eqs. (6)–(8). Some parameters are calculated using the following equations:

$$\vec{D}_{\text{alpha,wol}} = \left|\vec{C}_{\text{co1}} * \vec{Z}_{\text{alpha,wol}} - \vec{Z}\right| \tag{9}$$

$$\vec{D}_{\text{beta,wol}} = \left|\vec{C}_{\text{co2}} * \vec{Z}_{\text{beta,wol}} - \vec{Z}\right| \tag{10}$$

$$\vec{D}_{\text{delta,wol}} = \left|\vec{C}_{\text{co3}} * \vec{Z}_{\text{delta,wol}} - \vec{Z}\right| \tag{11}$$

Here, the coefficient vectors are represented by Cco_1, Cco_2, Cco_3 Cco, in Eqs. (9)–(11).

4. Attack against the prey:
As the prey stops moving, at that time grey wolf attack against the prey quickly. The enclosing purpose of wolf is accomplished as the value of |a| is decreased. The parameters |A| and |a| depend on each other. Here, the value of parameter a is evaluated using this formula: $|a| = [2 - 2 * ((1)/\text{Maxiteration})]$. The alpha wolf attacks against the prey if the following condition is true:
$|A| < 1$. Otherwise, the group of wolf cannot attack against the prey.

3 Proposed Randomized Neighbour Grey Wolf Optimizer

The exploration capacity of GWO is based upon the three best solutions, and in the next iteration the wolf upgrades their positions. Somewhere the position of three best solutions is overlapped, so some problems may occur. Due to this problem the efficiency and exploitation capability of algorithm is dismissed [2]. The following modified explore procedure overcomes this problem (12):

$$\vec{Z}(\text{it}+1) = \frac{\vec{Z}_{1*} - \vec{A}_1 + \vec{Z}_{2*} \rightarrow \vec{A}_2 + \vec{Z}_{3*}\vec{A}_3 + \vec{Z}_{4*}\vec{A}_4}{4^*(C_{1*}C_{2*}C_{3*}C_4)} \tag{12}$$

Here, $\vec{Z}_4$ represents the position vector of randomly elected agent wolf in explore area that is not equal to the $z_{\text{alpha,wol}}$, $z_{\text{beta,wol}}$ and $z_{\text{delta,wol}}$. By using the following equations the value of z_4 is evaluated (13 and 14):

$$\vec{D}_{\text{delta,wol}} = |\vec{C}co_4 * \rightarrow \vec{Z}_{\text{neigh,wol}}(\text{it}) - \vec{Z}_{\text{wol}}(\text{it})| \tag{13}$$

$$\vec{Z}_4 = \vec{Z}_{\text{neigh,wol}}(\text{it}) - \vec{A} * \vec{D}_{\text{neigh,wol}} \tag{14}$$

Here, C_{co4} C_{co} and $z_{\text{neig,wol}}$ are the position vectors of the randomly elected fourth best solution.

When adding the randomly selected neighbour with coefficient the efficiency of algorithm is increased. This proposed method is named as randomized neighbour grey wolf optimizer (RNGWO). The communal style of RNGWO can be depicted in Fig. 3.

As per the above description, the pseudo code of the proposed RNGWO is illustrated in following Algorithm 1.

4 Empirical Results

For validating the efficiency of proposed RNGWO, consider the 12 distinct optimization functions f_1 to f_{12}, as shown in Table 1. A comparative study is done with test problems between grey wolf optimizer (GWO) [2], randomized neighbour grey wolf optimizer (RNGWO), artificial bee colony (ABC) [7], fully informed grey wolf optimizer [8] and power low-based local search in spider monkey optimization (PLSMO) [9] (Figs. 2 and 3).

Table 1 Test problems

S. No.	Function-name	Explore Range	Dimens	AE
1	Easom's function	[−100, 100]	2	1.0E−13
2	Rastrigin	[−1, 1]	30	1.0E−01
3	Alpine	[−10, 10]	30	1.0E−05
4	Brown3	[−1, 4]	30	1.0E−5
5	Saloman problem	[−100, 100]	30	2.0E−1
6	Pathological function	[−1, 1]	30	1.0E−01
7	Sum of different powers	[−1, 1]	30	1.0E−01
8	Quartic fun	[−1.28, 1.28]	30	1.0E−5
9	Inverted cosine wave	[−5, 5]	10	1.0E−5
10	Rotated hyper-ellipsoid function	[−65.53, 65.53]	30	1.0E−5
11	Kowalik	[−5, 5]	4	1.0E−05
12	Shifted Griewank	[−600, 600]	10	1.0E−5

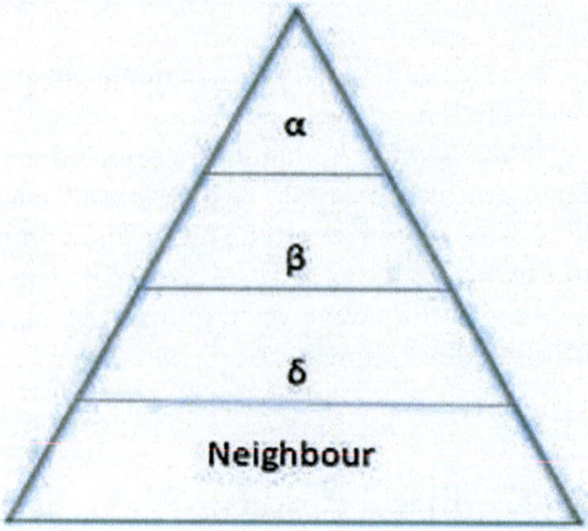

Fig. 2 Communal style of RNGWO

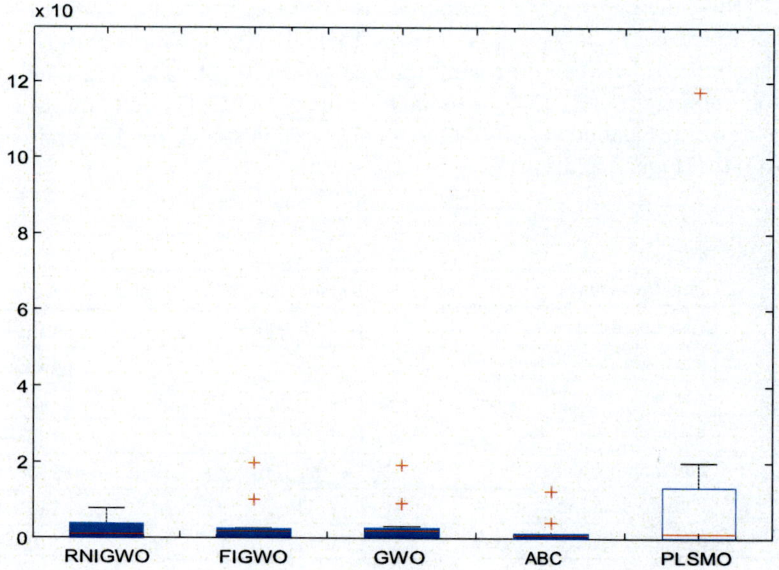

Fig. 3 Boxplot

```
Assume the zp is populance of grey wolf here p is 1 to n
while it <max-num of emphasis
  do for every searching agent
    Generate the current position of agent by equation 1 and 2
    Combine the neighbor (Zneig,wolf) as equation 13 and 14
```

$$\vec{Z}(it+1) = \frac{\vec{Z_1} \cdot \vec{A_1} + \vec{Z_2} \cdot \vec{A_2} + \vec{Z_3} * \vec{A_3} + \vec{Z_4} \cdot \vec{A_4}}{4*(C_1 \cdot C_2 * C_3 * C_4)}$$

```
  end for
  The fitness of all searching wolves is calculate.
  Upgrade Zalpha,wol, Zbeta,wol,
  Zdelta,wol and Zneig,wol
  it = (it + 1)
end while
return Zalpha,wol
```

Algorithm 1: Randomized neighbour grey wolf optimizer (RNGWO)

The experimental setup is as follows:

- Absolute figure of runs = 30
- Size of grey wolve's population = 50
- Gross number of emphasis = 5000
- Max function calculation = 200000

The mathematical calculations are represented in the mode of success run rate, standard deviation and average function calculation. Table 2 justifies the functions on which this planned method gets hit value.

The boxplot graph is developed to accomplish comparison of the RNGWO with distinct algorithm exposed to justify the result. The AR test and Mann-Whitney U rank-sum test [10] are also used to test and expose the results.

The boxplot graph of RNGWO with distinct consider algorithms are presented in Fig. 4. It is clear from the boxplot that RNGWO is good in terms of AFE as compared with other considered algorithms.

The acceleration rate is calculated by the following formula:

$$A_r = \frac{AFE_{(consider\ algo)}}{AFE_{(RNGWO)}} \quad (15)$$

The calculated acceleration rate (A_r) through equation 15 of RNGWO with considered algorithms is represented in Table 3.

Table 2 Analysis of conclusions of RNGWO, FIGWO, GWO, ABC, PLSMO algorithms; TF: Test Function

TP	Algorithm	SR	AF E	SD
f_1	RNGWO	30	8	0
	FIGWO	30	95	0.038526282
	GWO ABC PLSMO	30	95	0.038526287
		30	19,434.66667	4.05E-06
		30	10,342.9	2.15E-06
f_2	RNGWO	30	1722.933333	5.57748E−06
	FIGWO	30	8260	8.50007E−06
	GWO ABC PLSMO	30	8146.666667	8.29229E-06
		30	93,630	1.30E-06
		29	25,610.79	1.67E-01
f_3	RNGWO	30	1824.80	6.64459E−06
	FIGWO	30	10,296.67	9.17641E−06
	GWO ABC PLSMO	28	21,898.33	1.51706E-05
		30	43,613.33	4.08E-06
		30	50,570.87	3.19E-06
f_4	RNGWO	30	1256.40	6.02476E−06
	FIGWO	30	3345	8.56122E−06
	GWO ABC PLSMO	30	3366.67	8.26258E-06
		30	80,860	8.82E-06
		30	23,699.6	5.87E-07
f_5	RNGWO	30	1782	0.13919
	FIGWO	30	3836.67	0.1966
	GWO ABC PLSMO	30	3996.67	0.1979
		30	42,050	1.92E-06
		30	15,136.87	7.52E-07
f_6	RNGWO	30	2263.07	0.0586
	FIGWO	1	194,133.33	3.63636611
	GWO ABC PLSMO	2	189,610	3.4534
		30	12,876.67	1.32E-02
		30	5334.63	2.40E-01
f_7	RNGWO	30	672.53	4.40317E−06
	FIGWO	30	1261.67	6.13267E−06
	GWO ABC PLSMO	30	1270	6.38139E-06
		30	20,976.67	0.00E + 00
		30	16,922.17	0.00E + 00
f_8	RNGWO	5	18,709.87	7.3103E−05
	FIGWO	0	200,000	9.06641E−05
	GWO ABC PLSMO	1	198,185	7.70104E−05
		30	126,212.2	7.27E-06
		30	78,461.17	2.30E-06
f_9	RNGWO	30	401.87	6.34229E−06
	FIGWO	29	14,363.33	0.0794
	GWO ABC PLSMO	29	24,970	0.0898
		30	1,174,841.97	2.01E-07
		30	103,017.80	7.34E-07

(continued)

Table 2 (continued)

TP	Algorithm	SR	AF E	SD
f_{10}	RNGWO FIGWO GWO ABC PLSMO	30 30 30 30 29	1680.27 4361.67 4481.67 37,306.67 12,728.86207	5.69624E−06 8.10706E−06 8.105E−06 7.95E− 1.86E-02
f_{11}	RNGWO FIG WO GWO ABC PLSMO	30 30 30 30 30	14.27 60 60 7613.37 15,152.5	−0.0001 −0.000201932 −0.0002 2.14E-05 2.75E-05
f_{12}	RNGWO FIGWO GWO ABC PLS MO	30 30 30 30 30	7.47 50 50 32,420 9420.2	−456.2549 −512.220091 −506.0144 1.57E−06 1.12E−06

Table 3 AR of RNIGWO compared to the FIGWO, GWO, ABC and PLSMO

T P	F I GW O	GW O	ABC	P LSM O
f_1	11.87	11.87	215.7	1399.86
f_2	4.74	4.72	54.33	14.86
f_3	5.64	12.00	23.90	27.71
f_4	2.66	2.67	64.35	18.86
f_5	2.15	2.24	23.57	8.49
f_6	85.78	71.19	12.62	2.35
f_7	1.87	1.88	31.19	25.16
f_8	10.69	10.59	6.74	4.19
f_9	35.74	62.13	2923.46	7.57
f_{10}	2.59	2.66	22.20	7.57
f_{11}	4.20	4.20	533.64	1062.09
f_{12}	6.69	6.69	434.19	1261.64

The proposed method RNGWO is efficient for some optimization functions as shown in Table 1.

5 Conclusion

In this prospective effort, the solution is refurbished by transforming the information, from randomly chosen neighbour wolf to other wolf population. This method increases the capability of GWO algorithm. With the help of standard optimization test functions, ratify the fidelity of RNGWO method. Empirical outcomes expose that the planned method is truely good.

References

1. David WC, Alice ES (1996) Reliability optimization of series- parallel systems using a genetic algorithm. IEEE Trans Reliabil 45(2):254–260
2. Seyedali M, Seyed MM, Andrew L (2014) Grey wolf optimizer. Adv Eng Softw 69:46–61
3. Priyanka M, Harish S, Nirmala S (2019) Neighbourhood- inspired grey wolf optimizer. In: Proceedings of international conference on communication and computational technologies. Springer, pp 123–136
4. Raju P, Himashu M, Avinash P, Mukesh S (2016) Beecp: Biogeography optimization-based energy efficient clustering protocol for hwsns. In: Contemporary computing (IC3), 2016 ninth international conference on. IEEE, 1–6
5. Nitin M, Urvinder S, Balwinder SS (2016) Modified grey wolf optimizer for global engineering optimization. Appl Computat Intell Soft Comput 8
6. Xu H, Liu X, Su J (2017) An improved grey wolf optimizer algorithm integrated with cuckoo search. In: 2017 9th IEEE international conference on intelligent data acquisition and advanced computing systems: technology and applications (IDAACS), vol 1. IEEE, 490–493
7. Pei-Wei T, Jeng-Shyang P, Bin-Yih L, Shu-Chuan C (2009) Enhanced artificial bee colony optimization. Int J Innovative Comput Informat Control, 5(12):5081–5092
8. Sharma H, Meiwal P, Sharma N (2019) Fully informed grey wolf optimizer algorithm. In: International conference on information management & machine intelligence. Springer, 497–512
9. Sharma A, Sharma H, Bhargava A, Sharma N (2017) Power law-based local search in spider monkey optimisation for lower order system modelling. Int J Syst Sci 48(1):150–160
10. Patrick EM, Julius N (2010) Mann-whitney u test. The Corsini encyclopedia of psychology, pp 1–1

Chapter 20
Reverse Engineering National Cognition Impairment: A PGF-Mediated Approach

B. Malathi, M. Sankeerthana, D. Aishwarya, Mohammad Sana Afreen, and K. Chandra Sekharaiah

1 Introduction

Cognition is an important feature of human consciousness. It is the process of pursuing knowledge and understanding by thinking and through experience. Cognitive processes use existing knowledge and generate new knowledge.

Natural intelligence (NI) or human intelligence (HI) is the potential to grasp from experience, understand and use the knowledge, and adapt to new situations. Human cognition means the faculty of knowing. Cognitive science is the study of mind and intelligence under multiple disciplines, like psychology, artificial intelligence, neuroscience, linguistics etc.

Machine intelligence (MI) or artificial intelligence (AI) enables a machine to interact with an environment in an intelligent way. Making a brilliant computer system to solve complex problems just like humans do and enabling them to replicate the behavior of humans is possible only through artificial intelligence. Machine cognition gives a machine the same cognitive abilities that humans have.

Thus, the information processing machines showing intelligence by meeting the requirements are both the human brains and computers. The sections of the paper are organized as follows: Sect. 2 presents the overview of web2.0, web intelligence and features of intelligent web information systems. Section 3 presents the review and outline on cognitive informatics. In Sect. 4, we briefly discuss the web intelligence light on cognitive informatics and brain informatics. We have inspected for the national cognition of the students in the posts and comments made by them in google plus community of PGF founder and presented a few samples in Sect. 5. Section 6 includes the conclusion.

B. Malathi (✉) · M. Sankeerthana · D. Aishwarya · M. S. Afreen · K. Chandra Sekharaiah
JNTUH, Hyderabad, Telangana, India

N. Marriwala et al. (eds.), *Soft Computing for Intelligent Systems*, Algorithms for Intelligent Systems, https://doi.org/10.1007/978-981-16-1048-6_20

2 Web Intelligence

Two features of the web, the size and complexity, make the WI useful and necessary. The surplus amount of mutually joined web documents or web pages is dumped into the web. Owing to this, the challenges that pose are the problems in the storage, management, and efficient and effective retrieval of information, varied collection of web documents like structured, unstructured, semi-structured, interrelated and distributed. To deal with such challenges and the huge size and complexity of the web, the existing intelligent information systems need to be combined and extended to web-based information systems.

Web2.0 is World Wide Web (www) that highlights user-generated content. Social networking sites, video and image sharing sites, wikis, web applications and blogs are some of the examples of web2.0. Way back machine can be used to understand clear distinction between web1.0 and web2.0. An intelligent web information system or advanced WI system should be able do the tasks such as perform reasoning, learning and self-improvement, make dynamic recommendations to web user, perform automatic modification of websites content and organization, combine web usage data with marketing data and perform online behavioral analysis [1, 2].

$$\text{Web Intelligence} = \text{Artificial Intelligence} + \text{Information Technology}$$

Social web has become a crucial and necessary part of web2.0. It is a platform for sharing the experiences, emotions, opinions, feelings and sentiments. The applications here are very interactive. The vital features/characteristics of web2.0 in the manner of participation of user are freedom, receptivity/openness, cooperation, collaboration, coordination etc. [3, 4].

Social intelligence is fusion of human and machine intelligence for building social networks containing the different communities of people, organizations and other social entities.

$$\text{SI} = \text{MI} + \text{HI}$$

Data intelligence is to learn and gain deeper understanding of the information processing system which is achieved by the fusion of artificial intelligence (AI), machine learning (ML), deep learning (DL) and cognitive science.

3 Cognitive Informatics

Cognitive informatics, in order to investigate the human information processing mechanism, encompasses the study of information sciences, cognition and application of IT. It is also the study of internal information processing mechanisms of the natural intelligence (NI) or human intelligence (HI) and its application in computing and ICT using advanced information technologies. The basic issues that pose in informatics, psychology, artificial intelligence, ICT can be tackled by cognitive informatics.

Cognitive informatics accommodates the important forms of human/natural intelligence or the classes of cognitive information such as skill, knowledge, experience and wisdom. Applying the proper perception and reasoning, the instruction is transformed into an action/behavior by wisdom.

Based on the evolution of cognitive computing (CC), the latest and intelligent technologies will be developed for mining and processing knowledge, learning autonomously and machine-supported problem-solving. The conceptual base for world wide web plus (www+) and cognitive computing is cognitive informatics.

4 Web Intelligence Light on Cognitive Informatics

In the previous sections we have seen definitions of human cognition, human intelligence, machine cognition, machine/artificial intelligence, web intelligence, cognitive informatics and computing. Strengthening of web intelligence, its application, and its development can be possible by understanding the human intelligence using the discipline brain sciences. Also, the techniques of web intelligence provide strong latest platform for brain sciences. The union of web intelligence and brain informatics promotes the study and comprehension of data, information, knowledge, intelligence and wisdom. Thus, for achieving human-level web intelligence, WI transforms IT and AI.

Cognitive informatics is the inquiry of internal information processing mechanisms of natural/human intelligence and its application in computing and ICT using advanced information technologies.

$$\text{CI} = \text{HI} + \text{IT}$$

Usage of advanced WI-centric IT and cognitive neuroscience together to study human information processing mechanism is an emerging field called brain informatics.

$$\text{BI} = \text{WI centric IT} + \text{Cognitive science}$$

The relation of WI and BI is bidirectional. WI-based technologies like data mining, semantic web, wisdom web, multi-agent systems etc. provide a new stage for cognitive and brain sciences. Discovery and proper understanding of the human intelligence models such as neuroscience, cognitive science and brain informatics in brain sciences lead to new WI development. So, there is a great importance for understanding the three intelligence-related areas, which are machine intelligence, human intelligence and social intelligence for WI development. To develop true human-level WI, revelation of new cognitive models and computational models is important.

$$\text{WI} = \text{AI} + \text{HI} + \text{SI} + \text{IT}$$

5 Reverse Engineering National Cognitive Impairment in Students of JNTUH

National cognition means awareness/enlightenment on their duty toward their nation and perceiving national consciousness and integrity. National cognition in students has been misdirected in the academic premises of JNTUH. The details of the cyber-crime case study and the follow-up work done for cyber policing are presented in [5–16]. The students with national cognitive impairment lead to a regressive society. Hence the reverse engineering of national cognitive impairment in students is a necessity for transforming regressive society to progressive society. Our research work aims for it.

As part of our research methodology, we have considered the student community of a Google + account that belongs to the PGF founder, Dr. K. ChandraSekharaiah "https://sites.google.com/site/sekharaiahk/apeoples-governanceforumwebpage". The student community named "profchandswsnsubjectclass-group" comprises nearly 70 students of M.Tech and MCA from JNTUH University. The discussions in the community include topics on awareness regarding Twin Big Data Cybercriminal Organizations (TBDCOs), remedial approaches by Peoples' Governance Forum (PGF), cyber ethics, cyber policing methods etc.

The students who have passed out in a cybercriminal environment have become knowingly or unknowingly the victims by such cybercriminal organization and have national cognitive impairment. Guidance and counselling to the students of JNTUH through various social media and PGF is taken up by the PGF members to reverse engineer the national cognitive impairment that appeared in them directly or indirectly. The posts and comments of the students in the community are sampled for checking the national cognition in students and their cognition on the TBDCOs that prevailed in JNTUH during the time of their senior batches. We have made a study on the posts and comments of the students which indicates their Mother India Consciousness (MIC) as impressive. Our focus was on MIC generation in students. A sample of the posts and discussions by the students in the community group are illustrated in Figs. 1, 2, 3 and 4 and it indicates that the national cognitive impairment in them is reverse engineered by continuous guidance and counselling to the students in the JNTUH academic environment by the PGF founder and members (Table 1).

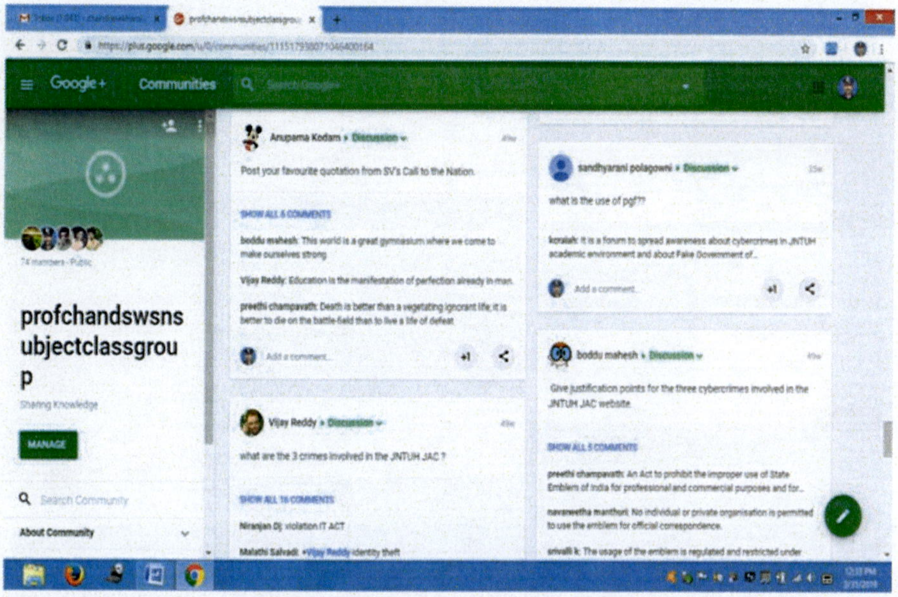

Fig. 1 A snapshot of the post made by a student Vijay Reddy "what are the 3 crimes involved in the JNTUH JAC?"

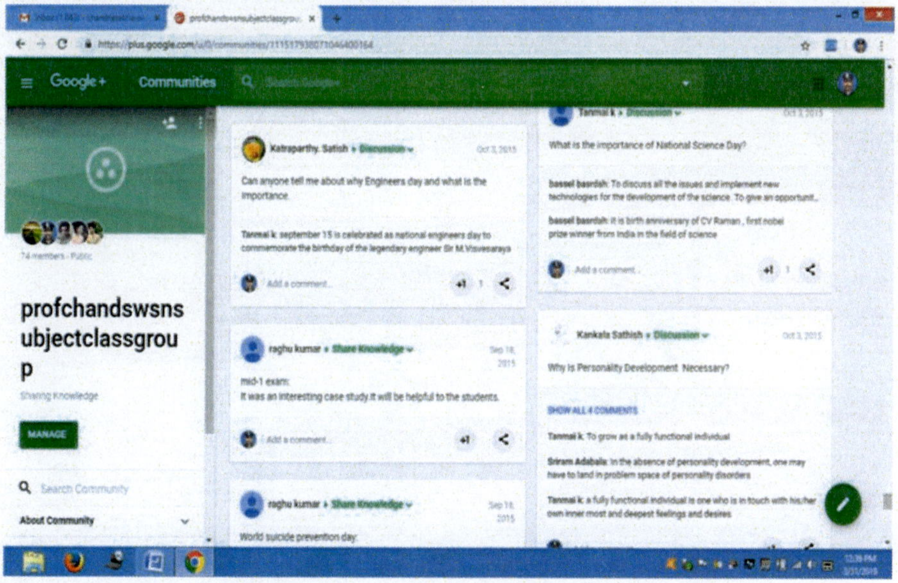

Fig. 2 A snapshot of the post made by a student Raghu Kumar which indicates that the mid1 exam conducted to them on the awareness of the TBDCOs is interesting and helpful

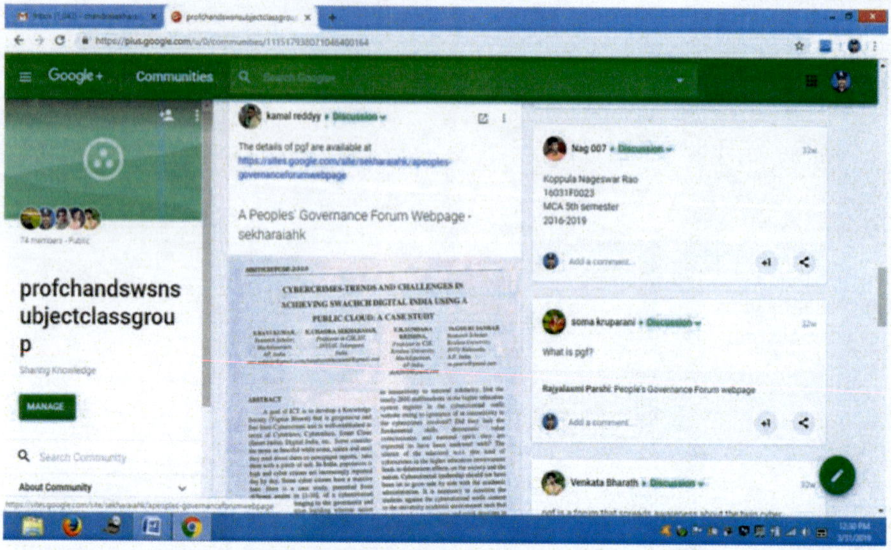

Fig. 3 A snapshot of the post made by a student Kamal Reddy on the remedial forum PGF to let know the other students of the community reg. it

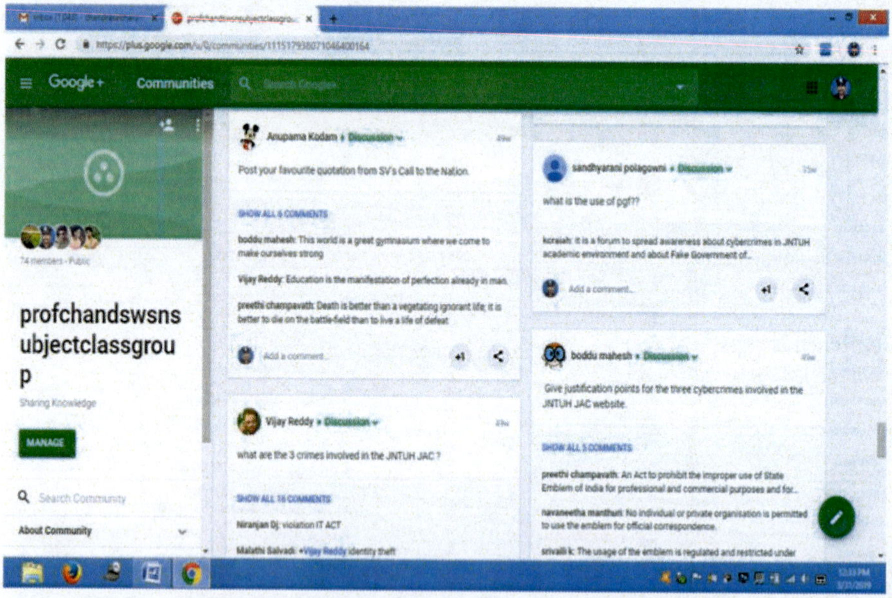

Fig. 4 A snapshot of the post made by a student Anupama Kodam regarding SV's Call to the Nation

Table 1 Some similar posts made by other students of the community

1. How can each one of us be an internal soldier to the nation by curbing sedition in India?
   Shared to the community profchandswsnsubjectclassgroup - Public
   Shireesha J - we should bring awareness people
   Teja Swi - by awareness camps and programs
   shilpa 8 - by bringing awareness in the people
   Swarna Latha - by providing awareness camps and programs in the people.
   Sreeja Pula - by learning anti cybercrime laws
2. Did you see the "Useful Links" webpage in my website?
   Shared with: Tanmai k, Maheswari Kotha
   Tanmai k - Yes sir, I have seen the useful links page in your website.
   Tanmai k - The page was very interesting and I have learned many useful things, especially about "Anti Corruption awareness efforts" in JNTU initiated by you. I am very glad to visit your website. Thank you sir.
3. Shravan Konakala - the crimes involved in cybercriminal website according to me are:
   1: Sedition law violation. 2: The cybercriminal website got around 2000 registrations, so it indicates a BIG DATA CYBER CRIME.
   comment if u find any other crime involved?
   Shared to the community profchandswsnsubjectclassgroup - Public
4. BHRUGUBANDA JAHNAVI - What are the crimes involved in the cybercriminal website
   Shared to the community profchandswsnsubjectclassgroup - Public
5. Shilpa 8 - What is a wayback machine??
   Shared to the community profchandswsnsubjectclassgroup - Public
6. Sreeja Pula - What is cyberforensics?
   Shared to the community profchandswsnsubjectclassgroup - Public
7. Shireesha J - What is internet achieving?
   Shared to the community profchandswsnsubjectclassgroup - Public
8. Teja Swi - What is IPC?
   Shared to the community profchandswsnsubjectclassgroup - Public
9. Teja Swi - What is JAC?
   Shared to the community profchandswsnsubjectclassgroup – Public
10. Sreepadi yashaswi - What is jurisprudence? ?
    Shared to the community profchandswsnsubjectclassgroup - Public
11. Sravani Busani - Advantages with WayBack Machine
    Shared to the community profchandswsnsubjectclassgroup – Public
12. arshiya pgf - Can cyber forensics captured serve the trial of the cybercrime?
    Shared to the community profchandswsnsubjectclassgroup – Public
13. Nagaraju Potharaju - So you agree that www.jntujac.com is a culprit organisation
    Shared to the community profchandswsnsubjectclassgroup – Public
14. Sandhyarani polagowni - What is the use of pgf??
    Shared to the community profchandswsnsubjectclassgroup – Public
15. Somshetty Teju - Pgf is a forum that spread awareness about the twin cybercriminal organizations. fake government of Telangana and jntuh jac.
    Shared to the community profchandswsnsubjectclassgroup - Public

6 Conclusion

The perspectives of the human cognition and machine cognition with the application of natural intelligence and artificial intelligence, respectively, are presented in this paper. The exploration of advanced information technologies is especially important in the field of web intelligence and cognitive informatics along with the utilization of the machine intelligence and human intelligence. The stand of brain informatics (BI) and web intelligence (WI) is bidirectional, i.e., BI for WI and WI for BI is presented in the paper. The concept of national cognition is to be introduced into the students for a progressive society and if national cognition impairment is caused in them due to any circumstances, proper guidance and counselling has to be provided for reverse engineering the cognition impairment. In the aforementioned case study, the PGF founder plays major role in introducing Mother India Consciousness (MIC) as the national cognition has been impaired in the students in the academic environment of JNTUH. The samples of google plus data after analysis indicate the same. Thus, our research work focus on transforming a regressive society to a progressive society by establishing national consciousness among the student community.

References

1. Malathi B, Chandrasekharaiah Jayasree GPL (2016) Analytical trends and practices of web intelligence. WIR 16:121–125. ACM, New York, NY, USA ©2016. ISBN: 978-1-4503-4278-0
2. Malathi B, Hurshan R, Mayeekar J, Chandrasekharaiah K (2017) Text opinion analyzer and classifier using word of support. In: Proceedings of the 11th INDIACom, INDIACom-2017. IEEE Conference ID: 40353, ISSN 0973-7529; ISBN 978-93-80544-24-3, 2017 4th international conference on computing for sustainable global development, 01st- 03rd March, 2017, BVICAM. New Delhi (INDIA)
3. Malathi B, Chandra Sekharaiah K (2018) Web intelligent information systems: a PGF-mediated social media evaluation perspective. Int J Manage Technol Eng (IJMTE) 8(XI):684–684 ISSN NO: 2249-7455
4. MalathiB, ChandraSekharaiah K (2019) Comparative evaluation of SMMD values of popular social media sites: PGF-a high SMMD case. In: Proceedings of the AICTE sponsored 2nd international conference on advances in computing and communication technologies (ICACCT-2019). CMRCET, Hyderabad, 16th–17th December 2019, pg. 11. Springer Nature https://www.springer.com/gp/book/9789811579608 pg 721–732). ISBN 978-981-15-7960-8
5. Ravi Kumar S, Chadra Sekharaiah K, Sundara Krishna YK (2019) Convergence potential of IT and RTI Act 2005: a case study. In: Proceedings of 2nd EAI international conference on big data innovation for sustainable cognitive computing, 12th–13th Dec 2019. Coimbatore, India
6. Pavana Johar K, Malathi B, Ravi Kumar S, Srihari Rao N, Madan Mohan K, Chandra Sekharaiah K (2018) India-abusive Government-of-Telangana (GoT2011): a constitutional IT (an SMI) solution. In: Proceedings of international conference on science, technology & management (ICSTM-2018), 12th August 2018. Punjab University Campus, Chandigarh, India, 978-93-87433-34-2, pp 330–337
7. Pavana Johar K, Malathi B, Ravi Kumar S, Srihari Rao N, Madan Mohan K, Chandra Sekharaiah K (2011) India-abusive Government-of-Telangana (GoT2011): a constitutional IT (an SMI) solution. Int J Res Electron Comput Eng (IJRECE) 6(3):1118–1124

8. Pavana Johar K, Gouri Sankar M, Ravi Kumar S, Madan Mohan K, Punitha P, Santhoshi N, Malathi B, Ramesh Babu J, Srihari Rao N, Chandra Sekharaiah K (2018) Cyberpolicing the multifaceted cybercriminal, fake Government of Telangana: what is sauce for the goose is sauce for the gander. In: Proceedings of 3rd international conference on research trends in engineering, applied science and management (ICRTESM-2018). Osmania University Centre for International Programmes, 4th November 2018, pp 321–330 & Universal Review, vol 7(XI), pp 256–264.
9. Santhoshi N, Chandra Sekharaiah K, Madan Mohan K, Ravi Kumar S, Malathi B (2018) Cyber intelligence alternatives to offset online sedition by in-website image analysis through webcrawler cyber forensics. In: Proceedings of international conference on soft computing & signal processing (ICSCSP 2018), June 22–23, pg 187–199
10. Srihari Rao N, Chandra Sekharaiah K, Ananda Rao A (2018) Janani Janmabhoomischa Swargaadapi Gareeyasi. Int J Eng Technol (IJET) 7(3, 29):225–231
11. Tirupathi Kumar B, Chandra Sekharaiah K, Mounitha P (2015) A case study of web content mining in handling cybercrime. Int J Adv Res Sci Eng 04(01). www.ijarse.com. ISSN 2319-8354
12. Madan mohan K, Chandra Sekharaiah K, Premchand P (2018) Impact of RTI act with in public authority organization toward employee employer engagement: a case study. School of Law, Pondicherry University, Pondicherry
13. Madan Mohan K, Chandra Sekharaiah K, Santhoshi N (2018) ICT approach to defuse the cybercriminal sedition dimension of Telangana Movement. Accepted for publication in Int J Eng Technol (IJET) 7(3, 29)
14. Gouri Shankar M, UshaGayatri P, Niraja S, Chandra Sekharaiah K (2016) Dealing with Indian jurisprudence by analyzing the web mining results of a case of cybercrimes. In: Proceedingss of ComNet 2016 International Conference, 20–21. Ahmedabad, India
15. PGF Website, https://sites.google.com/view/peoplesgovernanceforum/
16. PGF, https://sites.google.com/site/sekharaiahk/apeoples-governanceforumwebpage

Chapter 21
Analytical Breakthrough of Pennes' Bioheat Model in Malignant Tissues Exposed to Thermal Radiation: In Silico Investigation with Fractional-Order Three-Phase Lag

Sharduli, Iqbal Kaur, and Kulvinder Singh

Nomenclature

T_b	Blood temperature
ω_b	Rate of blood perfusion
t	Time
I_0	Intensity of the laser
c	Tissue's specific heat
c_b	Blood's specific heat
ρ	Tissue mass density
C_1, C_2, k_1, k_2	Functions of diffuse reflectance R_d
K	Thermal conductivity of the tissue
K^*	Material constant characteristic
τ_q	Heating flux's phase lag
τ_t	Temperature gradient's phase lag
τ_θ	Phase lag for the gradient of thermal displacement
$U(t)$	Unit step function
B	Factor of frequency
μ_a	Coefficient of absorption

Sharduli (✉)
Department of Biotechnology, University Institute of Engineering & Technology (U.I.E.T), Kurukshetra University, Kurukshetra 136119, Haryana, India

I. Kaur
Government College for Girls, Palwal, Kurukshetra 136131, Haryana, India

K. Singh
Department of Computer Science & Engineering, University Institute of Engineering & Technology (U.I.E.T), Kurukshetra University, Kurukshetra 136119, Haryana, India
e-mail: ksingh2015@kuk.ac.in

N. Marriwala et al. (eds.), *Soft Computing for Intelligent Systems*, Algorithms for Intelligent Systems, https://doi.org/10.1007/978-981-16-1048-6_21

Q_{ext}	Heat generated per unit volume of tissues
ρ_b	Blood mass density
δ	Penetration depth
χ	Time delay parameter
μ_s	Scattering coefficient
τ_p	Exposure time of the laser
g	Factor of anisotropy
R	Universal gas constant
Q_m	Metabolic heat generations in living tissues
Q_b	Thermal source of blood perfusion
E_a	Activation energy
θ_0	Incident heat flux intensity

1 Introduction

Thermal processes and bioheat transfer mechanisms are the embodiment of existing life forms which are essential for understanding life. Transfer of heat in tissues (both healthy and malignant cells) is influenced by the blood flow and the bad structure of the vascular tissues. These tissues are heterogeneous in nature and on occasion anisotropic with complex warm properties. It additionally delivers heat as a major aspect of active metabolism. Maintenance of a fairly constant body temperature in a range of thermal conditions infers that there is a nonstop trade of energy between deep, surface tissues and the surrounding environment. In order to carry out such investigations, the Pennes' interpretation of the impact of blood perfusion is regularly utilized as a first effort to demonstrate heat transfer in living tissue. The focal point of the arrangement improvement introduced here is to exploit the existing distributed arrangements of unadulterated thermal conduction. It is indicated that the basic arrangement of the transient thermal conduction might be effectively controlled for comparing bioheat transfer. Moreover, the current examinations of bioheat demonstrate that the consequence of bioheat velocity outlines in the thermal contaminated and tumor regions is more delicate to the rate at which thermal heating takes place and veins' size. In this research paper, a thorough attempt has been made to utilize the bioheat model dependent on the Pennes' bioheat transfer equation that would shape a vastly improved thermal infected region. It, therefore, works best on small to medium blood veins with healthier coverage to the tumors and blood vessels by the thermal radiations. In spite of this, when the bioheat model is applied in malignant cells exposed to laser or thermal radiations, it presents clinical breakthrough for the comprehensive treatment of tumor (T) cells in an effective way. Therefore, it is scientifically derived that future will definitely rely upon the combination of bioheat-based gene-encoding chimeric antigen receptor (CAR) T-cells therapy as presented in Fig. 1 for T-cell treatment. The CAR T-cell is a type of immunotherapy which additionally includes gene therapy (basically gene therapy treatment utilizes

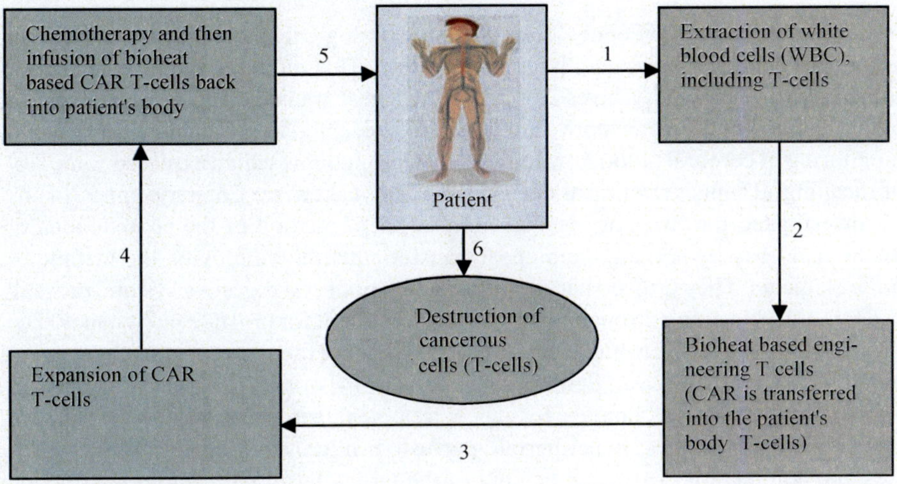

Fig. 1 Bioheat based gene encoding chimeric antigen receptor (CAR) T-cells therapy for next generation cancer treatments

an extraordinary transporter, generally an infection, to place RNA or DNA into living cells). In CAR T-cell therapy, the oncologists take immune cells from the blood and add genes to transform them so that they can more readily spot and kill malignant growth cells. After that, a lot of such cells are developed and put them back into the body.

This type of recently adopted immunotherapy is duly approved by the US Food and Drug Administration (USFDA), which is called CAR T-cell treatment. This sort of natural treatment, or biotherapy, utilizes the immune system to fight the malignant growth of cells. It helps the immune system to mark cancer cells and kill them more easily but biological investigation is going on in combination of chemotherapy and then infusion of bioheat-based CAR T-cells is placed back into patient's body. Finally, it can be inferred that bioheat model in malignant tissues can also be a fruitful perspective to model and treat any tumor but the size of malignant cells may present some thermal challenges.

2 Review of Important Work

This section presents the vital analysis of research work related to the bioheat and thermal heat transfer and other related properties to model the biological tissues which may be skin cells, healthy or malignant body cells of any living being. In this regard, Baish [1] proposed an innovative model based on the flow rate of blood and the vascular geometry of tissue transport (including oxygen, nutrient, and drug transport) for stable heat transport to the perfused tissue. The results show that, except for some anomalies, the Pennes' formula of the biological heat transfer equation can accurately predict the average tissue temperature. This model is still being used for some kind of elementary investigations of cell heat transfer, and also extended to other methods of cell transport. Similarly, Borgos [2] also presented a laser-based monitoring of cerebral blood flow for neurological patients which explored some sort of meaningful input for clinical decision-making. Likewise, Chatterjee and Adams [3] investigated the two-dimensional finite element method of the prostate area of the human body by solving the bioheat transfer equation employing the method of finite elements. They utilized nerve, muscle, skin, bone, fat digestive system, prostate and tumor tissue going through hyperthermia treatment for prostate malignancy. The results expressed that how the thermal heating affected the tumor regions by varying blood flow rates. Moreover, Sethi and Chakarvarti [4] critically described the use of various hyperthermia techniques for cancer treatment using ultrasonic hyperthermia, radio frequency devices, hyperthermic perfusion, microwave, magnetic nanoparticles and application of thermal heating to the target T-cell site. Additionally, Tian et al. [5] analyzed the membrane protrusions traits of rat brain vascular cells by various levels of hypoperfusion and found that when the level of hypoperfusion is promoted, the apoptosis of rats' brain microvascular endothelial cells became more neurological damage. In a similar concern, Crezee and Lagendijk [6] conducted experiments to verify the biological heat transfer by plotting temperature curves relative to the temperature difference between the blood vessels and organs under consideration, and measured the temperature curves around large artificial blood vessels in the perfused tissue. Likewise, Hase et al. [7] in their embolism hyperthermia research conducted experiments by induction heating of ferromagnetic particles injected into tumor tissue and liver cells to treat liver tumors. In another study, Hongliang et al. [8] in their research observed the effects and feasibility of heat and bioheat on surrounding organs. The method is to implant VX2 cancer of the liver with alternating magnetic field after embolization of magnetic nanoparticles to treat VX2 tumors, while the effect on the surrounding normal liver has no changes in the organization. On similar pattern, Hilger et al. [9] carried out the experiment of electromagnetic heating to eliminate breast tumors by utilizing an alternating magnetic field, and they further concluded that bioheat would be effective for the futuristic radiological treatments of breast tumors. Furthermore, Hilger et al. [10] also examined the treatment of breast cancer by adopting two distinct methodologies of magnetic heating on existing tumors with iron oxide and exposed to an alternate magnetic field to generate bioheat. In an alternate way, Hu et al. [11] convincingly used a bioheat model to consider the energy balance in healthy and malignant breast tumors under forced convection. They further concluded that by eliminating the impacts of the thermal field and noise, the level of the malignancy on the skin surface was fundamentally improved. Weinbaum and Jiji [12] presented a 3-D bioheat equation to illustrate the impact of blood stream on heat transfer. This bioheat theory explains the effect of heat transfer in areas with a blood vessel diameter less than 70 μm. Zhu et al. [13] explored the possible levels of bioheat equation for the thermal activity among the pair of blood vessel and their adjacent cells and they demonstrated that the temperature varieties in large blood vessels rely upon the fanning design and the neighborhood blood perfusion rate. Khaled and Vafai [14] analyzed the function of heat transport in biological tissues by decreasing the possibility of the flow instabilities in microorganisms. Similarly, P. R. Sharma et al. [15] probed the conventional Pennes' equation and the thermal wave prototype of biological heat transport to explore the numerical learning of heat dissemination in living tissues related to instant heating. They found that the deviation among the thermal wave representation of biological heat transport and the Pennes' representation greatly affects the heat transfer behavior, thereby providing realistic predictions for the temperature distribution in alive tissues. Tzu-Ching Shih et al. [16] analyzed the Pennes' biological heat transfer equation due to the sinusoidal heat flux on the surface of the skin by using the Laplace transformation. They suggested that the Pennes' bioheat equation would be appropriate for defining the transient temperature reaction of the tissue over the entire time range. Likewise, Lipkin and Hardy [17] built up the technique for estimating the temperature of human tissues. In the same way, Giering [18] depicted the various warm qualities of organic tissues. Shrivastava and Vaughan [19] proposed a nonexclusive bioheat move model which incorporate assurance of energy, pace of move of warmth from veins to tissue and relevant to a tissue. Additionally, Gardner et al. [20] proposed the flux rate and escape function in the case of one-dimensional optical switching in the organization using the Monte Carlo method. In a similar strategy, Cheng and Plewes [21] established a technique to determine the thermal properties associated with patients and their organs or tumors. Othman et al. [22] utilized the Pennes' bioheat transfer equation for getting the characterization of temperature variation in body tissues. Besides this, Kengne et al. [23] predict the dispersion of temperature in a finite biological tissue with spatial heating and oscillatory surface. In the same way, Youn and Lee [24] assessed the light dispersal and penetration depth in skin tissue utilizing high-intensity energy light sources. In another study, Deng and Liu [25] portrayed the hypothetical ways to solve 3-D bioheat transfer problems in the existence and absence of phase transition. In the same way, Kujawska et al. [26] investigated the ultrasonic approach to decide the thermal conductivity of animal tissues using pulsed centered ultrasound by developing a novel ultrasonic strategy to decide the thermal conductivity of certain animal body tissues, and on the other side, Kumar et al. [27, 28] depicted the exchange of heat in skin tissue of finite domain using metabolic heat. Youssef and Alghamdi [29] proposed a numerical model of one-dimensional thermo elastic skin tissue of small thickness using dual-phase-lag heat law. Ezzat et al. [30, 31] examined the thermal reactions of skin tissue utilizing a fragmentary model of bioheat condition along sinusoidal heat flux applied on the surface of the skin. Hobiny et al. [32] investigated the bioheat model to get to the thermal harm of living tissue by laser light using the fractional-order derivative. In spite of these, several other researchers such as Mahmoud et al. [33], Marin et al. [34], Zhang and Fu [35], Abbas and Marin [36], Bhatti et al. [37], Marin [38, 39], Lata and Kaur [40–43], Ezzat MA et al. [44–46] and Hobiny et al. [47] also worked profoundly on the various hypothesis of thermo elasticity for bioheat transfer. From this literature it is concluded that bioheat can be applied in plenty of areas, particularly for the treatment of malignant cells. Therefore, on the basis of work done by a number of researchers, it is concluded that bioheat is one of the most promising technique to cure the malignant cells but the size of malignancy may present some challenges while curing using bioheat and thermal energy. Therefore, on the basis of this literature survey, the following mathematical equation and problem are formulated and answer is given to know the possible temperature supplies in biological tissues consisting of healthy and malignant cells.

3 Mathematical Equations for Modeling

The novel model of fractional-order three-phase lag Pennes' bioheat transfer equation (FOTPLPBE) with instant surface heating due to laser irradiation is provided by the equation as underneath (1):

$$\left[K\left(1+\frac{(\tau_t)^\alpha}{\alpha!}\frac{\partial^\alpha}{\partial t^\alpha}\right)\frac{\partial}{\partial t}+K^*\left(1+\frac{(\tau_v)^\alpha}{\alpha!}\frac{\partial^\alpha}{\partial t^\alpha}\right)\right]\nabla^2 T$$
$$=\left(\frac{\partial}{\partial t}+\frac{(\tau_q)^\alpha}{\alpha!}\frac{\partial^{\alpha+1}}{\partial t^{\alpha+1}}+\frac{(\tau_q)^{2\alpha}}{2\alpha!}\frac{\partial^{2\alpha+1}}{\partial t^{2\alpha+1}}\right)\left(\rho c\dot{T}-Q_b-Q_m-Q_{ext}\right), \quad (1)$$

where

$$\theta(x,t)=T(x,t)-T_0, \quad (2)$$

$$Q_b=\omega_b\rho_b c_b(T_b-T), \quad (3)$$

$$Q_{ext}(x,t)=I_0\mu_a\left[U(t)-U\left(t-\tau_p\right)\right]\left[C_1 e^{-\frac{k_1}{\delta}x}-C_2 e^{-\frac{k_2}{\delta}x}\right]. \quad (4)$$

Note that the generation of metabolic heat is a function of local tissue temperature, which can be expressed as:

$$Q_m=Q_{mo}\times 2^{\beta\left(\frac{T-T_0}{10}\right)} \quad (5)$$

where T_0 is the starting temperature of the local tissue, Q_{mo} is the referring metabolic heating means, and β is the persistent relative metabolism. For taking all realistic goals into consideration, the dependence on metabolic heat can be approached as a linear function of local tissue temperature as described by the equation underneath (6):

$$Q_m=Q_{mo}\left(1+\beta\left(\frac{T-T_0}{10}\right)\right). \quad (6)$$

Following Jacques [48] C_1, C_2, k_1 and k_2 are given by

$$C_1 = 3.09 + 5.44R_d - 2.12\exp(-21.5R_d) \tag{7}$$

$$C_2 = 2.09 + 1.45R_d - 2.09\exp(-21.5R_d) \tag{8}$$

$$k_1 = 1 - 0.423\exp(-20.1R_d) \tag{9}$$

$$k_2 = 1.53\exp(3.4R_d) \tag{10}$$

and penetration depth of exposed T-cells is expressed by the Eq. (11)

$$\delta = \frac{1}{\sqrt{3\mu_a(\mu_a + \mu_s(1-g))}} \tag{11}$$

4 Problem Formulation and Solution

The temperature dissemination in a semi-infinite biological tissue with instantaneous surface heating and with the laser thermal source of 1-D model of fractional-order three-phase lag Pennes' bioheat transfer equation (FOTPLPBE) in a finite medium is considered. The 1-D form of Eq. (1) is written in terms of Eq. (12):

$$\begin{aligned}
&\left[K\left(1+\frac{(\tau_t)^\alpha}{\alpha!}\frac{\partial^\alpha}{\partial t^\alpha}\right)\frac{\partial}{\partial t}+K^*\left(1+\frac{(\tau_v)^\alpha}{\alpha!}\frac{\partial^\alpha}{\partial t^\alpha}\right)\right]\frac{\partial^2\theta}{\partial x^2}\\
&=\left(\frac{\partial}{\partial t}+\frac{(\tau_q)^\alpha}{\alpha!}\frac{\partial^{\alpha+1}}{\partial t^{\alpha+1}}+\frac{(\tau_q)^{2\alpha}}{2\alpha!}\frac{\partial^{2\alpha+1}}{\partial t^{2\alpha+1}}\right)\left(\rho c\frac{\partial\theta}{\partial t}\right.\\
&\quad-\omega_b\rho_b c_b(T_b-\theta-T_0)-Q_{mo}\left(1+\beta\left(\frac{T-T_0}{10}\right)\right)\\
&\quad\left.-I_0\mu_a\left[U(t)-U\left(t-\tau_p\right)\right]\left[C_1e^{-\frac{k_1}{\delta}x}-C_2e^{-\frac{k_2}{\delta}x}\right]\right)
\end{aligned} \tag{12}$$

It is considered that heat flux $\to 0$ deep inside the T-cell. The starting constraints are:

$$\theta(x,0) = \dot{\theta}(x,0) = 0, 0 \le x \le d. \tag{13}$$

Consider the semi-finite field of biological tissue with a thickness of L, and the skin tissue of malignant cell is exposed to instantaneous surface heating. We consider that heat flux $\to 0$ deep inside the malignant tissue. The applicable boundary conditions are:

$$\theta(0,t)=\theta_0, -\left[K\left(1+\frac{(\tau_t)^\alpha}{\alpha!}\frac{\partial^\alpha}{\partial t^\alpha}\right)\frac{\partial}{\partial t}+K^*\left(1+\frac{(\tau_v)^\alpha}{\alpha!}\frac{\partial^\alpha}{\partial t^\alpha}\right)\right]\frac{\partial\theta}{\partial x}=0,$$
$$\text{for } x=L, t>0. \tag{14}$$

For simplifying the solution, the following nondimensional quantities are given by

$$\theta'=\frac{\theta}{T_0}, T_b'=\frac{T_b-T_0}{T_0}, t'=\frac{1}{\rho cL^2}t, \tau_0'=\frac{1}{\rho cL^2}\tau_0, \tau_p'=\frac{1}{\rho cL^2}\tau_p,$$
$$\tau_t'=\frac{1}{\rho cL^2}\tau_t, \tau_q'=\frac{1}{\rho cL^2}\tau_q, \tau_v'=\frac{1}{\rho cL^2}\tau_v, x'=\frac{x}{L}, k_1'=Lk_1,$$
$$k_2'=Lk_2, R_b=\rho_b\omega_b c_b L^2, R_m=Q_{mo}\left(1+\beta\left(\frac{\theta T_0}{10}\right)\right),$$
$$R_r=\frac{L^2 I_0\mu_a}{T_0}, Q_m'=\frac{L^2}{T_0}Q_m \tag{15}$$

Using these nondimensional quantities defined in (15), the governing Eq. (12) boundary and initial constraints (13) and (14) can be written as (by ignoring dashes) Eq. (16).

$$\left[K\left(1+\frac{(\tau_t)^\alpha}{\alpha!}\frac{\partial^\alpha}{\partial t^\alpha}\right)\frac{\partial}{\partial t}+K^*\left(1+\frac{(\tau_v)^\alpha}{\alpha!}\frac{\partial^\alpha}{\partial t^\alpha}\right)\right]\frac{\partial^2\theta}{\partial x^2}$$
$$=\left(\frac{\partial}{\partial t}+\frac{(\tau_q)^\alpha}{\alpha!}\frac{\partial^{\alpha+1}}{\partial t^{\alpha+1}}+\frac{(\tau_q)^{2\alpha}}{2\alpha!}\frac{\partial^{2\alpha+1}}{\partial t^{2\alpha+1}}\right)\left(\frac{\partial\theta}{\partial t}-R_b(T_b-\theta)\right.$$
$$\left.-R_m-R_r\left[U(t)-U(t-\tau_p)\right]\left[C_1e^{-\frac{k_1}{\delta}x}-C_2e^{-\frac{k_2}{\delta}x}\right]\right). \tag{16}$$

Laplace transform is given by

$$\mathcal{L}[f(x,t)]=\int_0^\infty e^{-st}f(x,t)dt=\overline{f}(x,s). \tag{17}$$

with basic properties

$$L\left(\frac{\partial f}{\partial t}\right)=s\overline{f}(x,s)-f(x,0), \tag{18}$$

$$L\left(\frac{\partial^2 f}{\partial t^2}\right)=s^2\overline{f}(x,s)-sf(x,0)-\left(\frac{\partial f}{\partial t}\right)_{t=0}. \tag{19}$$

$$L(U(t-a))=\frac{e^{-as}}{s}, L(U(t))=\frac{1}{s}. \tag{20}$$

where

$$U(t-a)=\begin{cases}0, & \text{if } \ 0\le t<a\\ 1, & \text{if } \ t>a\end{cases} \tag{21}$$

Thus, we get

$$D^2\overline{\theta}=G\left(s\overline{\theta}-\frac{R_bT_b}{s}+R_b\overline{\theta}-\frac{R_m}{s}-\frac{R_r}{s}\left[1-e^{-\tau_p s}\right]\left[C_1e^{-\frac{k_1}{\delta}x}-C_2e^{-\frac{k_2}{\delta}x}\right]\right), \tag{22}$$

where

$$D=\frac{d}{dx},\mathrm{G}=\frac{\left(s+\frac{(\tau_q)^\alpha}{\alpha!}s^{\alpha+1}+\frac{(\tau_q)^{2\alpha}}{2\alpha!}s^{2\alpha+1}\right)}{\left[K\left(1+\frac{(\tau_t)^\alpha}{\alpha!}s^\alpha\right)\frac{\partial}{\partial t}+K^*\left(1+\frac{(\tau_v)^\alpha}{\alpha!}s^\alpha\right)\right]}, \tag{23}$$

It can be further simplified as

$$(D^2-\eta^2)\overline{\theta}=\zeta_1+\zeta_2e^{-\frac{k_1}{\delta}x}+\zeta_3e^{-\frac{k_2}{\delta}x}, \tag{24}$$

where

$$\eta^2=G(s+R_b), \tag{25}$$

$$\zeta_1=-G\left(\frac{R_bT_b}{s}+\frac{R_m}{s}\right), \tag{26}$$

$$\zeta_2=-G\frac{R_r}{s}\left[1-e^{-\tau_p s}\right]C_1 \tag{27}$$

$$\zeta_3=-G\frac{R_r}{s}\left[1-e^{-\tau_p s}\right]C_2 \tag{28}$$

And the boundary conditions after the application of Laplace transform (17) takes the form of the following Eq. (29) as:

$$\overline{\theta}(0,s)=\frac{\theta_0}{s},\ \left.\frac{d\theta}{dx}\right|_{x=0,L}=0,0\le x\le d,Re(s)>0. \tag{29}$$

By using the boundary conditions defined in Eq. (29) in Eq. 24, the exact solution is obtained as:

$$\bar{\theta}(x,s) = \frac{\theta_0}{s}\frac{\cosh\eta(x-d)}{\cosh\eta d} + \zeta_4\frac{\cosh\eta x}{\cosh\eta d} + \zeta_5\frac{\sinh\eta x}{\eta\cosh\eta d} - \frac{\zeta_1}{\eta^2} + \frac{\zeta_2 e^{-\frac{k_1}{\delta}x}}{k_1^2-\eta^2} + \frac{\zeta_3 e^{-\frac{k_2}{\delta}x}}{k_2^2-\eta^2} \tag{30}$$

where

$$\zeta_4 = \frac{\zeta_1}{\eta^2} - \frac{\zeta_2}{k_1^2-\eta^2} - \frac{\zeta_3}{k_2^2-\eta^2} \tag{31}$$

$$\zeta_5 = \frac{\zeta_1}{\eta^2} - \frac{\zeta_2 e^{-\frac{k_1}{\delta}d}}{k_1^2-\eta^2} - \frac{\zeta_3 e^{-\frac{k_2}{\delta}d}}{k_2^2-\eta^2}$$

The thermal damage, i.e., evaluation of burn caused by laser radiation, following Jasiński [49], Askarizadeh and Ahmadikia [50] is given by the Eq. (32) as:

$$\Omega = \int_0^t B e^{-\frac{E_a}{RT}}\,\mathrm{d}t \tag{32}$$

5 Results and Discussion

Generally, it is sufficient that the heat convection by blood flow is at least in the blood vessel. So, to complete this novel exploration on hypothetical and useful systems, a well-known thermal modeling in biological processes based on Pennes' bio-heat transfer equation (PBHTE) has been employed to predict temperature dissemination in living tissue by leading an arrangement of examinations estimating heat swap among blood flow and solid tissues (T-cells) by running a series of simulations. The outcomes exhibited and revealed that the temperature and thermal damage of blood vessels and tumors exposed to laser radiation heating in the physical structure are very well determined and obtained. The following Askarizadeh & Ahmadikia [50] parameters and their related values are used for numerical results:

$\rho_b = 1060\text{kgm}^{-3}$,	$c = 4187\text{Jkg}^{-1}\text{K}^{-1}$,
$c_b = 3860\text{Jkg}^{-1}\text{K}^{-1}$,	$\rho = 1000\text{kgm}^{-3}$,
$\omega_b = 1.87 \times 10^{-3}\text{s}^{-1}$,	$k = 0.628\text{Wm}^{-1}\text{K}^{-1}$,
$T_b = 37\ °\text{C}$,	$\tau_p = 10\text{s}$,
$g = 0.9$,	$L = 0.03\text{m}$,
$\mu_s = 12000\text{m}^{-1}$,	$\mu_a = 40\text{m}^{-1}$,
$R = 8.313\text{J/mol}\cdot\text{K}$	$T_0 = 37\ °\text{C}$
$Ea = 6.28 \times 105\text{J/mol}$	$B = 3.1 \times 1098\text{s}^{-1}$

The numerical results are simulated using Python software (Anaconda platform) installed on Windows 8.1 operating system with 8 GB RAM and Intel Core i3

(1.70 GHz) processor. The impact of laser source on the surface of malignant cells is incorporated. The proposed mathematical models depend on the bioheat transfer found and suitable boundary conditions. The conducting heat source, metabolic and perfusions are used in the formulations of the mathematical model. Numerical results are presented graphically from Figs. 1, 2, 3 to study the influence of fractional-order derivative parameter α, the laser exposure time τ_p, and the thermal relaxation time τ_0 on the temperature and the thermal deterioration. The malignant tissue is considered to be 0.03 m thick, and the referring temperature equals to the normal malignant temperature, that is, $T_0 = T_b = 37^{\circ} C$.

Figure 2 exhibits the deviation of temperature with the separation x by keeping the values of $\tau_0 = 5$ s and $\tau_p = 10$ s with various values of fractional-order derivative parameter α. It is seen that the temperature begins rising with the distance as the blood perfusions in malignant cells increase. Figure 2 demonstrates the thermal damage w.r.t. time t, keeping the values of $\tau_0 = 5$ s and $\tau_p = 10$ s with different values of fractional-order derivative parameter $\alpha = 0.3$. It has the most noteworthy impact on the thermal damages on the T-cells.

Figure 3 presents the change of temperature along the distance x by keeping the values of $\tau_0 = 5$ s and $\tau_p = 10$ s with different values of fractional-order derivative parameter α. It is very well understood that the temperature starts from the most extreme value and declines rapidly.

Figure 4 reveals the thermal damage w.r.t. time t keeping the values of $\tau_0 = 5$ s and $\alpha = .3$ with various values of τ_p. It is seen that the thermal damage to T-cell

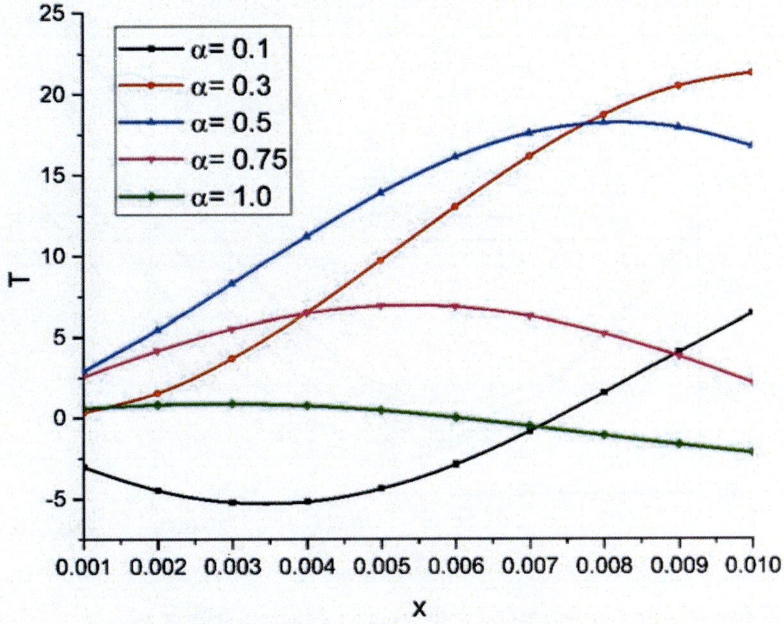

Fig. 2 Temperature variations w.r.t. T-cells depth for diverse values of α.

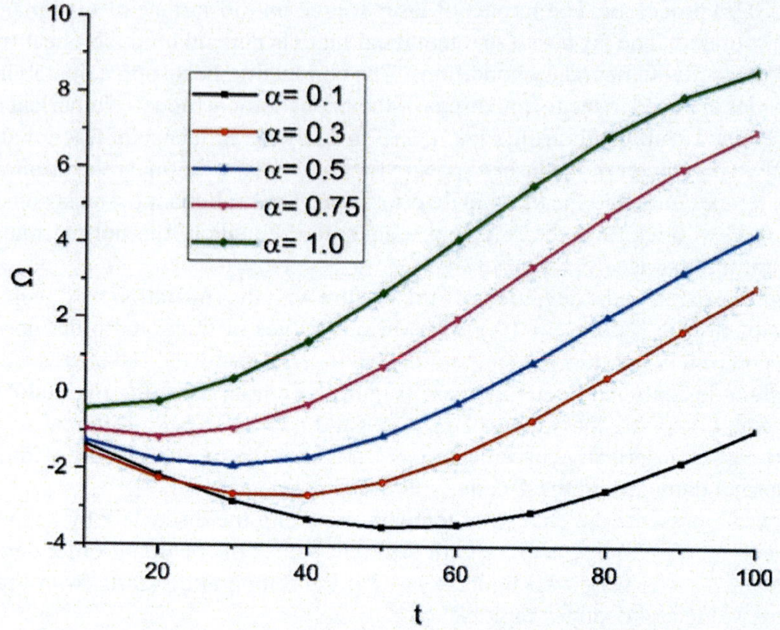

Fig. 3 The change of thermal deterioration w.r.t. diverse values of α in T-cells

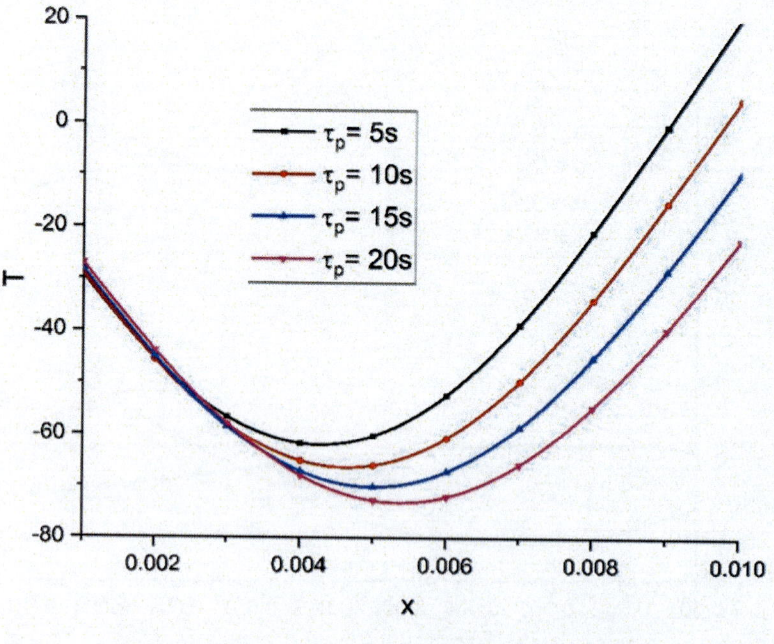

Fig. 4 Temperature distributions w.r.t. T-cells depth for different values of τ_p

starts from the most extreme value and decreases rapidly. It is observed that $\tau_p = 5\text{s}$ has the most noteworthy effect on the thermal damage to the T-cells.

6 Conclusion

The major objective of this research work is to analyze a new bioheat mathematical model centered on FOTPLPBE, which involves the time delay parameter α of the Pennes' biological heat transfer equation, and is employed to scrutinize the thermal characteristics of malignant tissues when exposed to thermal or laser radiations. It is very well concluded that the thermal changes and their damages to the malignant cells can be measured in a better way with the FOTPLPBE model. Blood perfusion prevents tissue damage by exerting a cooling function. In this research, the FOTPLPBE involving fractional-order parameter α becomes a new measure of efficiency for bioheat transfer in the malignant cells. These results may be beneficial in the study and further improvements in the applications of thermotherapy or bioheat therapy in malignant tissues. It will present new harbinger in the field of clinical and practical oncology in the long run.

The future research directions will be entirely concentrated on the appropriateness of the Pennes' modeling equation and other related transformation, particularly related to stem cells. Because stem cells are cells in blood and bone marrow that haven't developed into their last structure. The specialist utilizes them to replace cells in the bone marrow that different treatment kills which is the best for the kind of malignancy. It is widely proved that a number of times stem cells may find and kill malignant cells. In today's scenarios, the stem cell therapy is used in the treatment of various hereditary illnesses, tumors and blood disorders, including thalassemia and all sort of blood cancers. Blood reaped from the umbilical cord during delivery is a rich source of stem cell and if preserved, the patient and others with blood disorders problems, for example, extreme sickle cell anemia and thalassemia, just as some essential immunodeficiency sicknesses and certain metabolic issues, can be cured. So, it is intuitively defended that to treat any malignancy in the future will surely be based on the stem cells replication together with appropriate bioheat models like Pennes' modeling equation in situ to any therapy, including the treatment of diseases like anemia, arthritis, autism, brain disorders, blindness, diabetes, genetic disorders for cerebral palsy, myeloma, heart disease, kidney damage, leukemia, liver disease, retinal degeneration and other diseases.

Conflict of Interest All authors confirm no conflict of interest or funding for this study.

Transparency Declaration The authors certify that the manuscript is reported clearly, truthfully and precisely with no omission of any important aspect.

Authors' Contributions Statement Sharduli: Idea formulation, Conceptualization, Effective literature review, Methodology, Formal analysis, Validation, Software, Writing review and editing, Visualization.

Iqbal Kaur: Conceptualization, Formulated strategies for mathematical modelling, methodology refinement, Writing- review & editing.

Kulvinder Singh: Experiments and Simulation, Investigation, Methodology, Software, Supervision, Validation, Visualization, Writing - original draft.

References

1. Baish JW (1994) Formulation of a statistical model of heat transfer in perfused tissue. ASME J Biomech Eng 116(4):521–527
2. Borgos J (1996) Laser doppler monitoring of cerebral blood flow. Neurol Res 18:251–255
3. Chatterjee I, Adams RE (1994) Finite element thermal modeling of the human body under hyperthermia treatment for cancer. Int J Comput Appl Technol 7:151–159
4. Sethi M, Chakarvarti SK (2015) Hyperthermia techniques for cancer treatment: a review. Int J Pharm Tech Res 8(6):974–4304
5. Tian S, Wu J, Li C, Zhang B, Jiang C, Liu G, Chen Y, Qin W, Li F, Bai Y, Wu Y, Zhu Y (2017) Apoptotic cell characteristics of rat brain microvascular endothelia induced by different degrees of hypoperfusion. Int J Clinical Exp Pathol 1,10(11):11360–11368
6. Crezee J, Lagendijk JJW (1990) Experimental verification of bioheat transfer theories: measurement of temperature profiles around large artificial vessels in perfused tissue. Phys Med Biol 35(7):905–923
7. Hase M, Sako M, Fujii M, Ueda E, Nagae T, Shimizu T, Hirota S, Kono M (1989) Experimental study of embolohyperthermia for the treatment of liver tumors by induction heating to ferromagnetic particles injected into tumor tissue. Nippon Acta Radiologica 49:1171–1173
8. Sun H, Linfeng Xu, Fan T, Zhan H, Wang X, Zhou Y, Yang R-J (2013) Targeted hyperthermia after selective embolization with ferromagnetic nanoparticles in a VX_2 rabbit liver tumor model. Int J Nanomed 8:3795–3804
9. Hilger I, Andra W, Hergt R, Hiergeist R, Schubert H, Kaiser WA (2001) Electromagnetic heating of breast tumors in interventional radiology: in vitro and in vivo studies in human cadavers and mice. Radiology 218:570–575
10. Hilger I, Hergt R, Kaiser WA (2005) Towards breast cancer treatment by magnetic heating. J Magn Magn Mater 293:314–319
11. Hu L, Gupta A, Gore JP, Xu LX (2004) Effect of forced convection on the skin thermal expression of breast cancer. J Biomech Eng 126(2):204–211
12. Weinbaum S, Jiji LM (1985) A new simplified bioheat equation for the effect of blood flow on local average tissue temperature. J Biomech Eng 107:131–139
13. Zhu L, He Q, Xu LX, Weinbaum S (2002) A new fundamental bioheat equation for muscle tissue: part ii—temperature of SAV Vessels. ASME J Biomech Eng 124:121–132
14. Khaled ARA, Vafai K (2003) The role of porous media in modeling flow and heat transfer in biological tissues. Int J Heat Mass Trans 4989–5003
15. Sharma PR, Sazid A, Katiyar VK (2009) Numerical study of heat propagation in living tissue subjected to instantaneous heating. Indian J Biomech 205–209
16. Shih T-C, Yuan P, Lin W-L, Kou H-S (2007) Analytical analysis of the pennes bioheat transfer equation with sinusoidal heat flux condition on skin surface. Med Eng Phys 29:946–953
17. Lipkin M, Hardy JD (1954) Measurement of some thermal properties of human tissues. J Appl Physiol 7:212–217. https://doi.org/10.1152/jappl.1954.7.2.212
18. Giering K, Minet O, Lamprecht I, Muller G (1995) Review of thermal properties of biological tissues. SPIE Opt Eng Press 044:45–65. https://doi.org/10.1007/978-3-319-95432-5_4
19. Shrivastava D, Vaughan JT (2009) A generic bioheat transfer thermal model for a perfused tissue. J Biomech Eng 131:1–5. https://doi.org/10.1115/1.3127260

20. Gardner CM, Jacques SL, Welch AJ (1996) Light transport in tissue: accurate expressions for one-dimensional fluence rate and escape function based upon Monte Carlo simulation. Lasers Surg Med 18:129–138. https://doi.org/10.1002/(SICI)1096-9101(1996)18:2%3c129::AID-LSM2%3e3.0.CO;2-U
21. Cheng HLM, Plewes DB (2002) Tissue thermal conductivity by magnetic resonance thermometry and focused ultrasound heating. J Magn Reson Imaging 16:598–609. https://doi.org/10.1002/jmri.10199
22. Othman MIA, Ali MGS, Farouk RM (2011) The effect of relaxation time on the heat transfer and temperature distribution in tissues. World J Mech 01:283–287. https://doi.org/10.4236/wjm.2011.16035
23. Kengne E, Lakhssassi A, Vaillancourt R (2012) Temperature distribution in living biological tissue simultaneously subjected to oscillatory surface and spatial heating: analytical and numerical analysis. Int Math Forum 7:2373–2392. https://doi.org/10.4438/1988-599X-RE-2017-381-387
24. Youn J-I, Lee S (2012) Diffuse reflectance simulation in tissue for high-intensity laser therapy. Int Conf Biol Biomed Pharm Sci 1:90–93
25. Deng Z-S, Liu J (2012) Analytical Solutions to 3-D bioheat transfer problems with or without phase change. Heat Transf Phenom Appl. In Tech. https://doi.org/10.5772/52963
26. Kujawska T, Secomski W, Kruglenko E, Krawczyk K, Nowicki A (2014) Determination of tissue thermal conductivity by measuring and modeling temperature rise induced in tissue by pulsed focused ultrasound. PLoS ONE 9:e94929. https://doi.org/10.1371/journal.pone.0094929
27. Kumar R, Vashisth AK, Ghangas S (2017) Analytical solution of bioheat transfer equation with variable thermal conductivity in skin. Int J Eng Res Dev 14:11–17
28. Kumar R, Vashishth AK, Ghangas S (2019) Phase-lag effects in skin tissue during transient heating. Int J Appl Mech Eng 24:603–623. https://doi.org/10.2478/ijame-2019-0038
29. Youssef HM, Alghamdi NA (2020) Modeling of one-dimensional thermoelastic dual-phase-lag skin tissue subjected to different types of thermal loading. Sci Rep 10:3399. https://doi.org/10.1038/s41598-020-60342-6
30. Ezzat MA, Alsowayan NS, Al-Muhiameed ZIA, Ezzat SM (2014) Fractional modelling of Pennes' bioheat transfer equation. Heat Mass Transf Und Stoffuebertragung 50:907–914. https://doi.org/10.1007/s00231-014-1300-x
31. Ezzat MA, El-bary AA, Al-sowayan NS (2016) Tissue responses to fractional transient heating with sinusoidal heat flux condition on skin surface. Anim Sci J 87:1304–1311. https://doi.org/10.1111/asj.12568
32. Hobiny A, Alzahrani F, Abbas I, Marin M (2020) The effect of fractional time derivative of bioheat model in skin tissue induced to laser irradiation. Symmetry (Basel) 12:602. https://doi.org/10.3390/sym12040602
33. Mahmoud SR, Marin M, Al-Basyouni KS. Effect of the initial stress and rotation on free vibrations in transversely isotropic human long dry bone. Analele Univ "Ovidius" Constanta Ser Mat 23:171–84. https://doi.org/10.1515/auom-2015-0011
34. Marin M, Vlase S, Ellahi R, Bhatti MM (2019) On the partition of energies for the backward in time problem of thermoelastic materials with a dipolar structure. Symmetry (Basel) 11:863. https://doi.org/10.3390/sym11070863
35. Zhang J, Fu Y (2012) Pull-in analysis of electrically actuated viscoelastic microbeams based on a modified couple stress theory. Meccanica 47:1649–1658. https://doi.org/10.1007/s11012-012-9545-2
36. Abbas IA, Marin M (2017) Analytical solution of thermoelastic interaction in a half-space by pulsed laser heating. Phys E Low-Dimensional Syst Nanostruct 87:254–260. https://doi.org/10.1016/j.physe.2016.10.048
37. Bhatti MM, Ellahi R, Zeeshan A, Marin M, Ijaz N (2019) Numerical study of heat transfer and Hall current impact on peristaltic propulsion of particle-fluid suspension with compliant wall properties. Mod Phys Lett B 33:1950439. https://doi.org/10.1142/S0217984919504396

38. Marin M (2010) Lagrange identity method for microstretch thermoelastic materials. J Math Anal Appl 363:275–286. https://doi.org/10.1016/j.jmaa.2009.08.045
39. Marin M (1994) The Lagrange identity method in thermoelasticity of bodies with microstructure. Int J Eng Sci 32:1229–1240. https://doi.org/10.1016/0020-7225(94)90034-5
40. Kaur I, Lata P (2020) Stoneley wave propagation in transversely isotropic thermoelastic medium with two temperature and rotation. GEM Int J Geomathematics 11:1–17. https://doi.org/10.1007/s13137-020-0140-8
41. Kaur I, Lata P, Singh K (2020) Effect of Hall current in transversely isotropic magneto-thermoelastic rotating medium with fractional-order generalized heat transfer due to ramp-type heat. Indian J Phys. https://doi.org/10.1007/s12648-020-01718-2
42. Kaur I, Lata P (2019) Effect of hall current on propagation of plane wave in transversely isotropic thermoelastic medium with two temperature and fractional order heat transfer. SN Appl Sci 1. https://doi.org/10.1007/s42452-019-0942-1
43. Kaur I, Lata P (2019) Transversely isotropic thermoelastic thin circular plate with constant and periodically varying load and heat source. Int J Mech Mater Eng 14. https://doi.org/10.1186/s40712-019-0107-4
44. Ezzat MA, El-Karamany AS, El-Bary AA (2014) Generalized thermo-viscoelasticity with memory-dependent derivatives. Int J Mech Sci 89:470–475. https://doi.org/10.1016/j.ijmecsci.2014.10.006
45. Ezzat MA, El-Karamany AS, El-Bary AA (2015) A novel magneto-thermoelasticity theory with memory-dependent derivative. J Electromagn Waves Appl 29:1018–1031. https://doi.org/10.1080/09205071.2015.1027795
46. Ezzat MA, El-Karamany AS, El-Bary AA (2016) Modeling of memory-dependent derivative in generalized thermoelasticity. Eur Phys J Plus 131. https://doi.org/10.1140/epjp/i2016-16372-3
47. Hobiny A, Alzahrani F, Abbas I, Marin M (2020) The effect of fractional time derivative of bioheat model in skin tissue induced to laser irradiation
48. Jacques SL (1998) Light distributions from point, line and plane sources for photochemical reactions and fluorescence in turbid biological tissues. Photochem Photobiol 67:23–32. https://doi.org/10.1111/j.1751-1097.1998.tb05161.x
49. Jasiński M (2015) Modelling of thermal damage in laser irradiated tissue. J Appl Math Comput Mech 14:67–78. https://doi.org/10.17512/jamcm.2015.4.07
50. Askarizadeh H, Ahmadikia H (2016) Skin burn evaluation under the effects of the dual-phase-lag bio-heat transfer skin burn evaluation under the effects of the dual-phase-lag bio-heat transfer. In: 24th annual international conference of mechanical engineering. Azad University, Yazd, Iran, pp 1–6

Chapter 22
Various Swarm Optimization Algorithms: Review, Challenges, and Opportunities

Sachin Dhawan, Rashmi Gupta, Arun Rana, and Sharad Sharma

1 Introduction to Artificial Intelligence

The term "*Artificial Intelligence*" or "*Artificial Life*" refers to the hypothesis of reenacting human behavior through computation. It includes planning such computer frameworks which can execute errands that require human intelligence. For instance, earlier people could just perceive the speech of an individual. But now, this process is possible by any advanced gadget. This has become possible through artificial intelligence. Different instances of human knowledge may incorporate dynamic, language interpretation, visual observation, and so forth. Also, the internet of things (IoT) might probably slip into the vague pit of fashionable speech. Artificial intelligence also cascades into a specific snare, especially through the advent of modern words such as "AI," "profound science," "hereditary predictions," and then some [1–5]. AI gives various security techniques related to steganography in which we can achieve optimization using swarm intelligence with the comparison of various steganography techniques which can be useful for selecting the security technique and making the system more intelligent using swarm intelligence [6, 7]. Different methods make it conceivable. These strategies to actualize *artificial intel* into PCs are prominently known as approaches of *AI*.

Artificial intelligence can be categorized into three parts. Computational intelligence is one of the three parts of artificial intelligence. This paper will present a detailed review of computational intelligence.

S. Dhawan (✉)
Ambedkar Institute of Advanced Communication Technologies & Research, Geeta Colony, New Delhi, India

R. Gupta
NSUT, East Campus, New Delhi, India

A. Rana · S. Sharma
Maharishi Markandeshwar (Deemed to be University), Mullana, Haryana, India

N. Marriwala et al. (eds.), *Soft Computing for Intelligent Systems*, Algorithms for Intelligent Systems, https://doi.org/10.1007/978-981-16-1048-6_22

1.1 Classification of Artificial Intelligence

Artificial intelligence can be classified into three ways. Classifications are shown in Fig. 1.

2 Computational Intelligence

Any of the three approaches can be used to apply computational intelligence:

- Artificial neural network
- Fuzzy logic
- Evolutionary computation

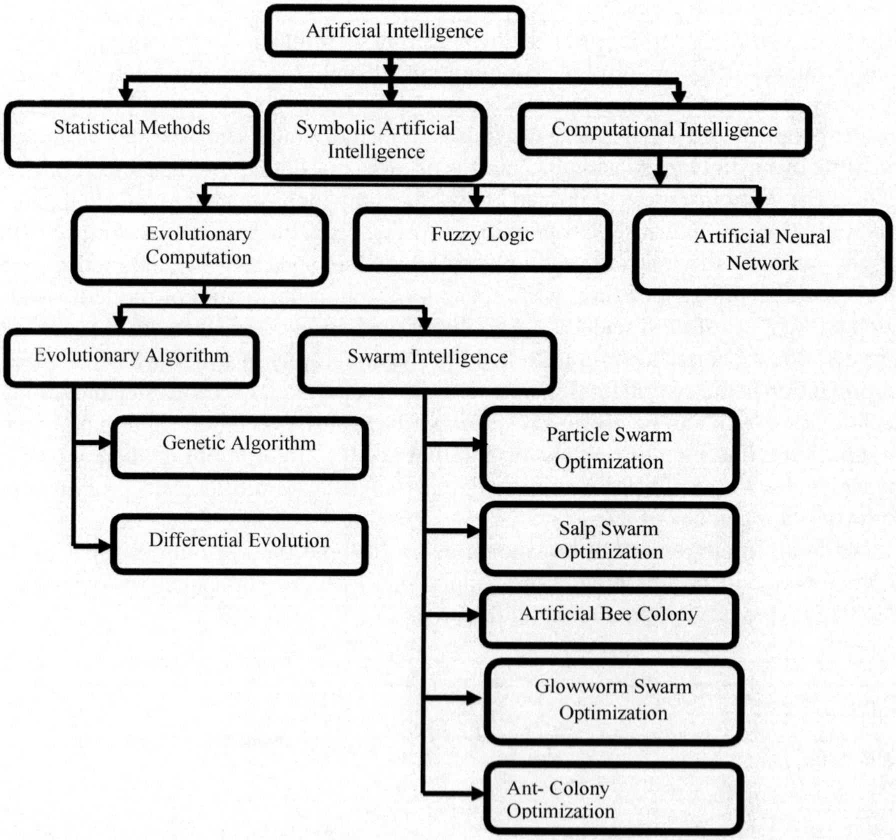

Fig. 1 Classification of artificial intelligence

2.1 Artificial Neural Network

Being a least mind-boggling structure, an artificial neural network (ANN) is a pantomime of the human cerebrum [8]. A human psyche can learn new things and acclimate to another in developing conditions. The psyche has the most dazzling ability to explore lacking and obscure, fluffy information, and makes its judgment out of it. For instance, we can perceive other's penmanship anyway, but the way where they form may not be exactly equivalent to how we make. We can separate a realized individual even from a foggy photograph. An adolescent can perceive that the condition of both a ball and an orange is a circle. In reality, even two- or three-day-old newborn child can see its mother from the touch, voice, and smell. The cerebrum is a significantly baffling organ that controls the entire body. Its ability is not just controlling the physical bits of the body, yet also of progressively complex activities like thinking, imagining, dreaming, imagining, learning, etc., practices that can't be depicted in physical terms. A fake systems machine is still past the breaking point of the most evolved supercomputers.

2.1.1 Artificial Neuron

An artificial neural system comprises preparing units called neurons. A counterfeit neuron attempts to repeat the structure and conduct of the common neuron. A neuron comprises data sources (dendrites) and one yield (neurotransmitter using axon). The neuron has a capacity that decides the enactment of the neuron [9] (Fig. 2).

$x_1 \ldots x_n$ are the contributions to the neuron. A predisposition is additionally added to the neuron alongside inputs. Generally, inclination esteem is initialized to 1. $W_0 \ldots W_n$ are the loads. Weight is the association with the sign. The result of weight

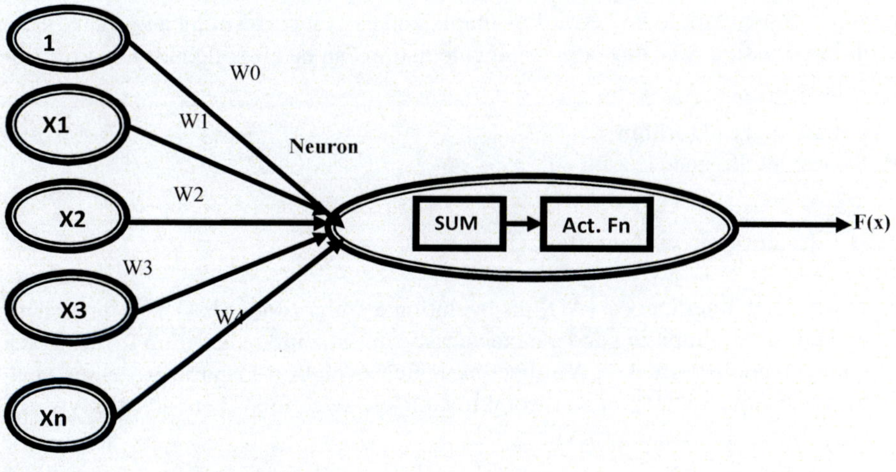

Fig. 2 Model of an artificial neuron

and info invigorates the sign. A neuron gets numerous contributions from various sources and has a solitary yield. There are different capacities utilized for initiation. One of the most usually utilized enactment work is the sigmoid capacity, given by

$$F(x) = \frac{1}{1 + e^{-xwn}}$$

2.2 *Fuzzy Logic*

Lotfi A. Zadeh, Professor of Computer Science at the University of California in Berkeley, discovered fuzzy logic in 1965 [10]. Fuzzy logic owes the meaning of the computer. It offers more computer knowledge. A logic based on the two "True and False" values of truth is sometimes inadequate when describing human reasoning [11]. Fuzzy logic uses the entire 0 False to 1 Real range to explain human reasoning.

2.3 *Evolutionary Computation*

The transformative calculation is a territory of software engineering that utilizes thoughts from organic development to take care of computational issues. Numerous such issues require looking through an enormous space of potential outcomes for arrangements, for example, among countless conceivable equipment circuit designs for a setup that produces wanted conduct, a lot of conditions will anticipate the high points and low points of a money-related market, or for an assortment of decision that will control a robot as it explores its condition. Such computational issues frequently require a framework to be versatile—that is, to keep on performing admirably in an evolving situation. The developmental calculation can be classified into two sorts of calculation:

- Evolutionary algorithm
- Swarm intelligence

2.3.1 Evolutionary Algorithm

- In artificial intelligence (AI), an evolutionary algorithm (EA) is a branch of evolutionary computing and a metaheuristic optimization algorithm focused on a common population. An EA utilizes biological evolution-inspired processes, such as replication, development, recombination and selection. There are two types of algorithms of evolution (EA).

   - Genetic algorithm
   - Differential evolution

- **Genetic Algorithm**
  The genetic algorithm (GA) presented by John Holland in 1975 [12] is an inquiry enhancement calculation dependent on the mechanics of the common choice procedure. The fundamental idea of this calculation is to impersonate the idea of "natural selection"; it reproduces the procedures saw in a characteristic framework where the solid will in general adjust and endure while the frail will in general die. GA is a populace-based methodology in which individuals from the populace are positioned dependent on their answers' wellness [13].
- **Differential Evolution**
  The differential evolution (DE) calculation is a populace-based calculation that can be viewed as GA since it utilizes comparable administrators; hybrid, transformation, and determination. The fundamental contrast among DE and GA is in building better arrangements, where DE depends on transformation activity while GA depends on the hybrid activity. This calculation was presented by Storn and Price in 1997. Since this calculation depends on transformation activity, it uses the change as a pursuit system and exploits the determination activity to coordinate the hunt toward the potential locales in the inquiry space.
  A man-made reasoning (AI) strategy dependent on aggregate conduct in decentralized, self-sorted out frameworks. Swarm intelligence (SI) is the method that pulled in numerous analysts associated with different fields for the improvement of any issue identified with their field. Bonabeau characterized SI as "The emanant aggregate knowledge of gatherings of basic operators" [14].
- **Swarm Intelligence Algorithms**
  This segment presents a few SI-based calculations, featuring their remarkable variations, their benefits and negative marks, and their applications. These calculations incorporate genetic algorithms (GA), ant colony optimization (ACO), particle swarm optimization (PSO), differential evolution (DE), artificial bee colony (ABC), glowworm swarm optimization (GSO), and cuckoo search algorithm (CSA).

2.3.2 Swarm Intelligence

(a) Ant Colony Optimization

Ant colony optimization (ACO) is a meta-heuristic method inspired by Marco Dorigo's 1992 Ant System (AS) idea in his Ph.D. proposition [14, 15]. This is inspired by real bees looking for behavior. This measure consists of four simple segments (insect, pheromone, daemon operation, and decentralized control) that contribute to the overall meaning. Ants are fanciful operators used to imitate the search of inquiry and abuse. All things considered a pheromone, which is a synthetic substance that ants disperse along the way they fly, and its strength varies after a while due to dissipation.

Applications of ACO.
The on-going research of ACO is in diverse engineering applications such as here , or .

- Network steering
- Data mining
- Discounted incomes in venture booking
- Grid work process booking issue
- Image preparing
- Intelligent testing framework
- System recognizable proof
- Particle swarm optimization

Particle swarm optimization (PSO) was introduced in 1995 by Kennedy and Eberhart as a method of optimization [16]. To direct the particles in their quest for optimal global solutions, it uses a simple approach to mimic swarm behaviors in flocking birds and in fish schooling. Del Valle and its co-authors defined PSO, as shown in Fig. 3, with three main habits of separation, teamwork, and cohesion [17]. Separation is the practice for preventing the alignment of the local mates as they follow the average path of the local mates. Cohesion is the driving force behind the conventional role. We have implemented the code of particle swarm optimization in MATLAB 2019 for two problem statements.

In the first problem statement we choose optimization problem; it is the minimization of the value of a function called sphere.

- problem.CostFunction = @(x) Sphere(x); % Cost Function
- problem.nVar = 5; % Number of Unknown (Decision) Variables

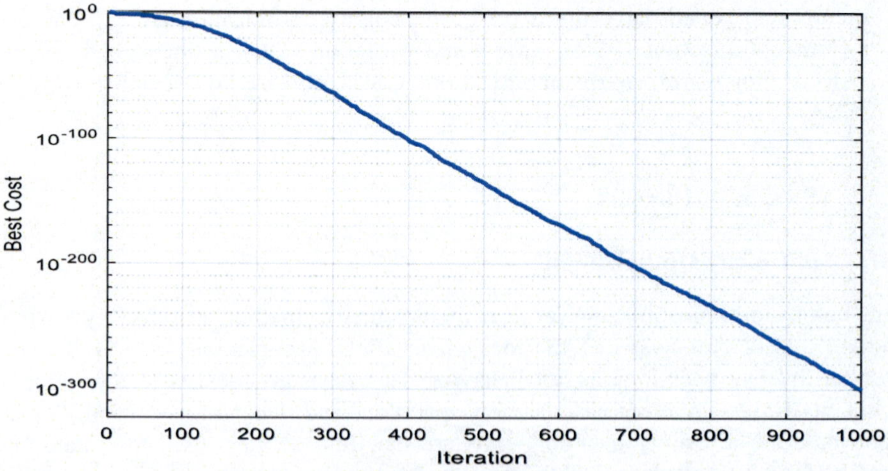

Fig. 3 Cost Function vs. Number of Iterations for PSO of Problem 1

- problem.VarMin = −10; % Lower Bound of Decision Variables
- problem.VarMax = 10; % Upper Bound of Decision Variables

In this implementation, I have used the following parameters, which are used in this problem.

params.MaxIt = 1000; % Maximum Number of Iterations.
params.nPop = 50; % Population Size (Swarm Size).
params.w = 1; % Inertia Coefficient.
params.wdamp = 0.99; % Damping Ratio of Inertia Coefficient.
params.c1 = 2; % Personal Acceleration Coefficient.
params.c2 = 2; % Social Acceleration Coefficient.
params.ShowIterInfo = true; % Flag for Showing Iteration Information.

Figure 3 shows the graph between the best value and no. of iterations; as the no. of iterations increases, the best value decreases.

We have also implemented the same problem with different parameters. Figure 4 shows the implementation of this problem.

problem.CostFunction = @(x) Sphere(x); % Cost Function.
problem.nVar = 20; % Number of Unknown (Decision) Variables.
problem.VarMin = −20; % Lower Bound of Decision Variables.
problem.VarMax = 20; % Upper Bound of Decision Variables.

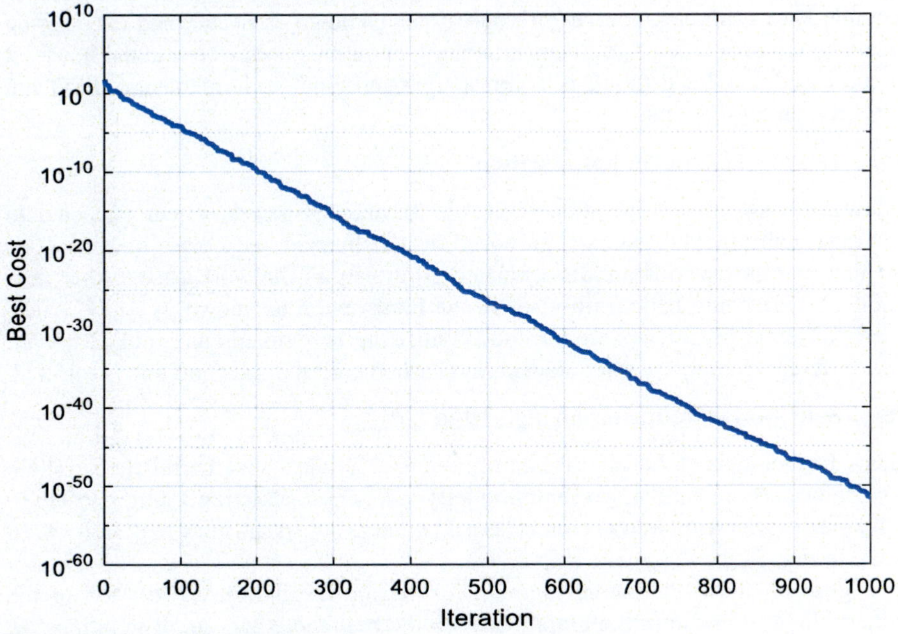

Fig. 4 Cost Function vs. Number of Iterations for PSO of Problem 2

In this implementation, I have used the following parameters, which are used in this problem.

```
% Constriction Coefficients.
kappa = 1;
phi1 = 2.05;
phi2 = 2.05;
phi = phi1 + phi2;
chi = 2*kappa/abs(2-phi-sqrt(phi^2-4*phi));
params.MaxIt = 1000; % Maximum Number of Iterations.
params.nPop = 50; % Population Size (Swarm Size).
params.w = chi; % Inertia Coefficient.
params.wdamp = 1; % Damping Ratio of Inertia Coefficient.
params.c1 = chi*phi1; % Personal Acceleration Coefficient.
params.c2 = chi*phi2; % Social Acceleration Coefficient.
params.ShowIterInfo = true; % Flag for Showing Iteration Information.
```

(c) **Artificial bee colony**

Artificial bee colony (ABC) is one of the most current estimates on swarm awareness. It was suggested by DervisKaraboga in 2005 [18]; in 2007, ABC's analysis was dissected, and it was believed that ABC conducts contrasted with a few specific methodologies all over. This estimate is inspired by the wise conduct of genuine bumble bees in finding sources of nourishment, known as nectar, and exchanging data about that source of nourishment among numerous honey bees in the house. It is professed that this estimate is as transparent and quick to implement as PSO and DE limit target function.

(d) **Glowworm swarm optimization**

Glowworm swarm optimization (GSO) is another SI-based system expected to upgrade multimodal capacities, proposed by Krishnanad and Ghose in 2005 [19]. At first, the glowworms are disseminated arbitrarily in the workspace, rather than limited areas being haphazardly set in the hunt region as shown in ACO. Afterward, different parameters are introduced utilizing predefined constants. However, like different techniques, three stages are rehashed until the end condition is fulfilled.

(e) **Salp swarm optimization algorithm** [20]

Salps are members of the Salpidae family and have a translucent, barrel-shaped shell. Their muscles are similarly resembling jelly fish. These also travel quite closely to jelly fish, in which the water is forced as a movement across the body to push forward [21]. In Fig. 5a the shape of a salp is seen.

Figure 6a shows unimodal function for getting the optimized solution of the sphere. Figure 6b shows the graph between the best score and no. of iterations. In

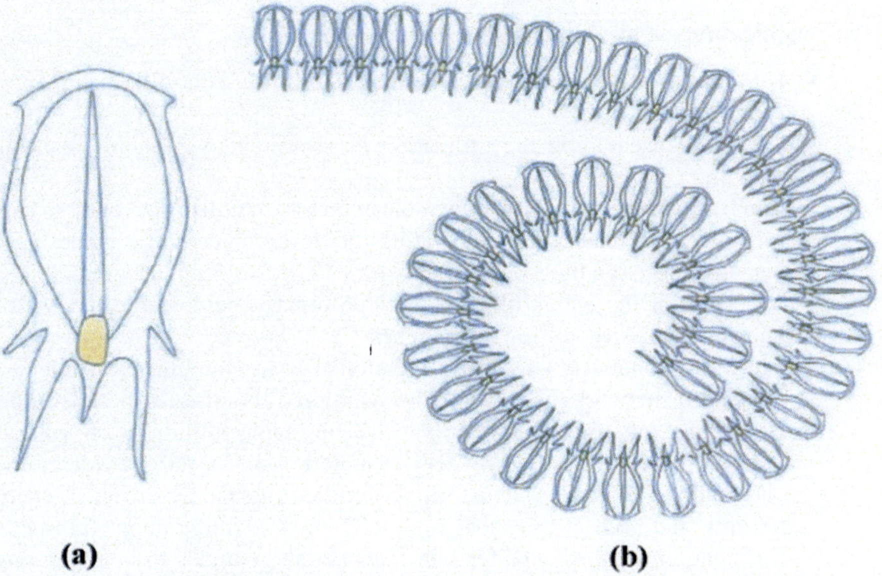

Fig. 5 Structure of Individual Salp & Swarm of Salps **a** Individual Salp **b** Swarm of Salps [20]

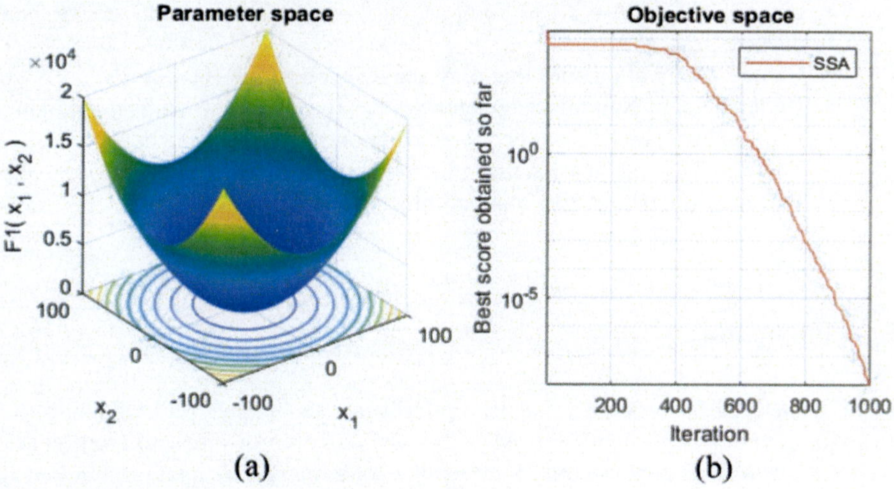

Fig. 6 **a**) Parameter Space **b**) Objective Space

this figure, it can be observed that unimodal test functions have only one optimum, and no local optima are present. For checking the speed of convergence and exploitative behavior, these types of search spaces are suitable.

(f) **Applications of swarm intelligence**

- U.S. military is researching swarm systems for controlling unmanned vehicles.
- NASA is researching the utilization of swarm innovation for planetary mapping.
- Tim Burton's Batman Returns was the main film to utilize swarm innovation for rendering, reasonably delineating the developments of a gathering of penguins utilizing the Boids framework.
- The Lord of the Rings film set of three utilized comparable innovation, known as Massive, during fight scenes.
- Clustering conduct of ants: The utilization of swarm intelligence in telecommunication networks has likewise been inquired about, as ant-based routing. Essentially this uses a probabilistic steering table fulfilling/fortifying the course effectively crossed by every "subterranean insect" (a little control parcel) that floods the system. Support of the course in the advances switch heading and both at the same time have been inquired about: in reverse fortification requires a Hilter kilter system and couples the two bearings together; advances fortification rewards a course before the result is known.

(g) **Applications of artificial intelligence**

- Finance: Banks use AI to compose tasks, put resources into stocks, and oversee properties.
- Games: There are games that play ace level chess for 100 dollars.
- Understanding Natural Language: Neither succession of words nor parsing is sufficient. The PC is to be given a comprehension of area content.
- The medicine-medical facility utilizes AI to arrange bed schedules, make staff revolution, and give clinical information.

3 Conclusion

In this paper, we have defined various optimization techniques for evaluating the behavior of the swarm algorithm. Also, the classification of artificial intelligence has been given. This paper mainly focuses on evolutionary computation. Further, it gives a detailed review of evolutionary algorithms and swarms intelligence. From the detailed review, it can be concluded that the swarm intelligence algorithm can be used for various problems, forgetting the optimized results like a wireless sensor network, image processing, steganography, or various security-related issues.

References

1. Kumar A, Salau AO, Gupta S, Paliwal K (2019) Recent trends in IoT and its requisition with IoT built engineering: a review. In: advances in signal processing and communication. Springer, Singapore, pp 15–25
2. Rana AK, Sharma S (2019) Enhanced energy-efficient heterogeneous routing protocols in WSNs for IoT Application. IJEAT 9(1). ISSN: 2249–8958
3. Kumar K, Gupta ES, Rana EAK Wireless sensor networks: a review on "challenges and opportunities for the future world-LTE
4. Rana AK, Krishna R, Dhwan S, Sharma S, Gupta R (2019) Review on artificial intelligence with internet of things-problems, challenges and opportunities. In: 2019 2nd international conference on power energy, environment and intelligent control (PEEIC). IEEE, pp 383–387
5. Rana AK, Sharma S Contiki Cooja Security Solution (CCSS) with IPv6 routing protocol for low-power and lossy networks (RPL) in internet of things applications. In: mobile radio communications and 5G Networks. Springer, Singapore, pp 251–259
6. Dhawan S, Gupta R (2020) Analysis of various data security techniques of steganography: a survey. Informat Sec J A Global Perspect 1–25
7. Dhawan S, Gupta R (2019) Comparative analysis of domains of technical steganographic techniques. In: 2019 6th international conference on computing for sustainable global
8. Jain AK, Mao J, Mohiuddin KM (1996) Artificial neural networks: a tutorial. Comput (Long. Beach. Calif) 29(3):31–44
9. Punyani P, Gupta R, Kumar A (2019) Neural networks for facial age estimation : a survey on recent advances, no. 0123456789. Springer Netherlands
10. Rana AK, Sharma S Industry 4.0 manufacturing based on IoT, cloud computing, and big data: manufacturing purpose scenario. In: Advances in communication and computational technology. Springer, Singapore, pp 1109–1119
11. Dernoncourt F (2013) Introduction to fuzzy logic
12. Holland JH (2005) Genetic algorithms, John H. Holland understand genetic algorithms, pp 12–15
13. Thengade A (2018) Genetic algorithm—survey paper genetic algorithm—survey paper
14. Ilamaran A, Ganapathiram S, Kumar RA, Uthayakumar J (2014) Swarm intelligence : an application of ant colony optimization, vol 4, pp 63–69
15. Farich M (2016).Artificial intelegence algoritma A*(A Star) Sebagai pathfinding enemy attack Pada Game Trash Collection (Doctoral Dissertation, University Of Muhammadiyah Malang)
16. Huang X, Sun N, Liu W, Wei J (2007) Research on particle swarm optimization and its industrial application, no. 973
17. Bai Q (1998) Analysis of particle swarm optimization algorithm, vol 3(1), pp 180–184
18. Yu X, Chen W, Zhang X (2018) An artificial bee colony algorithm for solving constrained optimization problems. In: Proceedings of 2018 2nd IEEE advance information management, communicates, electronic and automation control conference IMCEC 2018, pp 2663–2666
19. Krishnanand KN, Ghose D (2009) Glowworm swarm optimization for simultaneous capture of multiple local optima of multimodal functions. Swarm Intell 3(2):87–124
20. Mirjalili S, Gandomi AH, Mirjalili SZ, Saremi S, Faris H, Mirjalili SM (2017) Salp Swarm Algorithm: a bio-inspired optimizer for engineering design problems. Adv Eng Softw 114:163–191
21. Faris H, Mirjalili S, Aljarah I, Mafarja M, Heidari AA (2020) Salp swarm algorithm : theory, literature review, and application in extreme learning machines. Springer International Publishing

Chapter 23
Simulation and Analysis of Optical Communication System Using SMF for Different Wavelength Bands with NRZ Modulation

Swarnjeet Kaur and Kamal Malik

1 Introduction

Optical fiber communication system is currently the easiest and simplest way of communication as compared to other types of communication system [1]. The optical fiber cable is used to transmit the optical signal from source (T_x) to destination (R_x) [2, 3]. The traditional twisted and coaxial cables are replaced with latest optical fiber cables [4]. The fiber cables can be categorized according to their mode as well [5]. Light rays travel through the fiber like an electromagnetic pulse. The two components form patterns throughout the fiber: the electrical field and the magnetic field [6]. These patterns are called transmission modes. A fiber mode refers to the number of paths within the cable for the light rays. Optic fibers can be classified into two modes: single-mode fiber and multimode fiber [5, 7, 8].

In this paper, Sect. 2 consists of methodology, Sect. 3 discusses simulation setup, Sect. 4 provides the results based on the comparison of continuous wave laser frequency, power, bit rate, Q-factor and BER. The comparison is made in the tabular form for various optical bands.

S. Kaur (✉)
Electronics and Communication Engineering (ECE), CT University, Ludhiana, India

K. Malik
Computer Science and Engineering (CSE), CT University, Ludhiana, India

N. Marriwala et al. (eds.), *Soft Computing for Intelligent Systems*, Algorithms for Intelligent Systems, https://doi.org/10.1007/978-981-16-1048-6_23

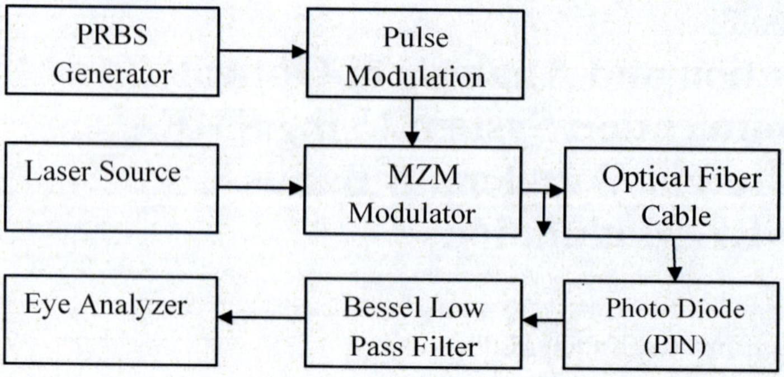

Fig. 1 Block diagram of optical fiber communication system using single-mode fiber

2 Methodology

The methodology shown below explains the optical fiber communication system in which laser source is used at the transmitter end. The digital data from NRZ pulse is combined at MZM and transmitted to the PIN diode after passing through low-pass Bessel filter. The signal is verified by checking the Q-factor and BER analyzer values (Fig. 1).

3 Simulation Setup

Figure 2 shows the simulation model of the paper. In this design, continuous wave (CW) laser having wavelengths of 1310, 1410 and 1510 nm and NRZ pulse generator are combined at the Mach–Zehnder modulator. The single-mode fiber cable of 1, 5,

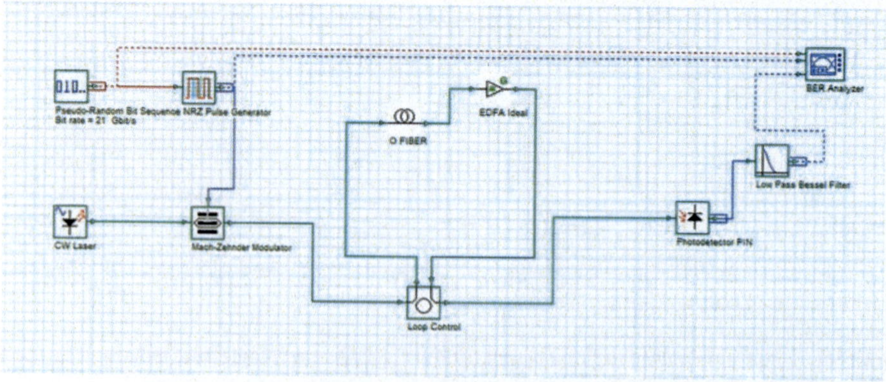

Fig. 2 Simulation layout of CW laser using single-mode fiber

Table 1 Various parameters used in optical communication system

Name	Value	Units
Bit rate	21	Gb/s
Power	10	mW
CW laser frequency	1310, 1410 and 1510	nm
Fiber length	1, 5, 7.5 and 10	km
Filter type	Bessel	–
Photodetector	PIN	–
Modulator	Mach–Zehnder	–
Amplifier	EDFA	–
Attenuation	0.3, 0.2	dB/km
Fiber type	Single-mode fiber	–

7.5 and 10 km, respectively, along with EDFA is used in this setup. The PIN detector at the receiving end is used to switch the optical signal back to the original one. The Bessel filter allows the low-frequency signals by rejecting the other signals. The BER analyzer with different fiber lengths is used to verify the Q-factor and BER and eye diagrams (Table 1).

4 Results and Comparison

The optical communication system using single-mode fiber is designed to transmit the data of 21 Gb/s over the distance of 1, 5, 7.5 and 10 km, respectively. The wavelengths 1310, 1410 and 1510 nm are used in this simulation model. The output of the simulation model can be represented in the form of EYE diagram to verify the Q-factor and BER values, which are given below in this section.

Figure 3 shows the simulative setup of SMF at wavelength 1310 nm. The data

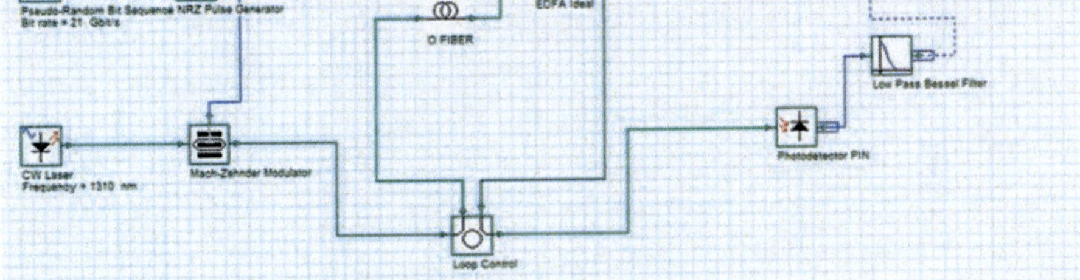

Fig. 3 Simulative setup of SMF at 1310 nm

rate is transmitted over the distances of 1, 5, 7.5 and 10 km, respectively.

Figure 4 explains the eye diagram with 1 km transmission distance using optical wavelength 1310 nm with Q-factor 5.63 and BER of $8.85e^{-009}$.

Figure 5 shows the eye diagram after 5 km transmission using optical wavelength 1310 nm with Q-factor 5.32 and BER value of $5.00e^{-008}$.

Figure 6 describes the eye diagram with 7.5 km transmission distance using optical wavelength 1310 nm with Q-factor 4.99 and BER of $2.90e^{-007}$.

Figure 7 explains the eye diagram after 10 km transmission distance using optical wavelength 1310 nm with Q-factor 4.99 and BER of $2.02e^{-006}$.

Figure 8 shows the simulative setup of single-mode fiber at wavelength 1410 nm. The data rate is transmitted over the distance 1, 5, 7.5 and 10 km, respectively.

Figure 9 describes the eye diagram after 1 km transmission using optical wavelength 1410 nm with Q-factor 5.62 and BER of $9.19e^{-009}$.

Figure 10 describes the eye diagram after 1 km transmission distance using optical wavelength 1410 nm with Q-factor 5.17 and BER of $1.21e^{-007}$.

Figure 11 describes the eye diagram after 1 km transmission using optical wavelength 1410 nm with Q-factor 4.78 and BER of $8.38e^{-007}$.

Figure 12 describes the eye diagram after 10 km transmission distance using optical wavelength 1410 nm with Q-factor 4.20 and BER of $1.26e^{-005}$.

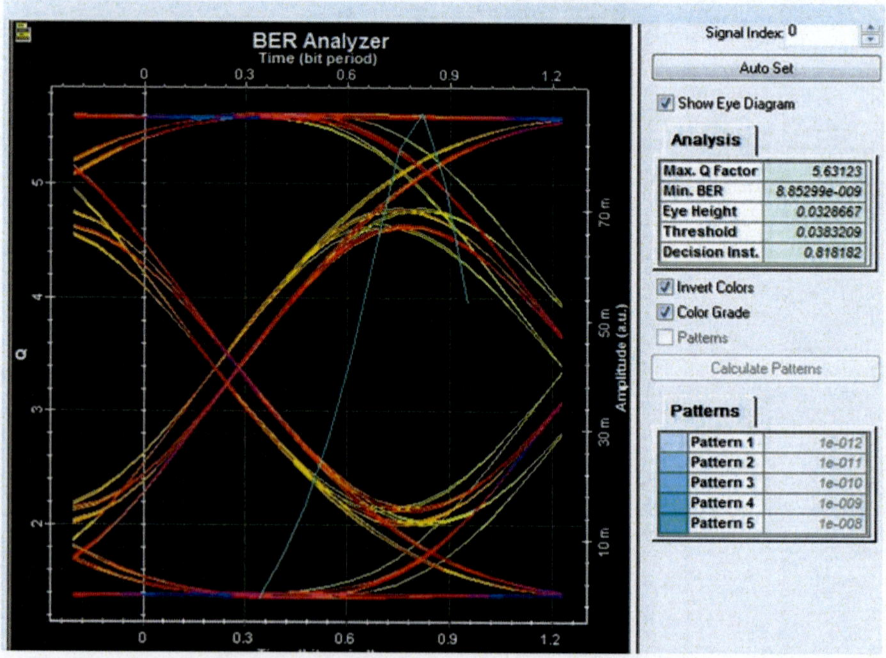

Fig. 4 Eye diagram with 1 km distance using optical wavelength 1310 nm

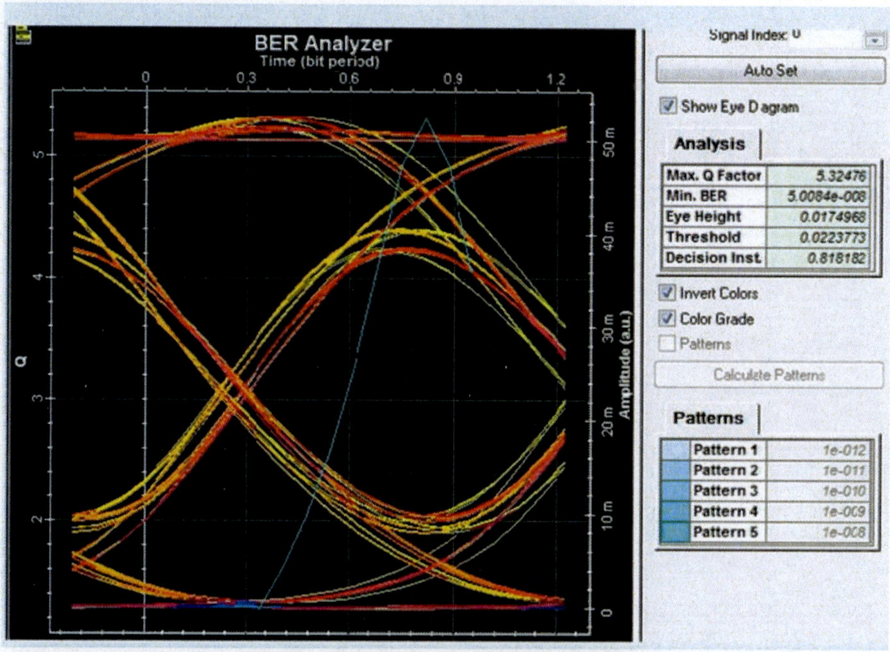

Fig. 5 Eye diagram after 5 km transmission using optical wavelength 1310 nm

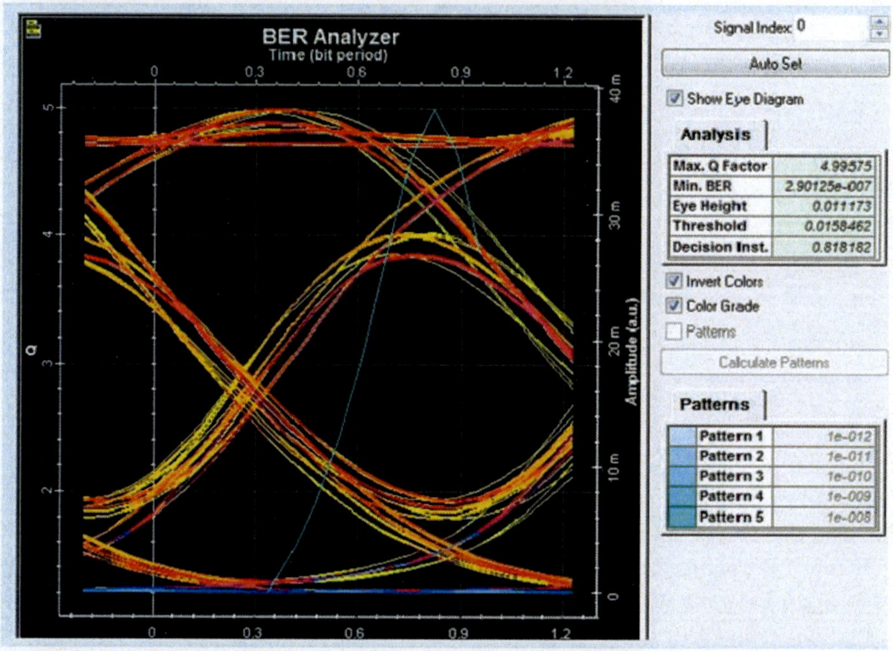

Fig. 6 Eye diagram after 7.5 km transmission using optical wavelength 1310 nm

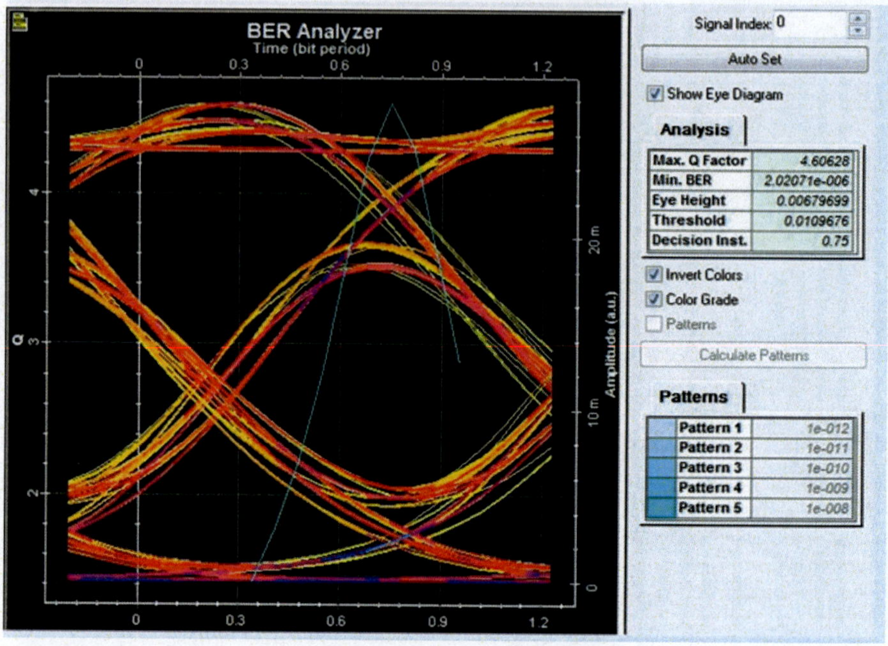

Fig. 7 Eye diagram after 10 km distance using optical wavelength 1310 nm

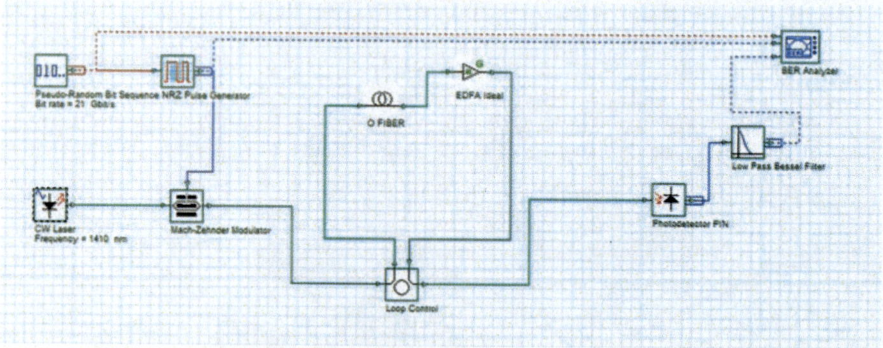

Fig. 8 Simulative setup of SMF at 1410 nm

Figure 13 shows the simulative setup of single-mode fiber at wavelength 1510 nm. The data rate is transmitted over the distance 1, 5, 7.5 and 10 km, respectively.

Figure 14 explains the eye diagram after 1 km transmission distance using optical wavelength 1510 nm with Q-factor 5.59 and BER of $1.06e^{-006}$.

Figure 15 shows the eye diagram after 5 km transmission using optical wavelength 1510 nm with Q-factor 5.01 and BER of $2.57e^{-007}$.

Figure 16 explains the eye diagram after 7.5 km transmission distance using

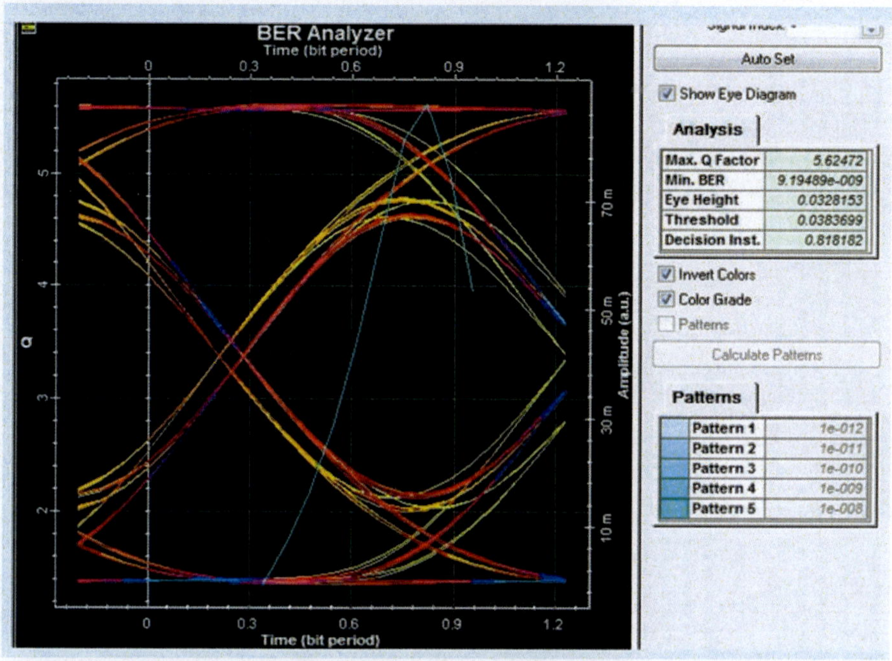

Fig. 9 Eye diagram with 1 km transmission using optical wavelength 1410 nm

optical wavelength 1510 nm with *Q*-factor 4.51 and BER of $3.07e^{-006}$.

Figure 17 explains the eye diagram after 10 km transmission distance using optical wavelength 1510 nm with *Q* factor 3.86 and BER of $5.49e^{-005}$. The comparison of CW laser using wavelengths 1310, 1410 and 1510 nm at power level 10 mW is done for the transmission of data of 21 Gb/s with the use of SMF and compared BER and *Q* factor in the tabular form in Tables 2, 3, 4 and 5. Further the comparison of different wavelengths along with Q-factor values are also shown in graphical form given below (Fig. 18).

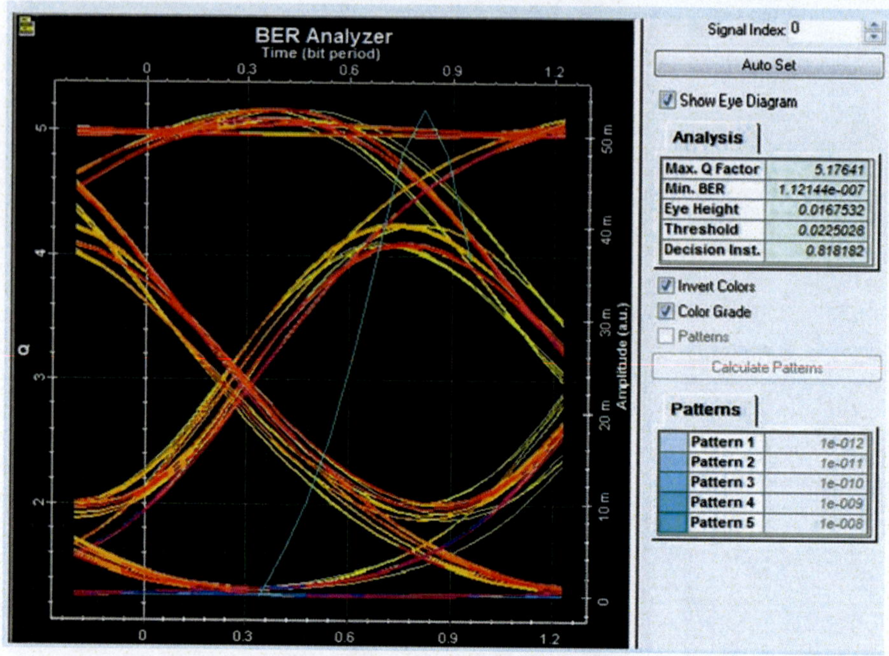

Fig. 10 Eye diagram after 5 km transmission using optical wavelength 1410 nm

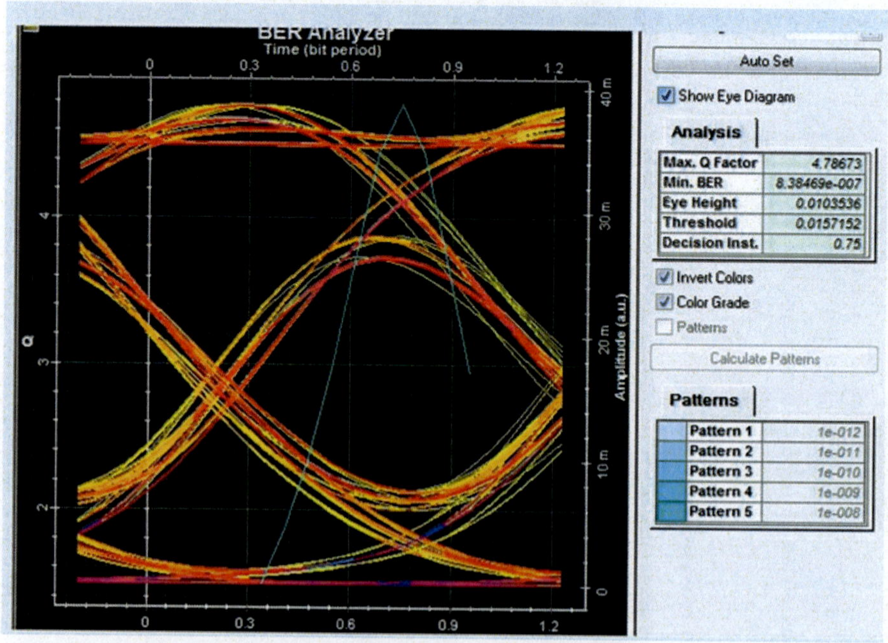

Fig. 11 Eye diagram after 7.5 km transmission using optical wavelength 1410 nm

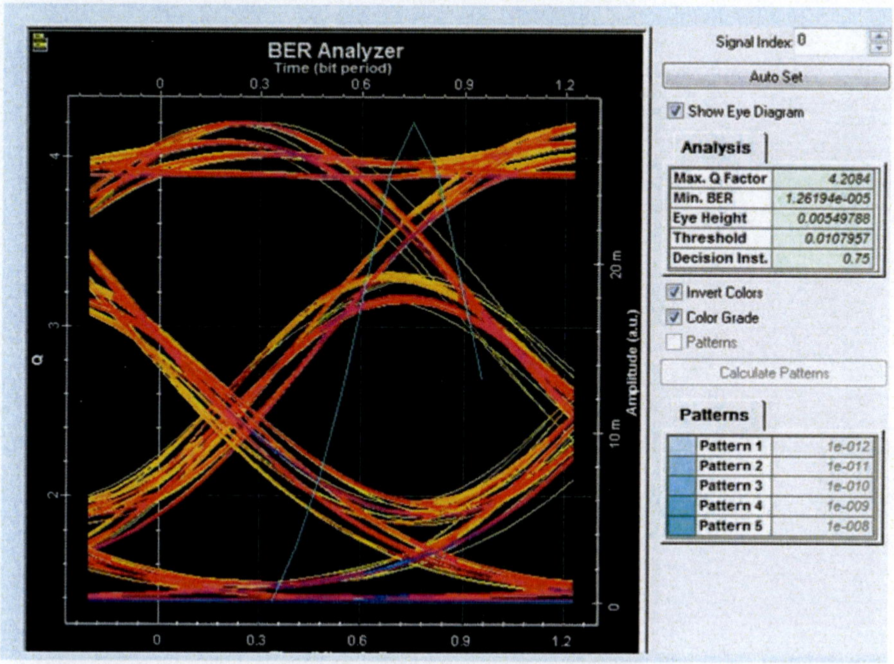

Fig. 12 Eye diagram after 10 km transmission distance using optical wavelength 1410 nm

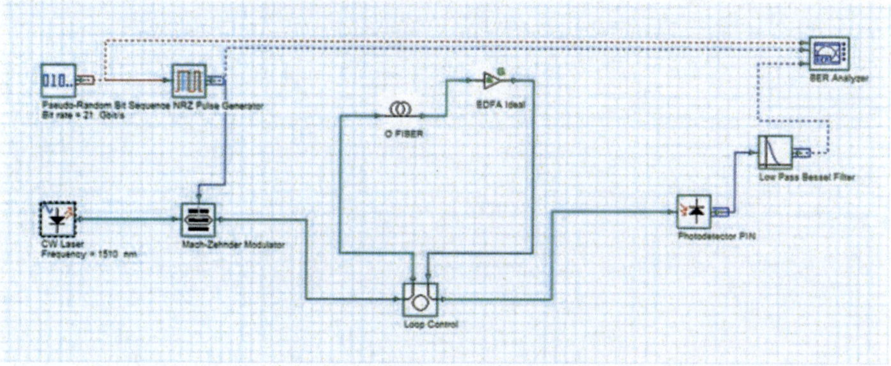

Fig. 13 Simulative setup of SMF at 1510 nm

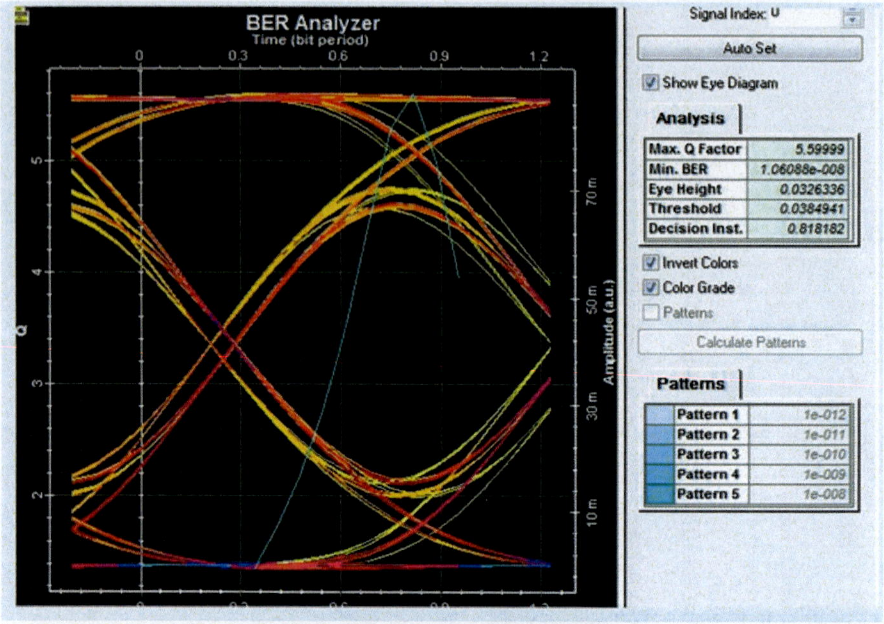

Fig. 14 Eye diagram after 1 km transmission distance using optical wavelength 1510 nm

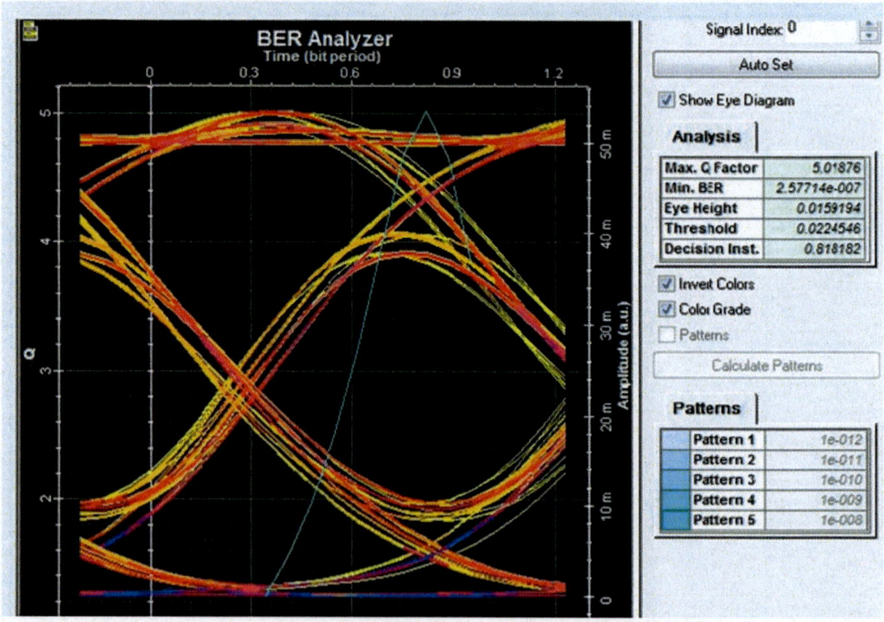

Fig. 15 Eye diagram after 5 km transmission using optical wavelength 1510 nm

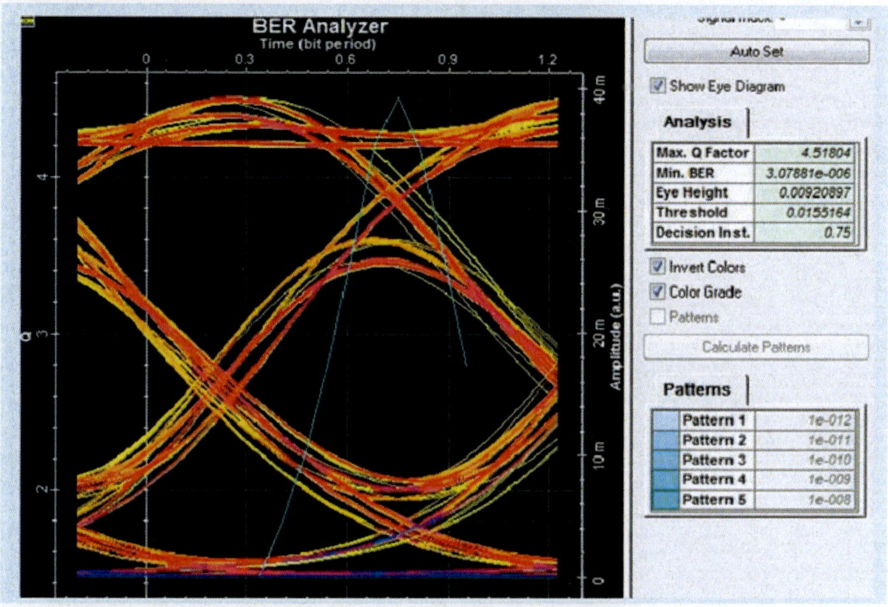

Fig. 16 Eye diagram after 7.5 km transmission distance using optical wavelength 1510 nm

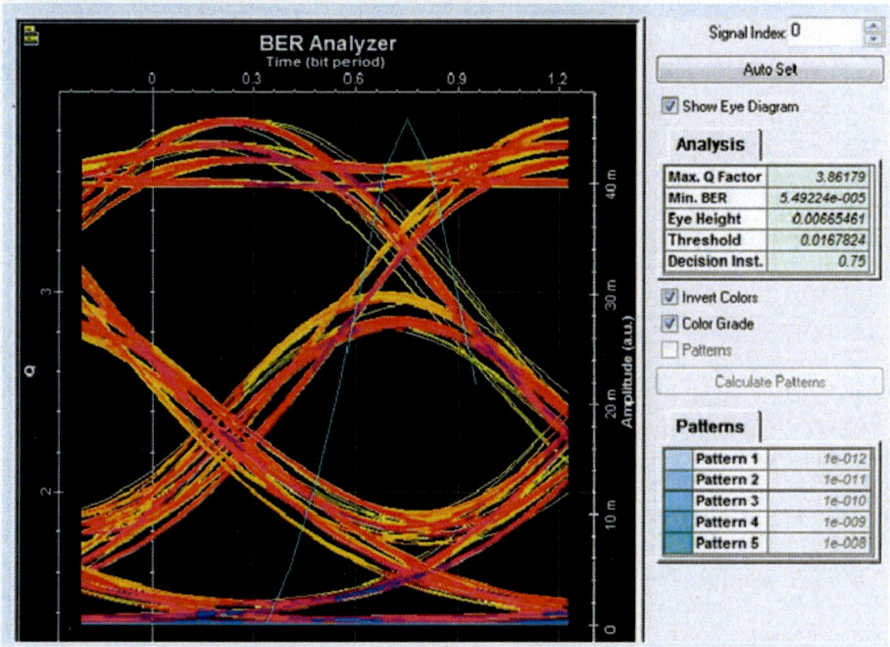

Fig. 17 Eye diagram after 10 km transmission distance using optical wavelength 1510 nm

Table 2 Optical signal is transmitted at various wavelengths for a distance of 10 km

CW laser frequency (nm)	Power (mW)	Bit rate (Gbps)	Single-mode fiber length (km)	Attenuation (dB/km)	Photodetector	BER	*Q*-factor
1310	10	21	10	0.3	PIN	2.02 e^{-006}	4.60
1410	10	21	10	0.3	PIN	1.26 e^{-005}	4.20
1510	10	21	10	0.2	PIN	5.49 e^{-005}	3.86

Table 3 Optical signal is transmitted at various wavelengths for a distance of 7.5 km

CW laser frequency (nm)	Power (mW)	Bit rate (Gbps)	Single-mode fiber length (km)	Attenuation (dB/km)	Photodetector	BER	*Q*-factor
1310	10	21	7.5	0.3	PIN	2.90 e^{-007}	4.99
1410	10	21	7.5	0.3	PIN	8.38 e^{-007}	4.78
1510	10	21	7.5	0.2	PIN	3.07 e^{-006}	4.51

Table 4 Optical signal is transmitted at various wavelengths for a distance of 5 km

CW laser frequency (nm)	Power (mW)	Bit rate (Gbps)	Single-mode fiber length (km)	Attenuation (dB/km)	Photodetector	BER	*Q*-factor
1310	10	21	5	0.3	PIN	5.00 e^{-008}	5.32
1410	10	21	5	0.3	PIN	1.26 e^{-007}	5.17
1510	10	21	5	0.2	PIN	2.57 e^{-007}	5.01

Table 5 Optical signal is transmitted at various wavelengths for a distance of 1 km

CW laser frequency (nm)	Power (mW)	Bit rate (Gbps)	Single-mode fiber length (km)	Attenuation (dB/km)	Photodetector	BER	*Q*-factor
1310	10	21	1	0.3	PIN	8.85 e^{-009}	5.63
1410	10	21	1	0.3	PIN	9.19 e^{-009}	5.62
1510	10	21	1	0.2	PIN	1.06 e^{-008}	5.59

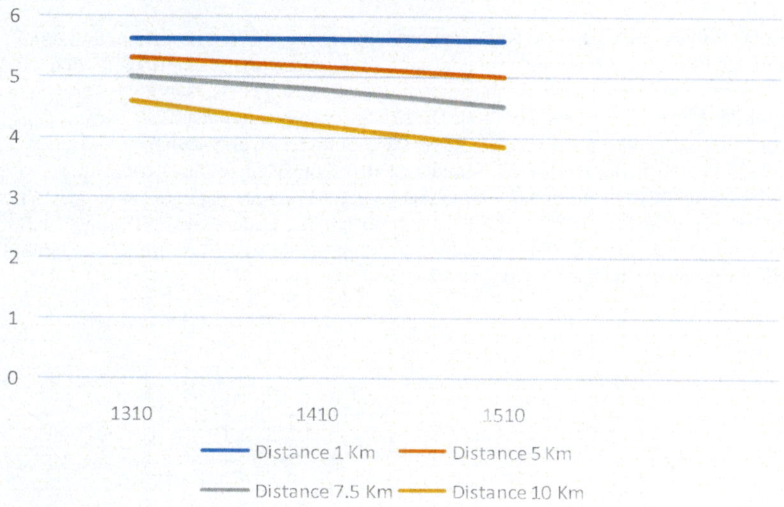

Fig. 18 Graph shows the comparison among the wavelengths 1310, 1410 and 1510 nm with their Q-factor values

5 Conclusion

The simulation design of CW laser using single-mode fiber is implemented with Optisystem software. By using EDFA amplifier, Bessel filter, NRZ and Mach–Zehnder modulator, 21 Gb/s data rate is sent over the distance of 1, 5, 7.5 and 10 km, respectively, at optical frequency 1310, 1410 and 1510 nm. We found the maximum Q-factor at the 1310 nm optical frequency and minimum BER $8.85e^{-009}$.

The future scope of this paper is to send the higher bits of data by using the different band of optical wavelength for the long distance communication with the help of SMF.

References

1. Szilagyi L, Belfiore G, Henker R, Ellinger F (2015) A high-voltage DC bias architecture implementation in a 17 Gbps low-power common-cathode VCSEL driver in 80 nm CMOS. In: 2015 IEEE international symposium on circuits and systems (ISCAS)
2. Larisch G, Rosales R, Lott JA, Bimberg D (2019) Energy-efficient VCSELs for 200+ Gb/s optical interconnects. In: Conference on lasers and electro-optics
3. Li K, Chen X, Mishra SK, Hurley JE, Stone JS, Li M-J (2020) Modal delay and modal bandwidth measurements of bi-modal optical fibers through a frequency domain method. Opt Fiber Technol 55:102145
4. Gheni HM, Abdullah M, Omar KA, Qasim AA, Abdulrahman AM, Fakhri MN, Dawood A (2019) Radio over fiber (RoF) implementation using MZM for long distance communication. In: 2019 international conference on information science and communication technology (ICISCT)

5. Larisch G, Juarez AA, Chen X, Li K, Bimberg D, Li M-J (2020) 910 nm single-mode VCSELs and its application for few-mode transmission over graded-index single-mode fibers. In: 2020 22nd international conference on transparent optical networks (ICTON)
6. Larisch G, Moser P, Lott JA, Bimberg D (2017) Large bandwidth, small current density, and temperature stable 980-nm VCSELs. IEEE J Quantum Electron 53(6):1–8
7. Li M-J, Chen X, Li K, Hurley JE, Stone J (2019) Optical fiber for 1310nm single-mode and 850nm few-mode transmission. In: Broadband access communication technologies XIII
8. Puerta R, Agustin M, Chorchos L, Tonski J, Kropp JR, Ledentsov N, Turkiewicz JP (2017) Effective 100 Gb/s IM/DD 850-nm multi- and single-mode VCSEL transmission through OM4 MMF. J Lightwave Technol 35(3):423–429

Chapter 24
Investigation of Free Space Optical Transmission System for Various Bands of Wavelength in Clear Weather Condition

Kulwant Singh and Kamal Malik

1 Introduction

In free space optical transmission system the data is transmitted in free space where air is used as a medium [1, 2]. The FSO system provides the free space connection for offices, campus and other established buildings [3]. It also provides the license-free bands with no signal interference [4]. The demerit of FSO transmission system is line of sight (LOS), which should be proper between the transmitter and the receiver, otherwise the data may be lost [5]. The FSO system is highly vulnerable to attenuation which occurs due to bad weather conditions [6, 7, 8]. It leads to loss of light beam signal which further limits the performance of FSO transmission system.

In this paper, the best suited wavelength for transmission is selected from S-band, C-band and L-band of optical communication system. The wavelengths selected from these bands are used for data transmission in clear weather condition which includes small dust particles or little water droplets during morning and evening conditions. Further, these wavelengths are simulated to get the best results for Q-factor, minimum BER and to analyze the eye diagram.

2 Materials and Methods

The FSO system is discussed in this paper, which comprises optical transmitter, medium (free space) and a receiver (photodetector). A PRBS generator is used to send the data rate in Gbps and passed through RZ pulse. The RZ pulse and laser source is modulated using Mach–Zehnder modulator (MZM). The optical signal on

K. Singh (✉)
Electronics and Communication Engineering (ECE), CT University, Ludhiana, India

K. Malik
Computer Science and Engineering (CSE), CT University, Ludhiana, India

N. Marriwala et al. (eds.), *Soft Computing for Intelligent Systems*, Algorithms for Intelligent Systems, https://doi.org/10.1007/978-981-16-1048-6_24

Table 1 Various parameters used in the setup of FSO system

Name	Values	Units
Bit rate	06	Gbps
Power	10	mW
CW laser frequency	1510 (S-band), 1550 (C-band), 1600 (L-band)	nm
Free space distance	1, 1.5, 2.5, 2.8 respectively	km
Filter type	Low-pass Gaussian	–
Photodetector	PIN diode	–
Modulator	Mach–Zehnder modulator (MZM)	–
Attenuation	1 (Clear weather condition with small dust particles)	dB/km

reaching destination is received by the PIN photodetector which converts the signal into electrical form and the signal is further passed through low-pass Gaussian filter. The performance of filtered signal is analyzed and checked by its Q-factor and BER. The values of Q-factor and BER analyzer along with distance values decide whether the signal quality is good or poor (Tables 1, 2, 3, 4 and 5).

Table 2 Optical signal is transmitted for the given wavelengths at a distance of 1 km

CW laser wavelength (nm)	Power (mW)	Bit rate (Gbps)	Free space distance (km)	Attenuation for clear weather condition small dust particles (dB/km)	Photodetector	Q-factor
1510	10	06	1	1	PIN	8.64
1550	10	06	1	1	PIN	8.57
1600	10	06	1	1	PIN	8.48

Table 3 Optical signal is transmitted for the given wavelengths at a distance of 1.5 km

CW laser wavelength (nm)	Power (mW)	Bit rate (Gbps)	Free space distance (km)	Attenuation for clear weather condition small dust particles (dB/km)	Photodetector	Q-factor
1510	10	06	1.5	1	PIN	8.25
1550	10	06	1.5	1	PIN	8.42
1600	10	06	1.5	1	PIN	8.49

Table 4 Optical signal is transmitted for the given wavelengths at a distance of 2.5 km

CW laser wavelength (nm)	Power (mW)	Bit rate (Gbps)	Free space distance (km)	Attenuation for clear weather condition small dust particles (dB/km)	Photodetector	Q-Factor
1510	10	06	2.5	1	PIN	7.02
1550	10	06	2.5	1	PIN	7.26
1600	10	06	2.5	1	PIN	7.32

Table 5 Optical signal is transmitted for the given wavelengths at a distance of 2.8 km

CW laser wavelength (nm)	Power (mW)	Bit rate (Gbps)	Free space distance (km)	Attenuation for clear weather condition small dust particles (dB/km)	Photodetector	Q-factor
1510	10	06	2.8	1	PIN	6.74
1550	10	06	2.8	1	PIN	6.53
1600	10	06	2.8	1	PIN	6.53

3 Methodology

The methodology described in Fig. 1 shows the FSO transmission system in which the laser is used as optical source and the digital data from RZ pulse is combined at MZM and transmitted to the PIN photodetector. After passing through low-pass Gaussian filter the signal is analyzed with BER analyzer.

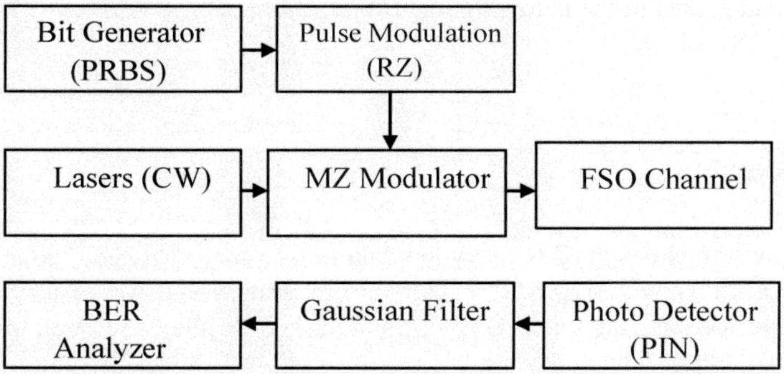

Fig. 1 Block diagram of FSO transmission system

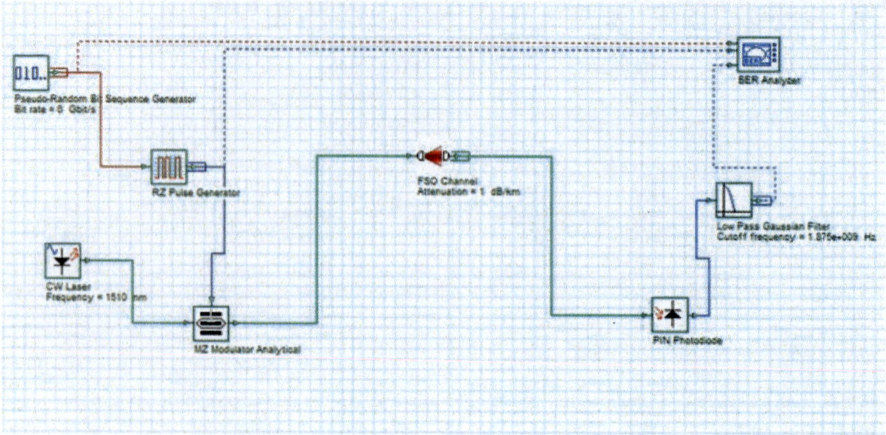

Fig. 2 Simulative setup of FSO system

Figure 2 shows the simulative setup in which wavelengths 1510, 1550 and 1600 nm are used to transmit the data over a distance up to 2.8 km. The results of successful transmission are checked with Q-factor and BER value.

4 Experimental Results

The free space optical (FSO) transmission system is aimed to transmit the data of 6 Gbps in free space having line of sight (LOS) with photodetector at the receiver. This data in gigabits are transmitted over the free space at distances of 1, 1.5, 2.5 and 2.8 km in clear weather condition. The wavelengths used for the transmission of data are 1510 nm (short band), 1550 nm (conventional band) and 1600 nm (long band). The outputs of transmitted data after passing through photodetectors are checked with Q-factor and BER through simulative set up figures, eye diagram figures and graph and discussed in the below figures (Figs. 3, 4, 5, 6, 7, 8, 9, 10, 11, 12, 13, 14, 15, 16, 17, 18 and 19).

5 Conclusion

The FSO system is designed to carry the data in free space between the transmitter and the receiver using the Optisystem software. With the simulative results discussed in the above section, it is found that with increase in distance the Q-factor reduces and BER increases. The wavelength 1550 nm in C-band gives the uniform results at all simulative distances. The maximum Q-factor and BER for data rate of 6 Gbps

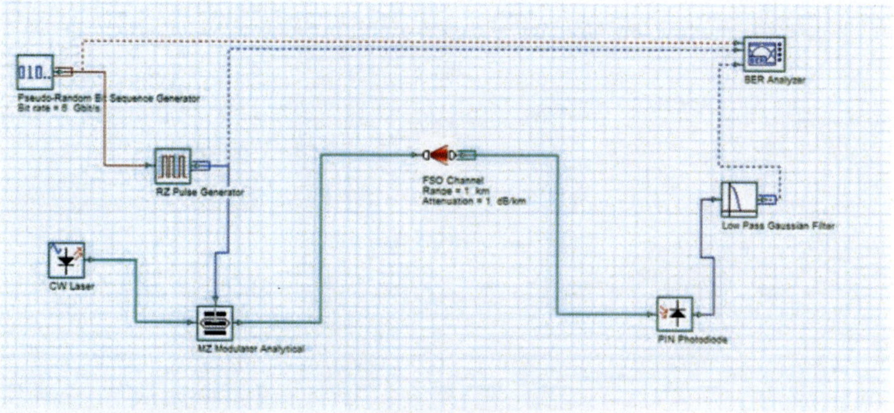

Fig. 3 Simulative setup of FSO system for free space distance of 1 km

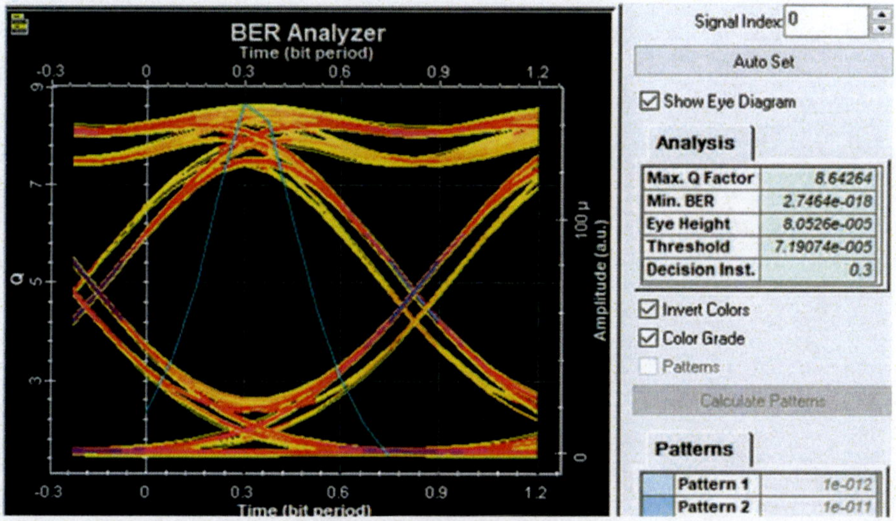

Fig. 4 Eye diagram with Q-factor 8.64 at wavelength 1510 nm and free space distance of 1 km

using wavelength 1550 nm are 8.57 and $4.85e^{-018}$, respectively, over the distance of 1 km.

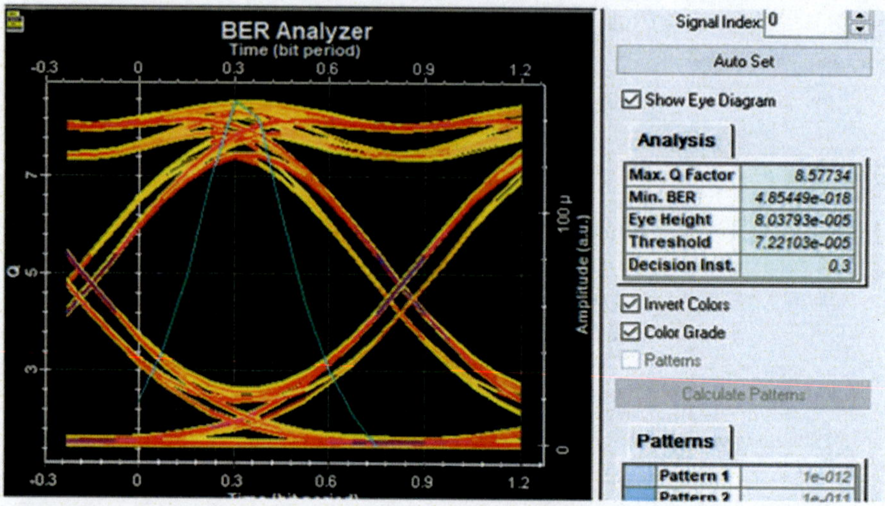

Fig. 5 Eye diagram with Q-factor 8.57 at wavelength 1550 nm with free space distance of 1 km

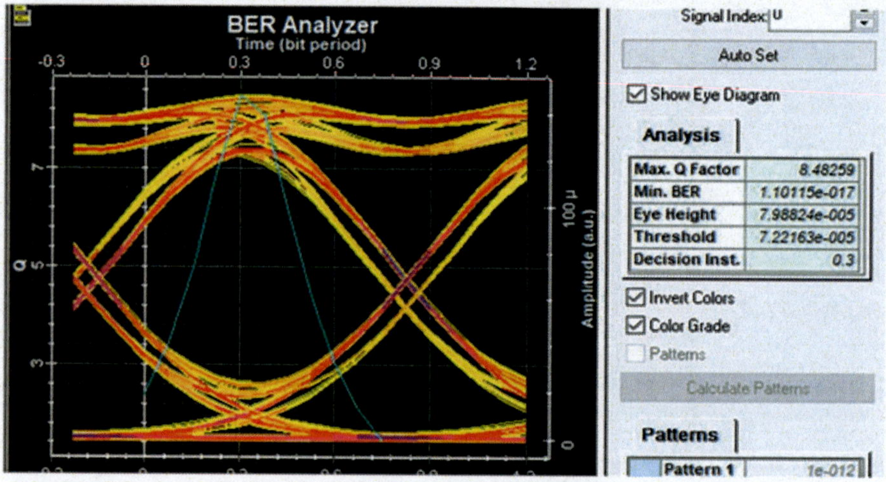

Fig. 6 Eye diagram with Q-factor 8.48 at wavelength 1600 nm and free space distance of 1 km

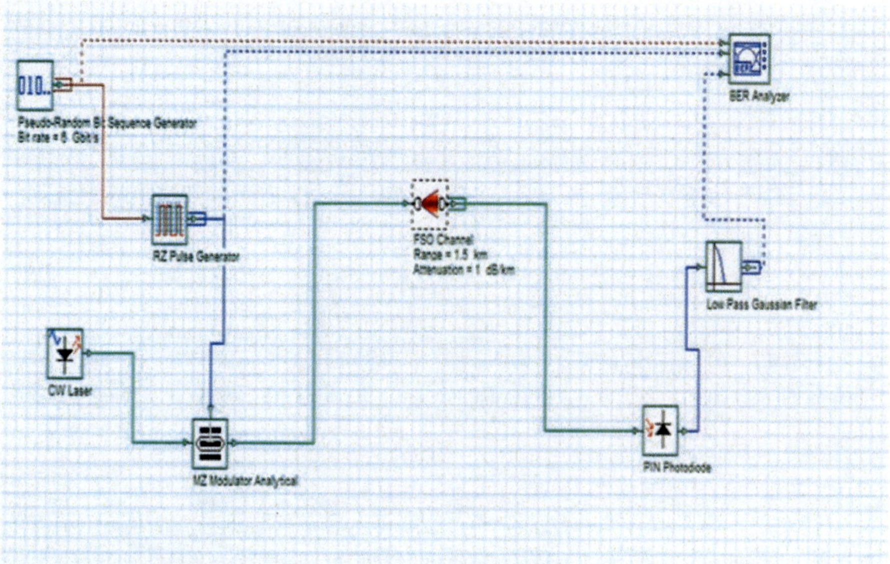

Fig. 7 Simulative setup of FSO system for free space distance of 1.5 km

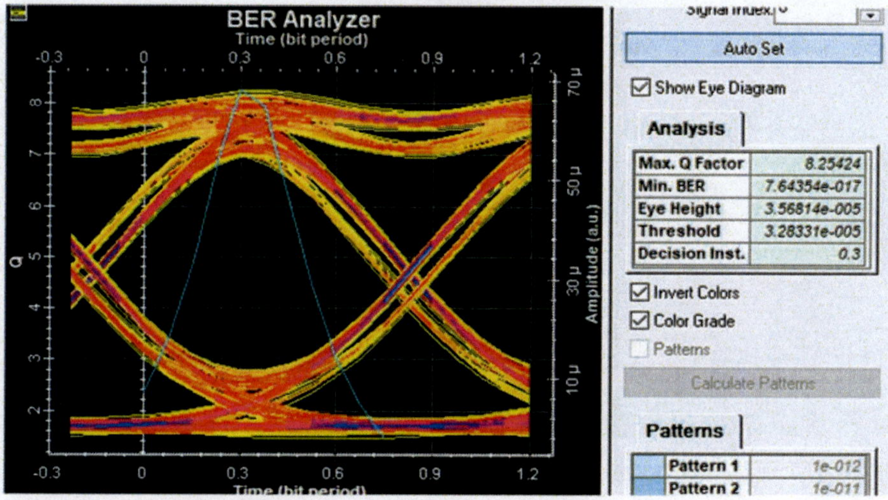

Fig. 8 Eye diagram with Q-factor 8.25 at wavelength 1510 nm and free space distance of 1.5 km

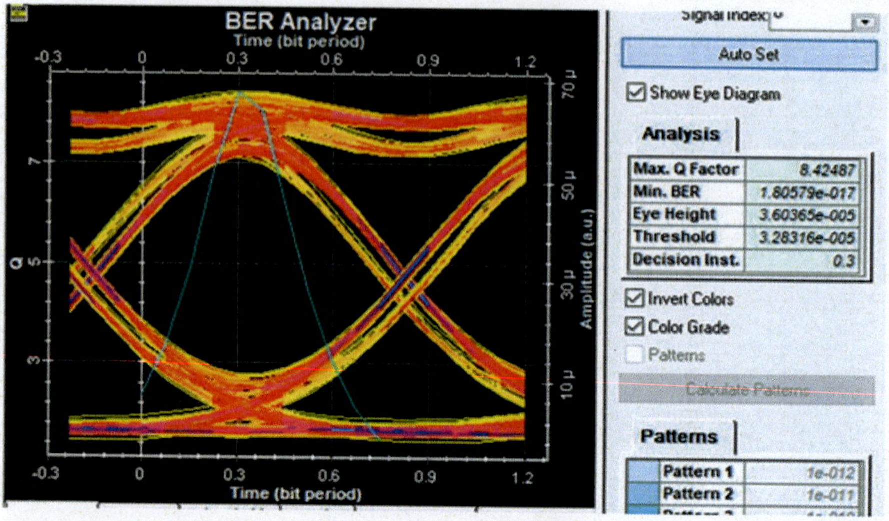

Fig. 9 Eye diagram with Q-factor 8.42 at wavelength 1550 nm and free space distance of 1.5 km

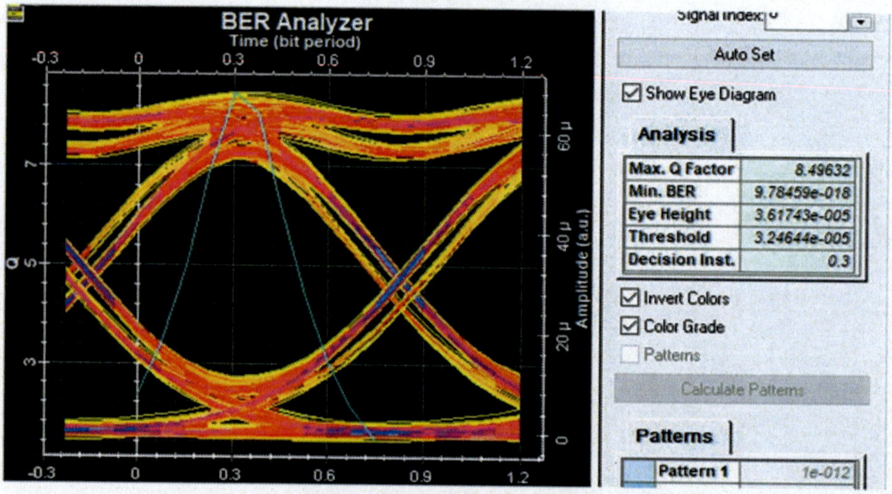

Fig. 10 Eye diagram with Q-factor 8.49 at wavelength 1600 nm and free space distance of 1.5 km

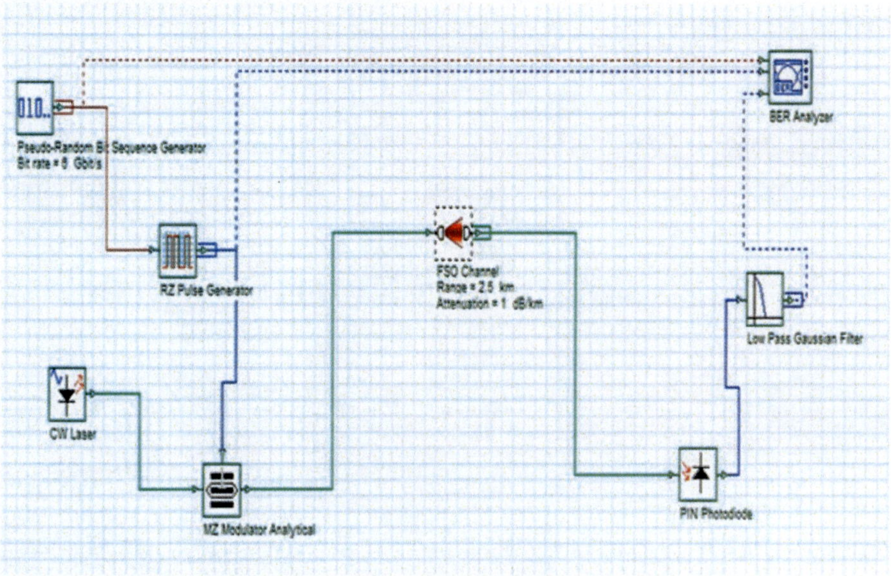

Fig. 11 Simulative setup of FSO system for free space distance of 2.5 km

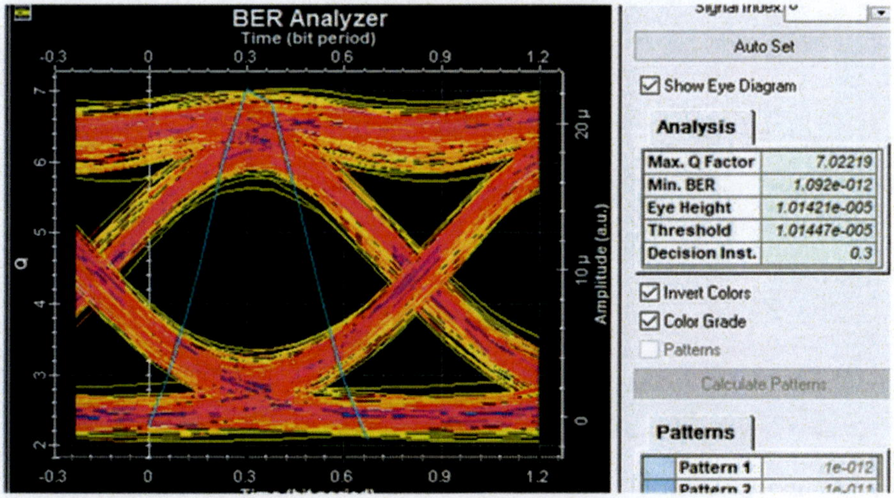

Fig. 12 Eye diagram with Q-factor 7.02 at wavelength 1510 nm and free space distance of 2.5 km

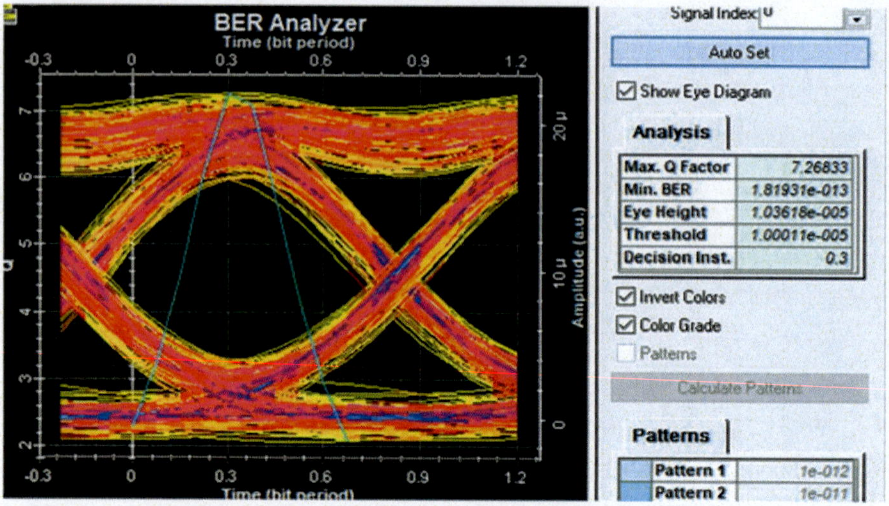

Fig. 13 Eye diagram with Q-factor 7.26 at wavelength 1550 nm and free space distance of 2.5 km

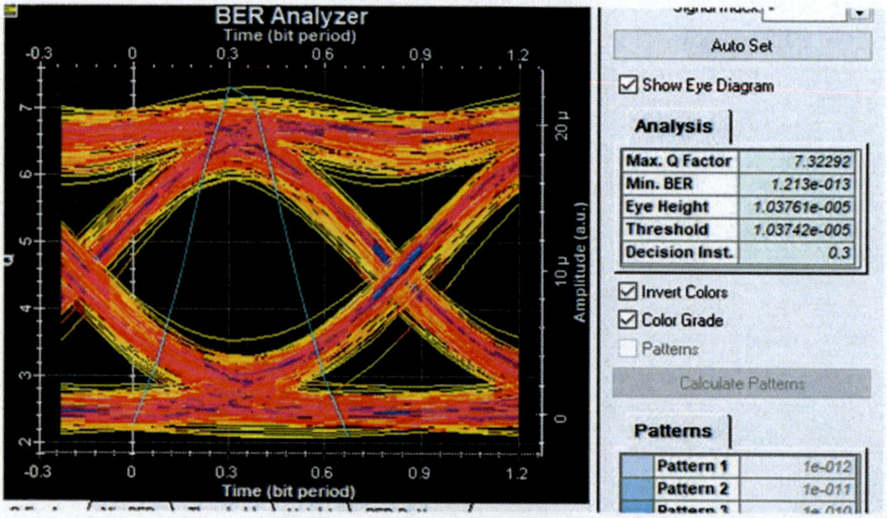

Fig. 14 Eye diagram with Q-factor 7.32 at wavelength 1600 nm and free space distance of 2.5 km

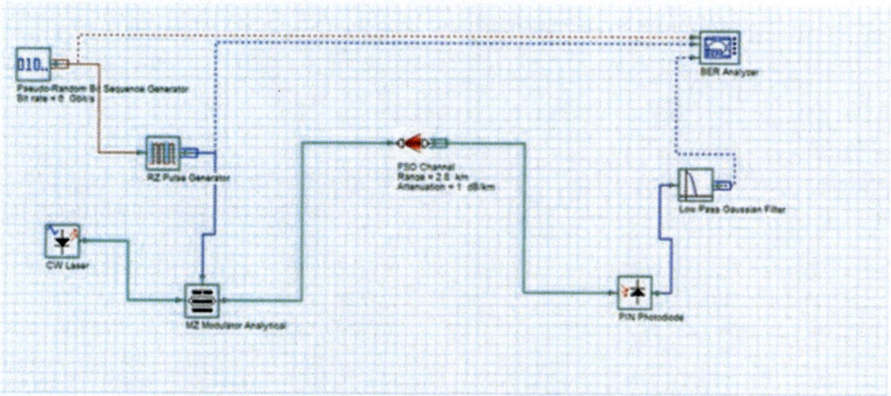

Fig. 15 Simulative setup of FSO system for free space distance of 2.8 km

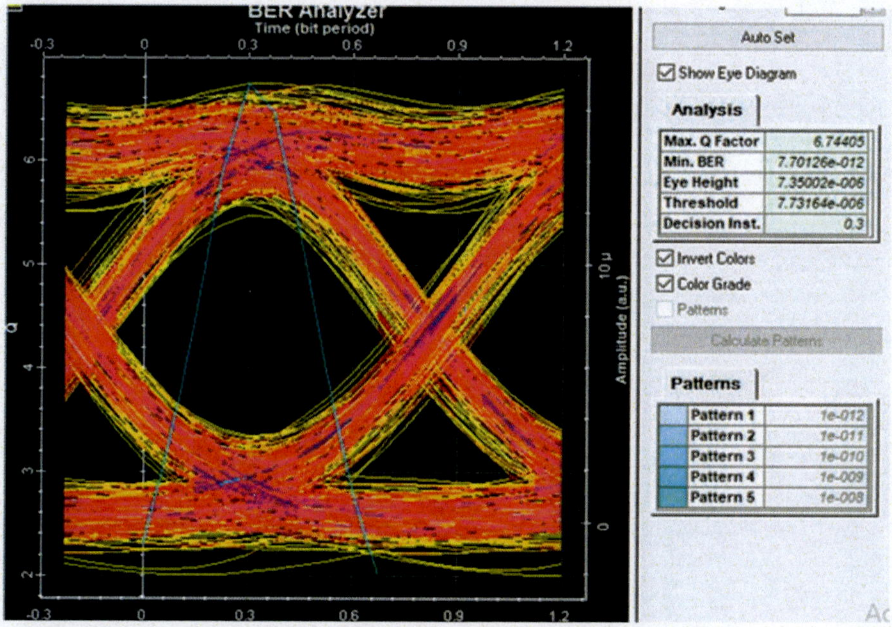

Fig. 16 Eye diagram with Q-factor 6.74 at wavelength 1510 nm and free space distance of 2.8 km

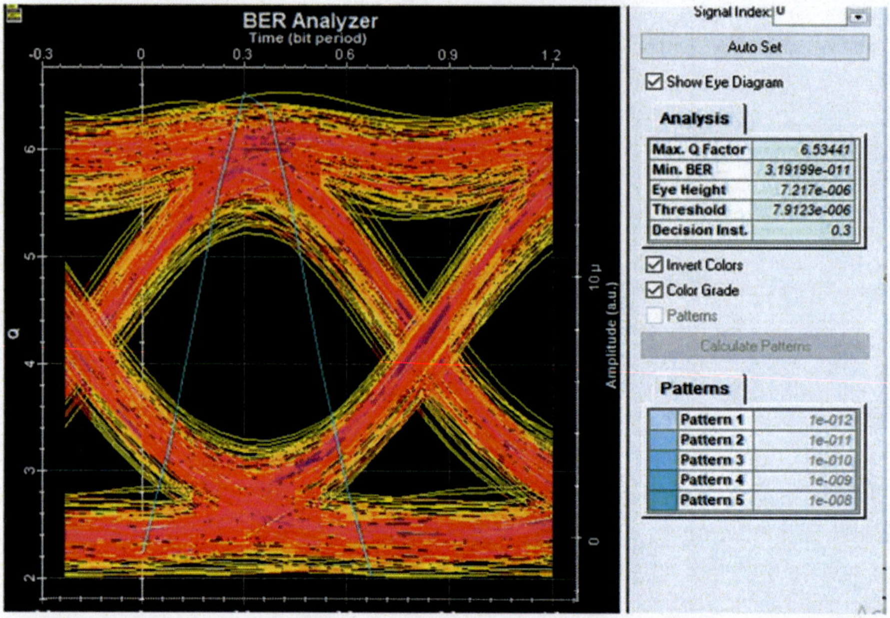

Fig. 17 Eye diagram with Q-factor 6.53 at wavelength 1550 nm and free space distance of 2.8 km

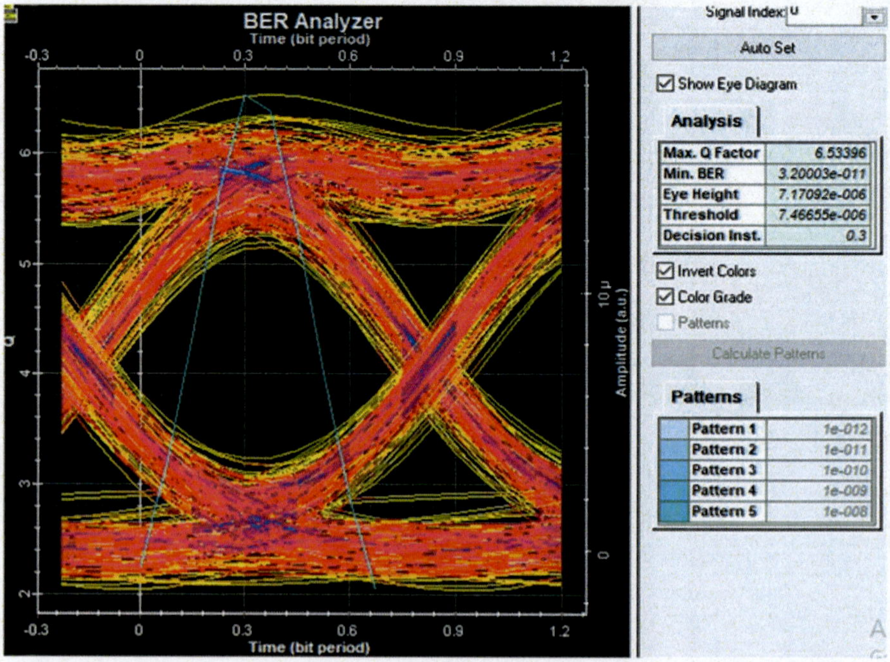

Fig. 18 Eye diagram with Q-factor 6.53 at wavelength 1600 nm and free space distance of 2.8 km

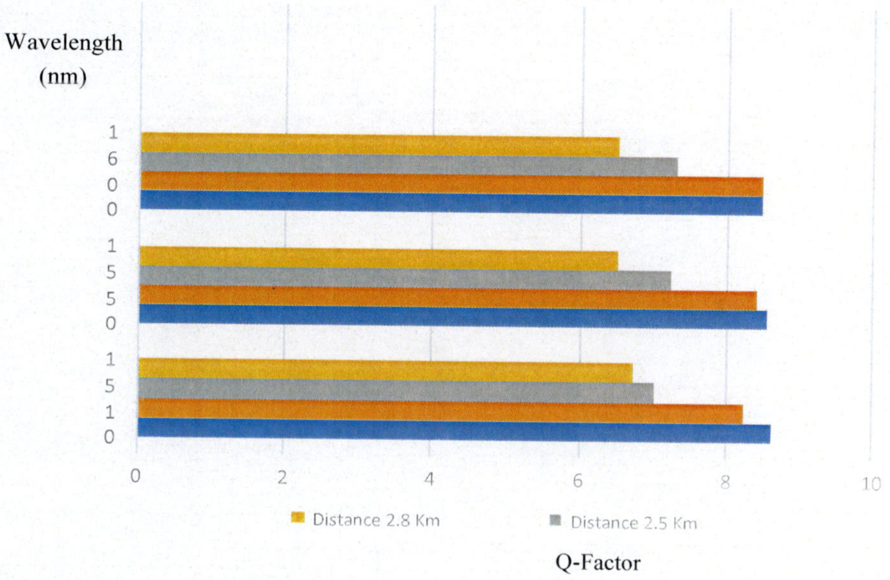

Fig. 19 Graph shows the comparison among the wavelengths 1510, 1550 and 1600 nm with their Q-factor values

References

1. Kaymak Y, Fathi-Kazerooni S, Rojas-Cessa R (2019) Indirect diffused light free-space optical transmissions for vehicular networks. IEEE Transm Lett 23(5):814–817
2. Li R, Chen T, Fan L, Dang A (2019) Performance Analysis of a multiuser dual-hop amplify-and-forward relay system with FSO/RF links. J Opt Commun Networking 11(7):362
3. Li R, Zhang J, Dang A (2018) Cooperative system in free-space optical communications for simultaneous multiuser transmission. IEEE Commun Lett 22(10):2036–2039
4. Jain D, Mehra R (2017) performance analysis of free space optical transmission system for S, C and L band. In: international conference on computer, transmissions and electronics, pp 183–189
5. Huang X, Li C, Lu H, Su C, Wu Y, Wang Z, Chen Y (2018) WDM free-space optical transmission system of high-speed hybrid signals. IEEE Photonics J 10(6):1–7
6. Kaushal H, Kaddoum G (2015) Free space optical transmission: challenges and mitigation techniques. arXiv:1506.04836v1 [cs.IT]
7. Li X, Zhao X, Zhang P, Tong S (2020) BER performance of FSO transmission system with differential signaling over correlated atmospheric turbulence fading. China Transm 17(4):51–65
8. Mahajan S, Prakesh D, Singh, H. (2019). Performance analysis of free space optical system under different weather conditions. In: 2019 6th international conference on signal processing and integrated networks (SPIN)

Chapter 25
Forest Fire Analysis

Shubh Gaur, Swati Chaturvedi, and Rohit Tanwar

1 Introduction

Forest fire and climate change are interrelated aspects, as shown in Fig. 1. As the number of forest fire increases the change in climate is also observed. The diagram depicts well that change in climate results in less rainfall, more evapo-transpiration or higher temperature, which leads to dry season; thus resulting in large number of forest fire which thereby results in more CO_2 emission, which is a really harmful gas and one of the reasons of climate change and global warming as well. Forest fires are mainly caused either due to natural reasons, which is called natural forest fire, or due to human interruption.

The change in weather also plays a big role like in summer season if the amount of precipitation is low for a couple of months in that case forest is full of dry leaves, and if there occurs even a slight spark it may lead to burn in flames. Even with the slightest of spark due to lightning to a very small percentage, there occurs a spontaneous combustion of leaves or any other dry fuel [1]. Volcanic eruptions can also lead to forest fire [2].

However, when considering human forest fire there could be numerous reasons, like smoking (negligently discarded cigarettes), campfires which are left unattended, burning of debris, equipment use and malfunctions [3] (Fig. 2).

Human-caused forest fire constitutes in a greater scale compared to naturally caused, but they lead to a greater forest area burned. Human-caused fire are detected earlier mostly, so are more controllable compared to the natural fires as they are detected later and hence more uncontrollable.

Now, a big question arises that how a forest fire burns. It contains three major elements—heat, oxygen and fuel, and this is also known as fire triangle.

S. Gaur · S. Chaturvedi · R. Tanwar (✉)
Department of Systemics, School of Computer Science, University of Petroleum and Energy Studies, Prem Nagar, Dehradun, India

N. Marriwala et al. (eds.), *Soft Computing for Intelligent Systems*, Algorithms for Intelligent Systems, https://doi.org/10.1007/978-981-16-1048-6_25

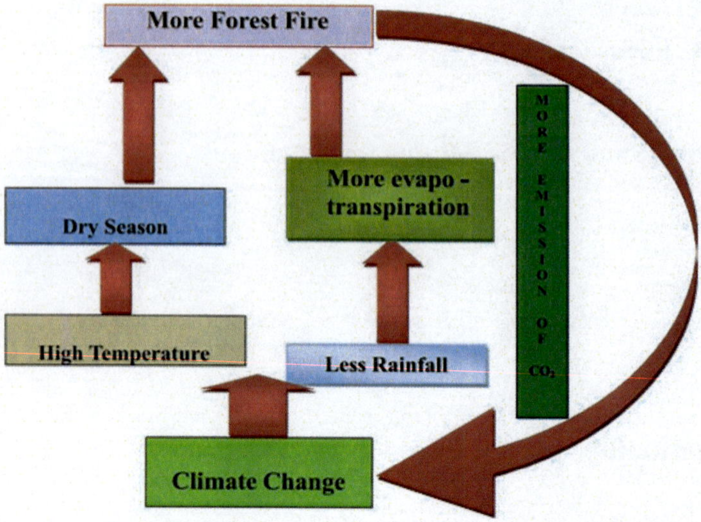

Fig. 1 Forest fire and climate change [2]

Fig. 2 Forest fire triangle [2]

1.1 Types of Combustion in Forest Fires

There are mainly three major types of combustion:

1. Smoldering fire: This fire doesn't emit flame but mainly smoke and is rarely self-sustained.
2. Flaming combustion: This type of fire consists of flames also. It may lead to formation of charcoal in case of absence of oxygen.
3. Glowing combustion: It is a slow and last stage of combustion which contains blue flames [4].

Table 1 Types of forest fires and its causes

Forest fire	Damage	Caused by
Surface fire	Moderate	Senescent leaves and twigs and dry grasses
Underground fire	Moderate	Consuming the organic matter beneath
Ground fire	High	Subsurface organic fuels
Crown fire	High	Senescent leaves, twigs, dry grasses, trees and shrubs burn
Firestorm fire	Very High	Subsurface organic fuels, Senescent leaves, twigs, dry grasses, trees and shrubs burn

1.2 Types of Forest Fires

1. Ground fires: These types of fires occur in the lower side or can say ground-like leaves area.
2. Surface fires: These fires occur on the surface which could be up to 1.3 m high.
3. Crown fires: These types of fire are the fastest spreading and the most dangerous as they occur on the top of trees. It can jump from one crown of tree to other and these are independent and very destructive.
4. Underground fires: The fires are mainly due to the organic matter inside the ground and it is lower in intensity and hence burns up to some meters on the surface.
5. Firestorms: This type of fire is the most rapid spreading, and spreads intensely over a large area. This makes fire to spread violently like storm [2].

Forest cover distribution depends on various geographical factors, like soil type, geology, altitude and climate conditions [5] (Table 1).

1.3 Facts About Forest Fires

1. In 1825, one of the largest fires occurred which burned around three million acres of forest in the areas of Maine and New Brunswick, Canada
2. There are only 10–20% fires which leads to forest fire; rest it strikes earth surface to 100,000 times in a day [6] (Fig. 3).

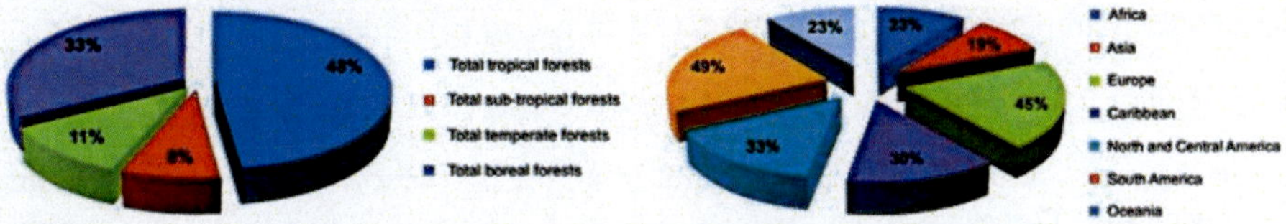

Fig. 3 Distribution of forests on the basis of major ecological zones [2], Fig. 4: The forest shares by regions [2]

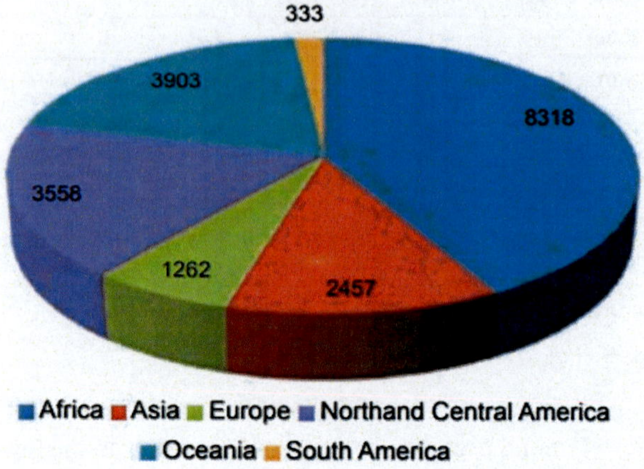

Fig. 4 Annual effect of forest areas in average by regions (area in thousand hectares) [2]

There is 2% average loss in the forest area in the period of 1975–1985 according to the Food and Agriculture Organization-U.N. "Production Year Book" but in GFRA report it was observed that the situation showed a bit progress after the year 1990. Apart from Europe where there was an increase in forest area, the rest of forest areas regions like in Asia has been reduced. There was reduction in forest area during the years 1990–2000 but increased in the years 2000–2010. The largest decrease in forests was in the Brazil which was around 2.6 million hectares per year on average and largest increase in forests were in China which was around 3.0 million hectares per year on average [2] (Fig. 5).

Let's have a look at the scale of forest fire in Brazil in the figure [3] which had the largest decrease in the world till 2010.

1.4 Major Effects of Forest Fires

1. Vegetation

Fire is responsible for killing various plants or part of plants. It mainly depends on how long and at how much temperature was the plant being exposed to fire. Also, the thickness of the branch and the diameter of stem play a major role to influence the tolerance of plant toward fire. For example, pine trees have three inches or more diameter, so pine needles survive up to 130 °F for up to 5 min.

2. Soil

There are some of the specific effects not in all areas forest fires soil but in middle coastal plains. There are huge chances of soil erosion, so to avoid that we should maintain moisture in the soil as it helps to prevent soil from consuming the humus in the soil. To do that the soil should be wet and also it should be ensured that an

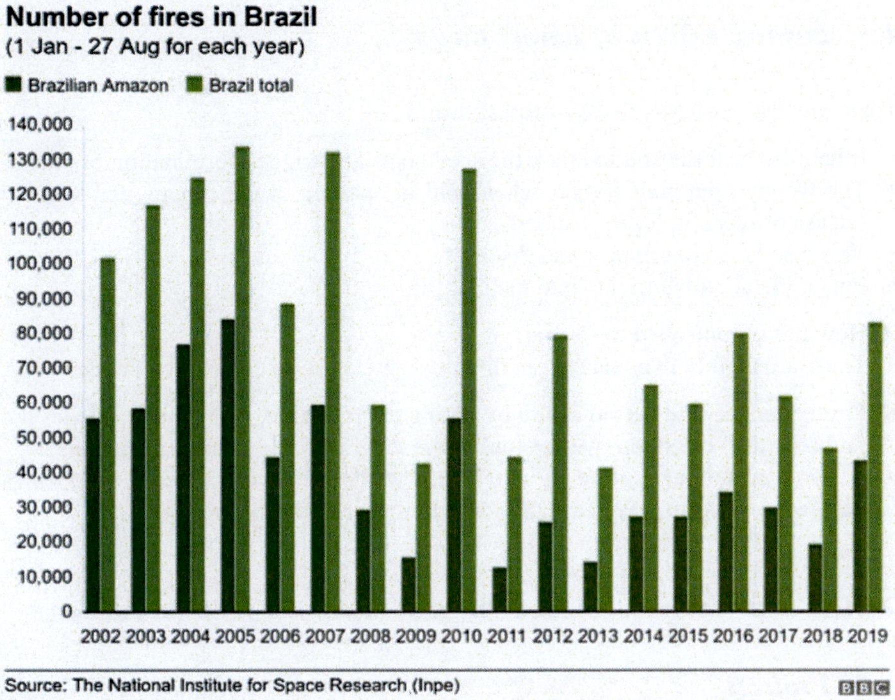

Fig. 5 The scale of forest fire in Brazil [3]. *Source* BBC (https://www.bbc.com/news/world-49515462)

organic layer will remain after a recommended burn. Moisture not only protects the duff layer adjacent to the soil, but also prevents the fire from consuming soil humus.

3. Air Quality

There could be temporary changes in the air quality due to forest fire. There are more problems to the local people living near that affected area as large quantity of smoke is produced during a short span of time.

4. Human Health and Welfare

The production of high concentration of smoke due to such fire can lead to very serious situations, mainly regarding respiratory issues which in some cases can lead to impaired alertness and judgement due to exposure of CO gas.

5. Wildlife

There could be huge loss to wildlife due to trapped in fire or due to destruction of forest there could be unavailability of food for many herbivore animals leading to overall effect in the food chain [7].

1.5 Positive Effects of Forest Fire

There are also positive effects of forest fire:

- It helps to heat the soil to crack the seed coats and trigger germination of plants.
- The woody seed pods trigger when held in opening of the canopy and leads to release of seeds in fertile ash beds.
- This also helps to kill pest and diseases.
- Puts mineral and nutrients into the soil [8].

How fire is controlled in forests?
There are mainly two techniques the fire fighters use to control forest fire:

- Firefighters help to get rid of fire by letting the fuel burn, and when it reaches the firebreak they often remove the fuel in the long line where fire is advancing.
- Air drop technique is used in which certain chemicals and water is dropped in large quantities from the airplanes or helicopter to control fire [9].

2 Related Work

2.1 Projects

1. **Near Real-Time Monitoring of Forest Fires**

This monitoring system was implemented by FSI. This system generates forest fire alerts with the help of SMS. The keyhole markup language helps to generate the forest fire alerts for the active fire areas. Its advantage is that it is compatible with Google-Earth. In this scenario SFD will be very helpful in locating the hotspots by giving the very precise location of the hotspot [2] (Fig. 6).

2. **Advanced Forest Fire Fighting (AF3)**

Advanced forest fire fighting (AF3) has main focus on innovative active and passive countermeasures. It enables remote sensing which helps in the early detection of fires and monitor it through advanced public information channels and integrated crisis management. This is the diagram of the whole solution (Fig. 7).

3. **Forest Fire Monitoring and Detection System (with SMS Alerts)**

In this project a tree inside a forest can be planted and used for monitoring the change in values with the help of sensors. If there is change in values above the threshold like temperature, smoke etc. it will send to nearest node and from there it will send to the head terminal node which uses GSM module to pass information (Fig. 8).

This whole procedure is done in three stages. In the first stage it senses and converts the signals. In the second stage this sensor reads the value and sends to the next nearest node and continues till it does not reach to the final node and the third

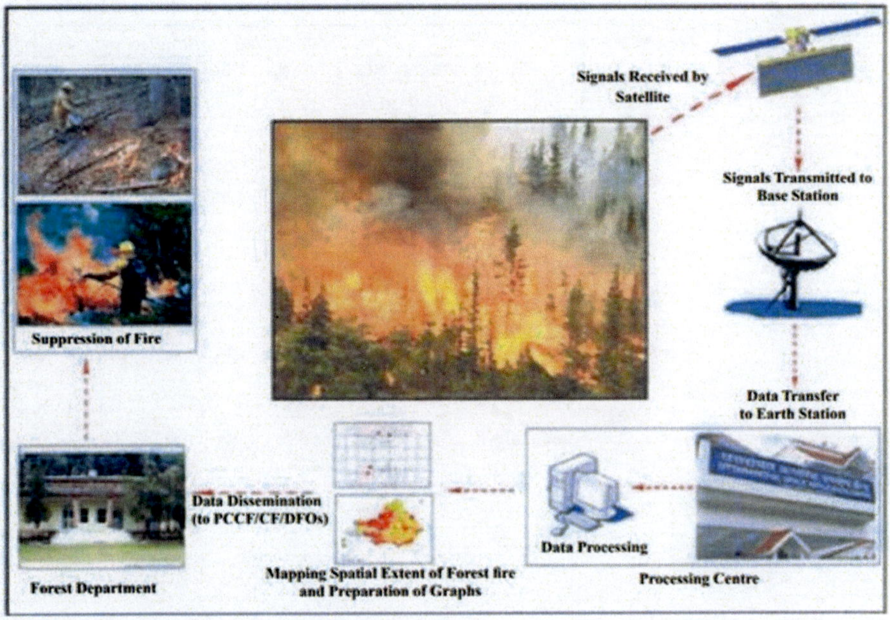

Fig. 6 Near real-time monitoring of forest fires [2]

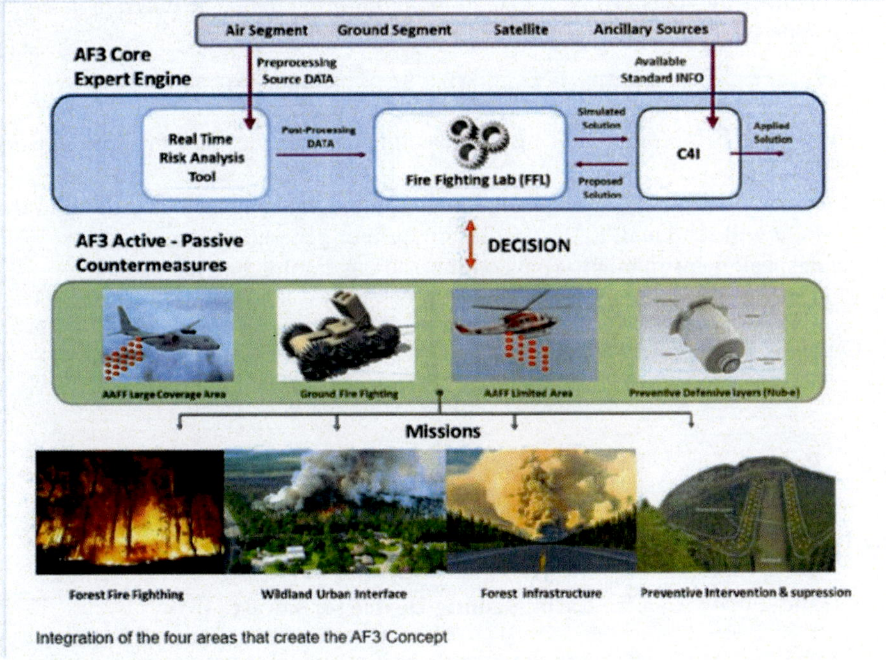

Fig. 7 AF3 Concept [10]

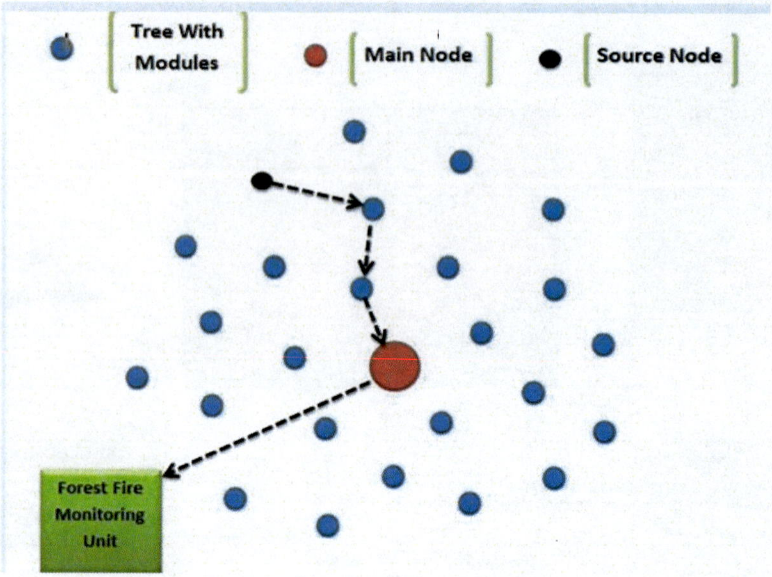

Fig. 8 Working of node [11]

stage involves the transferring of data to the monitoring unit. The diagram in Fig. 9 explains its working and flowchart [11].

4. **Forest Fire Detection and Prediction Using NodeMCU with IoT**

- This system helps for monitoring and detecting fire with the help of various sensors of IOT.
- If the sensor values received are greater than the values which are stored in the cloud, it will send mail to the user in the form of generating an alert.
- Thingspeak helps in monitoring continuously and uploading the values.
- Figure 10 depicts the block diagram of project [12].

Outputs of this project are given in Fig. 11.

2.2 *Patents*

2.2.1 Forest Fire Apparatus

1. **Hand-Throwing Fire Extinguishing Device for Forest** [13]

The forest fire extinguisher which sealed with a chemical container with super dry fine powder fires extinguishing chemical and a formal gas container with a foaming

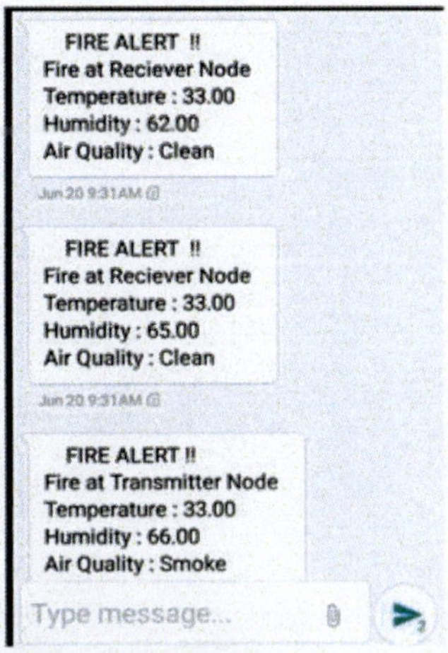

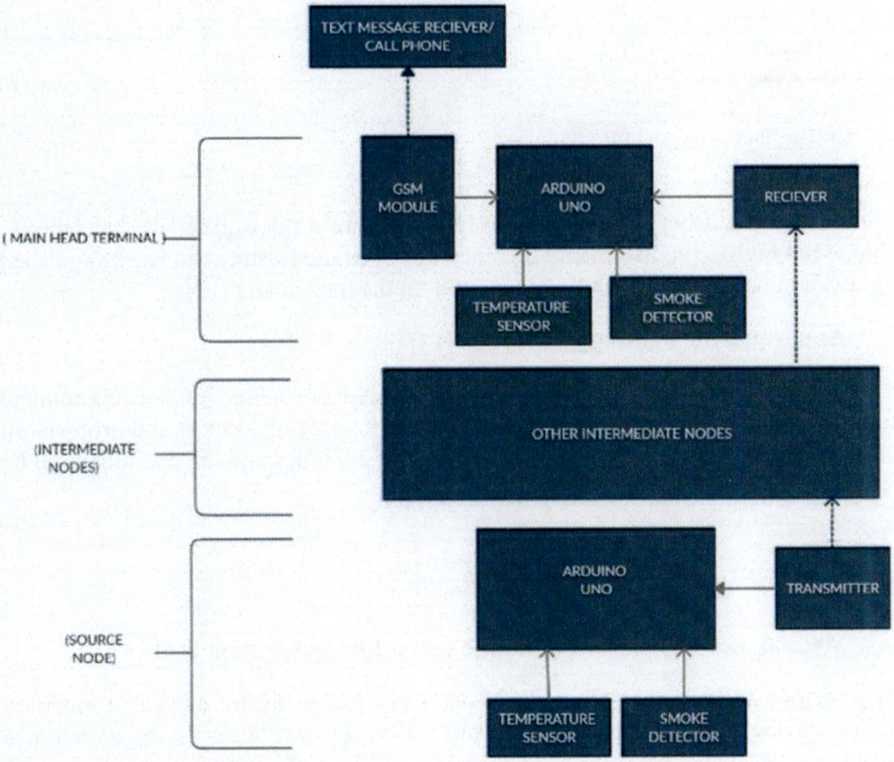

Fig. 9 Working of model [11]

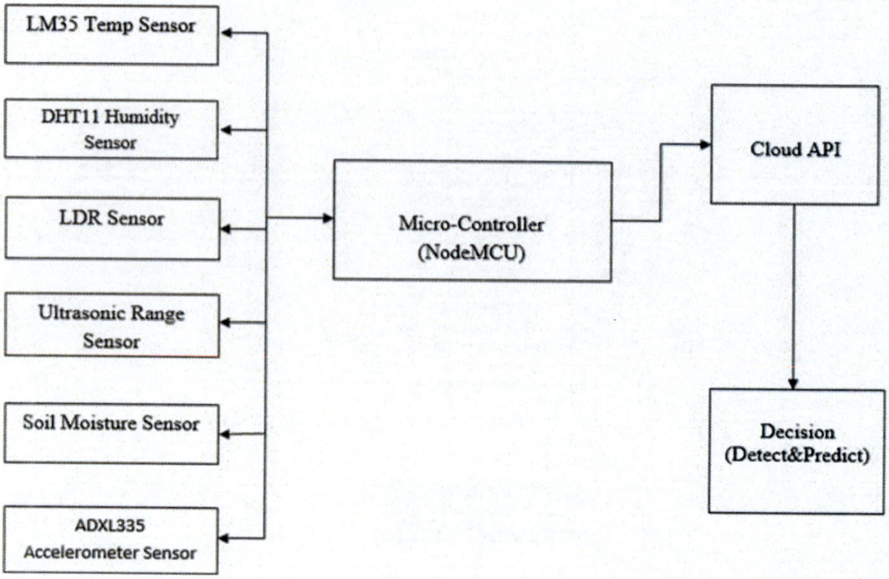

Fig. 10 Diagram of model [12]

agent, the extinguisher provides various fumaroles are at sidewall and it is hand throwable, it has high extinguishment efficiency low fired moderately can be extinguished lightweight capable for heavily transported on the hazard site [13].

2. **Apparatus for fighting forest fires** [14]

An apparatus called housing unit has two parts that define a fire-smothering chemical stored inside the apparatus. It is transported to the area by aircraft and dropped on the target area. So when it touches the ground it explodes and the chemical will be released from the apparatus [14] (Fig. 12).

2.2.2 Early Detection and Optical Visuals

1. **Method and system for automatic forest fire recognition** [15]

This method follows the technique of detecting forest fire by moveable mounted optical device that evaluates an image and trigger a local alarm. Taking a reference image and determine horizon. Then normalizing and performing filters compare the recording with at least a reference image. If we find distortion in the image and if it is above the threshold and smoke in at least one cluster of images, then it triggers the alarm [15].

Fig. 11 Outputs of model [12]

2. **Forest fire control system** [16]

It is a system for locating and suppressing a forest fire at the initial stage by finding forest fire location by one or more unmanned drones. This is a process and method to control the drone and configure according to the need. The system is used to a located forest fire and also provides other fire patterns. This system needs to develop a fire suppression system and control an extinguished chemical delivery such as GPS-guided missile system and fire suppression by missile chemicals [16].

3. **Forest fire smoke detection method using random forest classification** [17]

A forest fire smoke detection method is using random forest classification at the initial stage. At first, set a reference value which is set as a candidate block. Continuously capturing images, in frames, each block of pixels is compared with the previously captured frame which checks whether the value is greater or equal to the reference value. Every frame is compared by at least one previous frame. It extracts

Fig. 12 Apparatus [14]

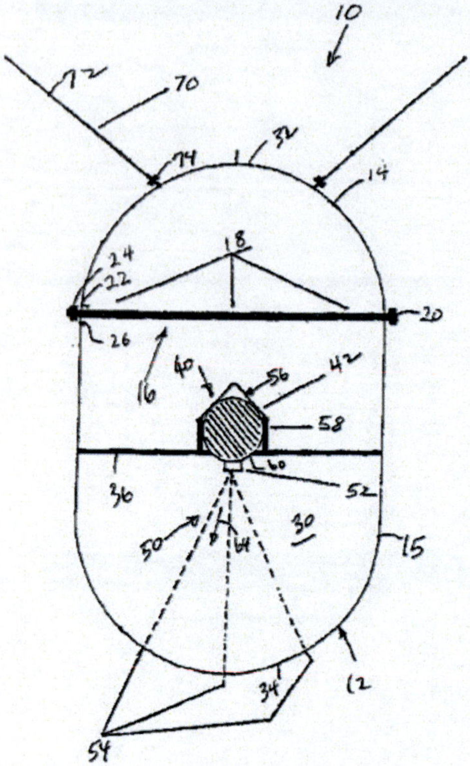

the information by candidate blocks, and then various random forest algorithms are applied to extract information. Then probability output is generated in the form of a histogram. First histogram is cumulative and then the second histogram is generated which detects the state of each pixel by comparing it with the candidate block [17] (Fig. 13).

4. **Forest surveillance and monitoring system for the early detection and reporting of forest fires** [18]

Forest surveillance and monitoring system is an early reporting system for a forest fire. In this, several numbers of remote devices are connected to the centralized system. Each remote comprised an IR sensor and a video camera is mounted on it that can be controlled remotely. It collects several data of every season and analyzed data. All remote devices are communicated via radio. Centralized system has a monitoring control of all the devices. When it seems fire can be caught there it can send forest fire personnel there to reduce the fires chance [18].

5. **Forest fire-fighting real-time early warning system** [19]

This invention is related to the early warning of forest fire detection. An early warning system has a hydrogen balloon in the air which consists of an infrared sensor on the

Fig. 13 Images pixels [17]

lower side. It monitors and detects the fire in the particular range. It alarms when the fire occurs [19].

6. **System for extinguishing and suppressing fire in an enclosed space in an aircraft** [20]

A system for extinguishing and suppressing a fire space includes a halon storage container and nitrogen generator that are connected to an enclosed duct. Here the extinguisher agent from the container archives rapid-fire suppression and uses nitrogen as a second extinguishing agent enriched in environmental air. It continually generates nitrogen through which we can archive a rapid extinguishing system [20].

2.3 Publication

1. **Blazing Heritage: A History of Wildland Fire in the National Parks** [21]

The national park has an important and major role in the development of wildlife management system on American public land with various missions and a powerful impact on the public. The nation has become the battleground between proponents of management and tool used as proponents of fire suppression. This book explains how the federal fire management has been shaped by the national park (1872–1916) which developed the fire management structure, the new deals and fire policies, post-war policies, Yellowstone and Cerro Grande [21].

2. **Ecology, Climate and Human Activities Conspire to Set the World on Fire** [22]

From the starting earth is a fire planet. It has been providing fuel for more than 420 million years. Few scientists even think that before a human fire is made fire flourished oxygenated atmosphere on earth and shaped the entire landscape, thus cleaning dense forests to make way for grassland. In recent years fire has extended their reach to earth's surface. Moreover, wildfires often become more severe and are changing the fire pattern that transforms ecosystem, reduces biodiversity and even a climate change. In recent years researchers have understood why fire is spreading and how to fight with science [22].

3. **Challenges of Predicting Wildfire Activity** [23]

The researcher has predicted that wildfire may increase its severity and several effects are predicted in the area of climate change. Two new studies show wildfire is becoming more complex and their future prevalence is less predictable than a commonly assumed "We hearing a lot that the entire planet will be burning in the future, but there's no good evidence to back this up, says Max Moritz, co-director of the Center for Fire Research and Outreach at the University of California, Berkeley" [23].

4. **Living in the Line of Fire: An Ecologist Advises Humans to Work with Forest Fires, Not against Them** [24]

It is time when we need to re-evaluate our forest fire policies, says the University of Wyoming ecologist, William Baker. Warning about forest fire from the existing method is costly and inefficient. It would not help us because the climate is changing more severely and the human population make forest fire more intense and common in the century. In his book, he tells that current fire suppression method is inefficient and costs around 1 billion dollars to US government and has a bad track record. The burning areas are increasing and keep expending forest fire [24].

5. **The Law of Fire: Reshaping Public Land Policy in an Era of Ecology and Litigation** [25]

Forest fire plays a certain role in Western public land, whatever the cause of the fire. Fire cause is increasing nowadays and there are certain debates, meetings and political engender that placed an enormous recurrent drain on the federal treasury. Seasonal fire causes burned acres, homes are destroyed, several people died every year due to wildfire and now it has become common in the states of New Mexico, Colorado, Oregon, Southern California and Arizona since 2002. Because of that forest policies have come under scrutiny but today debates are different and focus on wildland-urban interface problem, forest ecology, catastrophic fires, and legal gridlock concerns [25].

6. **The Healthy Forests Initiative: Unhealthy Policy Choices in Forest and Fire Management** [26]

In recent years forest fires are increasing severely and federal forest management is characterized by suppression of the forest fire. In summer 2002, President George W. Bush started a healthy forest initiative to improve forest health and reduce forest fire causes such as death of many people every year, private property and public resource to wildfire. It allows management agencies to take quick action and reduce the forest fire. To archive, several policies are made and help to timber harvests from public lands and make the forest healthier [26].

7. **Raging Wildfires: Climate Changes to Blame for Record Season** [27]

By observing steady climate condition, the team of researchers found the forests could retain their resilience by the wildfire if low-level fires were set every five or few years or so. They found the warmer climate causes forest burning and it only takes 20 years to make the climate hotter and slows down the growth of vegetation [27].

3 Summary

This paper broadly focused on the analysis of forest fires in various aspects of how it could be dangerous and useful, and at the same time various projects currently worked and policies made have also been studied. Flora and fauna is an essential part of forest lifecycle which has to be preserved but climate change is one of the major reason that is creating an overall imbalance in the cycle. Forest fires also play a role in increasing the global temperature in a certain way and drastic change in pollution levels as well. Many countries have made various national policies to control the rate of forest fires and investing lot of money for preventing forest fires, but they are inefficient compared to the cost invested for it. Wildfires are even more dangerous compared to forest fires as it sometimes affects the endangered species of the forest.

3.1 Challenges

1. The sensors used are of low accuracy and not cost-effective.
2. Video sensors wait for the threshold value to reach to generate request for action but till that fire increases.
3. Removal of dry trees and plants which can act as fire fuels. This is really essential to prevent forest fires but it is not possible for every forest.
4. There is a major challenge for infrastructure of remote areas for forest authorities to reach and control fires.

3.2 Future Scope

1. Pre–post analysis of forest fire should be done properly and ways and policies should be designed to prevent it.
2. Sensor designed shall be more accurate, advanced and cost-effective.
3. Certain robotic or sensor technologies shall be used for reaching the prevention of forest fires in remote areas.
4. People and young students should be more aware about forest fires and importance of forest preservation.
5. Closer monitoring of human activities inside forest activities, like mining, and residential development activities should be prohibited all over the world.
6. Sustainable development should be practiced.

4 Conclusion

In this paper we have done a deep analysis related to the aspects of forest fire, and forest fire can be very dangerous and can lead to significant health problems for nearby forest areas. Deforestation and global warming should be prevented to prevent climate change which is a big reason for forest fires.

Strict policies should be made and sustainable development should be practiced. The nearby homes built in forest area should be made of flame-resistant materials to prevent forest fires.

References

1. borealforest (2020) ForestFires. URL: https://www.borealforest.org/world/innova/forest_fire.htm#:~:text=Forest%20fires%20aways%20start%20by,to%20any%20number%20of%20reasons
2. Satendra, Kaushik AD (2020) Forest fire disaster Management. URL: https://nidm.gov.in/pdf/pubs/forest%20fire.pdf

3. National park service (2020) Forest fire. URL: https://www.nps.gov/articles/wildfire-causes-and-evaluation.htm#:~:text=Nearly%2085%20percent*%20of%20wildland,and%20intentional%20acts%20of%20arson.&text=Lightning%20is%20one%20of%20the%20two%20natural%20causes%20of%20fires
4. Shukla V (2020) Forest fire management. URL: www.vikasidea.com
5. Liu Z, Du J (2020) Forest fire control and fuels management. URL: https://www.iafss.org/publications/aofst/1/172/view/aofst.pdf
6. NSW Government (2020) How fire affects plants and animals. URL: https://www.environment.nsw.gov.au/topics/parks-reserves-and-protected-areas/fire/plants-animals-and-fire#:~:text=Positive%20effects%20of%20fire&text=We%20usually%20think%20of%20the,For%20example%2C%20fire%3A&text=creates%20hollows%20in%20logs%20and,animals%20for%20nesting%20and%20shelter
7. Academic (2020) Forestry wildlife. URL: https://www.auburn.edu/academic/forestry_wildlife/fire/effects.htm
8. doomething (2020) 11 facts about wildfires. URL: https://www.dosomething.org/us/facts/11-facts-about-wildfires
9. Ducksters (2020) Earth science for kids forest fires. URL: https://www.ducksters.com/science/earth_science/forest_fires.php
10. del Valle AE (2020) Advanced forest fire fighting (AF3) European Project, preparedness for and management of large scale forest fires. URL: https://www.researchgate.net/publication/282572570_Advanced_Forest_Fire_Fighting_AF3_European_Project_preparedness_for_and_management_of_large_scale_forest_fires
11. Ahmed A (2020) Forest Fire Monitoring & Detection System (with SMS Alerts). URL: https://create.arduino.cc/projecthub/ashad/forest-fire-monitoring-detection-system-with-sms-alerts-640a11
12. pantechsolutions. Forest fire detection and prediction using NodeMCU with IoT. URL: https://www.pantechsolutions.net/forest-fire-detection-and-prediction-using-nodemcu-with-iot
13. 宋明韬, 朱恒斌, 孙晓琳, 秦盼. URL: https://patents.google.com/patent/CN2885318Y/en
14. Appartus for fighting forest fires. URL: https://patentswarm.com/patents/US7261165B1
15. Behnke T, Hetzheim H, Jahn H, Knollen-berg J, Kührt E (2020) https://patents.google.com/patent/EP0984413A3/en
16. Ko BC, Nam JY, Kwak JY (2020) Forest fire smoke detection method using random forest classification. https://patents.google.com/patent/US8565484B2/en
17. Brogi G, Pietranera L, Frau F (2020) Forest surveillance and monitoring sys-tem for the early detection and reporting of forest fires. URL: https://patents.google.com/patent/US5734335A/en
18. Dagenhart WK (2020) Method and apparatus for containing wildfires. URL: https://www.freepatentsonline.com/y2016/0082298.html
19. "吴佑之", Early warning system for forest fires. URL: https://patents.google.com/patent/CN103456124A/en
20. Grabow T, Scheidt A, Kallergis K (2020) System for extinguish-ing and suppressing fire in an enclosed space in an aircraft. URL: https://patents.google.com/patent/US6676081B2/en?q=forest+fires&oq=+forest+fires
21. Rothman HK (2020) Blazing heritage: a history of wildland fire in the national parks. URL: https://academic.oup.com/jah/articleabstract/94/4/1305/902072?redirectedFrom=fulltext
22. Barazesh S (2020) Ecology, climate and human activities conspire to set the world on fire. URL: https://www.questia.com/magazine/1G1-203280827/ecology-climate-and-human-activities-conspire-to
23. Potera C (2020) Climate change: challenges of predicting wildfire activity. https://doi.org/10.1289/ehp.117-a293
24. Docksai R (2020) living in the line of fire: an ecologist advises humans to work with forest fires, not against them. URL: https://www.questia.com/magazine/1G1-229531531/living-in-the-line-of-fire-an-ecologist-advises-humans

25. Keiter RB (2020) The law of fire: reshaping public land policy in an era of ecology and litigation. URL: https://www.jstor.org/stable/43267312?seq=1
26. Davis JB (2020) The healthy forests initiative: unhealthy policy choices in forest and fire management. URL: https://www.jstor.org/stable/43267028?seq=1
27. "PeteThe Christian" Raging Wildfires: Climate Changes to Blame for Record Season, URL: https://www.questia.com/library/science-and-technology/environmental-and-earth-sciences/forest-fires

Chapter 26
Investigating the Need of Hybrid Integration of ERNN and BMO in Software Testing Effort Estimation

Bijendra Singh, Ankit Kumar, and Dheeraj Kumar Sahni

1 Introduction

Proper estimation of effort during software testing is playing significant role at the time of delivery of software. Estimation of effort during software testing is becoming a big problem because of complicated type of software. Effort estimation is a must during the termination of software project. Incomplete information as well as requirements that are not certain have been considered as major problem. However, several models to perform effort estimation are in existence from previous decade, but the level of accuracy is not satisfying enough. This research is presenting a new hybrid model. This model is a hybrid combination of Elman recurrent neural network (ERNN) and Barnacles mating optimizer (BMO), and is supposed to reach accurate effort estimation during software testing. BMO has been considered as novel optimization mechanism. It has been proposed in order to adjust the components of ERNN.

1.1 ERNN Software Testing Effort Estimation

Software testing is a practice to identify and resolve errors found in functionality or code of the software. The significance of testing might be considered by the fact that "Approximately thirty five percent of time and over fifty percent of total expense have been expending during testing application". The most general goal of testing is to achieve confidence that application is error-free and that it meets business and technical requirements. In other words, testing is executing a system

B. Singh (✉) · A. Kumar
Baba Mastnath University, Rohtak, Haryana 124001, India

D. K. Sahni
UIET, MDU, Rohtak, Haryana 124001, India

N. Marriwala et al. (eds.), *Soft Computing for Intelligent Systems*, Algorithms for Intelligent Systems, https://doi.org/10.1007/978-981-16-1048-6_26

in order to identify any errors, bug or missing requirements in actual requirements. Also, testing could be described as method of verifying and validating application. Test effort estimation is very essential and the management activity in application development approximates how long an operation would take to complete. The key pillars of application test effort estimation are:

- **Cost**: Every organization wants to earn profits in business. Hence cost of an application project is budget allocated during application testing. In simple words, it means how much money it takes to finish testing phase. Hence, it is very important to predict perfect budget of application testing life cycle.
- **Time**: Time is the key resource in an application project. Every project has a deadline to delivery. So, accurate estimation of time spent in testing process is also an important factor.
- **Resources**: The third important requirement is estimation of resources to carry out application testing of a project. Resources could be people, facilities, equipment or anything else required to complete application testing activity.
- **Human skills**: It means knowledge and experience of team members of any application testing project. They could badly affect your estimation. Let us say, if a team member has less application testing skill, then he/she would take more time to finish his/her work than another member who is expert in technical skills of application testing.

1.2 Software Testing Effort Estimation Techniques

There are numerous application test effort estimation techniques available for application developers to forecast test effort estimation. For large organizations where project size and team size is big, accurate and true test effort estimation could play an important role. A good quality test effort estimation technique could be beneficial for both test managers and application projects. Following are the key techniques for test effort estimation:

A. ***Use Case Point Estimation***
   Among various application test effort estimation approaches, the UCP approach is most important and commonly utilized technique. Test effort estimation using UCP is based on use cases. Use case is actually a system activity in various conditions. It presents system's functionality at various circumstances as this is responding to a request from stakeholders. They are named as primary actors. The key operation of use case point is to map the use cases to test cases.
B. ***Function Points/Test Point Effort Estimation***
   Test points are being utilized for test point analysis to find the test effort at the time of system and acceptance testing. Test point analysis only covers black-box testing. On the other hand, functional point analysis (FPA) never covers system and acceptance test cases. Therefore, both TPA and FPA are being merged together to calculate white and black-box application testing efforts.

This helps in determining the time required and risk (if any, after comparing TPA and pre-estimated hours) occupied in application projects. This method is useful when measurement of function point is available and also the previous data for development and testing are available.

C. ***Work Breakdown Structure Mechanism***
Complex application project is divided into modules in work breakdown structure mechanism. Then, these modules have been classified into sub-modules. Every sub-module has been classified as per its working. In other words, this technique works based on the method of dividing whole application project structure into smallest structures. Work breakdown structure has two types:

Functional Work breakdown structure: Application project is divided based on functions in software to be made. It has been considered useful to check system scale.
Activity Work breakdown structure: In this activity, application system is split according to the activities performed in system. These activities have been broken into tasks. This is a useful activity in estimating schedule and effort in application system.

D. ***Three-Point Software Testing Estimation***
The three-point estimation technique is one of the most important effort estimation techniques that could be utilized in management and information systems applications. The simplicity of three-point estimation technique is making a significant approach in case of project manager for predicting proper testing efforts in application projects.

E. ***Delphi Technique***
This is one of the most commonly utilized test effort estimation technique. It is a classic estimation technique based on surveys in which information is being collected from experts on their own earlier experience. In such type of mechanism, every task is assigned to every team member. Several rounds of survey are needed to be made. This is done until and unless real estimation is not found.

1.3 ERNN

Elman neural network has been considered as one of the well-known recurrent neural networks. ENN is having more inputs from hidden layer as compared to previous neural networks. These are making layer that has been known as context layer. Thus standard back-propagation mechanism is utilized in ENN. This mechanism is known as Elman back-propagation. Use of hidden layer increases the accuracy in Elman neural network as compared to traditional neural network.

1.4 *Barnacles Mating Optimizer*

The major idea of BMO has been derived from the mating activities of barnacles in nature. Barnacles have been considered as microorganisms that are hermaphroditic. These are having capabilities of male as well as female sex reproductions. These microorganisms are supposed to be fertilized by neighbor in order to produce new off-springs. These microorganisms are having unique feature that they have large penises. This is largest in microorganisms as compared to the size of their own body. Selection of barnacle's parents has been decided on a random basis to make new off springs.

Barnacles Mating Optimizer Algorithms

1. Set the population of barnacles Xi
2. Find fitness of every barnacle
3. Sorting to find best output at top of population
4. T is showing best solution
5. **Until** I is less than Max_Iter
   a. Set value of pl
   b. Perform selection using following equations
      barnacles_d = randperm(n)
      barnacles_m = randperm(n)
   c. **if** selection of Dad and Mum = pl
      i. **for** each variable
         Off spring generation wing equation:
         $X^{N\_new} - pX^{N}_{barnacles\_d} + qX^{N}_{barnacles\_m}$ for k<pl
      ii. **end for**
   d. **else if** selection of Dad and Mum >pl
      i. **for** each variable
         Off spring generation wing equation:
         $X^{N\_new} - rand() \times X^{N}_{barnacles\_m}$ for k>pl
      ii. end for
   e. end if
   f. Get recent barnacle back if this is going beyond boundaries
   g. Get fitness of each barnacles
   h. Update T if there is better solution
   i. 1=1+1
6. Return T

2 Existing Researches

There have been several researches in the field of software testing effort estimation. The existing researches have made use of neural network to train the network for efficient software testing effort estimation. The proposed work is considering integration of ERNN and BMO. The existing research-related ERRN has proved that ERRN is performing better as compared to traditional neural network. Thus the existing researches related to software testing effort, ERNN and BMO have been explained.

2.1 Researches in the Field of ERNN

Several researchers used ERNN to perform direct inverse control [1] that has been used for quadrotor attitude as well as altitude control mechanism. ERNN is presenting good performance as compared to traditional neural networks. In many researches comparison is made in back-propagation-based NN and ERNN mechanism [2] to perform research on altitude management of heavy lift hexacopter. This work is dependent on direct inverse control. Some researches have been performed for data rectification [3] that make use of Elman recurrent neural networks. ERNN is found to be capable of minimizing noise level at the time of measurements without having any information regarding nonlinear dynamics system. Elman neural network has been proposed with dynamic properties during identification in nonlinear dynamic system (Yuan-Chu Cheng [4]). Comparison of feed forward NN with RNN in case of find intrusion [5] has also been performed.

2.2 Researches in the Field of Barnacles

Many researchers consider application of barnacles optimizer in order to resolve economic dispatch issues [6], while some papers considered BMO as bio-inspired algorithm [7] to solve optimization issues. BMO has also been used to enhance gray-level image contrast [8] in many researches. BMO has been used to make evolutionary algorithm [9] in order to solve optimization issues in several researches.

2.3 Researches Related to Software Testing Efforts

Researchers have focused to propose cost-sensitive mechanism to increase the utilization of ML classifiers during estimation of efforts. Lot of researches have made use of self-learning machine-learning-based intelligent system in order to increase the reliability in prediction of software testing effort. Many researchers have proposed mechanisms that have reduced the expenses of effort estimation during testing [10], and on the other hand, various mechanisms to get effort estimation during testing in software industry are considered [11]. Issues along with pros and cons of different software test effort estimation mechanisms have been considered [12] in many researches. The comparison of accuracy and performance of different artificial intelligence mechanisms, such as genetic algorithm, neural network and SVM, has been made [13] in order to consider the need of BMO and ERNN. Many researchers have tried to make the system more intelligent [14], considering the challenges faced during test effort estimation [15].

2.4 Review of Literature

Table 1 shows the author of research and year of publication along with objectives of research. Moreover, the techniques used in research are explained with findings and results.

3 Testing Effort Estimation

Steps to find the test effort estimation are discussed below (Table 2):

A. **Classify Project Operation in Multiple Sub-tasks:**
Using work breakdown structure mechanism, complex project is categorized into modules, and these modules could be further categorized into sub-modules.

B. **Allocation of Each Operation to Team:**
Every operation has been assigned to team. Users may assign operation as follows (Table 3):

C. **Estimation of Effort during Operation**

There have been two mechanisms that could be applied to find the effort for operation:

1. Mechanism that is using functional point
2. Mechanism that performs estimation using three points

Mechanism 1: Mechanism used for Function Point.

During such mechanism, the test manager is considering scale, time and expenses for operations. Various steps included in this task are given as under:

Table 1 Literature review

S. No.	Author/year	Objective of research	Technique used	Findings
1	T. W. Karjala/1992	To perform rectification of content using ERNN	ERNN	ERNN is found capable of minimizing the noise level during measurement process. It has been performed without considering nonlinear dynamics of system
2	Yuan-Chu Cheng/2008	To propose Elman neural network with dynamic properties	ERNN	PID Elman network has been found prior to customized Elman network during identification in nonlinear dynamic system
3	N. Chowdhury/2008	To perform comparative analysis case of feed-forward neural network. The detection of intrusion is the main motive	Recurrent neural network	This research of ERNN provides efficient method during decision-making
4	Kerstner/2011	Mechanism for effort estimation of software	Test effort estimation	The challenges and pros and cons of different software test effort estimation mechanism have been considered
5	Amin Dastanpour/2014	Research is performed comparing GA on ANN and SVM to implement IDS	GA, neural network, SVM	The comparison of accuracy and performance of different artificial intelligence mechanisms, such as genetic algorithm, neural network and support vector machine, has been made

(continued)

Table 1 (continued)

S. No.	Author/year	Objective of research	Technique used	Findings
6	Kafle/2014	Research focused on effort estimation during software testing	Test effort estimation	Challenges faced during test effort estimation are considered
7	Adali/2017	Research has focused on estimation of effort during testing of software	Estimation of effort during testing	Different test effort prediction techniques used in software industry are considered
8	B. Y. Suprapto/2018	Research is performing comparison between back-propagation NN and ERNN during altitude management of hexacopter that is heavy lift	ERNN	The ERNN algorithm results in smaller MSE value
9	M. H. Sulaiman/2018	To use bio-inspired algorithm to solve optimization issues	BMO	Results show the effectiveness of BMO
10	M. H. Sulaiman/2018	Proposing evolutionary algorithm in order to solve optimization using BMO	BMO	The BMO has been found effective than other swarm algorithms
11	M. H. Sulaiman/2019	Research is considering application of BMO. The focus is to resolve problems related to economic dispatch	Barnacles mating optimizer	Minimized cost could be achieved without violating constraints with the help of BMO
12	In 2019, Moreira Nascimento	Research is making cost-sensitive approach. The objective is to impose the utilization of machine learning classifiers during effort estimation of software	ML classifier	Research has proposed mechanism that has reduced the expenses of testing effort estimation

(continued)

Table 1 (continued)

S. No.	Author/year	Objective of research	Technique used	Findings
13	F. Xiaorong/2019	Research has focused on intelligent network. It is considering auto software testing technology	Intelligent network	Research has opted to make the system more intelligent
14	B. Kamanditya/2020	This research focused to introduce direct inverse control in case of quadrotor attitude	ERNN	ERNN is showing better performance features as compared to previous back-propagation NN
15	S. Ahmed/2020	Research is focused to increase gray-level graphic contrast	BMO	Comparison is proving superiority of BMO mechanism over other optimization mechanisms

Table 2 Operations and sub-operations

Operation	Sub-operation
Analyze application requirement specification	1. To perform investigation during specification of soft requirement. 2. To make interview with developer as well as other stakeholders in order to consider website
Develop test specification	1. Making the scenarios to make testing 2. Producing the cases for testing 3. Review and revise test cases
Running test cases	1. Developing the environment for test 2. Implementation of test cases 3. Checking the results after test execution
Report faults	1. Develop fault reports 2. Report faults

Step A: Estimate size for operation

In Step A, the whole project operation is broken into small operation by using WBS method. Before starting actual estimating effort the functional points have been classified into categories such as **Complex, Medium and Simple**.

1. **Complex** system has several components interacting with each other.

Table 3 Operation and members

Operation	Person
Analyzing application needs and specification	Everyone
Product specification for testing	Analyst for testing
Generating environment for testing	Admin
Execution of test cases	Testing analyst and admin
Report faults	Testing person

Table 4 Representing weightage of groups

Group	Weightage
Complicated	05
Average	03
Easy	01

2. **Medium** system operates with limited components.
3. **Simple** performs operation with less components.

Test manager needs to provide appropriate **weightage** to each functional point depending on complicated application functions (e.g. Table 4).

More efforts are needed when there is more complicated function point. Website has been categorized into 12 function points that could check the complexity of every function point, as shown in Table 5.

Step B: Finding time for operation

Users need to guess the duration to test after categorization of complexity in case of function points. Timing is the time needed to complete an operation.

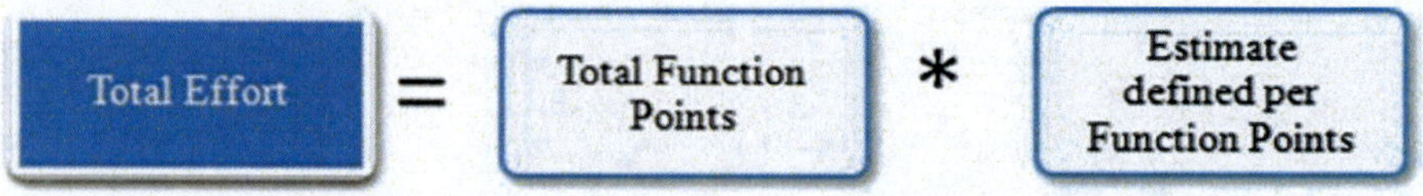

Effort has been considered as effort to overall test functions running on site.

Function Point presents the overall modules used in website.

Prediction of the defined points of function has been considered as average effort to fulfill function points.

Such value is dependent over productivity of member. These are the persons who would take in-charge of such operations. Users might predict the overall effort to test complete characteristics of website that are discussed below (Table 6):

Table 5 Table of module, applicable role and weightage

S. No.	Name of the module	Roles that are applicable	Module weightage
1	Enquiry of Balance	Administrator User	3
2	Fund Transfer	Administrator User	5
3	Mini Statement	Administrator User	3
4	Customized Statement	Administrator User	5
5	Change Password	Administrator User	3
6	New Customer	Administrator	3
7	New Account	Administrator	5
8	Edit Account	Administrator	1
9	Delete Account	Administrator	1
10	Delete Customer	Administrator	1
11	Deposit	Administrator	5
12	Withdrawal	Administrator	5

Table 6 Weightage, function points and total

	Weightage	# of function points	Total
Complicated	05	05	25
Average	03	04	12
Easy	01	03	03
Points used in function			40
Estimation of definition in per point			05
Effort estimated considering people hours			200

Step C: Considering the expense for operation

This step is helping to conclude the answer of question of user "What is the *cost of operation?*". *Let* on average team salary has been 200 rupees per hour. Then duration needed to develop test-specific operation is 200 h. Thus the cost for operation has been 200 × 200 = 40,000 Rupees.

Mechanism 2: Three-Point Estimation.

Such estimation has been considered as one of the mechanisms utilized to judge the operation. The simplicity of three-point estimation is making this as useful mechanism to project manager. **Three** values would be produced for every task depending on **prior experience** in case of three-point estimation:

Best Case is case when **120** man-hours.

Most Case presents the case when this operation is performed by **170** man-hours.
Worst Case is the case that present case to complete operation by **200** man-hours.

Set the value of each parameter as shown below

$$\mathrm{a} = 120$$
$$\mathrm{m} = 170$$
$$\mathrm{b} = 200$$

Effort considered to execute an operation might be found with the help of **double triangular distribution** equation that is presented in the following equation:

$$E = (a + 4m + b)/6$$
$$E = (120 + 4 * 170 + 200)/6$$
$$E = 166.6 \text{ (man-hours)}$$

In the above equation, parameter E has been considered as **Weighted Average. This is estimation of operation "*Develop Test Specification*".**

4 Problem Formulation

It has been observed that the existing researches in the field of software testing effort estimation are suffering due to lack of accuracy and performance. The traditional neural network and genetic algorithms are time-consuming. Moreover, there is a need to introduce cost-effective mechanism for software testing effort estimation. However, there are functional point method and three-point estimation mechanism for software testing effort estimation. There is a need to propose a mechanism that should provide the optimized and accurate results with high performance.

5 Proposed Work

The proposed work is supposed to study and analyze existing testing effort estimation techniques. During our research work, first we would concentrate on existing application test effort estimation techniques and analyze these based on numerous techniques. Next, we would use MATLAB tool to predict application test effort estimation. Hybrid combination of Elman recurrent neural network (ERNN) and barnacles mating optimizer (BMO) would be utilized for analysis, design and prediction of application test effort estimation. On the other hand, if required, we would also utilize other tools in our research work (Fig. 1).

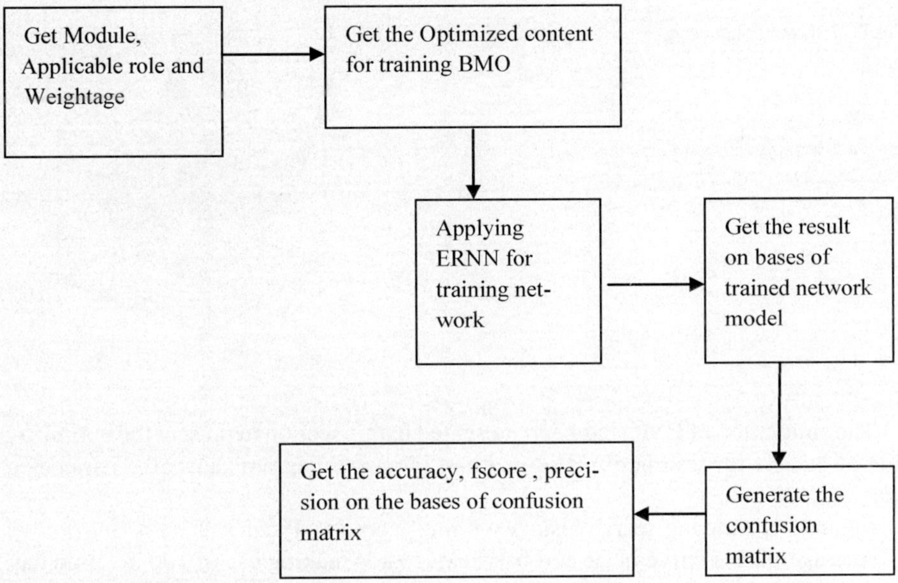

Fig. 1 Process flow of the proposed work

Elman neural network Algorithms with integration of BMO

Begin

1 Set the size of Barnacles population and ERNN structure

2 Setting data for training and testing

3 repeat step 4,5,6,7,8,9,10,11 Until MSE is less than stopping conditions

4 Set the Barnacles preys as solution to network

5 Feed forward network is running with help of optimized value that are set using **BMO**

6 Find the error

7 Reduce the error by setting network parameter with the support of **BMO**

8 Generate Barnacles (xj) by using Brownian motion from a certain position.

9 If error is detected, abandon the current position and move to a accurate solution

10 Consider the fitness of prey, select random Barnacle i

If

a. Xi > Xi Then

b. xi <— xj

c. xi <— xj

End if

11 Barnacle is finding best possible solution at each epoch until network is converged.

12 Get the output

13 End

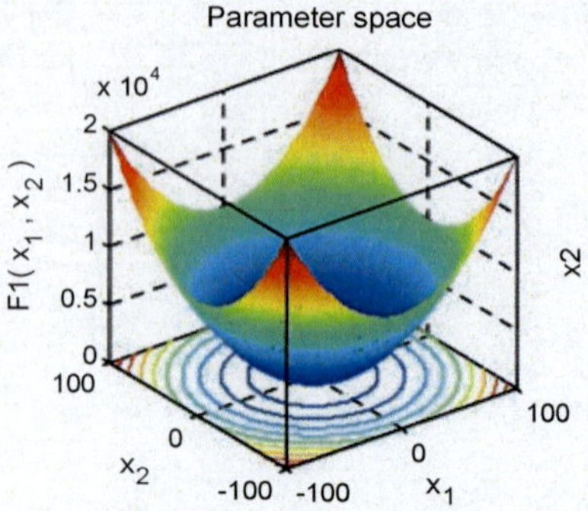

Fig. 2 Parameter space

The simulation of BMO has been presented in this section to present the optimization of dataset for testing effort estimation. Figure 2 diagram shows the parameter space.

Figure 3 is showing search history with respect to × 1 and × 2.

Average fitness curve in case of barnacles that is making use of 100 iteration has been presented in Fig. 4.

Convergence curve with respect to best score obtained so far has been shown in Fig. 5. After getting the best solution the convergence curve becomes stable.

Performance Parameters

The confusion matrix is produced presenting true positive, true negative, false positive and false negative (Fig. 6).

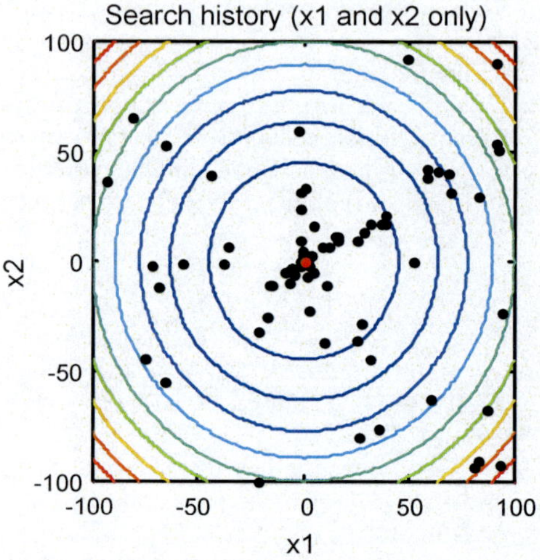

Fig. 3 Search history

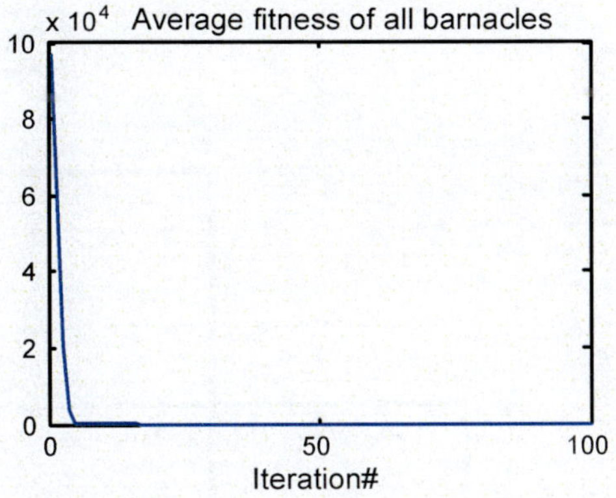

Fig. 4 Average fitness curve

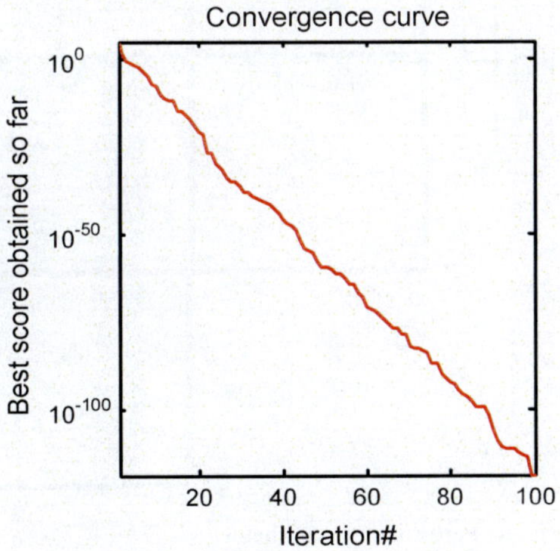

Fig. 5 Chart presenting best score at different iteration

Optimization Using BMO

The simulation of BMO has been presented in this section to present the optimization of dataset for testing effort estimation. Figure 7 diagram is showing the parameter space.

True positives: True positive has been considered as predicted positive data. It is the value of the actual class that is found valid, and the result of predicted class has also been valid.

True negatives: Valid but negatives have been known as forecasted negative values that are correct. This means the value of the real class has been found "no" and the result of forecasted class has also being "no".

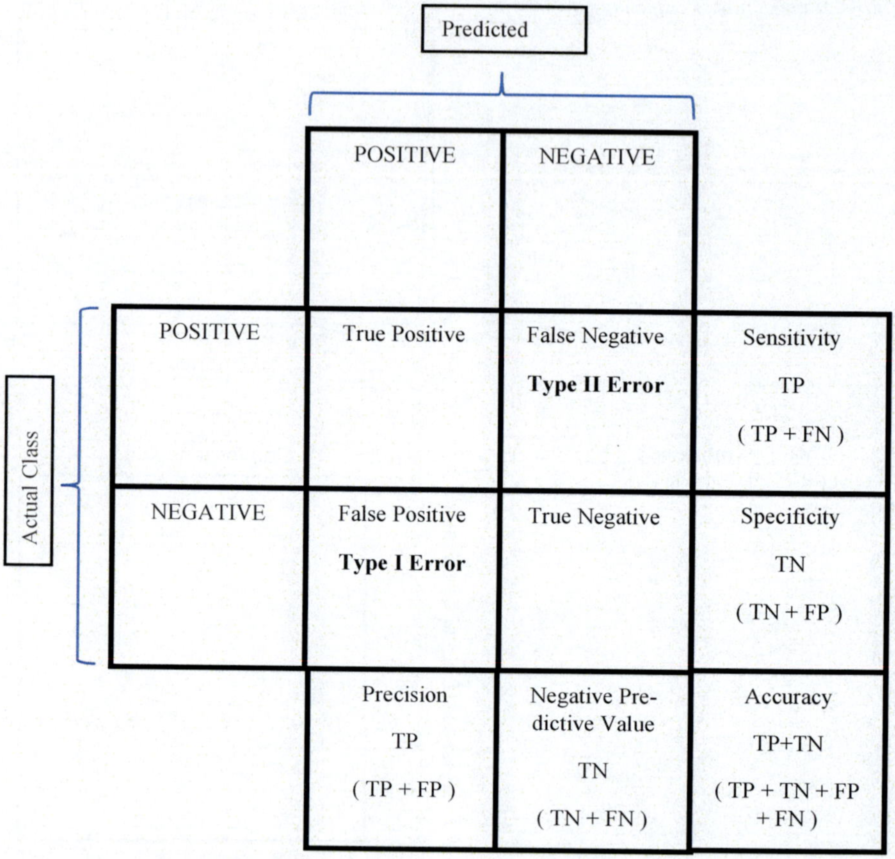

Fig. 6 Performance parameter

Invalid positives and Invalid negatives: Such data are occurring when the real category result is not matching with the result of the predicted class.

False positives: It is the case when the real class is "no" but the output of the predicted class has been found as "yes".

False negatives: It is the case where the real class is "yes"; on the other hand, predicted class is saying "no".

The parameters utilized for verifying results are precision, accuracy, f1 score and recall that are discussed below:

Accuracy: It is considered the most intuitive measurement of performance. This is known as the proportion of the right predicted findings to total findings. It has been considered that if the accuracy of model is more, then the model would be considered best. Accuracy has been considered as a good measure. The condition is that users should have symmetric datasets. Here values of invalid positive, as well as invalid negatives, are mostly equal.

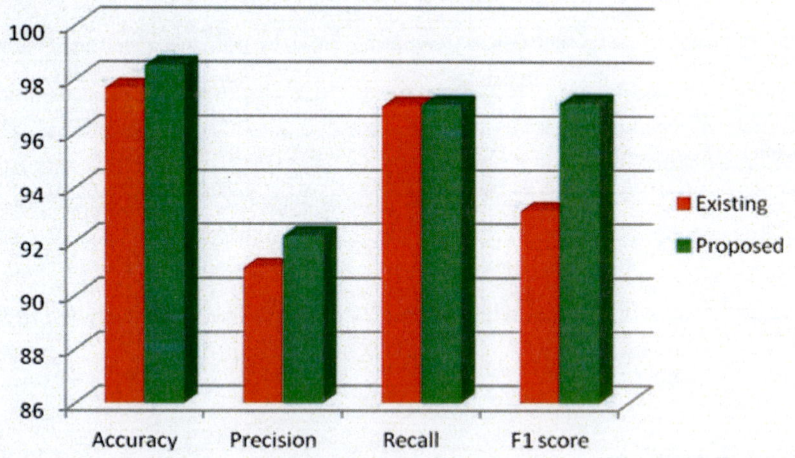

Fig. 7 Comparison chart for accuracy, precision, recall and f1 score in case of previous and proposed work

$$\text{Accuracy (A)} = \text{TP} + \text{TN}/\ \text{TP} + \text{FP} + \text{FN} + \text{TN}$$

Precision: It has been considered as a proportion ratio of right predicted positive results with overall predicted positive findings.

$$\text{Precision (P)} = \text{TP}/\text{TP} + \text{FP}$$

Recall: It is representing sensitivity and is considered as a proportion of right predicted positive findings to all findings in real class true.

$$\text{Recall(R)} = \text{TP}/\text{TP} + \text{FN}$$

F1 score: It has been considered as a weighted average of precision as well as recall. So, the score is considering false positives and false negatives both. The significance of F1 is considered more as compared to accuracy. It is playing important if there is an uneven class distribution. Thus accuracy is working well if the invalid and positives as well as invalid and negatives are having the same cost. This would be better to consider both precision and recall if the cost of invalid positives, as well as invalid negatives, is varying.

$$\text{F1 score} = 2 \times (\text{R} \times \text{P})/(\text{R} + \text{P})$$

Percentage wise increment in the proposed model as compared to previous model has been shown in Table 7.

The findings conclude that the accuracy in case of proposed work is more as compared to previous work. The precision, recall, F1 score are more, respectively, in

Table 7 Percentage wise presentation of increment in accuracy, precision, recall and f1 score

	Existing	Proposed
Accuracy	97.66	98.50
Precision	91.01	92.20
Recall	97.00	97.05
F1 score	93.10	97.1

case of proposed work. Chart-wise representation of above chart is shown in Fig. 7 (Table 8).

6 Conclusion

Research work has presented process of software testing effort estimation with support functional point method and three-point estimation. Hybrid combination of Elman recurrent neural network (ERNN) and barnacles mating optimizer (BMO) has been proposed in order to increase the reliability of result. Moreover, the proposed work is supposed to increase the performance. The accuracy has been increased as the proposed work is making use of ERNN that makes use of hidden layers. On the other hand, the results have been optimized using BMO.

7 Scope of Research

This research is supposed to play a significant role in estimating testing effort. The proposed work has integrated ERNN and BMO where ERNN has increased the accuracy and performance, while BMO has optimized the result. The research has been proven reliable, efficient and optimized. Such integration might be significant in another area where the decision making using intelligent system takes place such as AI, big data processing, IOT, health care, cloud computing and software development.

Table 8 Comparison between traditional testing effort estimation, fuzzy rule-based genetic algorithm, artificial neural network, BMO, ERNN

	Traditional testing effort estimation	Fuzzy rule-based	Genetic	Artificial neural network	BMO	ERNN
Training	N/A	N/A	N/A	Required	N/A	Required
Suitability	If there is limited cases	Algorithm for true false cases	Suitable for intermediate cases	Suitable for large set of cases	Suitable to variety of dataset	Large dataset is required for training
Time consumption	This model is time-consuming if several cases are available to find testing effort	Problem-solving control system mechanism takes average time	Algorithm is taking long time to get optimal solution	Takes time in learning but decision making is fast	The performance of BMO is remarkable	Moderate
Complexity	Easy to implement	Complexity depends on problem	The complexity depends of given problem	The implementation of ANN requires technical skill	Moderate	Moderate
Accuracy	Less	Less	Less	Less	Moderate	High

References

1. Kamanditya B, Kusumoputro B (2020) Elman recurrent neural networks based direct inverse control for quadrotor attitude and altitude control. In: 2020 international conference on intelligent engineering and management (ICIEM), London, United Kingdom, pp 39–43. https://doi.org/10.1109/ICIEM48762.2020.9160191
2. Suprapto BY, Kusumoputro B (2018) A comparison of back propagation neural network and elman recurrent neural network algorithms on altitude control of heavy-lift hexacopter based on direct inverse control. In: 2018 international conference on electrical engineering and computer science (ICECOS), Pangkal Pinang, pp 79–84. https://doi.org/10.1109/ICECOS.2018.8605240
3. Karjala TW, Himmelblau DM, Miikkulainen R (1992) Data rectification using recurrent (Elman) neural networks. In: IJCNN international joint conference on neural networks, Baltimore, MD, USA, vol 2, pp 901–906. https://doi.org/10.1109/IJCNN.1992.226873
4. Cheng Y-C, Qi W-M, Zhao J (2008) A new elman neural network and its dynamic properties. In: 2008 IEEE conference on cybernetics and intelligent systems, Chengdu, pp 971–975. https://doi.org/10.1109/ICCIS.2008.4670784
5. Chowdhury N, Kashem MA (2008) A comparative analysis of Feed-forward neural network & Recurrent Neural network to detect intrusion. In: 2008 International conference on electrical and computer engineering, Dhaka, pp 488–492. https://doi.org/10.1109/ICECE.2008.4769258
6. Sulaiman MH, Mustaffa Z, Aliman O (2019) An application of barnacles mating optimizer for solving economic dispatch problems. In: 2019 IEEE Jordan international joint conference on electrical engineering and information technology (JEEIT), Amman, Jordan, pp. 835–839. https://doi.org/10.1109/JEEIT.2019.8717377
7. Sulaiman MH et al (2018) Barnacles mating optimizer: a bio-inspired algorithm for solving optimization problems. In: 2018 19th IEEE/ACIS international conference on software engineering, artificial intelligence, networking and parallel/distributed computing (SNPD), Busan, pp. 265–270. https://doi.org/10.1109/SNPD.2018.8441097
8. Ahmed S, Ghosh KK, Bera SK, Schwenker F, Sarkar R (2020) Gray level image contrast enhancement using barnacles mating optimizer. IEEE Access 8:169196–169214. https://doi.org/10.1109/ACCESS.2020.3024095
9. Sulaiman MH, Mustaffa Z, Saari MM, Daniyal H, Musirin I, Daud MR (2018) Barnacles mating optimizer: an evolutionary algorithm for solving optimization. In: 2018 IEEE international conference on automatic control and intelligent systems (I2CACIS), Shah Alam, pp 99–104. https://doi.org/10.1109/I2CACIS.2018.8603703
10. Nascimento M, Vismari LF, Cugnasca PS, Camargo Júnior JB, Rady de Almeira Júnior J (2019) A cost-sensitive approach to enhance the use of ML classifiers in software testing efforts. In: 2019 18th IEEE international conference on machine learning and applications (ICMLA), Boca Raton, FL, USA, pp 1806-1813. https://doi.org/10.1109/ICMLA.2019.00292
11. Adali OE, Karagöz NA, Gürel Z, Tahir T, Gencel C (2017) Software test effort estimation: state of the art in Turkish Software Industry. In: 2017 43rd Euromicro conference on software engineering and advanced applications (SEAA). IEEE, pp 412–420
12. Kerstner M (2011) Software test effort estimation methods
13. Dastanpour A, Ibrahim S, Mashinchi R, Selamat A (2014) Comparison of genetic algorithm optimization on artificial neural network and support vector machine in intrusion detection system. In: IEEE conference on open systems (ICOS)
14. Xiaorong F, Shizhun J, Songtao M (2019) Research on intelligent network auto software testing technology. In: 2019 2nd international conference on safety produce informatization (IICSPI), Chongqing, China, pp 1–5. https://doi.org/10.1109/IICSPI48186.2019.9096016
15. Kafle LP (2014) An empirical study on software test effort estimation. Int J Soft Comput Artif Intell 2(2)

Chapter 27
Face Recognition Techniques, Challenges: A Review

Juhika Azmeen and Dibya Jyoti Borah

1 Introduction

Face recognition is a procedure of identifying or detecting a face that is either known or unknown from an enrolled user database. It is one of the most used biometric authentication techniques and can be considered as one of the most challenging tasks in image retrieval and computer vision. A face recognition system has various applications, such as in ATM, healthcare system, driving license system, passport authentication, criminal detection when there is no thumbprint on the scene [1], attendance system [2], etc. A brief process of a face recognition is shown in Fig. 1.

There are various biometric systems to identify a human face, fingerprint, palm print, speech, iris etc., but all these systems require human intervention. In such scenario, face recognition can be considered as the most convenient way than other biometrics as it requires no human intervention. Face recognition consists of face verification and face identification. The identification and recognition rate for face recognition in large database was seen to achieve more than 90% with the use of various face detection methods with controlled pose, illumination, and aging [3–6].

This paper broadly consists of four sections. Section 2 describes about the types of face recognition; Sects. 3 and 4 discuss about the methods and challenges of face recognition and comparison of the methods. Section 5 is the conclusion.

2 Face Detection

Face along with the different features of a face, such as fingerprint, eye, iris, and body structure, can play an important role in face recognition by identifying a face

J. Azmeen · D. J. Borah (✉)
Kaziranga University, Jorhat, India
e-mail: dibyajyotiborah@kazirnagauniversity.in; juhikaazmeen123@gmail.com

N. Marriwala et al. (eds.), *Soft Computing for Intelligent Systems*, Algorithms for Intelligent Systems, https://doi.org/10.1007/978-981-16-1048-6_27

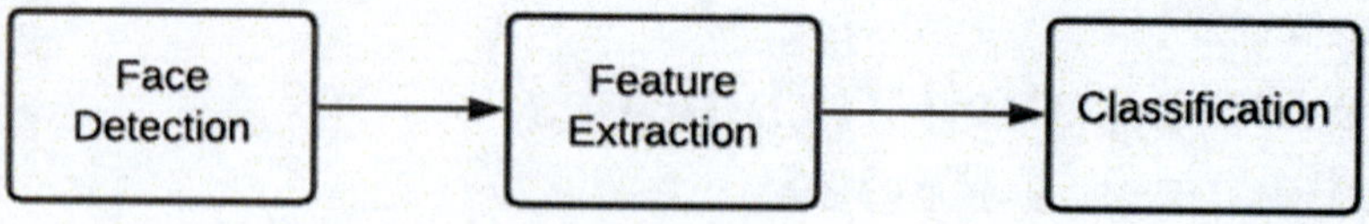

Fig. 1 Block diagram of face recognition, references [3]

directly from a vide or from a CCTV footage. The first step is to detect an image. A face has mainly 2D or a 3D dimension with different textures and facial expressions [7].

To remove the noise in the process of registration face processing is done in traditional FR systems. In the face recognition method, the input is compared with the database and is referred to as the gallery. The input image is known as probe. If it finds a match in the database, it gives a match report. In FR systems there are mainly three approaches. These are as follows:

1. *Feature-based Approach*: The input data for face detection in feature-based approach are mainly neck and eyes to ease the process of recognition.
2. *Holistic Approach*: Here the whole face is taken as an input for face detection. This makes detection easier.
3. *Hybrid Approach*: It is a combination of holistic and feature-based approach. In this kind of approach, the whole face or the local is used as an input for face detection system.

3 Techniques of Face Recognition

3.1 Linear Discriminant Analysis

In appearance-based methods, linear discriminant analysis acts as a powerful tool for feature extraction and data reduction. LDA method is considered as one of the best approaches since it optimizes the low-dimensional representation of the objects with focus on the most discriminant feature extraction. The optimality criteria are not directly related to the classification ability of the obtained feature representation. Moreover, the classification accuracy is affected by the small size problem encountered in FR task, which leads to the disadvantage of traditional LDA methods.

To overcome these two shortcomings, an efficient algorithm is proposed in [8], which combines the strength of D-LDA and F-LDA approaches. In the proposed framework, a new variant of D-LDA is implemented that results in a low-dimensionless face subspace keeping the discriminatory features preserved. To result in a set of optimal discriminate feature for face representation, a weighting function is introduced into the proposed variant of D-LDA so that a subsequent D-LDA step can be applied to carefully reorient the SSS-face subspace. The performance of the algorithm is overall superior to that of the traditional approaches such as fisherman faces, eigen faces, and DLDA methods.

3.2 Principal Component Analysis

Principal component analysis (PCA) algorithm is widely used in face feature extraction and recognition. Its main purpose is to transform a high-dimensional image into a low-dimensional one. In [1] an implementation of principal component analysis is used for face recognition in criminal identification and the results produced have an accuracy of 80%.

By combining the characteristics of PCA an improved method of PCA is proposed in [9] with the advantage of enlarging the classification distance between the different class samples. The improved experiments were done on ORL face database and the results obtained have 98% of correct recognition rate and is more efficient on the improved PCA method. GABOR-PCA, variant of GABOR wavelet transform, is best suited for face recognition task when used as a face extractor [9].

3.3 Convolutional Neural Network

Convolutional neural network (CNN) is one of the most commonly used deep learning approach in image and face detection. It is a kind of artificial neural network that uses convolutional methodology to extract the features from the input data so as to increase the number of features [10]. An improved architecture of CNN has been proposed in [11] by adding batch normalization process after two different layers in training step. The performance of the proposed software has been tested on Georgia Tech Face Database and the results show significantly improved performance.

3.4 Support Vector Machine

SVM is one of the powerful machine learning algorithms used in image processing for recognition and various pattern recognition problems. The various applications of SVM are image recognition. "The main aim of pattern recognition is to classify data based on previous knowledge or statistical information which are generally obtained from raw data" [12].

SVM uses the concept of constructing an optimal hyper plane, which is for classification for linearly separable problem. In SVM, the larger the margin size, the more accurately it can classify the patterns [12].

Hyperplane is represented by the following equation:

$$\text{Hyper plane: } ax + by = c$$

4 Challenges of Face Recognition

In face recognition algorithm the performance degrades when it undergoes changes in lightning, pose, and other factors. The main challenges of face recognition are listed below.

4.1 Large Variability in Facial Expression

A significant change in exposure, time, aperture, lens, observations, and sensor spectral response increase intrasubject variations. The variation in the same image due to illumination and viewing direction makes it hard to extract the important information of the face objects from their respective image [13].

4.2 Aging

As humans grow older they undergo significant change in its face. The law enforcement agencies require matching an investigated image with the individuals in the missing person database. For example, if the age of the image which has been investigated is 14 years and the image of the same person stored in the database is 5 years, then in such cases the algorithm may not yield better performance. If the database is regularly updated with recent templates or images, then this kind of challenges can be easily handled. However, this kind of approach may not always be feasible in situations such as missing person identification, border control, or homeland security. Many algorithms have been proposed by researchers to tackle the facial aging challenges in facial recognition, and a 2D log polar Gabor transform-based recognition algorithm is used for face images [14]. An improvement of 37.2% and 22.5% has been observed for age group of 1–18 years and 19–40 years, respectively, and beyond 41 years the face recognition performance is improved by 14.49% with proposed age transformation algorithm.

4.3 Change in Illumination

To detect and remove the illumination effect, wavelet-based face recognition has been proposed in [15]. The proposed algorithm has several advantages such as without any prior information it can be directly applied to single image. Due to multicase nature of the wavelet transform it has better edge-preserving ability in low-frequency illumination. The experiments were carried out on Yale B and CMU PIE face databases, respectively. A satisfactory recognition rate results are shown under varying illumination. Experimental results show that this method has achieved very high recognition rates by PCA recognition with the pre-processing.

4.4 Pose

The existing face recognition methods suffer from non-monotonic illumination variation, change in pose, random noise, age, and expression. In [16], a local directional pattern which overcomes the major drawbacks of LDP and is sturdier in face recognition is proposed by the author. As edge response magnitude is more stable than pixel intensity that provides more steadiness in the presence of noise, the proposed LDP appraises the edge response value in all different directions. Experimental results show that the performance of proposed LDP texture description is more robust than LDP and PCA.

5 Conclusion

Various face recognition techniques are used worldwide for recognizing a face. This paper attempts to review the various methods used for face recognition and the major challenges of face recognition system. The study reveals that the face recognition algorithms such as LDA, SVM, LDP, PCA and CNN may yield in better performance and you may choose any one of the algorithms as per the requirements. Many algorithms have been proposed to deal with the challenges of face recognition such as aging, pose, change in illumination, change in facial expressions, and the results show that the performance has significantly improved. The references enlisted will give a detailed idea of the various algorithms discussed.

6 Future Work

This paper attempts to review the various techniques and challenges of face recognition with few papers. However, the experimental results of these algorithms and the performance analysis with the trained datasets have been left for future work.

References

1. Abdullah NA, Jamri Saidi Md., Ab Rahman NH, Wen CC, Hamid IRA (2017) Face recognition for criminal identification: an implementation of principal component analysis for face recognition. In: AIP conference proceedings 1891, 020002
2. Bah SM, Ming. An improved face recognition algorithm and its application in attendance management system
3. Kakade SD (2016) A review paper on face recognition techniques. 2(2). ISSN : 2494-9150
4. Ahmad F, Najam A, Ahmed Z. Image-based face detection and recognition
5. Gan J, Zhou D, Li C (2005) A method for improved PCA in face recognition. Int J Inform Technol 11(11)
6. Singh S, Prasad SVAV (2018) Techniques and challenges of face recognition: a critical review. In: 8th international conference on advances in computing and communication (ICACC-2018)
7. Gorkberg B, Salah AA, Riccio D, Dugelay J-L. 3D face recognition
8. Lu J, Plataniotis KN, Venetsanopoulos AN (2003) Face recognition using LDA-based algorithms. IEEE Trans. Neural Networks 14(1) (2003)
9. Vinay A, Shekhar VS, Balasubramanya Murthy KN, Natarajan S. Face recognition using gabor wavelet features with PCA and KPCA—a comparative study
10. Parkhi OM, Vedaldi A, Zisserman A. Deep face recognition
11. Coşkun M, Uçar A, Yıldırım Ö, Demir Y. Face recognition based on convolutional neural network
12. Pradhan A (2012) Support vector machine—a survey. Int J Emerging Technol Adv Eng 2(8). ISSN 2250–2459
13. Li SZ, Jain AK. Handbook of face recognition
14. Singh R, Vatsa M, Noore A, Singh SK. Age transformation for improving face recognition performance
15. Zhanga T, Fanga B, Yuanb Y, Tanga YY, Shanga Z, Lia D, Langa F. Multiscale facial structure representation for face recognition under varying illumination
16. Jabid T, Hasanul Kabir Md., Chae O. Local directional pattern (L DP) for face recognition

Chapter 28
Novel Mechanism to Predict and Detect nCOVID-19 Using Deep Learning with Convolutional Neural Networks: An Holistic Approach

Sharduli, Amit Batra, Archit Sharma, and Kulvinder Singh

1 Introduction

The objective of this paper is to measure the functioning capability of convolutional neural networks which has suggested for the past few years to categorize medical images like X-ray and computerized tomography (CT). The testing of such networks is done for the categorization of two diverse plans of action, i.e., nCOVID-19 and normal; normal, viral and nCOVID-19. More particularly, a technique called transfer learning has been employed for the same purpose. With the assistance of transfer learning, the discovery of malformation in medical images data is the requisite objective usually resulting in accurate results. A different kind of nCOVID-19 eruption was discovered in December, 2019 which still needs utmost concern and awareness due to its global damage. nCOVID-19 results in a severe respiratory sign. The datasets employed in this scientific procedure are twofold. In the first fold, a set of 1426 X-ray images comprising 223 images with nCOVID-19, 701 images with usual pneumonia, and 502 images with usual features are analyzed. The second fold comprises 1493 X-ray images with 302 images of nCOVID-19, 672 images with

Sharduli
Department of Biotechnology, University Institute of Engineering and Technology, Kurukshetra University, Kurukshetra, Haryana, India

A. Batra
Department of CSE, University Institute of Engineering and Technology, Kurukshetra University, Kurukshetra, Haryana, India

A. Sharma (✉)
Faculty of Biotechnology, Department of Biotechnology, University Institute of Engineering and Technology, Kurukshetra University, Kurukshetra, Haryana, India

K. Singh
Faculty of CSE, Department of CSE, University Institute of Engineering and Technology, Kurukshetra University, Kurukshetra, Haryana, India
e-mail: ksingh2015@kuk.ac.in

N. Marriwala et al. (eds.), *Soft Computing for Intelligent Systems*, Algorithms for Intelligent Systems, https://doi.org/10.1007/978-981-16-1048-6_28

common pneumonia, and 519 images with normal features are considered. After feeding the above datasets in the machine, the proposed model is trained using deep learning with CNN. After investigating the output results of the applied datasets, the accuracy, sensitivity, and specificity metrics values have been found as 96.78, 98.66, and 96.46%, respectively. The applied set of images are categorized by employing diverse machine learning (ML) algorithms, and the performance of these algorithms is examined in terms of accuracy and loss. The algorithms may also be implemented using CNN, support vector machine (SVM), and random forest. Furthermore, in this research paper, a pragmatic effort is made to use the concept of CNN in two situations. In the first situation, CNN is employed to categorize the X-ray pictures, and in the second situation for drawing out the graphical features. As per the outcomes and results of performance, the CNN is found to be more accurate and precise method among the aforesaid methods with an accuracy rate of 95.2%. nCOVID-19 is a severe disease which can lead up to 4% of death rate. This pandemic disease can cause death because of alveolar harm and severe respiratory collapse [1]. Although the examination of the virus has now become a quick task, it incurs high financial cost to both countries and patients, particularly in countries where there are no specialized medical facilities. This forces us to examine the X-ray images and recognize possible signs that may result in automated diagnosis of the nCOVID-19. The objective of this research is to measure the efficacy of CNNs, pertaining their proficiency in the automated discovery and analysis of nCOVID-19 utilizing chest cum thoracic X-ray images. It is universally acceptable that the CNNs are advanced and complex models ensure large datasets to extract graphical features in order to categorize or classify the medical images. In order to acquire this, a group of 1426 and 1493 thoracic X-rays images are taken into consideration for data processing, train and test the CNNs. In spite of certain restrictions of the current examination, which are pertinent to the availability of the large datasets, it unlocks the path for more research appropriate to the nCOVID-19 virus in medical X-ray images. The outcomes and results are flourished enough and depict the effectiveness of deep learning by employing CNNs with transfer learning to the automated discovery of unusual X-ray images from small datasets, concerning to the nCOVID-19 virus.

2 Related Work

In the recent past, it is observed that most of the times, in the implementation of artificial intelligent-based techniques, deep learning models using CNN are employed to process and analyze the medical images dataset. Such type of artificially trained models can effectively assist and help clinicians and doctors to enhance diagnosis. The outcomes collected on the basis of CNN reveal that the radiologists can attain high specificity, where the term specificity can be outlined as the portion of real positives that are correctly recognized as such. The radiologists attain this high value of specificity in segregating nCOVID-19 from other cases of viral employing thoracic CT images. In this concern, a novel deep learning technique which is termed as

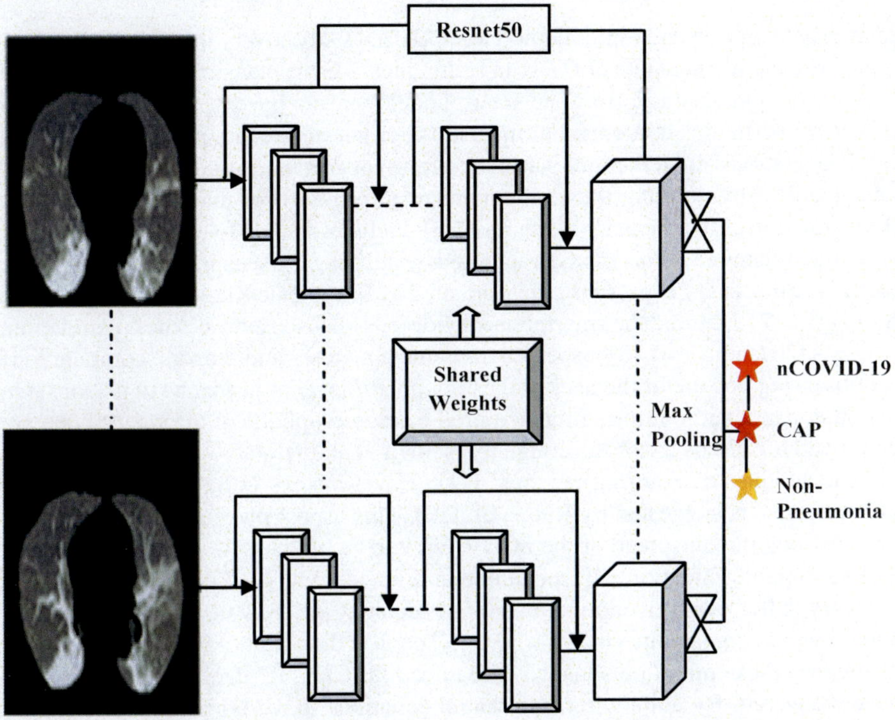

Fig. 1 Architecture for nCOVID-19 detection neural network (COVNet) model

coronavirus discovery network (COVNet) is recently incorporated by Li et al. [2] to discover nCOVID-19 centered on thoracic CT pictures. They experimented on three types of CT pictures, comprising nCOVID-19, community attained pneumonia (CAP), and non-pneumonia images, and these CT pictures are used to determine the effectiveness of the suggested technique as depicted in Fig. 1. Similarly, Hu et al. [3] employed a deep learning technique to predict the different patients with nCOVID-19 disease in China. Their team utilized a sequence of eight data points, i.e., eight days are utilized as network inputs. This changed network comprises four layers, namely, first hidden layer, second unseen layer, output layer, and input layer, having number of nodes 32, 4, 8, and 1, respectively. The hidden variables acquired from the second layer of the framework are undergone for processing through the single value decomposition technique before being supplied into different clustering techniques so as to categorize and classify the cases into states or cities to examine the degree of transference of the nCOVID-19 pandemic. Furthermore, an another type of prototype for nCOVID-19 discovery by applying data acquired from sensors within smartphones like temperature sensors, cameras, and microphones is suggested by Maghdid et al. [4]. Machine learning techniques are utilized for gaining information about the virus manifestation centered on the acquired dataset. This helps and assists by offering a relatively less price and rapid method to nCOVID-19 discovery in contrast to manually operating medical kits or CT scan techniques. This is plausibly reasonable since the information acquired from the smartphone's sensors have been used effectually in several particular implementations and the suggested technique combines these implementations altogether in a particular prototype. As an example, information and data acquired by utilizing the temperature-fingerprint sensor may be employed for the estimation of the intensity of the fever [5]. Pictures and audio-visuals captured by smartphone camera or information acquired by utilizing the sensors may be employed for fatigue discovery in persons as investigated by Babasaheb Karvekar [6]. Similarly, Story et al. [7] explored the smartphone's videos for nausea cum sickness estimation, whereas Lawanont et al. [8] inspected the camera pictures and sensors' computations for observing posture of the neck and estimation of level of headache in persons. On the other hand, audio information acquired from microphone of the smartphone are employed for discovery of the cough by Nemati et al. [9].

A technique to acquiring persons' basic travel history utilizing a phone-based online review is suggested by Rao et al. [10]. This aspect plays a fundamental part in mitigating the outspread of the nCOVID-19 to the vulnerable people population. The susceptible infection (SI) technique and its adjunct, i.e., susceptible infection recovery (SIR), are conventional pandemic techniques for estimating the evolution and growth of contagious viruses where "S" depicts the count of vulnerable persons, "I" depicts the count of persons who are infectious, and "R" depicts the people who have recovered. By employing differential equations to recognize the relation and association among "I", "S" and "R", the aforementioned techniques may be employed to estimate fortunately severe acute respiratory syndrome (SARS), as depicted by Ng et al. [11]. These characteristics are then fed as inputs to the long short-term memory (LSTM) deep learning technique by Hochreiter et al. [12] to amend and rewrite estimations of speed of infection of the SI technique as depicted in Fig. 2. A recognizable illustration of CT pictures of nCOVID-19 was also observed in earlier research by Kanne et al. [13]. They especially experimented on the distinction and dissimilarity with other cases of influenza-A and viral pneumonia by deep learning technique. Therefore, by considering the latest research potential of nCOVID-19, this research further promotes an innovative technology that uses some more obvious proposed algorithms to predict nCOVID-19, which are introduced in the next section.

3 Proposed Algorithm and Methodology

CNN is a mathematical setup that generally comprised three categories of layers. These three layers are: convolution, pooling, and entirely associated layers. A convolution layer has a significant job or part in CNN that comprised set of mathematical functions; for an instance, convolution—a type category of linear function. The first two layers, namely, convolution and pooling layers, are utilized and employed for drawing out the characteristics or features, while the last and third layer, an

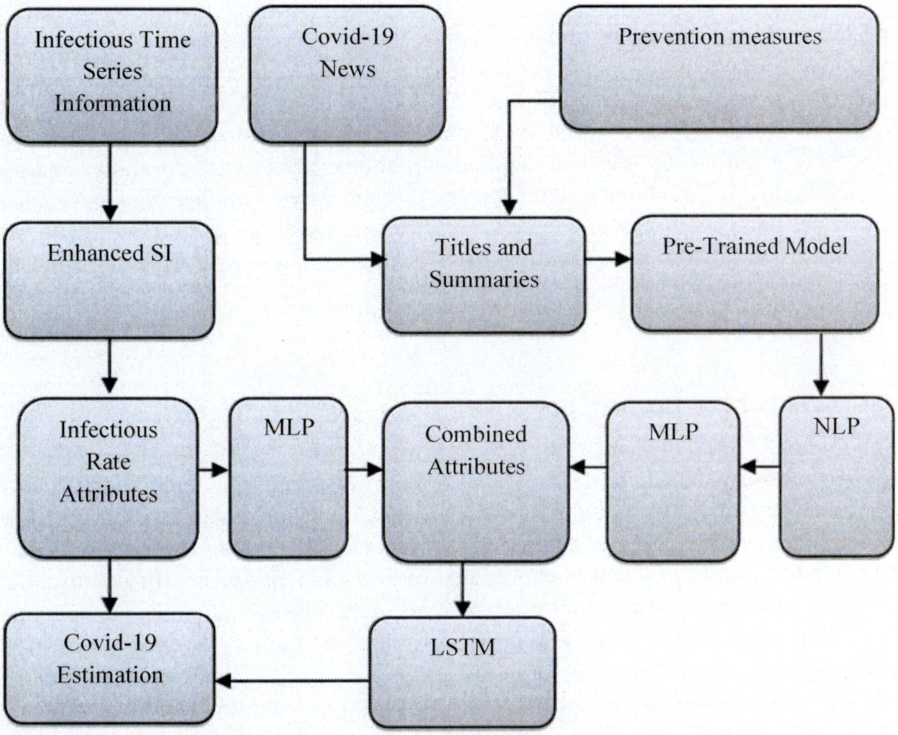

Fig. 2 An AI centered technique to coronavirus estimation that amalgamates SI technique, NLP, and ML

entirely associated layer, transforms the drawn characteristics or features into finalized outcome; for an instance, categorization. The task of optimization of values of the parameters like kernels is termed as training which is carried out in order to reduce the dissimilarity among outcomes and ground truth by utilizing an optimal technique termed as backpropagation and gradient descent. A number of various dissimilarities exist among such techniques and CNN. First, in CNN there is no need of manual characteristics withdrawal. Secondly, CNNs don't need partitioning or segmentation of tumors or body parts by human specialists. A particular structure comprises recurrence of heap of number of convolution layers and a pooling layer, accompanied with several entirely associated layers. Modern studies and examination of new techniques in the field of radiology employ manual characteristics extraction methods, as an instance in the analysis of text, accompanied by standard ML classifiers, such as SVM and random forests.

Although the convolution and merging functions as mentioned in this part of research paper are for two-dimensional CNN, but the same function can be performed for three-dimensional CNN. The CNN structure comprises various components; for instance, convolution, pooling, and entirely associated layers. The major characteristic of a convolution function is weight divide up or sharing throughout all the image locations. Another optional method to carry out down-sampling is a pooling function, as illustrated onwards. Weight sharing introduces the subsequent features of convolution functions: (a) allowing the local characteristic designs drawn out by kernels translation as invariable, because kernels make a journey throughout all the image locations and discover studied neighborhood designs, (b) understanding spatial hierarchies of characteristic designs through down-sampling in concurrence with a pooling function, which results in acquisition of huge area of sight, and (c) enhancing model efficacy by mitigating the count of parameters to understand in contrast with entirely associated neural networks.

3.1 Network Training

A model performance underneath typical kernels and weights is computed by a loss operation via forward propagation on a training information, and understandable limits; specifically, weights and kernels are modified as per the loss value via an optimization technique termed as backpropagation and gradient descent. In exercise, for grounds like memory restrictions, the gradients of the loss operation w.r.t the parameters are determined by employing a subdivision of the training information termed as mini-group, and is forced to apply to the modified parameters. It is scientifically observed that, in exercise, the speed of learning is the most significant hyperparameter which has to be fixed prior to the beginning of the training. This technique is termed as mini-batch gradient descent, also usually termed as stochastic gradient descent (SGD), and a mini-group measurement is a hyperparameter. Additionally, several enhancements on the gradient descent technique have been suggested and broadly employed, like SGD with momentum, RMSprop, and Adam. The following algorithm is successfully employed using Python software (Anaconda platform) installed on Windows 10 operating system with 12 GB RAM and Intel Core i7 (4.80 GHz) processor.

3.2 Algorithm Employed for Training

Training Algorithm

```
# The model is compiled as underneath:
print("The model is being compiled...")
optimizer = Adam_opt(lr=INIT_LR, decay=INIT_LR /
no_of_epochs)
model.compilation(loss="binary_cross_entropy", opt=optimizer,
measures=["accuracy"])
# The head of the network is trained
print("The head is being trained...")
Head = model.fitting_generation(
train.flow(trainingX, trainingY, size_of_batch=S),
steps=length(trainingX) // S,
valid_data=(testingX, testingY),
valid_steps=length(testingX) // S,
epochs=EPOCHS)
```

In the aforementioned algorithm, the network is compiled with decay in learning rate and the optimizer employed is Adam optimizer. The neural network is trained by calling fit_generator function and chest X-ray information is passed through data augmentation object.

4 Experimental Outcomes

Empirical studies show that although the manual RT-PCR can be employed to detect nCOVID-19 virus with more accuracy, but this proposed algorithm can help to detect the nCOVID-19 at a fast pace with some accuracy, confirmed by examining the chest X-ray images. It can be beneficial in cases where the number of cases are very large and especially in case of less resources. Therefore, the main advantage of these AI-centered techniques is to hasten the task of analysis and remedial of the nCOVID-19 which has spread as pandemic across the world. The accuracy and loss acquired by ResNet18 pre-trained network have been depicted in Fig. 3.

Moreover, the sensitivity for every infection category is investigated as shown in Table 1 and it depicts the efficiency of our proposed approach.

Fig. 3 Accuracy and loss curves

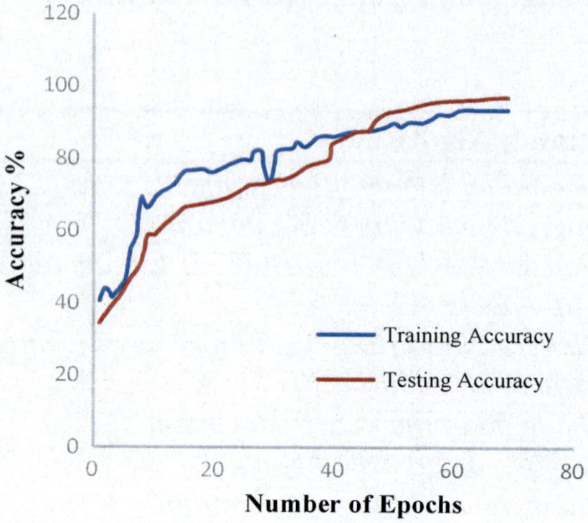

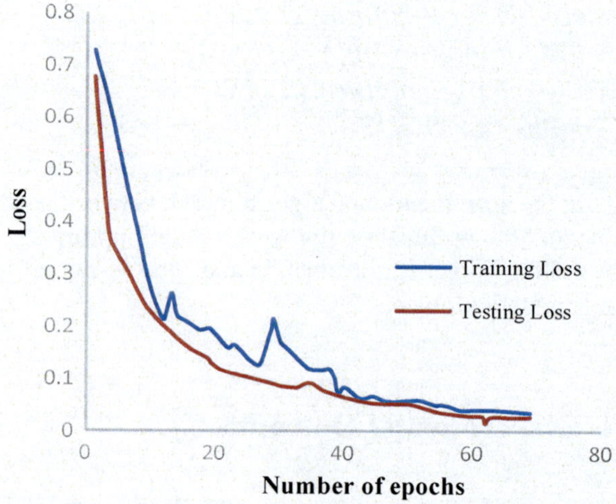

Table 1 Sensitivity for every infection category

Sensitivity (%)			
Model	nCOVID-19	Normal	Non-nCOVID-19
ResNet-50	96.1	99.8	98.6
Proposed Approach	96.9	99.99	99.1

5 Conclusion and Future Work

Today, the coronavirus disease has spread as pandemic throughout the world and the count of deaths and cases are multiplying on daily basis. Therefore, there is a need to take hard decisions against the spreading of this pandemic. If the future predictions are known, different authorities can take precautionary measures ahead of time and can help in preventing from this pandemic. Although the manual approach termed as RT-PCR is successfully being utilized to detect the coronavirus, but it takes quite sufficient amount of time. Our approach can help in detecting the same virus rapidly, but it exhibits less accuracy. The work is required to be carried out in future to increase the accuracy and to minimize training loss of these AI and deep learning techniques to overcome such disastrous situations.

Conflict of Interest All authors confirm no conflict of interest or funding for this study. **Authors' Contributions Statement**:

Transparency Declaration The authors certify that the manuscript is reported clearly, truthfully, and precisely, with no omission of any important aspect.

Authors' Contributions Statement Archit Sharma: Methodology, Objectives, Effective test cases, Editing and review.

Sharduli: Idea formulation, Conceptualization, Effective literature review, Methodology, Formal analysis, Software testing, Writing- review & final editing.

Amit Batra: Conceptualization, Formulated strategies for modelling, Software, Methodology, Writing- review & final editing.

Kulvinder Singh: Experiments and Simulation, Methodology refinement, Supervision, Validation, Visualization, Writing-original draft.

References

1. Xu Z, Shi L, Wang Y, Zhang J, Huang L, Zhang C, Wang FS (2020) Pathological findings of COVID-19 associated with acute respiratory distress syndrome. Lancet Respir Med 8(4):420–422
2. Li L, Qin L, Xu Z, Yin Y, Wang X, Kong B, Xia J (2020) Artificial intelligence distinguishes COVID-19 from community acquired pneumonia on chest CT. Radiology 284(2):200905
3. Hu Z, Ge Q, Li S, Jin L, Xiong M (2020) Artificial intelligence forecasting of COVID-19 in China, pp 1–20. Retrieved from https://arxiv.org/abs/2002.07112
4. Maghdid HS, Ghafoor KZ, Sadiq AS, Curran K, Rabie K (2020) A novel ai-enabled framework to diagnose coronavirus COVID 19 using smartphone embedded sensors: design study, pp 1–5. Retrieved from https://arxiv.org/abs/2003.07434
5. Maddah E, Beigzadeh B (2020) Use of a smartphone thermometer to monitor thermal conductivity changes in diabetic foot ulcers: a pilot study. J Wound Care 29(1):61–66
6. Babasaheb Karvekar S (2019) Smartphone-based human fatigue detection in an industrial smartphone-based human fatigue detection in an industrial environment using gait analysis environment using gait analysis. Retrieved from https://scholarworks.rit.edu/theses
7. Story A, Aldridge RW, Smith CM, Garber E, Hall J, Ferenando G, Hayward AC (2019) Smartphone-enabled video-observed versus directly observed treatment for tuberculosis: a multicentre, analyst-blinded, randomised, controlled superiority trial. Lancet 393(10177):1216–1224
8. Lawanont W, Inoue M, Mongkolnam P, Nukoolkit C (2018) Neck posture monitoring system based on image detection and smartphone sensors using the prolonged usage classification concept. IEEJ Trans Electr Electron Eng 13(10):1501–1510

9. Nemati E, Rahman MM, Nathan V, Vatanparvar K, Kuang J (2019) Poster Abstract: a comprehensive approach for cough type detection. In: Proceedings of 4th IEEE/ACM conference on connected health: applications, systems and engineering technologies (CHASE 2019), pp 15–16
10. Rao ASRS, Vazquez JA (2020) Identification of COVID-19 can be quicker through artificial intelligence framework using a mobile phone-based survey in the populations when cities/towns are under quarantine. Infection Control and Hospital Epidemiology, 1400
11. Ng WT, Turinici G, Danchin A (2003) A double epidemic model for the SARS propagation. BMC Infect Dis 3:1–16
12. Hochreiter S, Schmidhuber J (1997) Long short-term memory. Neural Comput 9(8):1735–1780
13. Kanne JP (2020) Chest CT findings in 2019 novel coronavirus (2019-nCoV) infections from Wuhan, China: key points for the radiologist. Radiology 295(1):16–17

Chapter 29
A Review on Surface Defect Detection of Solar Cells Using Machine Learning

Nitu Rana and Shaveta Arora

1 Introduction

In the last few decades, a lot of work has been done for surface defect detection of solar modules. Surface defects are the irregularities such as cracks, scratches, dislocations etc. visible on the surface of different materials that can badly affect the quality of the final product. The motivation for research in this area is due to the fact that the solar modules are gaining noteworthy recognition. In 2019, approximately 100 GW had been added to the solar energy capacity of the world, taking the total capacity of the solar energy systems in the world to 627 GW [1, 2]. This paper presents a review of surface defects in solar cells and the significant contribution of image processing and computer vision techniques to facilitate surface defect detection.

1.1 Classification of Solar Cells

Solar cells can be classified as: monocrystalline silicon and polycrystalline silicon, as shown in Fig. 1, on the basis of the production materials used for manufacturing of solar modules. The monocrystalline modules are formed from bars of single crystal silicon which is then cut into wafers to form monocrystalline modules. They are so-called because a single crystal is used. In contrast, the polycrystalline modules are formed by melting different pieces of silicon and forming wafers from it.

N. Rana (✉)
Department of Electrical Electronics and Communication Engineering, The Northcap University, Gurugram, India
e-mail: nitu19ecd004@ncuindia.edu

S. Arora
Computer Science Engineering Department, The NorthCap University, Gurugram, India
e-mail: shavetaarora@ncuindia.edu

N. Marriwala et al. (eds.), *Soft Computing for Intelligent Systems*, Algorithms for Intelligent Systems, https://doi.org/10.1007/978-981-16-1048-6_29

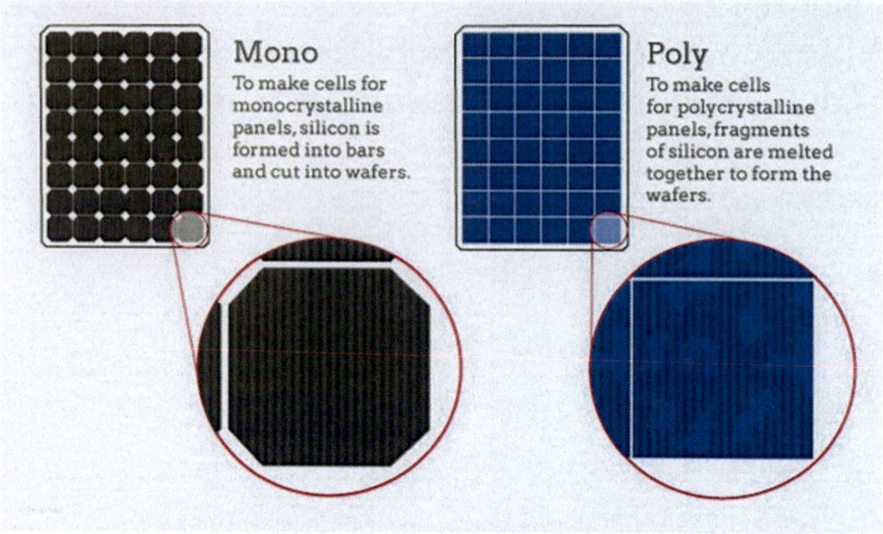

Fig. 1 Monocrystalline and polycrystalline solar modules [3]

1.2 Types of Defects

During the manufacturing process some surface defects appear in the solar cells, such as pasting spots, thick lines, scratches, chromatic aberrations, dirty cells, broken gates, etc. [4, 5]. Sometimes the faults also occur due to environmental conditions after installation of the solar modules [6–8]. Table 1 gives the different types of defects that occur in solar cell modules due to:

(A) Manufacturing process and
(B) Environmental conditions.

1.3 Detection and Classification of Defects

Different types of defects will severely affect the performance of the solar cells. Therefore, it is important to detect surface defects of solar cells so as to control the quality and efficiency of solar cells. Manual inspection of solar cells is not very effective, so automatic surface defect detection techniques using machine vision are implemented to improve the production quality and also the lifetime and efficiency of solar cells.

For analysis of solar cell modules, the first step is acquisition of images, as shown in Fig. 2. Images may be acquired by infrared (IR) imaging technique or electro-luminescence (EL) imaging. Infrared (IR) imaging is a non-destructive method for monitoring the quality of solar modules without taking any direct measurements.

Table 1 Different types of defects in solar cell modules

Defects	Causes
(A) *Manufacturing defects*	
Scratches, broken gates	Mechanical pressure, improper manipulation
Missing angle, breakage	Miscut, collision
Common spot, fingerprint, pasting spots	Hygiene problem, Manual errors
Blue spot	Chemical reaction is not uniform in the coating
Chromatic aberration	Firing problem with rainbow colors
(B) *Defects due to environmental conditions*	
Corrosion of frame and interconnects, deframing	Due to exposure to the atmosphere The adhesive and sealing between panel parts and frame gets detached
Cracks in the transparent glass	Mishandling
Deposition of soil on glass surface of the panel	Lack of maintenance, dusty environment
Yellowing	Exposure to high temperature, UV and water leads to degradation in the adhesive material between the glass and the cell
Discoloration, burn marks in bus bars	Oxidation of ARC, poor EVA quality, burn marks due to heating of module parts

UV Ultraviolet, *ARC* anti-reflective coating, *EVA* Ethylene vinyl acetate

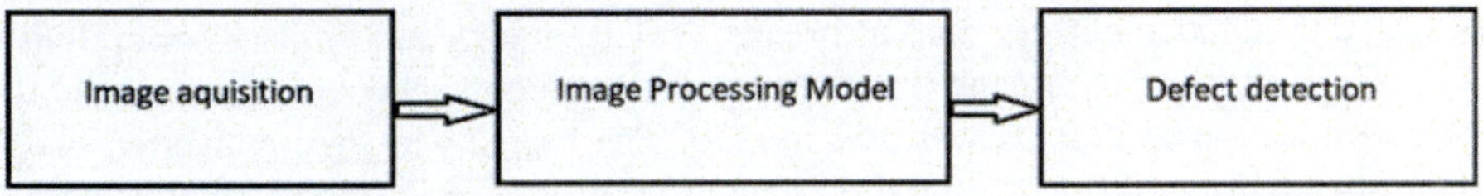

Fig. 2 Steps in analysis of a solar module

But the IR cameras have a very low resolution, due to which detection of small defects such as micro cracks is not feasible. Electroluminescence (EL) imaging is also a non-destructive method for monitoring of solar modules and it is better than IR imaging as the EL images of solar modules have a very high resolution which gives a detailed view of the modules and even micro cracks can be detected [9].

After image acquisition, image processing model is used, and different operations, like image enhancement, segmentation, restoration, rotation etc., are performed on the images to prepare the image dataset for further processing [10]. Thereafter, as shown in Fig. 2, different machine learning and deep learning approaches are used for classification and detection of surface defects. Machine learning is a subset of artificial intelligence that provides the capability of a machine to learn from the data provided and gradually improve its performance. Machine learning technology has varied applications today and finds its usage in areas like suggestions in web searches, object identification, object detection, speech transcription, recommendations on e-commerce websites, fraud detection in banks, drug discovery, genomics, medical diagnosis etc. [9].

Deep learning models are mathematical models consisting of several processing layers and are used for classification and detection of surface defects. Deep learning algorithms are used to find patterns in comparatively large datasets by using the back-propagation algorithm that facilitates the machine in changing its internal parameters which are then used to determine the representation in each layer from the representation in the previous layer [9].

The techniques used by different researchers are reviewed in the next section so as to get an inspiration for future work in the field of surface defect detection of solar modules.

2 Literature Review

The automatic detection of surface defects of solar cells can be carried out by using computer vision in a less time-consuming and efficient manner. Many researchers have proposed different methods and algorithms for solar cell surface defect detection.

Tsai et al. [11] used Fourier image reconstruction for detection of defects in solar cells. They used EL images of polycrystalline solar modules. The suggested method was used to identify defects such as breaks, (small) cracks and finger interruptions. Fourier transform of the image was taken and high-frequency coefficients related to bar and line-shaped defects were set to zero. The image was then converted back to the spatial domain from spectral domain. The difference in intensity between the original and the high-pass filtered image were identified as defects. The proposed method was not suitable for complex shape defects.

Zhang et al. [12] suggested a novel approach based on independent component analysis (ICA) reconstruction algorithm in 2013 for defect detection in solar modules. In this approach, a non-defective image was taken as the training image and then after applying the ICA in the training image, the de-mixing matrix and corresponding independent components were obtained. The de-mixing matrix was again formed by reordering the ICs and thereafter the defect image was reconstructed. The local anomalies were highlighted in the resultant image. The proposed method proved to be effective in the detection of defects in surfaces having periodic patterns.

Tseng et al. [4] devised a supervised classification technique to automatically detect finger interruptions in multi-crystalline solar cells. They obtained regions of interest (ROI) from the EL images. Then the features were extracted from interrupted fingers and non-interrupted fingers images. For detection of defects, the researchers performed binary clustering of features from the region of interest.

Tsai et al. [13] proposed an approach based on machine learning for automatic defect detection in which they utilized Haar-like feature extraction and binary tree clustering technique. They used non-defective images for training. For training, a binary-tree clustering method was proposed to partition non-defective samples from the defected ones. The intersample distances were computed for clusters and the location of the defects was detected.

In 2017 Qian et al. [5] presented a paper in which they briefly described the typical solar cell surface defects. The authors also explained different methods based on global, local and local–global schemes for classification of defects.

Bartler et al. [14] used an adapted VGG-16 architecture for classification of defective solar cells. For completely automated defect detection, they introduced a deep learning-based classification strategy for the EL images. This included image preprocessing for distortion correction, segmentation and perspective correction and a deep convolutional neural network for solar defect classification. The method was proposed considering a highly imbalanced dataset. The authors also explored the impact of data augmentation and minority oversampling on the accuracy of the system. The dataset used was highly imbalanced and the authors compared different methods to address the data imbalance problem.

Ying et al. [15] devised a multi-channel convolutional neural network (MCCNN) for detection of the defects in solar modules. Images were acquired at different scales and then features extracted from those multi-scale images. The image features thus extracted were used in the fully connected layer for training. For final classification the random forest classifier was used.

Qian et al. [16] suggested a new method for finding micro cracks based on self-learning features and low-rank matrix recovery. The authors applied the low-rank matrix recovery to the decomposition of self-learning feature matrix to obtain the initial detection result. Then the initial detection result was improved by applying the super pixel segmentation to find the location of defects. Finally, morphological post-processing was used to fine tune the result.

Deitsch et al. [17] proposed two approaches of automatic detection of defects in EL images of solar cells. They proposed two approaches for classification: one based on support vector machines (SVM) and the other based on convolution neural networks (CNN) which employed fine-tuned regression network centered on VGG-19. They explored several feature detector and descriptor algorithms. They analyzed a technique for dimensionality reduction known as t-distributed stochastic neighbor embedding (t-SNE) [18]. SVMs were trained with both a linear and a radial basis function (RBF) kernel. Experiments were performed and it was determined that the performance of both the classifiers was satisfactory for monocrystalline solar cells, but in case of polycrystalline solar cells CNN classifier performed better than the SVM classifier.

Akram et al. [19] presented a novel approach for automatic detection of defects from EL images of solar modules using a light convolutional neural network (CNN) model. They studied VGG-based deep architectures [20], i.e., VGG 11, VGG 13, VGG 16 and VGG 19. Isolated deep learning which means learning from scratch [21] was used for this study. In isolated learning previously learnt knowledge relevant to other tasks is not given due consideration. The researchers concluded that combining data augmentation, suitable model size, L2 weight regularization, batch normalization, and dropout produced outstanding output. Dropout increased the training time though it helped to reduce overfitting [22].

Balzategui et al. [23] created a dataset by augmentation, trained the neural network and then constructed a heatmap. Then they analyzed how the network performed at full image and at defect level by comparing the resulting heatmap with the ground truth annotated by the expert. The algorithm proposed here created a segmentation heatmap which helped to visualize the location as well as shape of the defects. The researchers attempted to assess the quality of the segmentation.

Liu et al. [24] proposed an algorithm consisting of three steps. In the first stage for extraction of features, they combined CNN with the feature pyramid network (FPN) [25] and performed fusion of features at different scales to extract strong semantic features. Then, guided anchor region proposal network (GA-RPN) [26] was used to obtain anchors matching the size and shape of the defects, and then the ROI pooling layer was used to process the generated anchors to obtain object proposals. Lastly, for classification of the fully connected layer, softmax classification and bounding box regression were used to get the final defect detection result.

Su et al. [27] proposed a new feature descriptor (FD)—center pixel gradient information to center-symmetric local binary pattern (CPICS-LBP) which can merge the CPICS-LBP by thresholding. By fusing the two, every pixel of the image was converted into binary code for classifying the defects from the heterogeneous background. The authors also presented bag of CPICS-LBP (BCPICS-LBP) scheme combined with CPICS-LBP and a pre-trained k-means model for extraction of global defect features from the image of solar module.

Chen et al. [28] proposed a multi-spectral CNN which could extract the multi-spectral features of solar cell surface, which thus improved the detection rate of most of the defects. The researchers also presented a comparison and analysis of proposed model with some typical machine learning algorithms: local binary pattern (LBP) + histogram of oriented gradient-support vector machines (HOG-SVM) and Gabor-SVM.

Zhang et al. [29] fused two distinct convolution neural networks, i.e., faster R-CNN (fast region-based CNN) [30] and R-FCN (region-based, fully-convolutional networks) [31], to improve position accuracy and detection precision. The authors also used two other schemes to improve the detection performance keeping in mind the attributes of the solar cell surface defects. Initially, the authors added the multi-aspect and multi-scale regions to set the anchor points of region proposal network (RPN). This was done to minimize the false negative rate. Thereafter, considering fine defects of solar cells the strategy used was the hard negative sample mining. Using this technique helped overcome the issue of low detection precision resulting from a large negative sample space. The method proposed here not only increased the accuracy of locations of defects, it also reduced the false positive rate and the false negative rate of a single network and the object recall rate was also improved.

A new structure-aware-based crack defect detection scheme (SACDDS) was suggested for detection of cracks in non-homogeneously textured background by Chen et al. in 2019 [32]. They used mathematical models for representation of the structural features of crack defect and crystal grains which are randomly distributed. Then they obtained the identification functions of the blob and linear structure in accordance with the Hessian eigenvalues of the mathematical models of the crack defect and crystal grains. The identification function of two structures was used to design structure similarity measure (SSM) function. This helped to suppress the crystal grains and highlight the crack defect at the same time. To reduce the non-uniform response of crack region and get the crack defects tensor voting-based non-maximum suppression (TV-NMS) method was proposed.

The comparative analysis of different methods used by the researchers is given in Table 2.

3 Performance Metrics

The performance metrics used commonly are precision, recall, F-measure or F-score, accuracy, FNR (false negative rate), FPR (false positive rate), mean average precision (MAP) and balanced error rate. However, if the dataset is imbalanced accuracy is not a good performance measure as it could give a false sense of high performance. Precision, recall and F-score are preferred by most of the researchers as they do not give any false performance even if the dataset is imbalanced.

4 Tools Used

The major software tools employed in the research papers are: Python with Caffe toolbox, OpenCV for image processing, Keras API with Tensorflow as backend in Ubuntu16.04, Deep learning acceleration library CUDNN7.0 and the MATLAB.

5 Conclusion

This review of solar cell surface defect detection methods gave some insights into the current research going on in this area and the future scope in automatic visual inspection of solar modules. Some key observations worth mentioning are:

- The detection speed of most of the algorithms satisfies the requirements of real-time production of solar modules.
- The proposed methods correctly located the position of defects.
- Histogram of oriented gradient (HOG) is not good for multi-class classification of EL images.
- The multi-spectral CNN exhibits poor feature extraction ability for linear defects and small area defects.

Table 2 Comparative analysis of different methods used

Ref. No.	Method used	Type of defects	Sample size	Performance
[17]	• SVM • CNN(VGG-16)	Micro cracks, short circuited cells, open interconnects, soldering failures	2624 Solar cell images[1]	SVM Accuracy = 82.44% CNN Accuracy = 88.42%
[14]	VGG-16 CNN	Micro cracks and large cracks	98,280 images	Balanced error rate = 7.73%
[28]	Multi-spectral CNN	Broken gates, scratches paste spots, thick lines, dirty cell, color difference	15,330 non defective and 5915 defective images	Accuracy = 94.30%
[19]	Light CNN and isolated deep learning approach	Micro cracks, short circuited cells, open interconnects, soldering failures	2624 Solar cell images[1]	Accuracy = 93.02%
[23]	CNN and heatmap reconstruction	Cracked defective images	542 images (145 defective and 397 defective)	Precision = 85% Recall = 92%
[24]	CNN combined with feature pyramid network (FPN)		5000 images	Mean average precision mAP = 94.62
[4]	• Supervised classification method • Spectral clustering method	Finger interruptions	60 multi-crystalline solar cell images	Accuracy = 99.07%
[29]	• Faster R-CNN and R-CNN • Hard negative sample mining strategy and • Online hard Example mining	Broken cell, crack, unsoldered area defects	1462 Solar cell surface images (1461 defective and 1 defect free)	Overall detection accuracy = 98.3%
[13]	• Haar-like feature extraction • Fuzzy C means clustering	Finger interruptions, micro cracks, breaks	50 images (31 defect free and 19 defective)	FNR = 5.2% FPR = 41.9%

(continued)

Table 2 (continued)

Ref. No.	Method used	Type of defects	Sample size	Performance
[32]	• Structure aware-based crack defect detection scheme • Structure similarity measure • Tensor voting-based non-max. suppression	Cracks	10,000 EL images	F-measure = 94.9
[15]	Multi-channel CNN	Hidden crack, open welding, broken grid, shadow, solid black	6945 images	F1 value = 0.9302 Accuracy = 96.76%
[27]	CPICS-LBP and BCPICS-LBP with k-means model for classification	Cracks, finger interruptions	3021 images	Average precision = 92.02% Average Recall = 93.33% F-measure = 92.81%
[16]	A technique utilizing self-learning features and low-rank matrix recovery for micro cracks detection	Micro cracks	120 images(80 defect free and 40 defective)	Precision = 0.2070 Recall = 0.7556 F-measure = 0.3250

[1]Dataset available at: https://github.com/zae-bayern/elpv-dataset

- For deep learning-based models large datasets are required and labeling those datasets is a challenge.

6 Future Work

- Multi-class classification can be done upon availability of large training data and results can also be improved by having a large size training data.
- Performing transfer learning from a network which is trained on a richer pixel-wise labeled dataset and comparing the results with existing approaches.
- Estimation of the geometric size of the defects.
- Automatic parameter setting by the model.
- Proposed methods correctly located the position of defects but more work needs to be done to identify the true shape and length of defects.

References

1. Mustafa RJ, Gomaa MR, Al-Dhaifallah M (2020) Environmental impacts on the performance of solar photovoltaic systems. Sustainability 12(2):608
2. Renewables 2020 Global Status Report. 2020. Available online: https://www.ren21.net/wp-content/uploads/2019/05/gsr_2020_full_report_en.pdf. Last accessed on 24 Aug 2020
3. https://www.energysage.com/solar/101/monocrystalline-vs-polycrystalline-solar-panels/. Last accessed 16 Oct 2020
4. Tseng D-C, Liu Y-S, Chou C-M (2015) Automatic finger interruption detection in electroluminescence images of multicrystalline solar cells. In: Mathematical problems in engineering
5. Qian X et al (2017) Solar cell surface defects detection based on computer vision. Int J Performability Eng 13(7)
6. https://www.electricalindia.in/effect-of-environmental-conditions-on-performance-of-solar-panels/. Last accessed on 25 Aug 2020
7. Mehta S et al. (2018) Deepsolareye: power loss prediction and weakly supervised soiling localization via fully convolutional networks for solar panels. In: 2018 IEEE Winter conference on applications of computer vision (WACV). IEEE (2018)
8. Abd Allah El-G, Gharabawy E (2018) Review on corrosion in solar panels Int. J. Smart Grid **2** (2018)
9. Le Cun Y, Bengio Y, Hinton G (2015) Deep learning. Nature 521(7553):436–444
10. Aghaei M et al (2015) IR real-time analyses for PV system monitoring by digital image processing techniques. In: International conference on event-based control, communication, and signal processing (EBCCSP). IEEE
11. Tsai D-M, Wu S-C, Li W-C (2012) Defect detection of solar cells in electroluminescence images using Fourier image reconstruction. Solar Energy Mater Solar Cells 99:250–262
12. Zhang X et al (2013)A novel method for surface defect detection of photovoltaic module based on independent component analysis. In: Mathematical problems in engineering
13. Tsai D-M et al (2015) Defect detection in multi-crystal solar cells using clustering with uniformity measures. Adv Eng Inform 29(3):419–430
14. Bartler A et al (2018) Automated detection of solar cell defects with deep learning. In: 2018 26th European signal processing conference (EUSIPCO). IEEE
15. Ying Z et al (2018) Automatic detection of photovoltaic module cells using multi-channel convolutional neural network. In: 2018 Chinese Automation Congress (CAC). IEEE
16. Qian X et al (2018) Micro-cracks detection of multicrystalline solar cell surface based on self-learning features and low-rank matrix recovery. Sens Rev
17. Deitsch S et al (2016) Automatic classification of defective photovoltaic module cells in electroluminescence images. Solar Energy 185: 455–468
18. van der Maaten L, Hinton G (2008) Visualizing data using t-SNE, The. J Mach Learn Res 9:2579–2605
19. Akram MW et al (2019) CNN based automatic detection of photovoltaic cell defects in electroluminescence images. Energy 189:116319
20. Simonyan K, Zisserman A (2015) Very deep convolutional networks for large-scale image recognition. In: Proceedings of international conference on learning representations (ICLR2015)
21. Chollet F (2018) Deep learning with python. Manning Publications Co., New York
22. Srivastava N, Hinton G, Krizhevsky A et al (2014) Dropout: a simple way to prevent neural networks from overfitting. J Mach Learn Res 15:1929–1958
23. Balzategui J et al (2019) Semi-automatic quality inspection of solar cell based on convolutional neural networks. 2019 24th IEEE international conference on emerging technologies and factory automation (ETFA). IEEE
24. Liu L et al (2019) Surface defect detection of solar cells based on feature pyramid network and GA-Faster-RCNN. In: 2019 2nd China symposium on cognitive computing and hybrid intelligence (CCHI). IEEE

25. Lin TY et al (2016) Feature pyramid networks for object detection. In: CVPR (2016)
26. Wang J et al (2019) Region proposal by guided anchoring. In: CVPR (2019)
27. Su B et al (2019) Classification of manufacturing defects in multicrystalline solar cells with novel feature descriptor. IEEE Trans Instrum Measure 68(12):4675–4688
28. Chen H et al (2020) Solar cell surface defect inspection based on multispectral convolutional neural network. J Intell Manuf 31(2):453–468
29. Zhang X et al (2020) Detection of surface defects on solar cells by fusing multi-channel convolution neural networks. Infrared Phys Technol 103334
30. Ren S, He K, Girshick R et al (2015) Faster r-cnn: Towards real-time object detection with region proposal networks. Adv Neural Inf Process Syst 91–99
31. Dai J, Li Y, He K et al (2016) R-fcn: object detection via region-based fully convolutional networks. Adv. Neural Inf. Process. Syst. 379–387 (2016)
32. Chen H et al (2020) Structure-aware-based crack defect detection for multicrystalline solar cells. Measurement 151:107170
33. van der Maaten L (2014) Accelerating t-SNE using tree-based algorithms. J Mach Learn Res 15:3221–3245
34. Géron A (2017) Hands-on machine learning with Scikit-Learn and TensorFlow. O'Reilly Media, USA
35. Goodfellow I, Bengio Y, Courville A (2016) Deep learning. MIT Press, Cambridge

Chapter 31
Design of Various Control Strategies for Electro-Hydraulic System

Divya Pandey, Shekhar Yadav, and Satanand Mishra

1 Introduction

We are analyzing control strategies of the electro-hydraulic system from different important techniques and understanding the various performances. In this paper, we are using two control methodologies, i.e., loop shaping with open-loop bandwidth of 5 Hz and internal model control approaches for designing the electro-hydraulic system. Electro-hydraulic system is typically nonlinear having unstable closed-loop system.

In future research work, fractional order controller is used in many control areas related to electro-hydraulic system having lower order system. Fractional order PID controllers have been noticed by both industrial and academic areas. In fact, they contribute more pliability and robustness in the implementation of controller design in comparison to standard PID controllers because they have five tuning parameters. However, the tuning of the FOPID controller can be much more complicated. In the literature, different techniques for the implementation of FOPID controller have been suggested in order to solve this difficulty [1].

In this paper, we are using journal bearing test rig as a electromechanical basis for the evaluation of performance in different load rolling [2]. The hydraulic system utilizes the hydraulic pump like pressure oil source. Hydraulic system is a highly nonlinear system. Flow coefficient, elastic modulus, and oil viscosity are uncertain [3]. The Conventional PID controller is having a wide application in electro-hydraulic

D. Pandey (✉) · S. Yadav
Department of Electrical Engineering, Madan Mohan Malaviya University of Technology, Gorakhpur, Uttar Pradesh 273010, India

S. Yadav
e-mail: syee@mmmut.ac.in

S. Mishra
CSIR AMPRI, Bhopal, AcSIR & CSIR- Advanced Materials and Processes Research Institute (AMPRI), Bhopal 462026, Madhya Pradesh, India

N. Marriwala et al. (eds.), *Soft Computing for Intelligent Systems*, Algorithms for Intelligent Systems, https://doi.org/10.1007/978-981-16-1048-6_31

loading system, which is simple in characteristics, easy to apply, and has understandable consequences. The above nonlinear factors are complicated to accomplish better performance. Therefore, some advanced techniques having controllers are preferred for the electro-hydraulic loading system. In reference to the electro-hydraulic shaking table, adaptive inverse control and internal model control strategies are used in [4]. Further, for the electro-hydraulic servo system, fuzzy internal model control is proposed in (Hong et al. 2008). Meta-heuristic techniques likes Particle Swarm Optimization applied in (Olranthichachat et al. 2011) is used for the loop-shaping control of the MIMO electro-hydraulic servo system.

The notion of fractional order PID controllers was suggested by Podlubny in 1997,1999 [5–10]. He indicated that a preferable response is produced in comparison with conventional PID controller for controlling the fractional order systems. In reference to frequency domain application, adopting FOPID controllers is also considered in (Vinagre et al. 2000). In (Monje et al. 2004), the tuning parameters of FOPID are complacent with pre-determined method specifications, and an optimization technique is introduced. Ziegler-Nichols tuning rules for FOPID are presented in (Valerio & Costa 2006). Further, research developments are running for the development of new tuning approaches and are exploring the different applications of FOPID. Karam and Elbayomy et al. introduced control strategy of electro-hydraulic servo system using the genetic algorithm based on PID controller and gave its effectiveness through different demonstration [11]. Guan and Pan suggested an adaptive sliding control method which is utilized for an electro-hydraulic system with nonlinear parameters [12].

The robust control techniques are used for evaluation of uncertain nonlinearities and vagueness in tuning parameters for the estimation of transient responses. In the literature [13, 14], the nonlinear version of quantitative feedback theory (QFT) is used for hydraulic system to construct a robust time-invariant controller. For the tracking of signals for real-time measurement, neural network adaptive inverse controller is utilized in an electro-hydraulic system suggested by Yao et al. [15]. For the robust $H\infty$ position, control synthesis of electro-hydraulic system utilizes the approaches of linear matrix inequalities introduced by Milíc et al. [16]. Feed-forward inverse model (FFIM) control is also proposed in the electro-hydraulic hybrid system for the greater robustness of the process [17–19]. Fractional order controller is utilized for the insulator fatigue test in electro-hydraulic system for tuning the parameters and to control the valves [5]. Adaptive robust control of Hydraulic manipulators is proposed for trajectory tracking in presence of uncertain nonlinearities and parametric uncertainties [6]. Furthermore, for improvement in the tracking performance for an electro-hydraulic servo system, Adaline neural network is used in adaptive inverse control of the system [7].

In the present scenario, the research techniques on the fractional order PID controller in this area are restricted. In this paper, we explored the applications of ant colony optimization based fractional order PID controller in electro-hydraulic loading system. In the literature, for the first time nature inspired algorithm (Ant colony algorithm) is used for tuning of FOPID variables for optimizing time domain and frequency domain. Therefore, the effectiveness of this method is analyzed with the comparative study of PID and FOPID controller. The results show that this method is more suitable over all standard methods in a probabilistic manner. This method used computational intelligence of nature inspired algorithm to minimize errors and to obtain optimal control variables (K_P, K_I, K_D, μ, and λ) for FOPID controller.

2 Methodology

In this paper, different control methods are examined for exploring the concepts of electro-hydraulic system. The electro-hydraulic loading system is applied to rolling force on the test bearing. The pressure applied to the system is 0–20 MPa and it is comparable to 0–90 tons load. The fractional order controller is a based on fractional calculus equation. For designing the fractional order controller, parameter tuning is important.

2.1 *System Description of EHLS*

2.1.1 Electro-hydraulic Loading System

In this system, for loading the test bearing, hydraulic cylinder is applied in the test rig. With the help of electro-hydraulic proportional relief valve, the hydraulic system adjusts the load [2]. The main components of the electro-hydraulic loading system consists of digital display temperature controller, axial piston pump, accumulator, proportional valve, liquid level relay, cooler, oil filter, etc. The compilation of real-time load is done through pressure sensors for effective results.

The test rig utilizes the hydraulic cylinder for loading of the test bearing. The various parts of electro-hydraulic loading system are controller, pressure sensor, proportional relief valve which acquires control and pressure of the system. For the employment of proportional relief valve, the control is executed by the proportional amplifier. The working principle of the electro-hydraulic loading system is shown in Fig. 1.

This system consists of two valves, namely, the main valve and the pilot valve. The main valve and pilot valve having the fluid pressure are utilized for the stabilization of spring force increment in the system. The pressure exerted is continuously maintained, and force is balanced. The overflow of the systems flows back to the tank.

The transfer function for the system is given as follows:

$$G(s) = \frac{-15.28s^5 + 59.92s^4 - 2659s^3 + 3.052 \times 10^4 s^2 - 6.682 \times 10^4 s + 7.804 \times 10^4}{s^6 + 19.55s^5 + 2331s^4 + 687.9s^3 + 1.396 \times 10^5 s^2 - 1.606 \times 10^6 s + 1.213 \times 10^7} \quad (1)$$

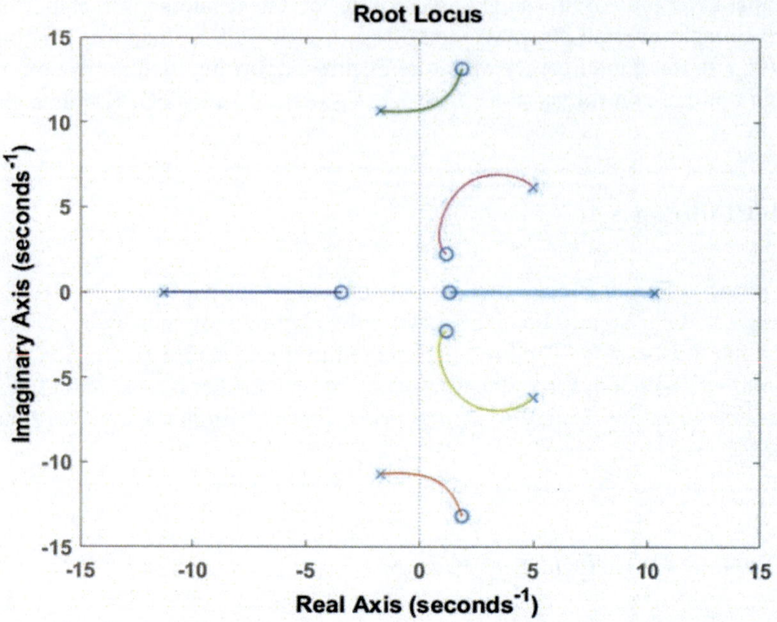

Fig. 1 Root locus plot of electro-hydraulic system

2.1.2 Controller Design for Electro-Hydraulic System

In this paper, we are utilizing two control approaches for electro-hydraulic system. Here, we study the nature of electro-hydraulic system by root locus techniques which is shown in Fig. 1. This paper represents the design of robust controller for an electro-hydraulic system in presence of nonlinearities and uncertainties. In this paper, we are using loop-shaping controller of target open-loop bandwidth of 10 Hz, and order of desired controller is 6.

The another technique for control strategy of electro-hydraulic loading system is internal model control (IMC) which is having the order of desired controller 10. The implementation of the loop-shaping controller is to compensate the control erection interaction and to reduce the consequences of resonance. We obtain the step responses by utilizing the two techniques and then the comparative analysis is done in time-domain characteristics in reference of rise time, settling time, % overshoot, peak and peak time (Figs. 2, 3 and 4; Table 1).

2.1.3 Fractional Order PID Controller

The general form of fractional order PID controller is the $PI^{\lambda}D^{\mu}$ (Podlubny, 1999a), involving an integer of order λ and a differentiator of order μ (λ and μ can be any real number) [20]. The performance and stability factor improves in general by using

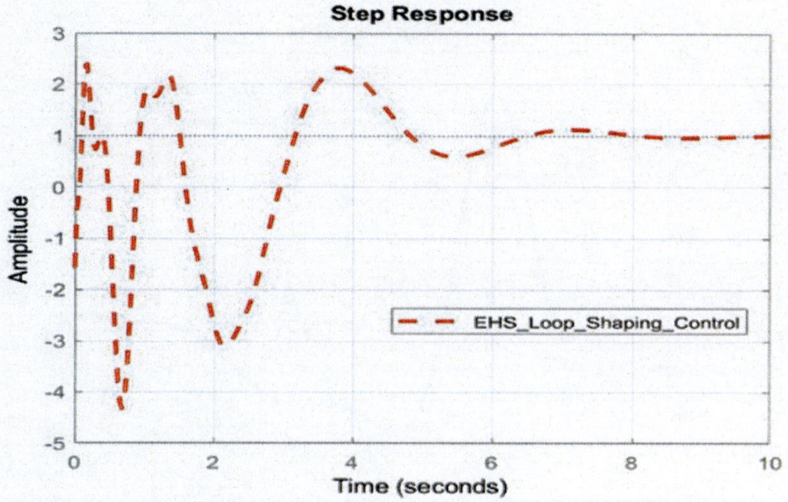

Fig. 2 Step response of loop-shaping controller for electro-hydraulic system

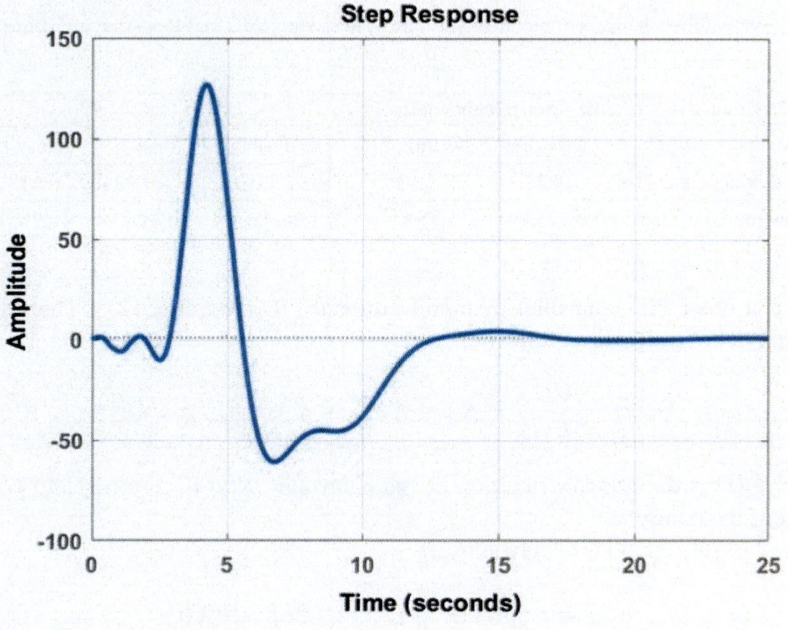

Fig. 3 Step response of internal model control for electro-hydraulic system

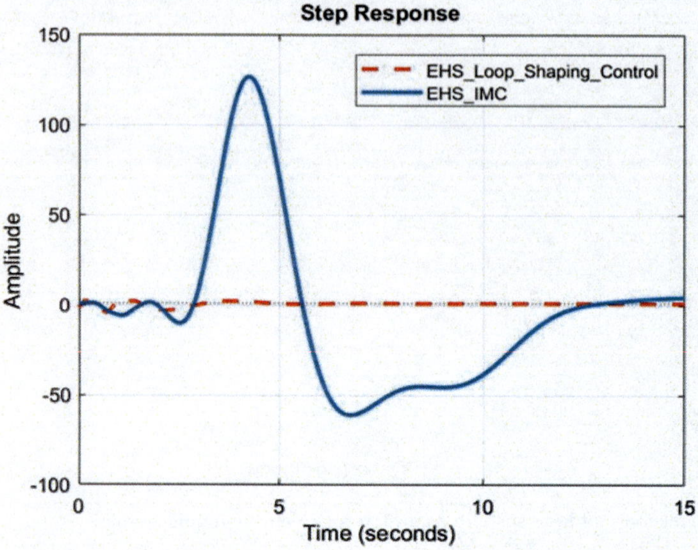

Fig. 4 Comparative analysis of electro-hydraulic system with IMC and loop-shaping controller

Table 1 Comparison of time-domain characteristics

Parameters	Rise time	Settling time	% overshoot	Peak	Peak time
Loop-shaping controller	0.0827	7.4105	139.5380	4.3350	0.6780
Internal model control	0.1239	15.7540	1.2580e+04	126.8038	4.2337

fractional order PID controller than conventional PID controller [21]. The transfer function of controller is as follows:

$$G_c(s) = \frac{U(s)}{E(s)} = K_p + K_I \frac{1}{s^{\lambda}} + K_D s^{\mu}, (\lambda, \mu > 0) \tag{2}$$

where $G_c(s)$ is the transfer function of the controller, $E(s)$ is an error, and $U(s)$ is output of the controller.

The control signal $u(t)$ is given by:

$$u(t) = K_p e(t) + K_I D^{-\lambda} e(t) + K_D D^{\mu} e(t) \tag{3}$$

The advantage of using this fractional order PID controller is to design fractional order models used in different real-time complex problems for robust control methodology of different dynamical systems and improved performances in implementation. To design fractional order controller, approximation algorithm can be developed [20]. Another advantage is that these controllers are less sensitive toward change of parameters of output. Due to five tuning parameters there is a better tuning of dynamical features of fractional order systems (Figs. 5, 6 and 7).

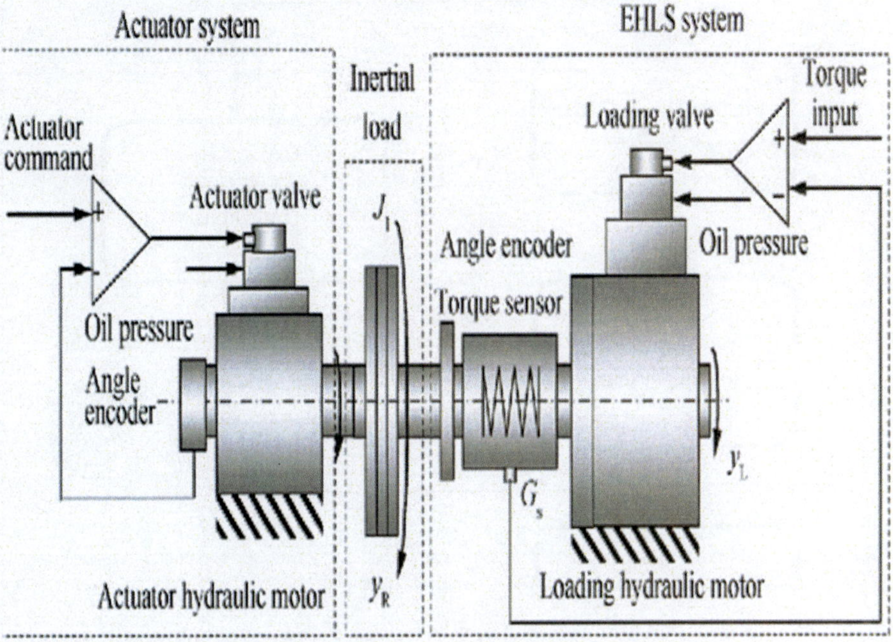

Fig. 5 Schematic diagram of electro-hydraulic loading system

Fig. 6 An overview of journal bearing

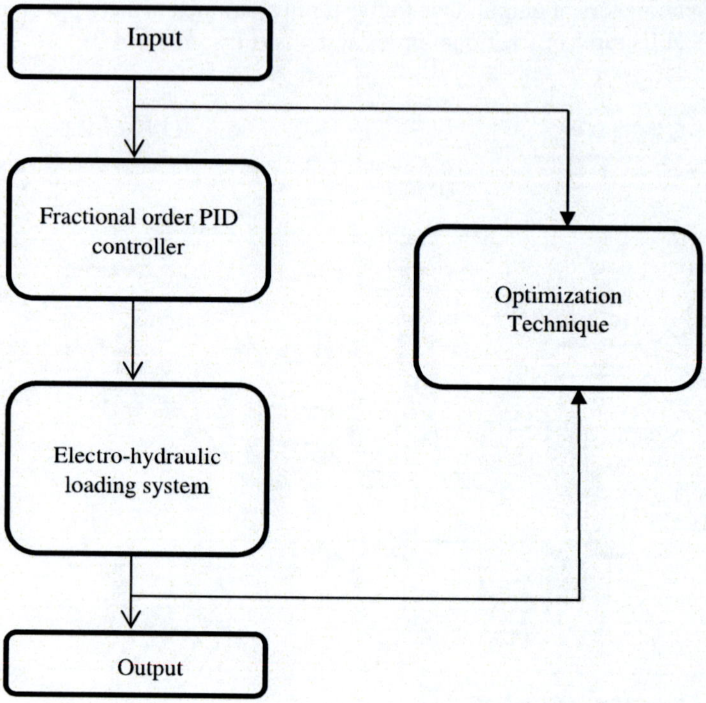

Fig. 7 Generalized block diagram for optimization technique used in electro-hydraulic system

The block diagram is as shown in Fig. 7. As shown in the flow chart, the input (i.e. pressure) is given to the fractional order controller or controlling the electro-hydraulic system. After that the optimization algorithm is applied for the optimization of parameter values. After the maximum iterations by ants and obtained the optimum parameters values in response.

The mathematical model of control strategy has been demonstrated. There are some assumptions that are taken for the modeling of procedures with small consequences. In this model, six-order transfer function is obtained. There are many variables in the mathematical model which cannot be evaluated. In this mathematical model, system identification is used in demonstration for validation and verification process. The model parameters are cited in reference [2]. The input and output variables of the model are collected as the variable source of system identification.

In this model, the system identification output represents effectively that six-order mathematical order system is more precise than the fifth- and seventh-order system. The six-order system is efficient up to 98.01%. The simulation model diagram of the system shows that the fractional order PID obtained better responses than the PID because of the unstability of the closed-loop system and also the complexity of the system.

3 Result and Discussion

In this research work, we are analyzing the performance of electro-hydraulic system by two techniques which are loop sharing and internal model control. The simulating model shows that this technique is efficiently used for optimizing FOPID controller in both the domain for electro-hydraulic loading system. In this paper, we are using non-minimum phase system by which different challenges due to nonlinearities and unstable closed-loop system are present. The analyzation is very important for the control strategies of such a complex system. In this step, responses are obtained by using loop-shaping controller and internal model control techniques and then understanding the time-domain characteristics produced by two control strategies.

4 Conclusion

In this, we are designing the controller for electro-hydraulic system by two methods. The responses obtained shows the unstable closed-loop behavior. The loop-shaping and internal model control technique used for higher order system and FOPID controller is suitable for lower order system. In this, the comparative analysis of step responses are done for the electro-hydraulic system with the help of both the controllers. In future research work, the fractional order PID controller is utilized for different applications in reference to lower order system. For optimal solution of the parameters of fractional order PID controller, different metaheuristic algorithms are applied in different fields. These metaheuristic algorithm approaches are applied in various difficult problems which do not obtain the optimal solution by deterministic approach. The different perspectives are represented here for envisaging the behavior of the electro-hydraulic system.

Acknowledgements This work was supported by Madan Mohan Malaviya University of Technology, Gorakhpur (U.P.).

References

1. Faieghi MR, Nemati A (2011) On fractional-order PID design. In: Applications of MATLAB in science and engineering. IntechOpen
2. Wang N, Wang J, Li Z, Tang X, Hou D (2018) Fractional-order PID control strategy on hydraulic-loading system of typical electromechanical platform. Sensors 18(9):3024
3. Viersma TJ (1980) Analysis, synthesis and design of hydraulic servosystems and pipelines (Studies in mechanical engineering S.), Elsevier (1980).
4. Shen G, Zhu Z, Tang Y, Zhang L, Liu G, Zhao J, Yang C, Han J (2013) Combined control strategy using internal model control and adaptive inverse control for electro-hydraulic shaking table. Proc Inst Mech Eng C J Mech Eng Sci 27(10):2348–2360

5. Zhao J, Wang J, Wang S (2013) Fractional order control to the electro-hydraulic system in insulator fatigue test device. Mechatronics 23(7):828–839
6. Mohanty (2011) Indirect adaptive robust control of hydraulic manipulators with accurate parameter estimates. IEEE Trans Control Technol 19(3),565–575
7. Yao J, Wang X, Hu S, Fu W (2011) Adaline neural network-based adaptive inverse control for anelectro-hydraulic servo system. J Vibr Control 17:2007–2014
8. Majhi SK, Pradhan JK, Pati BB (2019) Optimal fractional order PID controller design using Ant Lion optimizer
9. Yildirim B, Özdemir MT, İbrahim EK (2019) Design of integer order approximation fractional order controller with for automatic voltage regulation system. In: 2019 4th International conference on power electronics and their applications (ICPEA), pp 1–6. IEEE, Sep 25 2019
10. Nakata N (2013) Effective force testing using a robust loop shaping controller. Earthquake Eng Struct Dynam 42(2):261–275
11. Elbayomy KM, Zongxia J, Huaqing Z (2008) PID controller optimization by GA and its performances on the electro-hydraulic servo control system. Chin J Aeronaut 21:378–384
12. Guan C, Pan S (2008) Adaptive sliding mode control of electro-hydraulic system with nonlinear unknown parameters. Control Eng Pract 16(11):1275–1284
13. Niksefat N, Sepehri N (2000) Design and experimental evaluation of robust force controller for an electro-hydraulic actuator via quantitative feedback theory. Control Eng Pract 8(12):1335–1345
14. Nam Y, Hong SK (2000) Force control system design for aerodynamic load simulator. Control Eng Pract 10(5):549–558
15. Milíc V, Šitum Ž, Essert M (2010) Robust ∞ position control synthesis of an electro-hydralic servo system. ISA Trans 49: 535–542
16. Shen G, Zhu ZC, Han JW (2013) Adaptive feed-forward compensation for hybrid control with acceleration time waveform replication on electro-hydraulic shaking table. Control Eng. Pract 21:1128–1142. Sensors (2018)
17. Karer G, Music G, Igor Š (2011) Feed forward control of a class of hybrid systems using an inverse Model. Math Comput Simul 82:414–427
18. Shen G, Zhu ZC, Li X (2016) Real-time electro-hydraulic hybrid system for structural testing subjected to vibration and force loading. Mechatronics 33:49–70
19. Upadhyaya V, Singh A, Pandey P, Sharm R (2019) Genetic-algorithm-based optimization of ant colony controller for fractional-order systems. In: Computational intelligence: theories, applications and future directions, vol 1. Springer, Singapore, pp 441–452
20. Singh R, Kumar A, Sharma R (2016) Fractional order PID control using ant colony optimization, pp 1–6
21. Rajasekhar S, Das S, Suganthan PN (2012) Design of fractional order controller for a servo-hydraulic positioning system with micro Artificial Bee Colony algorithm. In: IEEE congress on evolutionary computation, pp 1–8

Chapter 32
Multi-Wave Mixing Process can be Used to Generate Single Photon Source for Quantum Information Processing

Priyanka and Savita Gill

1 Introduction

It is possible to describe a state as a nonclassical state that does not have any classical analogue [1–3]. If the "Glauber Sudarshan P-function of the field became negative than the delta function", a radiation field can be nonclassical. It is possible to calculate the nonclassicality of a system by sub-Poissonian photon statistics and antibunching. In the last three to four decades, these two nonclassical effects had been explored. But in the majority of cases, nonclassical effects were concentrated in the lowest order. Later, higher order nonclassical states have been introduced [4–7]. Higher order squeezing has been analyzed in depth out of these higher order nonclassical effects [4, 5, 8, 9], but higher order sub-Poissonian photon statistics and higher order antibunching have not yet been specifically studied. Lee developed the idea of HOA [6] and projected it in trio coherent state [10]. Kim [11] reported a general method for the analysis of second-order sub-Poissonian photon statistics.

Previous studies of these nonclassical effects have been mainly used in the academics [12], but its various applications in quantum information processing like communication [13], dense coding [14], quantum teleportation [15], quantum cryptography [16] are well-known for the time being. Many of the physical systems above are experimentally feasible and readily observed in nonlinear optics laboratory [17, 18]. We can obtain photon statistics experimentally using homodyne detection technique [19, 20].

The current work focuses on representing the higher order nonclassicality, i.e., higher order antibunching and higher order sub-Poissonian photon statistics in pump mode in multi-wave mixing nonlinear optical process. Section 2 emphasizes the criteria of nonclassicality of the nonlinear optical system. Furthermore,

Priyanka · S. Gill (✉)
University Institute of Engineering and Technology, Kurukshetra University Kurukshetra, Kurukshetra 136119, India
e-mail: savita2015@kuk.ac.in

N. Marriwala et al. (eds.), *Soft Computing for Intelligent Systems*, Algorithms for Intelligent Systems, https://doi.org/10.1007/978-981-16-1048-6_32

Sect. 3 provides a second-order solution of equation of motion of multi-wave mixing nonlinear optical process and indicates the occurence of HOA and HOSPS. In Sect. 4, we have compared the results obtained using graphs, and Sect. 5 emphasizes on conclusion.

2 Conditions of Nonclassicality of a Nonlinear Optical System

2.1 *Condition for Higher Order Antibunching (HOA)*

"HOA criteria was found by Lee" [6], which is

$$R(m,l) = \frac{\langle N_x^{m-1}\rangle\langle N_x^{l+1}\rangle}{\langle N_x^{m}\rangle\langle N_x^{l}\rangle} - 1 < 0 \tag{1}$$

where the number operator is represented by N. $\langle N^{(k)}\rangle = \langle N(N-1)(N-2)............(N-k+1)\rangle$, $N^{(k)}$ is kth factorial moment of number operator. Both integers (l) and (m) fulfilling the condition $(l \leq m \leq 1)$ and subscript x signify the specific mode. "$m = 1$ is chosen by Ba An" [10]. The criterion of lth order antibunching is given as

$$A_{x,l} = \frac{\langle N_x^{l+1}\rangle}{\langle N_x^{l}\rangle\langle N_x\rangle} - 1 < 0 \tag{2}$$

and

$$\langle N_x^{l+1}\rangle < \langle N_x^{l}\rangle\langle N_x\rangle \tag{3}$$

by simplifying Eq. (3) and obtaining criteria for $l - th$ order antibunching, which is

$$d(l) = \langle N_x^{l+1}\rangle - \langle N_x\rangle^{l+1} < 0 \tag{4}$$

Therefore, in quantum cryptography, we can say that single photon source to be utilized must fulfill the criteria given in Eq. (4) of HOA.

2.2 *Condition for Higher Order Sub-poissonian Photon Statistics (HOSPS)*

"Prakash and Mishra [4] give the criteria for $(l-1)$th order HOSPS", which are

$$D(l-1)=\sum_{k=0}^{l}\sum_{i=0}^{l-k}{}^{l}C_k(-1)^k S_2(l-k,i)\langle N^i\rangle\langle N\rangle^k$$
$$-\sum_{k=0}^{l}\sum_{i=0}^{l-k}{}^{l}C_k(-1)^k S_2(l-k,i)\langle N\rangle^{k+i}<0 \tag{5}$$

where Stirling number $S_2(l,k)$ is of second order. Criteria for the second-order sub-Poissonian photon statistics, for which $l=3$ is given as

$$D(2)=\langle N^3\rangle+2\langle N\rangle^3-3\langle N^2\rangle\langle N\rangle+3\langle N^2\rangle-3\langle N\rangle^2<0 \tag{6}$$

Equation (6) is the condition of second-order sub-Poissonian photon statistics.

3 Multi-wave Mixing Process

We selected eight wave mixing processes to analyze HOA and HOSPS in a manner that three photons of frequency ω_1 are absorbed with four photons of frequency ω_2 and one photon of frequency ω_3 being expelled. For this form, the Hamiltonian is given as

$$H=\omega_1 a^{\dagger}a+\omega_2 b^{\dagger}b+\omega_3 c^{\dagger}c+g(a^3 b^{\dagger 4}c^{\dagger}+a^{\dagger 3}b^4 c) \tag{7}$$

where coupling constant is represented by g, annihilation (creation) operators are $a(a^{\dagger})$, $b(b^{\dagger})$, $c(c^{\dagger})$, respectively. $A=a\exp i\omega_1 t$, $B=b\exp i\omega_2 t$, $C=c\exp i\omega_3 t$ are the slowly varying operators oscillating at frequencies ω_1, ω_2 and ω_3.

3.1 *Time Evolution of Pump Mode A*

Time evolution of operator A in pump mode is given as

$$\frac{dA}{dt}=\frac{\partial A}{\partial t}+i[H,A] \tag{8}$$

We obtain

$$\dot{A}=-3igA^{+2}B^4C \tag{9}$$

$$\dot{B}=-4igA^3B^{+3}C^+ \tag{10}$$

and

$$\dot{C} = -igA^3B^{+4} \tag{11}$$

by applying short-time approximation, expanding $A(t)$, in line with Taylor's expansion and bringing terms up to second order in g^2t^2 as

$$A(t) = A - 3igtA^{\dagger 2}B^4C + \frac{3}{2}g^2t^2(6A^\dagger A^2N_B^4N_C + 6AN_B^4N_C - 16A^{\dagger 2}A^3N_B^3N_C \\ -72A^{\dagger 2}A^3N_B^2N_C - 96A^{\dagger 2}A^3N_BN_C - 24A^{\dagger 2}A^3N_C - A^{\dagger 2}A^3N_B^4 - 16A^{\dagger 2}A^3N_B^3 \\ -72A^{\dagger 2}A^3N_B^2 - 96A^{\dagger 2}A^3N_B - 24A^{\dagger 2}A^3) \tag{12}$$

Using Eq. (12), number operator $N_A(t)$ is given as

$$N_A(t) = A^\dagger A - 3igt(A^{\dagger 3}B^4C - A^3B^{\dagger 4}C^\dagger) + 3g^2t^2(9A^{\dagger 2}A^2N_B^4N_C + 18A^\dagger AN_B^4N_C \\ +6B^{\dagger 4}B^4N_C - 16A^{\dagger 3}A^3N_B^3N_C - 72A^{\dagger 3}A^3N_B^2N_C - 96A^{\dagger 3}A^3N_BN_C - 24A^{\dagger 3}A^3N_C \\ -A^{\dagger 3}A^3N_B^4 - 16A^{\dagger 3}A^3N_B^3 - 72A^{\dagger 3}A^3N_B^2 - 96A^{\dagger 3}A^3N_B - 24A^{\dagger 3}A^3) \tag{13}$$

To study antibunching, initially we consider a quantum state which is a product of coherent state $|\alpha\rangle$ for pump mode A and vacuum state $|0\rangle$ for stokes mode B and signal mode C, i.e.

$$|\psi\rangle = |\alpha\rangle_A|0\rangle_B|0\rangle_C \tag{14}$$

Using Eq. (14) in Eq. (13), we get $\langle N_A(t)\rangle_\alpha$ as

$$\langle N_A(t)\rangle_\alpha = |\alpha|^2 - 72g^2t^2|\alpha|^6 \tag{15}$$

where $A|\alpha\rangle = \alpha|\alpha\rangle$. And value of $\langle N_A^2(t)\rangle$ and $\langle N_A^3(t)\rangle$ is given as

$$\langle N_A^2(t)\rangle_\alpha = \langle A^{\dagger 2}(t)A^2(t)\rangle = |\alpha|^4 - 144g^2t^2(|\alpha|^8 + |\alpha|^6) \tag{16}$$

and

$$\langle N_A^3(t)\rangle_\alpha = |\alpha|^6 - 36g^2t^2(6|\alpha|^{10} + 12|\alpha|^8 + 4|\alpha|^6) \tag{17}$$

We are now using Eqs. (15–17) on the condition of normal and higher order antibunching (Eq. (4)), we get

$$d_A(1)_\alpha = -144g^2t^2|\alpha|^6 \tag{18}$$

and

$$d_A(2)_\alpha = -144g^2t^2(3|\alpha|^8 + |\alpha|^6) \tag{19}$$

Equations (18) and (19) show that eight wave mixing processes fulfill the antibunching criteria of normal and higher order.

Furthermore, to study higher order antibunching, we have now taken the initial state which is a product of vacuum state $|0\rangle$ for pump mode A and $|\beta\rangle$ for mode B and vacuum state for mode C, i.e.,

$$|\psi\rangle = |0\rangle_A|\beta\rangle_B|0\rangle_C \tag{20}$$

Now taking the expectation values of $N_A(t)$, $N_A^2(t)$ and $N_A^3(t)$ and in pump mode A in relation to condition (20) is

$$\langle N_A(t)\rangle_\beta = 0 \tag{21}$$

$$\left\langle N_A^2(t)\right\rangle_\beta = 0 \tag{22}$$

$$\left\langle N_A^3(t)\right\rangle_\beta = 0 \tag{23}$$

Therefore, normal and higher order antibunching in mode A with respect to $|0\rangle|\beta\rangle|0\rangle$ is given as

$$d_A(1)_\beta = 0 \tag{24}$$

$$d_A(2)_\beta = 0 \tag{25}$$

Similarly, we can study the nonclassical effect (HOA) by assuming a quantum state which is the product of vacuum state for pump mode A and for stokes mode B and $|\gamma\rangle$ for signal mode C, i.e.,

$$|\psi\rangle = |0\rangle_A|0\rangle_B|\gamma\rangle_C \tag{26}$$

now taking expectation values of $N_A(t)$, $N_A^2(t)$ and $N_A^3(t)$ in pump mode A in relation to condition (26) is

$$\langle N_A(t)\rangle_\gamma = 0 \tag{27}$$

$$\left\langle N_A^2(t)\right\rangle_\gamma = 0 \tag{28}$$

$$\left\langle N_A^3(t)\right\rangle_\gamma = 0 \tag{29}$$

Therefore, normal and higher order antibunching in mode A with respect to $|0\rangle|0\rangle|\gamma\rangle$ is given as

$$d_A(1)_\gamma = 0 \tag{30}$$

$$d_B(2)_\gamma = 0 \tag{31}$$

From Eqs. (24, 25) and from Eqs. (30, 31), we can say that neither normal antibunching nor higher order antibunching is existing in mode A with respect to $|0\rangle|\beta\rangle|0\rangle$ and $|0\rangle|0\rangle|\gamma\rangle$, respectively.

Now using Eqs. (13, 14, 20, 26) in Eq. (6), we get

$$D(2)_\alpha = -576g^2t^2|\alpha|^6 \tag{32}$$

$$D(2)_\beta = 0$$

$$D(2)_\gamma = 0$$

Thus, we have obtained negative value which indicates that higher order sub-Poissonian photon statistics exists in pump mode A.

4 Result

The presence of HOA and HOSPS in pump mode A in multi-wave mixing nonlinear optical process is shown in Eqs. (18, 19 and 32), respectively. If we plot a graph between HOA say $d(1)$, $d(2)$ and HOSPS say $D(2)$ with number of photons in mode A, respectively, it is clearly evident that HOA and HOSPS increase nonlinearly with increase in $|\alpha|^2$. Further, it is evident from the figures that nonclassicality increases as we go toward the higher order of antibunching (Figs. 1, 2 and 3).

5 Conclusion

We have reported normal as well as higher order nonclassicality in terms of higher order antibunching and higher order sub-Poissonian photon statistics in pump mode in multi-wave mixing nonlinear optical process. We found that nonclassicality increases as we go toward the higher order of antibunching. We have also found that normal and higher order antibunching directly depends upon the number of pump photons existing in the system. Thus, we can conclude that multi-wave mixing

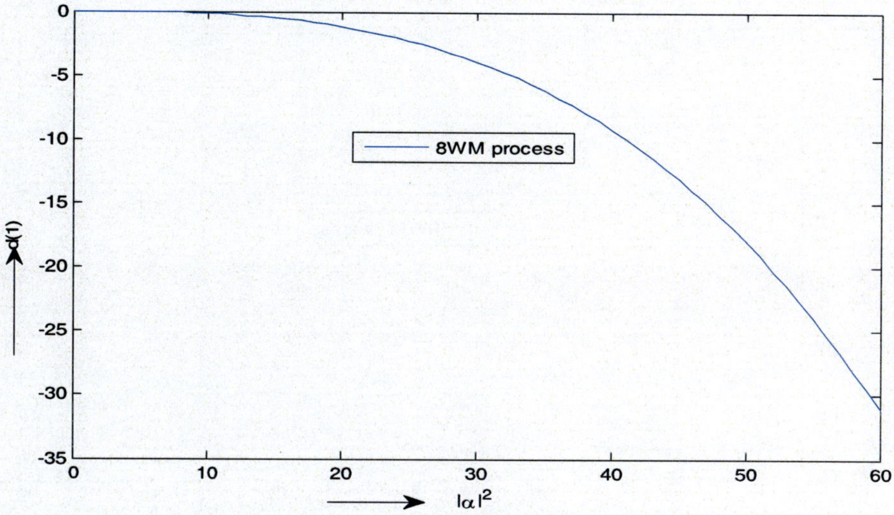

Fig. 1 Variation of normal antibunching $d(1)$ with $|\alpha|^2$ (taking $g^2t^2 = 10^{-4}$)

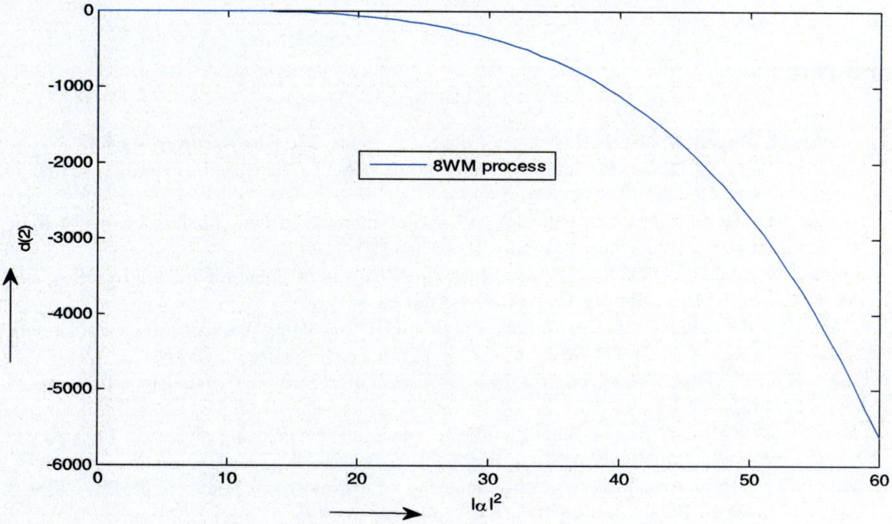

Fig. 2 Variation of higher order antibunching $d(2)$ with $|\alpha|^2$ (taking $g^2t^2 = 10^{-4}$)

nonlinear optical process shows higher order nonclassicality and that can be utilized to generate single photon source for quantum information processing.

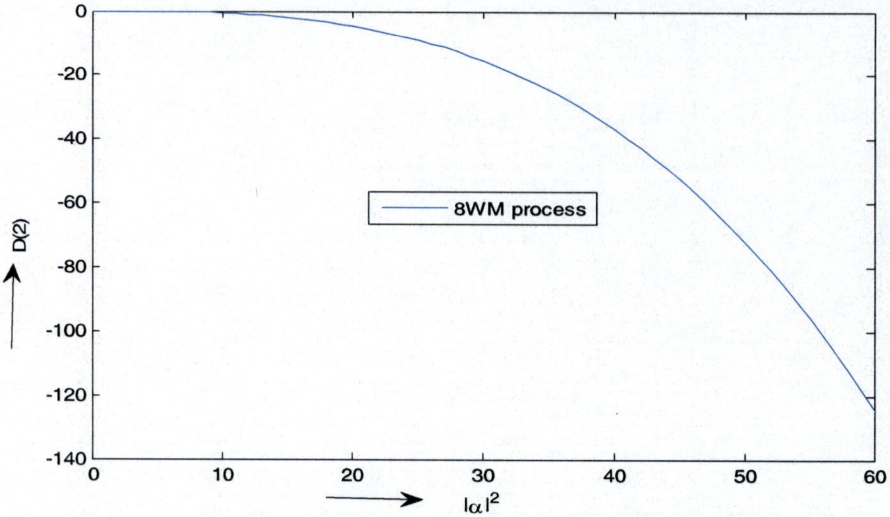

Fig. 3 Variation of higher order sub-Poissonian photon statistics $D(2)$ with $|\alpha|^2$ (taking $g^2t^2 = 10^{-4}$)

References

1. Meystre P, Sargent M (1991) Elements of quantum optics, 2nd edn. Springer, Berlin
2. Dodonov VV (2002) Nonclassical states in quantum optics: a 'squeezed' review of the first 75 years. J Opt B: Quantum Semiclassical Opt 4
3. Brown RH, Twiss RQ (1956) Correlation between photons in two coherent beams of light. Nature 177(4497):27–29. https://doi.org/10.1038/177027a0
4. Hong CK, Mandel L (1985) Higher order squeezing of a quantum field. Phys Rev Lett 54(4):323–325. https://doi.org/10.1103/PhysRevLett.54.323
5. Hong CK, Mandel L (1985) Generation of higher order squeezing of quantum electromagnetic fields. Phys Rev A 32(2):974–982. https://doi.org/10.1103/PhysRevA.32.974
6. Lee CK (1990) Higher order criteria for nonclassical effects in photon statistics. Phys Rev A 41:1721–1723
7. Lee CT (1990) Many photon antibunching in generalized pair coherent states. Phys Rev A 41(3):1569–1575. https://doi.org/10.1103/physreva.41.1569
8. Hillery M (1987) Amplitude squared squeezing of the electromagnetic field. Phys Rev A 36(8):3796–3802. https://doi.org/10.1103/PhysRevA.36.3796
9. Giri DK, Gupta PS (2003) Short time squeezing effects in spontaneous and stimulated six wave mixing process. Opt Commun 221(1):135–143. https://doi.org/10.1016/S0030-4018(03)01464-0
10. An BN (2002) Multimode higher order antibunching and squeezing in trio coherent states. J Opt B: Quantum Semiclassical Opt 4(3):222–227. https://doi.org/10.1088/1464-4266/4/3/310
11. Kim K (1998) Higher order sub-poissonian. Phys Lett A 245(1–2):40–42
12. Mollow BR, Glauber RJ (1967) Quantum theory of parametric amplification. Phys Rev 160(5):1076–1096. https://doi.org/10.1103/PhysRev.160.1076
13. Bennett CH et al (1993) Entanglement-assisted classical capacity of noisy quantum channels. Phys Rev Lett 83(15):3081–3084. https://doi.org/10.1103/PhysRevLett.83.3081
14. Braunstein SL, Kimble HJ (2000) Dense coding for continuous variables. Phys Rev A 61(4):042302(1–4). https://doi.org/10.1103/PhysRev.61(4).04202

15. Braunstein et al (2000) Universal teleportation with a twist. Phys Rev Lett 84(15):3486–3489. https://doi.org/10.1103/PhysRevLett.84.3486
16. Bennett CH et al (1992) Quantum cryptography without Bell's theorem. Phys Rev Lett 68(5):557–559. https://doi.org/10.1103/PhysRevLett.68.557
17. Bachor HA (1998) Chapters 8 and 10: A guide to experiments in quantum optics. Wiley-VCH, Weinheim
18. Vogel W, Welsch D, Wallentowitz S (2001) Chapter 6: Quantum optics: an introduction, 2nd edn. Wiley-VCH, Berlin. https://doi.org/10.1002/3527608524
19. Erenso D, Vyas R, Singh S (2002) Higher-order sub-Poissonian photon statistics in terms of factorial moments. J Opt Soc B 19(6):1471–1475. https://doi.org/10.1364/JOSAB.19.001471
20. Vyas R, Singh S (1989) Photon-counting statistics of the degenerate optical parametric oscillator. Phys Rev A 40(9):5147–5159. https://doi.org/10.1103/PhysRevA.40.5147

Chapter 33
Performance Analysis of Cluster-Based Energy-Efficient Routing Scheme for WSN

Satyanarayan Padaganur, Paramanand S. Patil, and Mallikarjun Deshmukh

1 Introduction

WSN are popular because they can be easily deployed and maintained when compared to earlier forms of networks, and also they have low cost. The three main key elements of WSN are monitoring, computing, and communication, whose combination in one sensor network provides plentiful of remote sensing applications [1]. WSN is a network of distributed independent devices that can sense and extract data from the various surrounding environment, and then transfer data to a base station for further processing through wireless communication. The various problems faced by the WSN are storage issues, communication failures, computational constraints, and insufficient power supply. The main upcoming task occurring in wireless sensor network is to save energy, and hence prolong the network lifetime [2].

To solve the energy consumption problem, we are using Clustering of sensor nodes. And to design and develop a cluster-based energy-efficient routing scheme with consideration of low cost, low power, and for longer distance communication as shown in Fig. 1. In clustering, clusters are formed to lessen the total transmission power, and to equalize the load among the nodes to perpetuate the network lifetime [2, 3].

In the proposed work, we have proposed clustering, which is implemented for Wireless Sensor Network that ensures an Energy-Efficient Routing scheme. There are different types of cluster head selection approach as shown in Fig. 2.

In WSN, the constrained abilities (battery control, transmission extend, preparing equipment and memory utilized, and so on.) of the sensor nodes joined with the extraordinary area based on the conditions met (not effectively got to all together energize the batteries or supplant the whole sensors) make the vitality productivity and the versatility factors considerably more significant [4]. Besides, the test of

S. Padaganur (✉) · P. S. Patil · M. Deshmukh
B.L.D.E.A's V. P. Dr. P.G. Halakatti College of Engineering and Technology, Vijayapur, Karnataka, India

N. Marriwala et al. (eds.), *Soft Computing for Intelligent Systems*, Algorithms for Intelligent Systems, https://doi.org/10.1007/978-981-16-1048-6_33

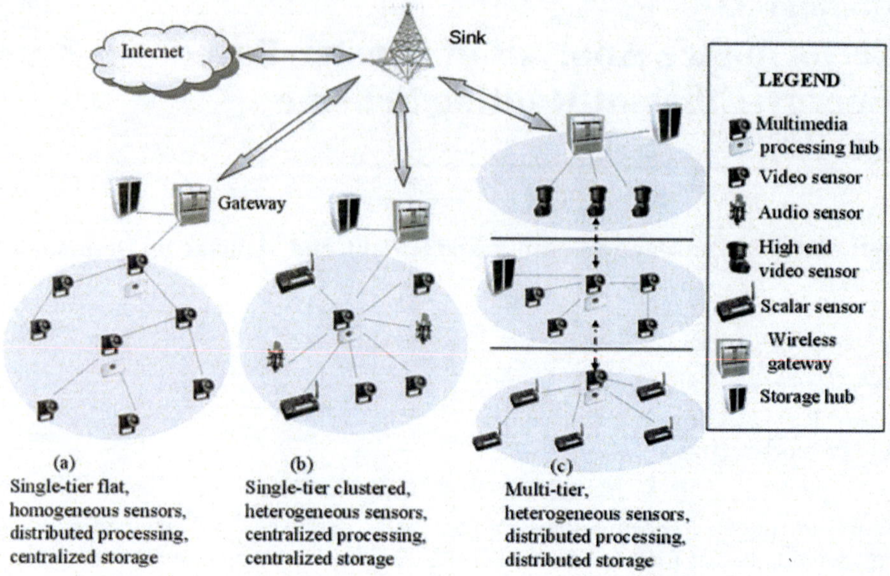

Fig. 1 Wireless sensor network

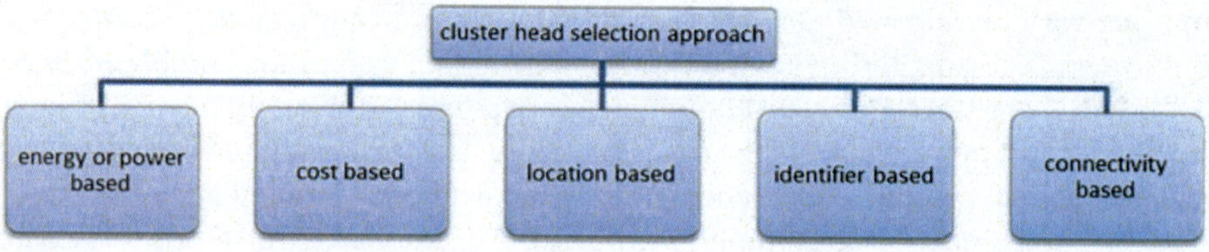

Fig. 2 Cluster head selection approach

drawing out system lifetime under the above confinements is hard to be met by utilizing just customary methods. The inconstancy of system topologies because of node disappointments, presentation of extra nodes, varieties in sensor area and changes to group portions in light of system requests requires the flexibility of basic system structures and activities. Battery control, memory, handling force, and lifetime are physical requirements that effect each part of WSNs. Individual sensors are problematic in unforgiving and eccentric conditions. The amount of information assembled may surpass the prerequisites of the system, and thus assessment of the information and transmission of just pertinent and satisfactory data should be performed.

2 Review Literature

Since all sensor nodes in remote sensor systems work by their own particular batteries that are implanted inside them, if a node comes up short on its battery, the sensor organizer can't work ordinarily [5]. In this circumstance, we should utilize the steering conventions which can devour the vitality of nodes effectively. Numerous conventions for vitality effective steering in sensor systems have been recommended, yet LEACH and PEGASIS are most outstanding conventions [6–8]. Anyway LEACH utilizes vitality intensely in the head nodes, and the head nodes tend to bite the dust early and PEGASIS—which is known as a superior vitality proficient convention that has a long exchange time from a source node to sink node, and the nodes near the sink node exhaust vitality forcefully since it makes a long bounce of information sending [9, 10].

3 Proposed Work

To setup the cluster-based network system, the following three phases has been adopted: Initial Phase, Implementation Phase, and Testing Phase.

We know that in wireless sensor networks, the sensors are deployed in various places covering a large area. The energy thus consumed by these nodes is drained out easily as it is not being used efficiently. In order to overcome this, we propose a clustering scheme. Wherein the clusters of the deployed sensor nodes based on their locations are made. The distance between the sensor nodes is considered as a parameter for the formation of cluster. A maximum distance limit is found within which the sensor nodes communicate easily and effectively and cluster is formed. Thus, the wireless sensor network with several single nodes is now converted into a hierarchical network with clusters. Once successfully the clustering process is carried out, the next task is Cluster-Heads selection. Cluster Head is one of the nodes in the cluster that is responsible for collecting the data from other nodes inside and initiates the communication. There are various parameters that can be considered while electing a particular node from a cluster as a cluster head. They are:

(1) Energy—When we use energy as a parameter for electing cluster head we select the sensor node that has highest energy level in the cluster of various nodes.
(2) Cost—The sensor node that is most cost effective among various other sensor nodes of a cluster is elected to be the cluster head.
(3) Location—As we know, in this scheme, the data that a cluster head aggregates is further sent to base station. Hence, node that is nearest to the base station is elected as a cluster head.

Our proposed work is based on Energy-Efficient Cluster-based scheme that is an extension to version of LEACH. To carry out the process of clustering, a maximum distance is found for which the sensor nodes can transmit and receive data effectively and efficiently as shown in Fig. 3. This distance is then considered and the nodes falling in that range are considered to be in one cluster and hence clusters are formed based on the distance between the sensor nodes. Thus, the wireless sensor network with several single nodes is now converted into a hierarchical network with clusters. For a network with several clusters there is need for proper management and coordination between these clusters and to serve this purpose we select one of them from the cluster as a Cluster Head. Hence, in the clustering process the next task that is to be performed is the selection of a Cluster Head. Thus, energy utilization in a clustered network is limited by using a hierarchical routing approach, i.e., forming clusters. Basically, there can be three modes of transmission in a cluster-based network in which following are the types of communication that will occur in the proposed topology. Intra-Cluster Communication—The communication within a cluster that is the transmission of sensed data from a cluster member to cluster head. In this type of transmission sensed data from various sensor nodes in a cluster is reported to the elected Cluster Heads.

Inter-Cluster Communication—Since the clusters are formed based on the distance for which the nodes can communicate effectively, the scenario where a

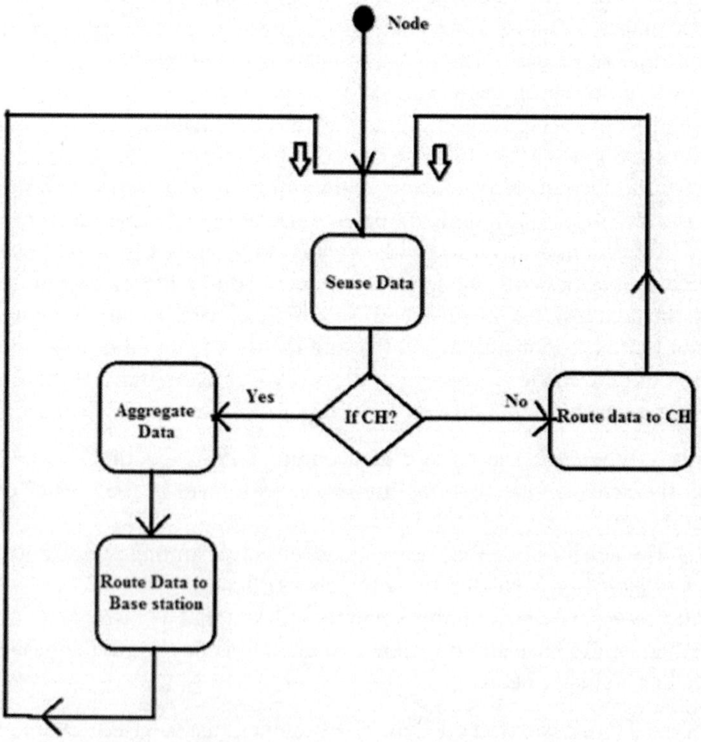

Fig. 3 Flow chart for cluster formation

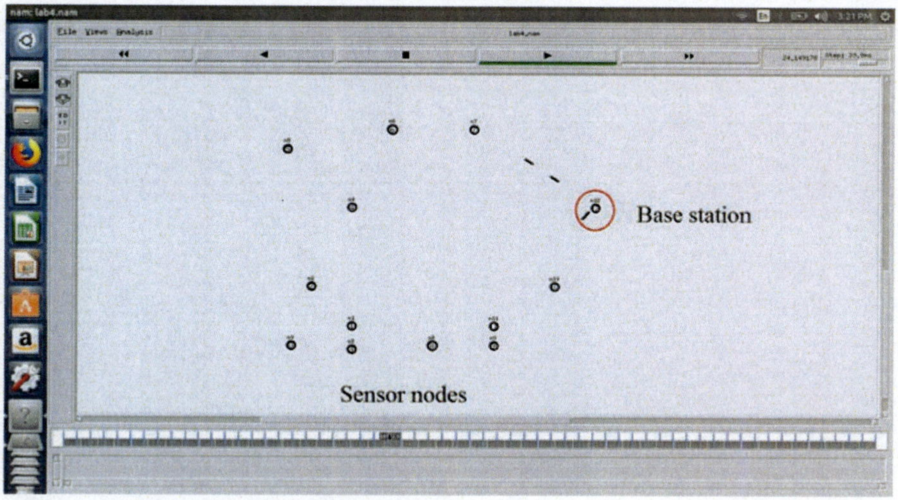

Fig. 4 Topology of nodes without clustering

cluster node has to communicate with other cluster node that is present in other clusters comes under this type of communication. For this type of communication to take place, a base station is used, which collects the sensed data aggregated in the sender cluster head and forward this data to the receiving cluster head. The topology of nodes without clustering is shown in Fig. 4.

Figure 4 shows a topology that depicts a randomly placed set of nodes that forms a wireless sensor network. The network consists of deployed sensor nodes and a base station wherein the sensor nodes individually transmit the data to the base station as and when they sense the data. After examining the operation of this network, results show that the energy consumed and the number of packet dropped is more, while the throughput of this type of network is less.

Nodes with clustering: The topology of nodes with clustering is shown in Fig. 5.

Clustering provides the following benefits

Increased resource availability: If one sensor node in a group comes up short, the other sensor nodes in the cluster can pick the workload. This keeps the loss of profitable time and data if a sensor node neglects to detect the information. Strategic asset use: We can circulate venture crosswise over nodes in whatever arrangement we incline toward. Increased performance: Multiple sensor nodes in a cluster provide greater sensing power and hence the system reliability. Greater scalability: As the number of sensor nodes increases clustering proves to be the best approach as the management of such large number of nodes as cluster becomes easy. Simplified management: Clustering rearranges the administration of expansive or quickly developing frameworks. Clustering enables to implement the following strategies. Load is calculated as the area that a sensor node has to sense and report. The load balancers in the form of Cluster head collect information from various other sensor nodes of that cluster.

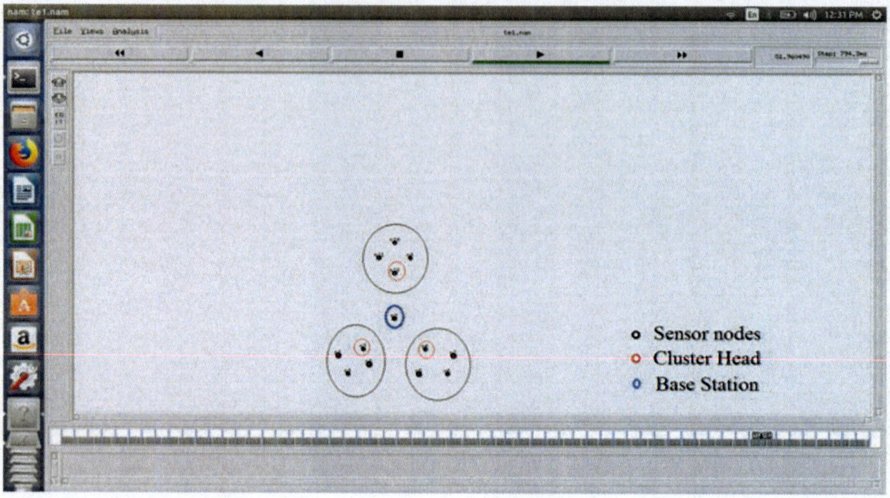

Fig. 5 Topology of nodes with clustering

The aggregated information is further sent to the base station. Hence, the clustering technique along with the load balancers, that is, the elected cluster head helps in the efficient usage of available energy.

4 Simulation and Results

The energy consumed in network with cluster and in network without cluster is compared in the following Table 2. It is evident from the table that the energy consumed by the cluster network is less compared to the network without cluster [2]. As the number of nodes increases, clustering proves to be a very effective approach as the total energy consumed by clustered network in that case will drastically be less than energy consumed by a network without cluster (Table 1).

The energy consumption of a clustered network and a network without cluster is depicted graphically in Fig. 6. The lines seem to close in the graph above, it is so because the number of nodes considered for our experiment is less. For a practical scenario where several hundreds and thousands of nodes are deployed in large areas, we can observe a greater difference in the energy consumption by both the networks.

Throughput: When nodes transmit, their neighboring nodes are prohibited from transmitting due to interference. Thus, a system throughput is limited by interference in the system.

Formula of throughput: Throughput = received size/(stop time−start time).

Table 1 Comparison of energy consumption of a clustered network and non-clustered network

No. of nodes	With cluster	Without cluster
	Total energy = 1440 J	Total energy = 1530 J
	Average energy (J)	Average energy (J)
1	1440	1530
2	720	765
4	360	382.5
6	240	255
8	180	191.25
10	144	153
12	120	127.5

Table 2 Comparison of packet drop with clustering and without clustering

	With cluster	Without cluster
No. of packets dropped	16	526
No. of packets sent	19,458	15,678
No. of packets received	19,668	17,569

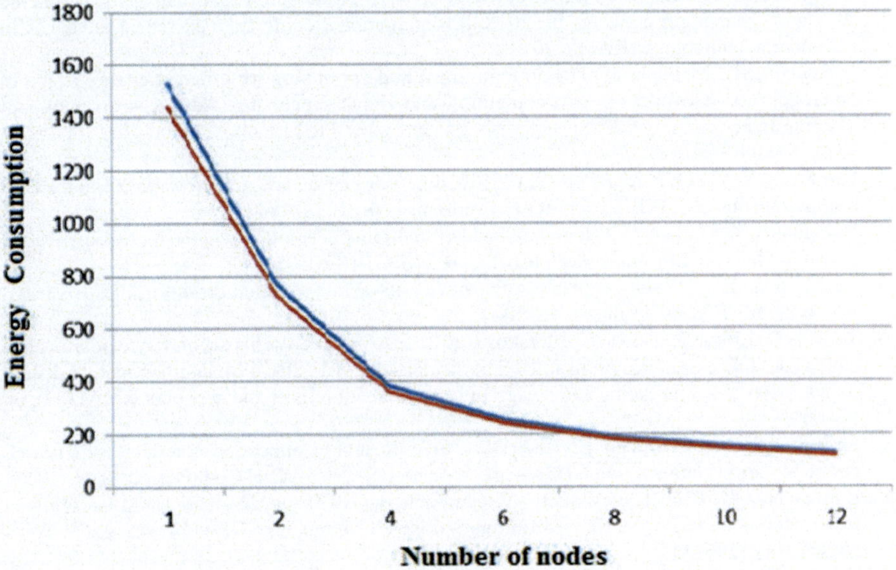

Fig. 6 Graphical representation of energy consumption

Formula of packet drop: Packet drop = (generated packet − received packet).

The transmission control convention (TCP) distinguishes parcels misfortune and performs retransmission to guarantee dependable informing.

5 Conclusions

In the proposed work, cluster-based network can improve the energy consumption, throughput, and packet drop of network. Nodes in two cases, i.e., with clustering and without clustering for a wireless sensor network. The results show that using clustering technique, we can reduce energy consumption of nodes, thus it can save energy and also lifetime of the nodes can be increased. Hence, the performance analysis shows that energy consumed in a clustered network is less than a network without cluster.

References

1. Padaganur SK, Mallapur JD (2018) A feed forward based smart link management scheme for future generation network. In: 2018 4th international conference on recent advances in information technology (RAIT), 2018
2. Padaganur SK, Mallapur JD (2019) Optimized node swapping for efficient energy usage in heterogeneous network. In: Soft computing and signal processing. Advances in intelligent systems and computing 900. Springer Nature Singapore Pte Ltd. https://doi.org/10.1007/978-981-13-3600-3_66
3. Fan J, Li GY, Yin Q, Peng B, Zhu X (2012) Joint user pairing and resource allocation for LTE uplink transmission. IEEE Trans Wirel Commun 11(8):2838–2847
4. Padaganur SK, Mallapur JD Neural embedded smart link generation scheme for heterogeneous network. Heliyon. https://doi.org/10.1016/j.heliyon.2018.e01089
5. Lei L, Zhong H (2012) Operator controlled device-to-device communications in LTE-advanced networks. IEEE Wirel Commun 96–104
6. Tsolkas D, Liotou E, Passas N, Merakos L (2012) A graph-coloring secondary resource allocation for D2D communications in LTE networks. In: 2012 IEEE 17th international workshop on computer aided modeling and design of communication links and networks (CAMAD), pp 56–60
7. Phunchongharn P, Hossain E, Kim DI (2013) Resource allocation for device to device communications underlaying LTE advanced networks. IEEE Wirel Commun 20(4):91–100
8. López-Pérez D, Chu X, Vasilakos AV, Claussen H (2013) On distributed and coordinated resource allocation for interference mitigation in self-organizing LTE networks. IEEE Trans Netw 21(4):1145–1158
9. Zheng K, Hu F, Wang W, Xiang W, Dohler M (2012) Radio resource allocation in LTE-advanced cellular networks with M2M communications. IEEE Commun Mag 50(7):184–192
10. Abdel-hadi A, Clancy C (2014) A utility proportional fairness approach for resource allocation in 4G-LTE. In: International conference on computing, networking and communications, pp 1034–1040

Chapter 34
Analytical Study of Virtual Machine Migration Techniques in Cloud Computing

Loveleena Mukhija, Rohit Sachdeva, and Mohanjeet Singh

1 Introduction

Cloud Computing is one of the eminent model in today's scenario. Nowadays, there is an exponential rise in the users of mobile and computer which has in turn given rise to whopping growth in the data. It is like centralizing the activities related to computing just analogous to electricity producing and dispensing by the virtue of grids.

Cloud Computing has now become imperative in IT sphere which provides its services that too as required by an application or offers sharing of computational resources which it provides accordingly on the basis of the usage. The services provided by this paradigm have evolved from storing huge data to intricate services. Thus, this emerging model has alleviated and proved to be as a boom in the current scenario of ICT and it has all been possible through pivotal concept behind this, which is virtualization.

In this paper, we focus on the existing techniques of the virtual machine migration, survey the literature for assessing these diverse ways and will uncover the various parameters of performance. This paper is structured as follows. Section 2 presents the overview discussing the various terminology involved in the cloud scenario. Section 3 presents a detailed description of VM migration and its various aspects and also details about live virtual machine migration technique. Section 4 presents a comparative analysis of the existing techniques for VM migration. We wrap up our work in Sect. 5 with future path.

L. Mukhija (✉)
Punjabi University, Patiala, India

R. Sachdeva
MM Modi College, Patiala, India

M. Singh
PMN College, Rajpura, India

N. Marriwala et al. (eds.), *Soft Computing for Intelligent Systems*, Algorithms for Intelligent Systems, https://doi.org/10.1007/978-981-16-1048-6_34

2 Overview

Data Center: Data center is the eminent part of cloud computing scenario, containing collection of servers on which business information is stored and applications run which in turn can significantly minimize the considerable variety of physical machine (hosts) which are operating in the data center.

Virtualization: Cloud Computing harness the beneficial effects like reducing monetary expenses, improved efficiency, and flexibility with the primal method of virtualization. Virtualization makes it possible to create numerous virtual machines, which in turn can share physical resources. It can be viewed as a fine level of software streaming between physical hardware and the operating system. It is also named as a virtual machine monitor or a hypervisor. This technology focuses to harness maximal from hardware resources which traditionally were getting less than optimal usage in the time span of high priced mainframe computers. A few examples of popular virtualization software are VMware ESX, ESXi, and Citrix.

Virtualization can seize many benefits, it can facilitate:

- Parallel execution of multiple application.
- Moving running application to another host.
- Managing resources dynamically.
- Utilization of resources in a cost-effective manner.
- Execution of application onto varied systems.
- Multiplexing of resources like hardware or software.
- Execution of VM in isolation.
- Fault tolerance by avoiding failure of one VM affecting on execution of other VMs or on the entire physical machine.
- Energy saving by clubbing several server loads to few servers and switching off the idle servers leading to reduced energy consumption.

Virtual Machine (VM): It can be termed as a computing resource. Virtual machines are, in fact, operating systems or software that impersonate the computing system with certain set of resources like memory capacity or CPU.

Virtual Machine Allocation: One of the integral aspect in the cloud computing environment is the mapping of virtual machines (VM) to the physical machines (PM). This is the initial phase of mapping. It initiates as soon as the user from application requests for provisioning of virtual machines. This phase is bifold where a former VM configuration phase maps the users' applications to the virtual machines based on the resources required by the application, and in the former phase which is called as VM placement phase or Virtual Machine Allocation virtual machines are actually mapped onto the physical machines based on some optimization objective (Fig. 1).

Server Consolidation: One of the conducive phase in the virtualized cloud data center environment is consolidation of server; the focus of this phase is to integrate numerous virtual machines on the fewer possible physical servers, and this facilitates

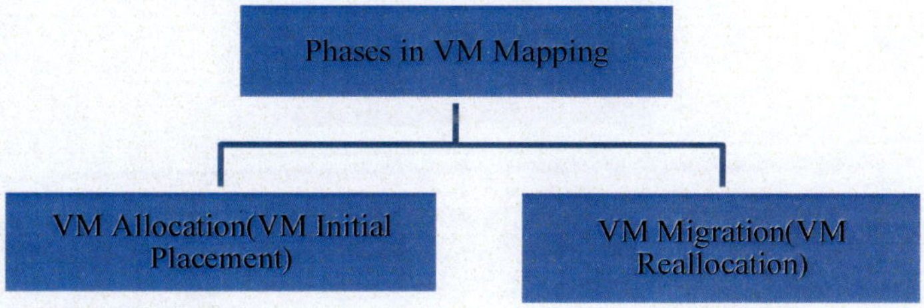

Fig. 1 Phases in VM mapping

in saving power by switching off the idle servers which can provide an overall benefit of reduced energy consumption and thus an increased rate in the availability of the system.

Server consolidation phase in turn gives rise to certain phases like:

- Detection of Overloaded host
- Detection of Underloaded host
- Selection of Virtual Machine
- Migration of Virtual Machine.

3 Virtual Machine Migration

Virtual Machine Migration: This is the former phase of mapping process, and it involves the reallocation of virtual machines to the physical machines.

It is the part of the server consolidation phase where while clubbing virtual machines to the physical machines certain virtual machines need to be migrated from one host to another so as to leverage energy conservation. It involves the reshuffling of the virtual machines to the physical machines. Migration of virtual machines always considers some optimization objective, while reshuffling, the optimizing objective may subject to reduction in energy consumption and increased utilization of resources which certainly revolves around this phase. This phase can be invoked based on some criteria which has triggered it or it may be due to any periodic reallocation.

Optimization Objectives:

See (Fig. 2).

Need of Virtual Machine Migration:

- Optimal Utilization of resources
- Cost efficiency
- Energy efficiency
- Fault tolerance.

Fig. 2 Optimization objectives with virtual machine migration

Criteria for Virtual Machine Migration:

- Migration Time: For the migration of virtual machine, it will be taken into account the minimum time taken by a particular virtual machine to complete migration.
- Arbitrarily chosen: VM to be migrated is selected according to an arbitrarily variable value which is set with value according to the index of allocated hosts.
- Interdependence: If the interdependence of an application in terms of resource usage is more on the server then that VM will be selected for the migration, considering the fact that sever may be getting overloaded.

Invoking of Virtual Machine Migration:

Virtual machine migration is based on some invocation scheme. Invocation can be:

- Based on Event
- Based on Interval
- Based on Threshold
- Hybrid.

Classification of Virtual Machine Migration Algorithm:

1. **Non-Live migration:** This technique refers to migration in which the virtual machine discontinues the application execution while migrating VM memory state from the source server to the target server. Non-live VM migration pattern does not initiate the VM (at target server) until complete VM is transferred.
2. **Live migration:** This technique refers to migration in which the virtual machine keeps operating while it has been transferred to the destination server. Live VM migration un-interruptedly services the running applications during the elapsed VM migration time.

3.1 Live Virtual Machine Migration Algorithms

Live VM migration helps in fulfilling the dynamic demands of resources of running applications. It facilitates the following features:

1. **Load balancing**: Having simultaneous live VM (s) migration helps in having optimum balance in load.
2. **Proactive fault tolerance**: Live migration facilitates handling of faults proactively so as to fault guarantee the availability and reliability of critical services.
3. **Power management**: VM live migration is a good technique for cloud power efficiency as it switches off the idle mode server to either sleep mode or off, which leads to great energy saving because.
4. **Resource sharing**: VM live migration helps in sharing of limited hardware resources which leads to cut down operational cost.
5. **Online system maintenance**: It is one of the requirements to enhance or upgrade or for servicing the physical systems, and thus it requires moving of all VMs to another server without any hindrance.

Live VM Migration Techniques:

- **Post-Copy**: In this approach, VM to be migrated is suspended first at the source machine then it copies its processor state to the destination, and thereafter VM resumes and initiates fetching memory pages from the source.
- **Pre-Copy**: This approach comprises two phases: Warmup and Stop and Copy. In the Warmup phase, the VMM copies all the pages from source to the destination while VM is in operation. And if in meantime some changes are done on the pages, then those pages will be treated as dirty pages. In the former Stop and Copy phase, thereafter dirty pages will be copied to the destination and operations will be resumed by the VM.
- **Hybrid**: This technique employs both the methods pre-copy and post-copy.
- **Multiple**: In this technique, multiple virtual machines are migrated simultaneously.

Challenges of Virtual Machine Migration:

Virtual Machine Migration is one of the key phases in the virtualized data environment. Migration must be coherent so as not to hinder the services for both user applications or cloud service providers. Extensive methods have been used for seamless services but still migration faces a few challenges like:

- Faults in the network
- Migration Downtime
- Page Missing
- Large Application Migration
- Availability of Resources.

4 Comparative Analysis of Existing Techniques

See Table 1.

Table 1 Comparison of VM migration techniques

Approach	Objective	Technique	Remarks
Nathu et al. [1]	VM live migration	To design an architecture considering energy management	Amalgamated various severs on a single server
Tolia et al. [2]	Short-term VM migration	To consolidate workloads	By distributing workloads, underutilized servers were put in the sleep mode
Lim et al.[3]	To compute latency of migration	Used regression method to get performance	Migration latency is dependent on availability of resources
Elmroth et al. (2009)	To predict quantitatively the rate of energy consumption and performance of migration estimation	By using high-level linear model for predicting migration energy used	Accuracy of prediction was around 90%, it also derived a dropped migration cost by a rate of around 72.9%, and also energy saving by 73.6%
Tarighi et al. [4]	Fuzzy algorithm	To discover server with peak load	(TOPSIS) By using this algorithm, maximal loaded server was discovered thus helped in taking decision of migration
Marzolla et al. [5]	Local VM consolidation	To design protocol for consolidation by applying consolidation at local level and migrating VM from smallest filled to greatest filled	The projected protocol not considered any energy saving
Mastroianni et al. [6]	Considered Bernoulli trials for VM placement and consolidation	To uncover by a probabilistic way, a solution for placing and consolidating VM	Considered energy and cost of migration
Goiri et al. [7]	Considered algorithm taking into account numerous objectives	To propose an algorithm based on profit entailed to place a virtual machine	The projected algorithm considered various performance parameters like energy conservation, violation of SLA, and overhead of virtualization

(continued)

Table 1 (continued)

Approach	Objective	Technique	Remarks
Belg et al. (2012)	Energy aware Heuristic	To dynamically allocate VM considering current resource allocation	Main idea was to lower violations of SLA by preserving free resources
Billa et al. (2012)	Partial migration	To partially migrate free VM of user to server	This approach helped in reducing overall energy consumption
Shekhar et al. (2013)	Greedy heuristic	To reduce energy consumption	Designed for energy aware live migration of VM
Vu et al. [8]	Algorithm for enhancing communication	To decrease overall cost of VM traffic and reducing energy	The algorithm proposed has ultimately achieved by increasing CPU utilization
Lu et al. [9]	Multiple VM migration	To optimally schedule the multi-tier virtual machines	This resulted in dropped rate in service downtime by 70% and improved application through by 52%
Babu et al. [10]	Pre-copy migration	To optimize pre-copy migration	By using this technique, instructions were sent to destination independently and execution started resulted in low downtime and low impact on application performance
Desai et al. [11]	Pre-copy migration CBC algorithm	To improve the total migration time and total downtime	By using this technique Modified existing pre-copy algorithm handle both low and high dirty page rate
Deshpande et al. [12]	Dropping traffic related to migration and reducing of contention of network by the VM traffic	Forwarded an approach based on traffic of migration of VMs which are co-located	Resulted in dropped rate of entire degradation and total migration time for an application

5 Conclusion and Future Work

The paper uncovered various algorithms and techniques for migrating the virtual machine in the environment of Cloud Computing. It analyzed the various things associated with migration of virtual machines type, content, and benefits which can be exploited. As of future direction, we are forwarding to have better techniques for VM migration thus obtaining further better diverse services and applications for a versatile cloud environment by incorporating certain security issues which are still a challenging aspect in VM migration in the cloud environment.

References

1. Nathuji R, Schwan K (2007) Virtual power: coordinated power management in virtualized enterprise systems. ACMSIGOPS Oper Syst Rev 41(6):265–278
2. Tolia N, Wang Z, Marwah M, Bash C, Ranganathan P, Zhu X (2008) Delivering energy proportionality with nonenergy-proportional systems—optimizing the ensemble. In: Workshop on power aware computing and systems (HotPower '08). San Diego
3. Lim MY, Rawson F, Bletsch T, Freeh VW (2009) PADD: power aware domain distribution. In: Proceedings of the 29th IEEE international conference on distributed computing systems (ICDCS'09). Montreal, pp 239–247
4. Tarighi M, Motamedi SA, Sharifian S (2010) A new model for virtual machine migration in virtualized cluster server based on fuzzy decision making. J Telecommun 1(1):40–51
5. Marzolla M, Babaoglu O, Panzieri F (2011) Server consolidation in clouds through gossiping. In: IEEE international symposium on the world of wireless, mobile and multimedia networks (WoWMoM). IEEE, pp 1–6
6. Mastroianni C, Meo M, Papuzzo G (2011) Self-economy in cloud data centers: statistical assignment and migration of virtual machines. In: Euro-Par 2011 parallel processing, pp 407–418
7. Goiri I, Berral JL, Fitó O, Julià F, Nou R, Guitart J, Gavalda R, Torres J (2012) Energy-efficient and multifaceted resource management for profit-driven virtualized data centers. Future Gener Comput Syst 28(5):718–731
8. Vu HT, Hwang S (2014) A traffic and power-aware algorithm for virtual machine placement in cloud data center. Int J Grid Distrib Comput 7(1):21
9. Lu H, Xu C, Cheng C, Kompella R, Xu D (2015) vHaul: towards optimal scheduling of live multi-VM migration for multi-tier applications. In: IEEE 8th international conference on cloud computing. IEEE, New York, pp 453–460
10. Babu BS, Savithramma RM (2016) Optimised pre-copy live VM migration approach for evaluating mathematical expression by dependency identification. Int J Cloud Comput 5(4):247
11. Desai MR, Patel HB (2016) Performance measurement of virtual machine migration using pre-copy approach in cloud computing. In: Proceedings of the second international conference on information and communication technology for competitive strategies. ACM, Udaipur, pp 1–4
12. Deshpande U, Kulkarni U, Gopalan K (2012) Inter-rack live migration of multiple virtual machines. In: Proceedings of the 6th international workshop on virtualization technologies in distributed computing date. ACM, Delft, The Netherlands, pp 19–26
13. Kikuchi S, Matsumoto Y (2011) Performance modeling of concurrent live migration operations in cloud computing systems using prism probabilistic model checker. In: IEEE 4th international conference on cloud computing. IEEE, DC, pp 49–56

14. Beloglazov A, Abawajy J, Buyya R (2012) Energy-aware resource allocation heuristics for efficient management of data centers for cloud computing. J Future Gener Comput Syst 28(5):755–768
15. Kurdi HA, Alismail SM, Hassan MM (2018) LACE: A Locust-Inspired scheduling algorithm to reduce energy consumption in cloud datacenters. IEEE Access 12(3):85–92
16. Das J, Dasgupta A, Ghosh SK, Buyya R (2019) A learning technique for VM allocation to resolve geospatial queries. In: Recent findings in intelligent computing techniques. Springer, Singapore, pp 577–584

Chapter 35
Comprehensive Survey of IDS Techniques in Mobile Ad Hoc Network (MANET)

Kitti Chawla, Jasmeen Gill, and Mohanjeet Singh

1 Introduction

Wireless Sensor Network is one of the networks which spans through air into energy. It can broadcast, dispatch, and settle for signals, such as microwaves radio or satellite signal. It scatters the signals into all the direction and, in fact, also comprises the nodes which are usually of low performance. In this network, the task is cumulatively performed by the nodes. One of the eminent part is the sensor node which monitors the varied environmental conditions such as pressure, temperature, and air pollution level, and organizing the collected data at a central location through the network. The advancement in the numerous electromechanical systems has also pushed the verge of connection with the detector nodes. Thus wireless sensor network has spanned to maturation and made WSNs into sensible feasibleness [1] (Fig. 1).

Advantages in WSN

- Wireless sensor network can be extended, i.e., any new node can be added any time.
- Wireless sensor network can operate smoothly even if any of the node fails as such possible by distributing tasks.
- Implementation cost of wireless sensor network is economical.
- In the wireless sensor network accessibility of any node can be done through centralized node.
- Wireless sensor network can be operated in any harsh hostile environment.

K. Chawla (✉) · J. Gill
RIMT University, Mandi Gobindgarh, India

M. Singh
P.M.N. College, Rajpura, India

N. Marriwala et al. (eds.), *Soft Computing for Intelligent Systems*, Algorithms for Intelligent Systems, https://doi.org/10.1007/978-981-16-1048-6_35

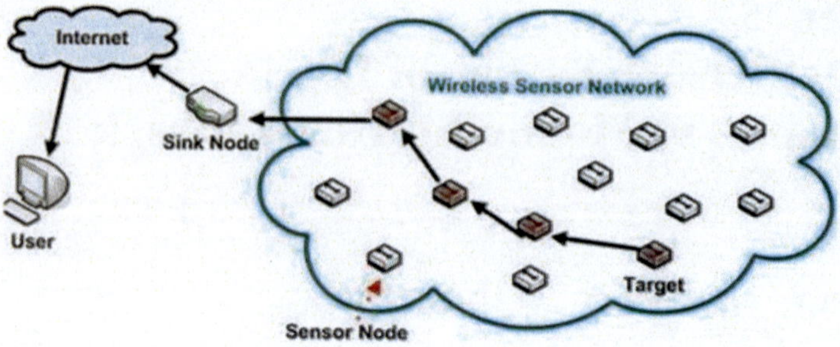

Fig. 1 Wireless sensor network architecture [2]

- Wireless sensor network can be applicable in various fields such as forest fire detection and monitoring of water.

2 Categories of WSN

2.1 Infrastructure Network

This network prerequisites physical device like base station/access point for providing the communication between wireless nodes. One node cannot communicate with another node directly in the network. Base station behaves as an intermediate between source and destination. If any node takes out the scope of one base station, it takes place under the scope of another base station. Cellular network is an example of infrastructure-based network. It is a consolidate system. The disadvantage of the system is that if central system break down, then the entire system will crash [1].

2.2 Infrastructure-Less Network

Another prominent alternative of wireless network is a network without basic infrastructure like base station, access point were servers etc. and is terms as infrastructure less. This network shares most important traits like dynamic topology, mobility in nodes self-configuration, etc. This additionally contains maintenance for the discovery of routes at the side of at the associated price. This is considered as the network which is infrastructure-based wireless network, providing the feature of routing which is the trustworthy node that the square measure referred to as access point. Configuration of access point square measure metrics gives dropped dynamicity in comparison with the other one having nodes of moving nature. The metric of square measure is very similar to the square metric of base station which considers various metrics like authentication of nodes, their association disassociation, and also the regulation of flow of traffic for the concerned users and their access points, and further it also considers the web; therefore, facilitating Internet access to its users.

3 MANET

It is one of the communication systems which lets the communication to be done remotely. It is a form of a network which is self-arranging in nature. Routers tend to move and join via remote connection with no entrance criteria. Each device can move liberally in all the varied direction, which in turn gave rise to another issue involved which is maintenance of knowledge of every device for critical issue of proper route (Fig. 2).

3.1 Types of MANET

i. **Vehicular Ad hoc Networks (VANETs)**: It is a mechanism that utilizes mobile cars like a node to develop a dynamic network for communication between vehicles with in outspread range. It is one of the best solution for preventing road accidents. It provides the traffic information to the vehicle. So that they can analyze the traffic environment and enhance the driving safety. It is a traffic management application.
ii. **Intelligent Vehicular Ad hoc Networks (In VANETs)**: It assists the vehicle to act in a smart manner throughout V2V crashes, collision, etc.

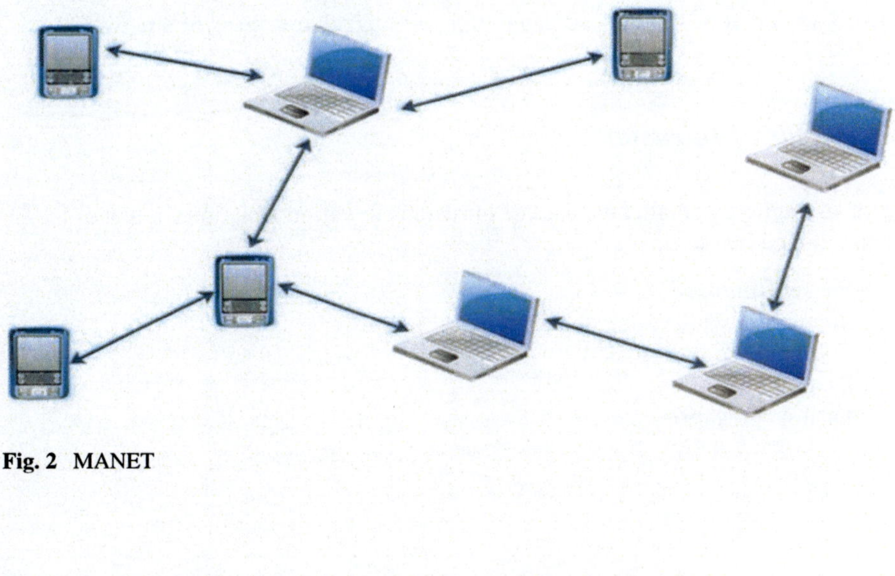

Fig. 2 MANET

Fig. 3 Types of MANET

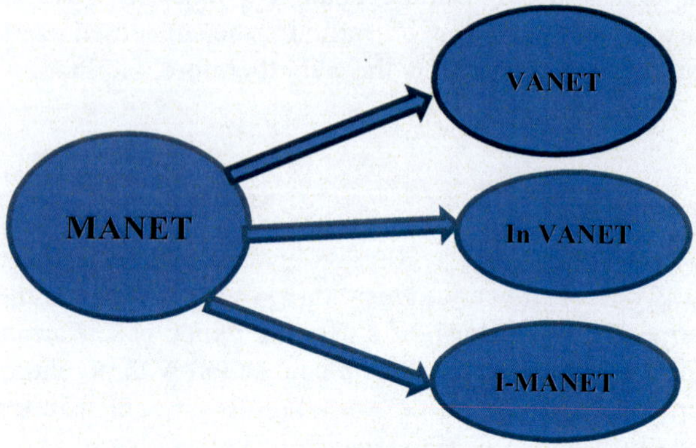

iii. **Internet-Based Mobile Ad hoc Networks (I-MANET)**: This network assists the nodes of dynamic nature and the nodes which are of stable Internet gateway nodes (Fig. 3).

3.2 Intrusion in Network

Security is one of the major concern in the ever-increasing communication system. Network can be peeked due to intrusion, and this can prove to be threatening in communication and acts as hazardous in (N2N) communication. It can be looked from a diverse perspective and each way considering its own criteria to be called as successful in intruding or attacking the network, and it can range from simple guessing of passwords to information impersonating. It can vary with its nature such as it can be an active or a passive attack and can hamper the proper functioning of network to denial of service and hampering various other aspects of a network.

3.3 Areas of Intrusion

There are numerous malicious uses of intrusions in varied situations, a few of them are mentioned below:

i. Packet Routing
ii. Tampering with Voting and Reputation Systems
iii. Fair Resource Allocation
iv. Distributed Storage
v. Data Aggregation.

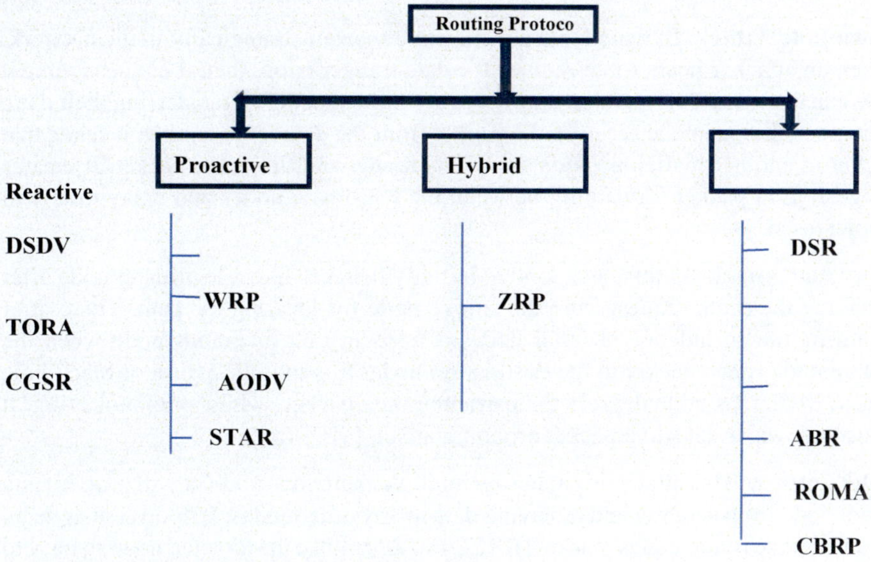

Fig. 4 Routing Protocols

3.4 Routing Protocol in MANET

Routing protocols are various terms of rules or regulations which may define a network issue. There are different routing protocols and their application depends according to the network. The next section hierarchically classifies various routing protocols [3] (Fig. 4).

3.5 Types of Attacks in Manet

Black Hole Attack: This is one of the most prevalent attack which uniformly attacks the network. Within the early stages, the root node does not have a coherent path to the destination. Once supply node sends (RREQ) route request for locating the path, then malicious node incorrectly announces the shortest route to the destination and supply node considers taking into account that route as the shortest path and places the packet for transfer to destination. After receiving the packet, the malicious node drops the packet. It works in slow pace and discards the informative packets and seizes the network.

Wormhole Attack: In which two attackers are located strategically in the network. When an attacker hears route request for data transmission, then it creates a tunnel between them and declares that they have the shortest path for transferring their data. When a malicious node receives the packet from the source node, then it sends that packet to another malicious node using this tunnel and drops the packet. It creates congestion in packet forwarding between the legitimate nodes and delay time will be increased.

Gray hole attack: In this, attack node initially behaves like a legitimate node after receiving the route request from the source node for locating the path. There after acquiring the confidence of other nodes it turns in to a malicious node when the source node sends packet to the destination node. It is modified form of black hole attack. Malicious node depress the particular data packets while communicating. It is also known as selective packet dropping attack [4].

DDOS attack: This attack originates by multiple malicious node. It will prohibit any target node from sending and receiving data to any other nodes. It floods the node by throwing fake route request packet (RREQ) so as to make the targeted node so busy to handle that request as a result that node cannot forward packets to its destination. The number of request exceeds the handling capacity of the node and that node falls in a deadlock condition and thereafter increasing the delay time. The aim of this attack to drop the packet and decrease the network bandwidth by blocking the legitimate user to access network resources. (Security challenges attack in manet) [5].

Replay attack: Supply node dispatches the packet to the target node.The associated degree mischievous node steals the packet, modifies the contents of packet and resends that packet to the destination node. Receiver can receive the packet from supply node and also the malicious node but he does not know which is the correct packet sent by actual sender because same packet is received twice. This kind of attack can make the information of packet redundant in nature and thus in turn overloading the network and making less efficiency of network.

Ad hoc networks like MANET and Mobile Wireless Sensor Network (M-WSN) face critical issues in security. It has been a major concern in both the fields but considering the complex nature of node architectures and their routing behavior, the task is not easy at all. As the nodes are mobile in nature which gives rise to another concern of tracing the intruder introspecting the network. So we require a robust mechanism for detecting the intruder.

4 Intrusion Detection System (IDS)

Due to the dynamic nature of MANET, a malicious attacker can access the network and hack the confidential data. So we require an efficient tool which can identify as well as prevent the common security attacks and save a network from various security loopholes which is called IDS. IDS analyzes the network traffic or recognizes the systems behavior and alerts the administrator after detecting the malicious node.

5 Sources of IDS

NIDS (Network based IDS): This IDS runs on the network as a full and additionally also analyzes congestion between nodes on the network. NIDS is capable of observing the external malicious activities that would be initiated from the associate in nursing external threat at an earlier section before the threat unfolds to a different system [6].

HIDS (Host-based IDS): This IDS runs on individual hosts/device on the network. It generates log files and stores on that host and analyzes the incoming and outgoing packet from the device and compare them with internal database. It always decides which user/process are involved in venomous activity. If any suspicious activity is found then it alerts the administrator. This IDS is used for maintaining the security of system. An HIDS is needed to be additional on each machine and specific configuration to the software system and software package framework. For distinguishing, the result of violation system programming is employed.

6 Categories of IDS

Signature-based intrusion detection systems (SIDS): This technique works on some rules which are predefined from the point of various different security attacks. The rules used are predefined and whenever a system deviates from these rules, the situation will be termed as an attack. In this scenario, whenever there is a deviation from the predefined rules, it is an intrusion. Therefore, one of the drawbacks, this method suffers from is that it cannot identify any attack which is not predefined in the database. The database needs to be continuously updated for the attack signatures. SIDS is unable to find zero day attack because no similar signature exists in the database until it takes out new intrusion signature and stored in database [6] (Fig. 5).

Anomaly-based intrusion detection system (AIDS): This technique is used for considering the system specifications at particular snapshot with its typical nature, and if any disparity is found from normal nature and if it exceeds pretended threshold, then it sparks the alarm. This task is classified into twofold: former is training and then detection. The system must be trained for typical behavior before it can be used in a detection model. And in the latter mode, i.e., detection of the system, differentiates from typical or normal behavior with abnormal behavior on the basis

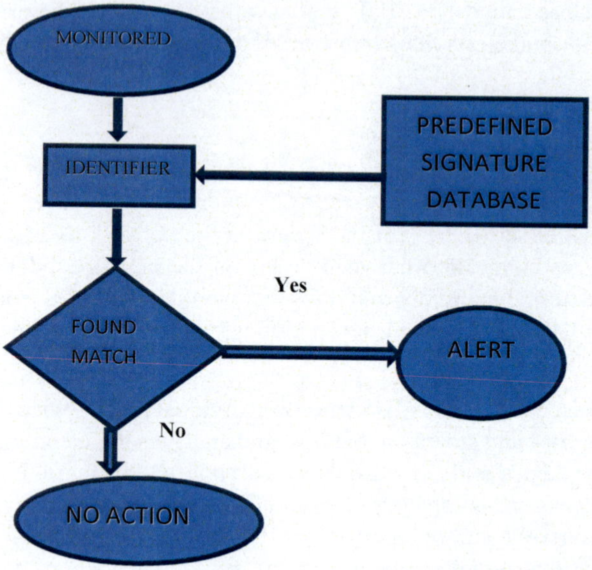

Fig. 5 Signature-based IDS

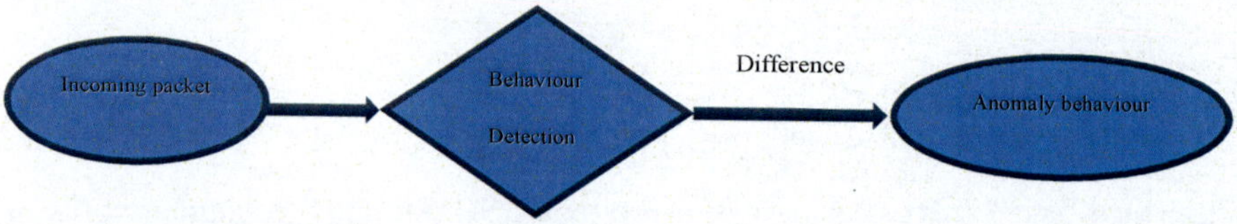

Fig. 6 Anomaly-based IDS

of heuristic technique or rules. This technique is capable of discovering unpredicted threats (attacks) (Fig. 6).

Hybrid-Based Intrusion Detection System: This technique is the combination of signature-based IDS as well as anomaly-based IDS. This technique takes over the basic property of both IDS systems. It is divided into two detection modules. The first module is used for conforming the known attacks by using signature-based IDS and the second module to analyze the overall network behavior which is different from normal behavior. This IDS is suitable for credible wireless sensor network. This technique will be able to identify more attacks, and the ratio of false detection attack is approximately less. (2016 shanti s,e. grajan attacks).

7 Literature Review

Mohammed [7]: This paper proposed an IDS based on watchdog and path rater methodology. The former technique is used to expose the malignant node on the basic of node behavior and the other technique basically aims to select the reliable path. The author used Netsim dataset and presented the performance using DSR and AODV routing protocol with or without considering the consequence of intrusion. AODV protocols are found to be more productive than DSR. The author has justified the proposed technique with some performance metrics parameters such as PDR ratio, throughput, and delay for analyzing the output.

Singh et al. [8]: This paper presented planning the construction trust primarily based intelligence intrusion detection system make use of cryptography scheme for diagnosing the three nature of attackers such as flooding attack, black hole attack selective packet dropping attack. Author employed N2 simulator for considering some parameters such as packet delivery ratio, throughput, packet loss, packet overhead and delay for analyzing the performance of techniques. The focus of the proposed technique of this multitier trust method is to find a secure path rather than the shortest path.

Vimala et al. [9]: This paper presents a novel IDS which categorizes the attacks which is done by neural network based (NNIDS), trail based support vector machine (TSVM IDS) and determinant fuzzy based approach (DF-IDS). This prospective method uses KDD Cup 1999 dataset for testing and training. These proposed techniques have been compared with Decision Tree, Naïve Bayes method used in the previous work for analyzing the performance of prospective algorithm using metrics such as accuracy and error rate and acquire better performance than the previous methods. NNIDS, TSVM, DF_IDS techniques are compared with each other for finding the optimal technique for detecting the intrusion in MANET using a graph. The performance of TSVM IDS is better than the other techniques.

Training time			Testing time		
NNIDS	TSVMIDS	DF_IDS	NNIDS	TSVMIDS	DF_IDS
280 s	264 s	320 s	58 s	42 s	61 s
Accuracy			Detection rate		
NNIDS	TSVMIDS	DF_IDS	NNIDS	TSVMIDS	DF_IDS
Below 80%	Above 80%	Below 80%	Below 20%	Above 30%	Above 30%

Chaudhary [10]: This paper presents comparison between Mamdani type and Sugeno-type fuzzy inference based IDS for finding the packet dropping attacks by considering malicious node on the basis of some ratings like true positive rate and false positive rate. The author has considered the QualNet simulator for evaluating the proposed technique and MATLAB for the purpose of evaluation, and it has been found that both fuzzy inferences are performing in a similar way. But Sugeno-type model performs better over Mamdani type model in the terms of defuzzification.

Bala et al. [11]: This paper presents network information theory (NIT) based model, and the proposed system will use the time-variant snapshot for detecting the routing attacks in MANET. Every node in the network gets information from NIT about the other nodes like location, displacement speed, and energy details. The author implements the same system with the NS2 simulator considering network parameters like: throughput, end-to-end delay, packet delivery ratio. And finally in this paper, the author presented the productive output for detecting the intrusion and enhancing the performance level.

Rodney et al. [12]: This paper presents effective security techniques based on machine learning that will identify the intruder node on the bases of their behavior either normal or abnormal. The author used support vector machine (SVM) and logistic regression (LR) methods and Iris dataset and SPSS tools for detecting the malicious node using PDER and PMMR parameters to analyze the performance of the models. Pycharm simulator with Python is used in this model. The target of this paper is to identify the perfect representation for detecting the packet behavior with higher accuracy. After comparing both techniques using matrices, the obtained result presented that the performance of LR method is more effective than SVM technique in terms of higher detection rate of malicious node in MANET.

Mohamed et al. [13]: This paper presented an IDS against gray hole and black hole attacks using Adaptive Neuro-Fuzzy Inference System (ANFIS) to automate the procedure of generating a fuzzy system and reform the system using genetic algorithm (GA). Packet Delivery Ratio (PDR) and Routing Overhead (ROH) parameters are used for analyzing the performance of suggested model in the five situations. It proves that improved network with an average of 36% in presence of black hole attack only, and with an average of 37.8% with presence of both attacks in case of low speed mobility by using the presented IDS in the PDR with an increase of 2.5% in the ROH. But network improved in case of high speed mobility with an average of 31% in the presence of black hole attack and with an average of 27% with the presence of both attacks in case of high mobility by using presented IDS in the PDR with an increase of 1.8% in ROH.

Zulfiqar et al. [5]: This paper presents DDBG technique to select IDS using Connected Dominating Set (CDS)method. This technique is used for detecting the black hole and gray hole attck. The author used NS2 simulator for analyzing the performance of the proposed technique and compare this technique with native AODV, LID-AODV, and HPSO-GA considering some parameters like detection rate, PDR ratio, throughput, routing overhead, average delay. The outcome of simulation exposes that our proposed technique gives 98% detecting rate, 97% PDR, 77% throughput, routing overhead 5787 bytes, and 0.105 less delay while detecting the malicious node.

Banerjee et al. [14]: This paper presents MANET and its susceptivity. The author described types of IDS and different IDS techniques which supported to detecting the intruder and protect the MANET. The author considers a security as an enormous challenge. At purpose IDS came into this scope so as to provide security in MANET to uncover there obtaining weak point.

Shah et al. [15]: This paper proposed an artifical neural network using AODV protocol to establish the density of hello message to enhance the performance of the MANET. The author outlines the design and pattern of an ANN using Verilog Hardware Descriptive Language (VHDL) application of hardware of ANN using FPGA. The author obtained better results of FPGA using neural controller after comparing with MATLAB result of ANN.

Otoum et al. [16]: This paper presents restricted Boltzmann machine based mostly clustered IDS (RBC-IDS). A capable deep learning based method where mostly IDS technique is employed for analyzing the essential infrastructure of the device network. The target of the paper is to seek out the potential of deep learning as a substitute to vigorous ML-based IDS. The author compares novel idea of deep learning based (RBC-IDS) with machine based (ASCH-IDS) considering parameters such as high accuracy rate, less negative rate and high detection rate, ROC curve, F_1 score curve using NS3 simulator. The author compares both techniques and finds that there is higher detection rate and accuracy rate of RBC-IDS than ASCHIDS method.

Sukriti et al. [17]: This paper proposed a whale optimization algorithm for finding the intruder in MANET. It is applicable on support vector machine (SVM) and used NSL KDD dataset. The author compares the results of whale optimization algorithm and genetic algorithm considering the three parameters like accuracy, sensitivity, and specificity. The author discussed four attacks like DOS, U2R, R2L, and probing attack. The comparison results show that WOA provide better accuracy than genetic algorithm. The objective of this paper is a comprehensive study on the WOA for feature reduction for enhancing better performance of IDS. The author proposed a comprehensive study on the application WOA (Whale Optimization Algorithm) optimization algorithm for feature reduction for making a better intrusion detection system (IDS).

Comparative Analysis of the Existing techniques.

Year and Author	Objective	Technique	Drawbacks
Hortelano [18]	The objective of this paper is to enhance the accuracy so as to detect intruder nodes which are behaving maliciously in the P2P networks	The author has used watchdog sensor technique and a Bayesian filtering and has employed NS-2 simulator	This technique employs more number of experiments. So as to apply Bayesian and watchdog technique

(continued)

(continued)

Year and Author	Objective	Technique	Drawbacks
Chang [19]	With the help of DSR protocol, the author has proposed to find the malicious node which can uncover attacks like black/gray hole attacks and as well as cooperative black hole attacks. And the simulation is being done in the QualNET	The author has used a novel technique cooperative bait detection system (CBDS)	The proposed method was not simple and not precise
Nadimi [20]	The aim of this paper was to monitor behavioral parameters found in the WSN and to increase reliability while communication is within a patch network. And finally, classifying the behavior of each into different groups	The author has proposed ANN system and implementation	The author has proposed this technique to implement on the active animals, which is a serious hazard
Sherif [21]	The focus of this paper is to uncover black hole attack and the various methods to detect and prevent them in AODV type MANET	The author has used various algorithms, FuzzyLogic, Genetic, Clustering Algorithms as well as the author has also proposed Mobile Agents for the purpose of detection of these attacks	Major drawback that this method suffers is from power consumption issues
Mohanapriya [22]	The author focused on detecting and preventing selective black hole attack in MANET	The author has employed Light-weight intrusion detection system (IDS)	The major drawback is that it suffers from low packet delivery ratio
Dhaka [4]	Author considers the attacks using NS-2 simulator and finds out secure and short route to the destination	Control sequence packets	Unable to identify gray hole attack

(continued)

(continued)

Year and Author	Objective	Technique	Drawbacks
Arathy [23]	The author proposed an optimized technique for identifying dual things like attacks such as black hole and single taking parameters such as reduced overhead of routing and computational	The author employs Destination Sequence Number (ADSN)	It suffers from high average sequence number which are not being detected by the proposed system
Tamilselvi [15]	Author proposed the system for improving different parameters like PDR in the presence of various attacks like black hole and employs the NS-2 simulator	Bait request	The system suffers from the delay if the route is long
Venu [24]	To identify the black and gray hole attacks	The author employs the frame checking sequence method	This procedure is complex
Gopinath [25]	Aim of the paper is to provide reliable and secure routing scheme for the purpose of secure transmission of data	secure location aware routing protocol (SCLARP) is developed	It suffers from overhead of extra hardware like antenna, GPS, etc

8 Conclusion

MANET follows dynamic topology, and due to the mobility in nature this network is suffering from numerous attacks and it is hard to identify an intruder. So security is the substantial provocation. To secure the MANET, an effective IDS is required. In this paper, a detailed survey of IDS methodologies have been studied. The techniques used with pros and cons have been uncovered. However, such techniques suffer from higher complexity level, more time consumption in low PDR, and less throughput. So in the future, the objective will be to develop an optimized IDS considering improving some critical parameters.

References

1. Ghorale MMG, Bang AO (2015) Wireless ad-hoc networks: types, applications, security goals. In: National conference convergence, vol 2015, pp. 28.
2. Hadi TH (2017) MANET and WSN: what makes them different. Int J Comput Netw Wirel Commun (IJCNWC) 7(6):23–26
3. Hinds A, Ngulube M, Zhu S, Al-Aqrabi H (2013) A review of routing protocols for mobile ad-hoc networks (manet). Int J Inf Educ Technol 3(1):1
4. Dhaka A, Nandal A, Dhaka RS (2015) Gray and black hole attack identification using control packets in MANETs. Procedia Comput Sci 54:83–91
5. Ali Zardari Z, He J, Zhu N, Mohammadani KH, Pathan MS, Hussain MI, Memon MQ (2019) A dual attack detection technique to identify black and gray hole attacks using an intrusion detection system and a connected dominating set in MANETs. Future Internet 11(3):61
6. Khraisat A, Gondal I, Vamplew P, Kamruzzaman J (2019) Survey of intrusion detection systems: techniques, datasets and challenges. Cybersecurity 2(1):20
7. Saifuddin KM, Ali AJB, Ahmed AS, Alam SS, Ahmad AS (2018) Watchdog and Pathrater based intrusion detection system for MANET. In: 2018 4th international conference on electrical engineering and information and communication technology (iCEEiCT). IEEE, pp 168–173
8. Singh O, Singh J, Singh R (2018) Multi-level trust based intelligence intrusion detection system to detect the malicious nodes using elliptic curve cryptography in MANET. Clust Comput 21(1):51–63
9. Vimala S, Khanaa V, Nalini C (2019) A study on supervised machine learning algorithm to improvise intrusion detection systems for mobile ad hoc networks. Clust Comput 22(2):4065–4074
10. Chaudhary A (2019) Mamdani and sugeno fuzzy inference systems' comparison for detection of packet dropping attack in mobile ad hoc networks. In: Emerging technologies in data mining and information security, vol 814. Springer, Singapore, pp. 805–811
11. Bala K, Jothi S, Chandrasekar A (2019) An enhanced intrusion detection system for mobile ad-hoc network based on traffic analysis. Clust Comput 22(6):15205–15212
12. Sebopelo R, Isong B, Gasela N (2019) Identification of compromised nodes in MANETs using machine learning technique. Int J Comput Netw Inf Secur 11(1):1
13. Abdel-Azim M, Salah HED, Eissa ME (2018) IDS against black-hole attack for MANET. IJ Netw Secur 20(3):585–592
14. Banerjee S, Nandi R, Dey R, Saha HN (2015) A review on different intrusion detection systems for MANET and its vulnerabilities. In: 2015 International conference and workshop on computing and communication (IEMCON). IEEE, pp 1–7
15. Shah SK, Vishwakarma DD (2012) FPGA implementation of ANN for reactive routing protocols in MANET. In: 2012 IEEE International conference on communication, networks and satellite (ComNetSat). IEEE, pp 11–14
16. Otoum S, Kantarci B, Mouftah HT (2019) On the feasibility of deep learning in sensor network intrusion detection. IEEE Netw Lett 1(2):68–71
17. Sukriti, Batla N (2018) An intrusion detection system in MANET using whale optimization algorithm (WOA). IJRECE 6(3)
18. Hortelano J, Calafate CT, Cano JC, De Leoni M, Manzoni P, Mecella M (2010) Black-hole attacks in p2p mobile networks discovered through Bayesian filters. In: OTM confederated international conferences on the move to meaningful internet systems, vol 6428. Springer, Berlin, Heidelberg, pp 543–552
19. Chang JM, Tsou PC, Chao HC, Chen JL (2011) CBDS: a cooperative bait detection scheme to prevent malicious node for MANET based on hybrid defense architecture. In: 2011 2nd international conference on wireless communication, vehicular technology, information theory and aerospace and electronic systems technology (Wireless VITAE). IEEE, pp 1–5
20. Nadimi ES, Jørgensen RN, Blanes-Vidal V, Christensen S (2012) Monitoring and classifying animal behavior using ZigBee-based mobile ad hoc wireless sensor networks and artificial neural networks. Comput Electron Agric 82:44–54

21. Sherif A, Elsabrouty M, Shoukry A (2013) A novel taxonomy of black-hole attack detection techniques in mobile Ad-hoc network (MANET). In: 2013 IEEE 16th international conference on computational science and engineering. IEEE, pp 346–352
22. Mohanapriya M, Krishnamurthi I (2014) Modified DSR protocol for detection and removal of selective black hole attack in MANET. Comput Electr Eng 40(2):530–538
23. Arathy KS, Sminesh CN (2016) A novel approach for detection of single and collaborative black hole attacks in MANET. Procedia Technol 25:264–271
24. Sandhya Venu V, Avula D (2018) Invincible AODV to detect black hole and gray hole attacks in mobile ad hoc networks. Int J Commun Syst 31(6):e3518
25. Gopinath S, Kumar KV, Sankar TJ (2019) Secure location aware routing protocol with authentication for data integrity. Clust Comput 22(6):13609–13618
26. Pandey MA (2015) Introduction to Mobile Ad Hoc Network. Int J Sci Res Publ 5(5):1–6
27. Islabudeen M, Devi MK (2020) A smart approach for intrusion detection and prevention system in mobile ad hoc networks against security attacks. Wireless Pers Commun 112:1–32
28. Tamilselvi P, Babu CG (2019) An efficient approach to circumvent black hole nodes in manets. Clust Comput 22(5):11401–11409
29. Fatima SK, Fatima SG, Sattar SA, Srinivasa Rao DD (2019) An analysis on cooperative attacks in manets. Int J Adv Res Eng Technol 10(2):245–253

Chapter 36
Research Works in Alkaloid Enhancement in Plants—A Brief Review

R. Jeyapackiaseeli and T. Deva Kumar

1 Introduction

2 Plant Metabolism

Everz and Mazene (2019) Metabolism is essential for cells for their sustainability. Metabolic reactions involve many chemical reactions which are useful for cell functions. Three main functions of metabolites are first, provide energy for biological functions; second, generation of nucleic acid, lipids that are building blocks of cells; third, functionality which is the destruction of unwanted content of a cell. Generally, metabolism is classified as primary and secondary metabolism (Fig. 1).

Lassen et al. [1] Primary metabolism include photosynthetic process and the pathway is required to produce chloroplast in a sufficient level. Photosynthesis process uses for the production of chemical energies such as ATP and NADPH which are used in primary metabolism.

Secondary metabolism consists of metabolic pathways and product of metabolism which are not required for survival of plants. It is mainly responsible for the production of bioactive compounds used medicinally. Secondary metabolites [2] are formed in the biosynthetic pathway (chemical reactions are carried out in the form of ATP and NADPH, and others) by cell and it is acquired from the primary metabolism.

R. Jeyapackiaseeli (✉) · T. Deva Kumar
Department of ECE, National Engineering College, Kovilpatti, India
e-mail: jeyaseeli16121998@gmail.com

T. Deva Kumar
e-mail: tdkece@nec.edu.in

N. Marriwala et al. (eds.), *Soft Computing for Intelligent Systems*, Algorithms for Intelligent Systems, https://doi.org/10.1007/978-981-16-1048-6_36

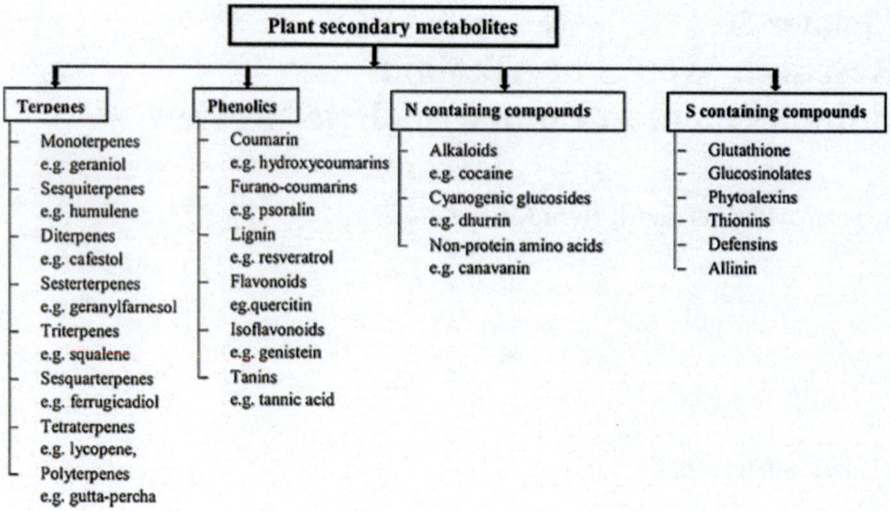

Fig.1 Types of plant secondary metabolites

Bioprocessing gives knowledge about phytochemicals production. Phytochemicals are the main constituents which contain many chemical organic compounds and it is ecology friendly.

Rosaza and Michel (1992) clearly show from Fig. 2 the VLB121 enormous structures of potential metabolic transformations possible with this alkaloid. The reactions

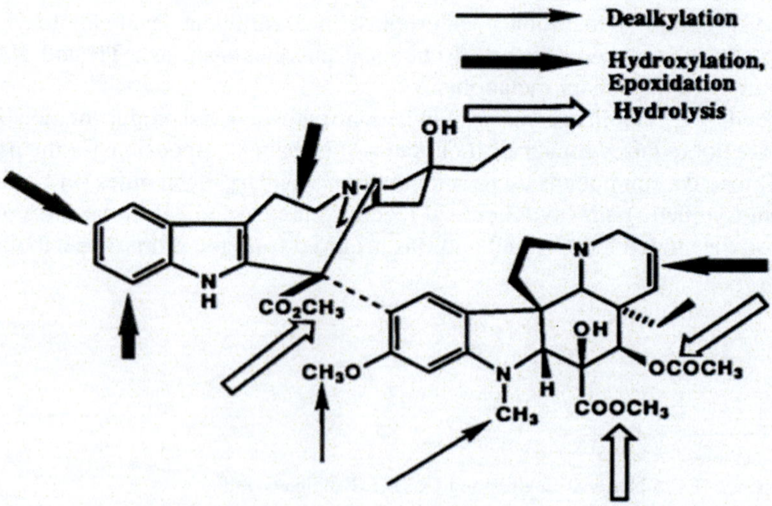

Fig. 2 Metabolic transformation of nitrogen atom in vinblastine

include hydroxylations (aromatic or aliphatic positions), ester hydrolyses, *N*- or 0-dealkylation reactions, and either hydration or epoxidation of the isolated double bond. All of these bio-transformation reactions are predictable based upon extensive knowledge of the metabolic disposition of other classes of drugs.

According to Seger and Sturm [2], some specific endogenetic metabolites are found and quantified by profiling of metabolics which is called metabolic profiling. Mass spectrometry is the platform for analyzing formation and selection of ions which are analyzed (that is analytes) in metabolic profiler.

3 Analytical Platform for Metabolites

Seger and Sturm [2] reviews the different analytical platforms for metabolites' structural and ionic variations.

3.1 Mass Spectrometry

Mass spectrometry is one of the analytical platform of molecules and it is used to calculate the ratio of molecular mass to ion ratios.

3.2 Mass Analyzers

Mass analyzer depends on mass of ion and these ions are separated according to their velocities.

3.3 NMR Spectroscopy

NMR is an analytical platform technique for molecules, and it is used to study the structural characterization of organic compound including molecular signatures and 3D structures of macro-molecules. In this method, the pH value of the sample and its equilibrium temperatures are deeply monitored for further measurements. In this approach, the effect of biological molecules using magnetic spectrum for an one-dimensional state of liquid is used. If the biological system is complicated, then the analytical platform is also modified to produce best results. For this, to analyze the complex biological system, the better way is using combination of both NMR and MS along with chromatography assay metabolic platforms.

4 Single Cell Metabolomics

According to Tom et al. [3], to analyze the secondary metabolites, concentration is needed in the molecules within the cell. Because the cell contains the molecules like DNA, RNA, proteins nucleic acids, and so on. So that the proteomics method is used for identifying the particular protein within the time point. This process is called single-cell proteomics Shabnam Akhtar et al. (2017). Several cell metabolomics techniques are analyzed as follows.

4.1 DI–MS Analysis

Terpenoid indole alkaloids are formed in the biosynthetic pathway. And this terpenoid indole alkaloid is the base of other alkaloids like Catharanthus. For this purpose, locating these alkaloids from stem of Catharanthus roseus is done by Direct Infusion Mass Spectrometry (DI–MS). This analysis is based on the sample mixture that is ionized by the mass spectrometer, and it utilizes the electrospray ionization (ESI). The chemical structure of vinorelbine slightly differs from VLB, which gives knowledge about the VLB as the base from which many alkaloids are derived synthetically by changing its chemical structure in CT moiety and their metabolite undergoes oxidation and cyclization (Fig. 3).

4.2 SI-MS Imaging Technique

The quantity of protein is analyzed by single-cell proteomics method and the chemical structure of molecule is analyzed by DI-MS. Similarly, metabolite measurement is also needed for that secondary ion mass spectrometry (SIMS) that is used to measure metabolites at a single-cell and sub-cellular resolution, but the drawback of this method is that it does not give structural information on Analytes [3].

4.3 MALDI

Ionization of molecules provides the alteration in their chemical structure which yields new alkaloid generation due to its changes in structures. For that, ionization of bio molecule is carried out by Matrix-assisted laser desorption/ionization (MALDI) method. This method uses low consumption of sample and its fragmentation offers high sensitivity. The limitation of this method is measuring the low molecular weight which is difficult because this method uses low sample (high weight molecules) consumption.

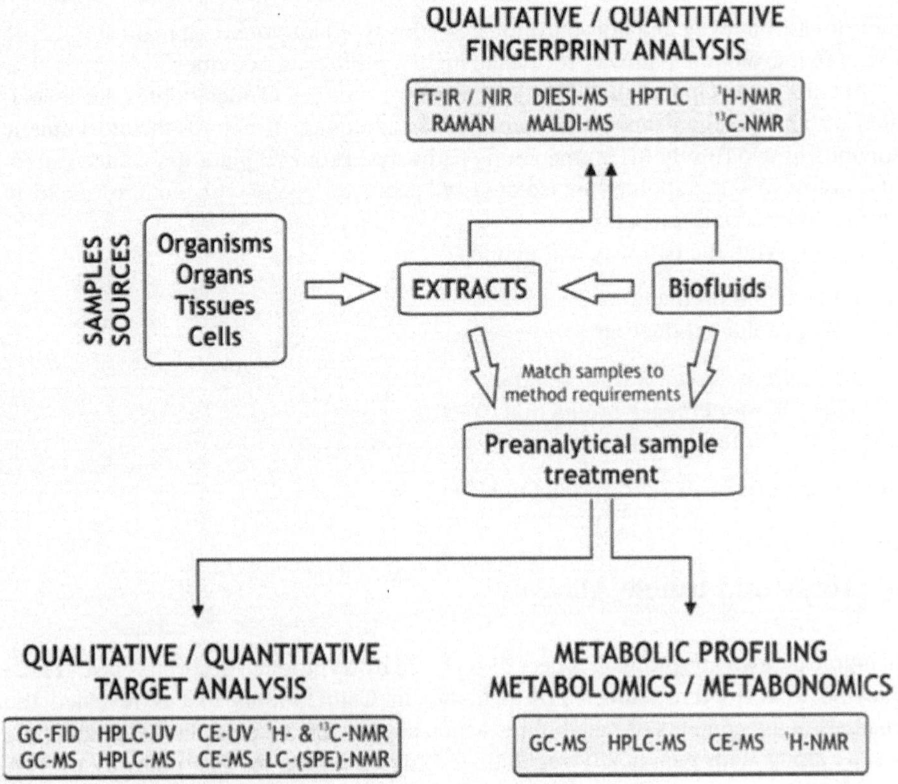

Fig. 3 Flowchart of analytical technologies in metabolome analysis

4.4 Super Critical Fluid Chromatography

Goetz et al. (2014) suggest that the super critical fluid chromatography was used to detect the intra-molecular hydrogen bonding with increasing membrane permeability and it confirms NMR chemical shift.

Mass cytometry (CyTOF) has screening record throughput for known protein among single cells. High resolution Gas (chromatography)-(High and low resolution) mass spectrometry (HG-MS) are used to calculate retention indices of compounds from their structural formulas for aliphatic alcohol and saturated hydro carbons.

5 Metabolic Path Way

Chemical reactions (metabolic reactions) are carried out in biosynthetic pathway and that reaction maybe broke or built by the moleculars enormously. This reactions are very important in cell process and its proper function. The pathway is classified as mevalonate pathway and non-mevalonate pathway. Eukaryotes and plant cytosol are found in mevalonate pathway for balancing the biological activities.

According to Gunatilaka (2006), in plants, a number of metabolites are associated with microorganisms which incorporate within their structures the biosynthetic subunits of two (rarely more) metabolic pathways. From 70 plant-associated microbial strains, 230 metabolites are isolated and characterized in which micro-organism has bioactive small molecules.

The biosynthetic pathway compounds,

i. Flavonoid production
ii. Terpenoids production
    a. Monoterpenoids production
    b. Sesquiterpenes production
    c. Carotenoids production.
iii. Plant-origin alkaloids production.

6 Terpenoid Indole Alkaloid

In indole pathway tryptophan is decarboxylysed by tryptophan decarboxylase (TDC) enzyme to form tryptamine. TIA pathway in Catharanthus roseus revealed the changes in the contents of metabolites which are in roots and leaves of Catharanthus roseus along with genetically regulating compartmentation. The pathway of TIA contains more steps and needs further illustrations. Gene bank database provides information about the sequence of genes involved from GPP to secologinin pathway of iridoid.

Yuko [4]: Effects of some naturally occurring indole alkaloid analogues such as dihydrocatharanthine and catharanthine were found to be inhibitors of TRPM8 activity, and their IC50 values were equivalent to that of BCTC, a potent TRPM8 antagonist.

Kingston [5]: In vinca alkaloids, Tubulin polymerization inhibitors are combretastatins, NPI-2358, the halichondrin B analogue eribulin, dolastatin 10, noscapine, hemiasterlin, and rhizoxin while the taxanes and the epothilones are the major classes of tubulin polymerization promoters along with treatments of discodermolide, eleutherobin, and laulimalide.

Tohuru et al. (2007) explains synthetic route of vinblastine alkaloid generation for a cell growth inhibitory activity of alkanyl moiety transformation, which is amenable to the preparation of a range of analogues at C-4¢.

For the production or creation of pathway as synthetically many alkaloid products are generated due to changes in their chemical structure of indole alkaloids. The changes in molecular chemistry that is breaking and combining the chemical bonds of atoms produce variety of alkaloid products (Fig. 4).

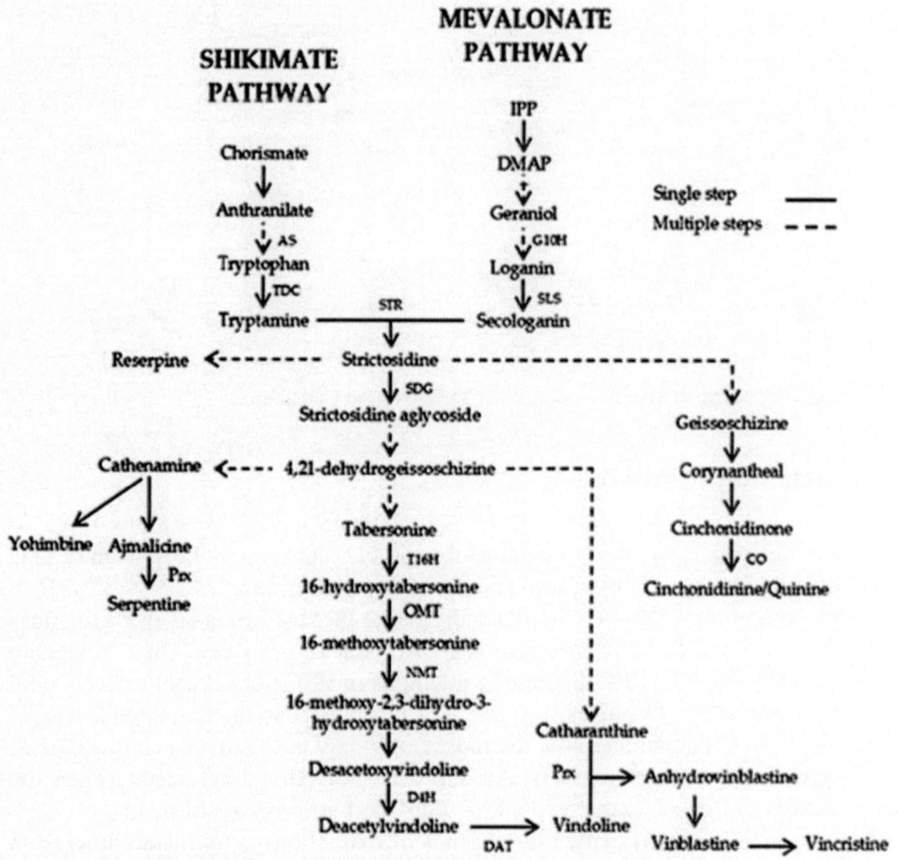

Fig. 4 Terpenoid indole alkaloids (TIAs) pathway

6.1 Alkaloids

Alkaloids are the naturally occurring organic compounds that contain one nitrogen atom and it causes alkalinity and acts as defence compound in plants against pathogens and predators due to their toxicity. These kind of alkaloids are extracted from the herbal plants which are used in treatment of disease. Vinca rosea is a medicinal plant which contains more number of alkaloids and also which are used for many treatments like cancer, diabetics, etc., and the extraction of alkaloids are discussed in this section. Before going to extraction, the samples are prepared from Catharanthus roseus. This plant acts as an anticancer agent and in antidiabetic activity. Vinca alkaloids is a molecular interaction between tubulin and vinca alkaloids such as indole alkaloids as vinblastine, vincristine, catharanthine, vindolinine, vinfluinine, and so on alkaloids which belong to vinca rosea (Fig. 5).

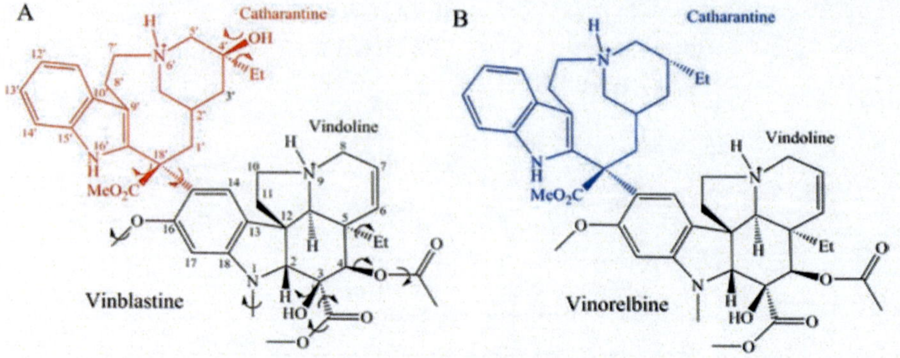

Fig. 5 Structures of **a** VLB and **b** vinorelbine. Derived from Catharanthus

7 Extraction Methods

The extraction process is used to separate the soluble plant metabolites, from sample plant material.The initial crude some needs further processing.

Poompavai et al. [6] proved that Pulsed Electric Field (PEF) is an emerging non-thermal process which helps in enhancing the extraction process. The extracts are analyzed and yield 135.12% increase in alkaloids in PEF-treated root extracts with electrical parameters (15 pulses, 100 μs, 3.125 kV/cm) than the untreated extracts.

Sathya et al. [7] concluded that the most commonly used solvents are methanol, ethyl acetate, and acetone extracts. Among these solvents, ethyl acetate gives the best antibiogram for the extracts like leaf, stem, and flower parts of plants.

Nakabayashi et al. [8] explan three kinds of extractions of N-metabolites from biological organs which are 1. LC – FTICR – MS Analysis, 2. LC-FTICR-MS/MS analysis, and 3. MALDI analysis, which were analyzed and improves efficiency of metabolites production.

Goetz et al. (2014) explains NMR method and ESPA methods are used for data interpretations of particular chemical structures and intra molecular hydrogen bonding. Alkaloids extracted from the cell and the extracted alkaloids are chromatographed by thin layer chromatography (TLC).

Microwave radiation interacts with molecular dipoles of polar materials (e.g., solvents and sample) which causes heating near the surface of the biological tissue and heat is transferred by conduction.

Mangeney et al. (1979): A bio genetic scheme for the formation of the major indole alkaloids of the vinblastine group; the precursors of ammonium salts are catharanthine and vindoline which would lead to anhydrovinblastine and followed by leurosine.

Potier (1980) proved that the variety of alkaloids are obtained from the synthesis of vinblastine-type alkaloids. The active compounds are selected from the interaction between these alkaloid compounds (known as "spindle poisons") and tubulin.

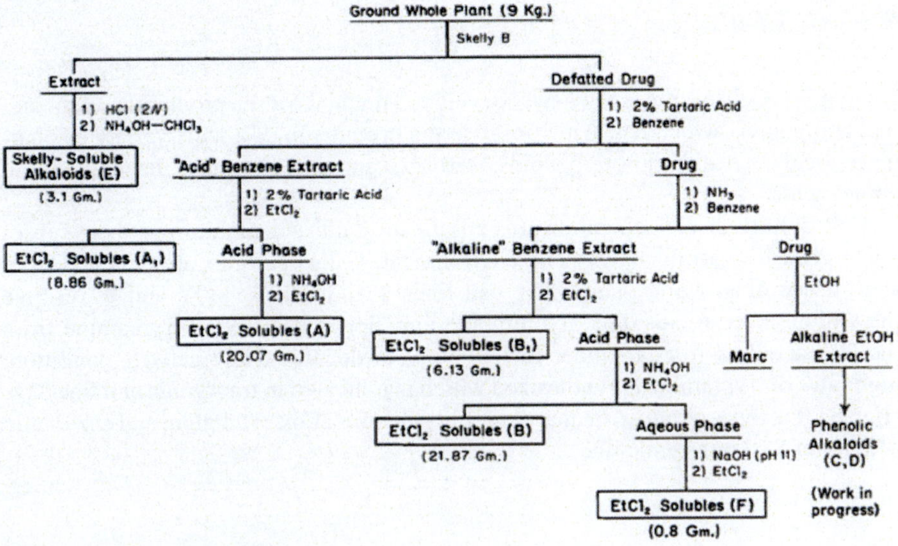

Fig. 6 Alkaloid extraction scheme

Kutney et al. [9]: The possible relationship between the Aspidosperma and Iboga alkaloids and the order of sequential formation of the indole alkaloids is Corynanthe → Aspidosperma → Iboga (Fig. 6).

Gordon and Norbert [10]: The numerous amounts of alkaloids are available in vinca rosea which deals with its medical property. Enhancing the migration of dissolved ions and promoting solvent penetration because of nitrogen atom is the reason for alkalinity of alkaloids. Phyto-chemical constitutes efficiency of the final extractions; they eventually have an effect on the final extracts. It can be concluded that no universal extraction methods is the ideal method and each extraction procedure is unique to the plants. Previously optimized methods can be used to lead in the selection of suitable methods. However, evaluation and selection of pre-extraction preparation and extraction methods are depend on the study objectives, samples, and target material.

8 Alkaloid Enhancement Techniques

Alkaloid enhancement in vinca rosea plant is analyzed through the enhancement techniques which are either synthetic or natural methods. Few such methods are plant cell culture, tissue culture, root culture, shoot culture.

8.1 Cell Culture

Kutny [11]: Cell culture is the newly developed method for the production of medicinal compounds which has made possible the production of a wide variety of pharmaceuticals like alkaloids, terpenoids, steroids, saponins, phenolics, flavonoids, and amino acids.

Cell culture is the seperation of cells from plant and monitoring their subsequent growth in artificial/synthetic environment. Callus cultures are obtained from stem tissue of growing plants. The cell lines 17BG, 299Y, 340Y and 951G give the vindolinines, trictosidine, gyohimbine ajmalicine, n-dimethyl tryptamine from the Catharanthus roseus. Under low tryptophan decarboxylate activity condition, high rates of tryptamine is synthesized which is indicated in transgenic cell line S10. High STR activity obtained or analyzed from accumulation of alkaloid and enzymatic activity of S10 transgenic line.

8.2 Organ Culture

Organ culture (Gandhi et al. 2014) has been explored for production of such phytochemicals. vinblastine, and vincristine that are derived from catharanthine of Catharanthus roseus (L.) is produced with better level.

8.3 Cell Suspension Culture

Gandhi et al. (2014): Cell suspension culture systems are used for extraction of secondary metabolites in large-scale culturing of plant cells. A suspension culture is developed under suitable conditions of aeration, agitation, light, temperature, and other physical parameters and by transferring the relatively friable portion of the callus into liquid medium and is maintained. Mishra [6] conclude that the cell suspension culture produces higher yield and production of alkaloid content than the agar gel medium.

8.4 Root Culture

Root culture gives the details about production of valuable secondary metabolites and recombinant proteins. Root culture is classified as transformed and non-transformed root culture. Non-transformed root culture does not show great variations but transformed root culture is used because of the increasing bio-mass and alkaloid contents and along with it has wider spectrum of secondary metabolites. And transformed root culture is also called as "hairy root culture". The advantages of Hairy roots cultures (Gandhi et al. 2014) are (1) growth rate is high, (2) stability of long cultural period in genotypic and phenotypic, (3) plant growth regulators are not needed, and (4) secondary metabolites are produced at higher level.

8.5 *Tissue Culture*

Tissue culture is the separation of cell tissue and growing in artificial environment. The challenges faced by tissue culture environment is the lack of skilled manpower. The herberine and sanguinarine, vincristine and indole alkaloid content is highly produced. Tissues have high rates of cell division and produce the growth regulating substances [12]. Full strength MS medium reduced root length and half strength MS medium gives better result of root length. In tissue culture the content cytokinin is the group of plant growth regulators. The tissue culture method gives the production of secondary metabolites, improved crop varieties, and genetic transformation. To improve productivity of tissue culture, the product combines the bioprocessing with genetics.

8.5.1 Technique of plant tissue culture

Micro Propagation

Micro-propagation is the propagation through tissue culture. Here, leaf, bud, meristem can be used as explants. To increase the production of number of shoots, a multiplication stage is required which is called shoot multiplication stage. For further increasing the number of shoots, implementing subculturing to these new shoots on to new medium produced more shoots. The four stages are (Fig. 7):

Establishment of explants
Shoot multiplication
Rooting
Acclimatization.

Establishment of Explants

According to the explants type the product will be produced. The contaminants in plant cell is removed by using chemical solutions which is called sterilization of explants.

Shoot Multiplication

Propagules number is increased

Fig. 7 Different stages of micro-propagation of vinca rosea. **a** shoot formation, **b** shoot elongation, **c** multiple shoot formation, **d** shoot profilteration, **e** early shoot multiplication, **f** shoot formation, **g** well acclimatized plant, **h** well growing plant

Rooting Stage

Multiplication of explants and development of strong root growth

Acclimatization Stage

This stage plants are weekend and hardened. During hardening it changes from low intensity to high intensity and from high humidity to low humidity.

8.5.2 Bio-Reactor

For the production of bioactive components in an active field, the application of bio-reactor system for large-scale cultivation of plant cells are widely used. In a bio-reactor process, the plant cells are placed in liquid suspension which offers a unique combination of physical and chemical environments.

Kargi et al. [13]: By the use of ajmalicine, the total alkaloid concentration was determined by absorbance measurement at 280 nm and subsequent comparison to a calibration curve is prepared. At optimized biofilm thicknesses for indole alkaloid titters the biofilm reactor was more effective than suspension cultures.

Table 1 Remarks of Enhancement Techniques

Enhancement techniques	Remarks
Cell culture	0.2 g/l of alkaloid content is increased
Tissue culture	Tissue culture is the best method for increasing production of alkaloids
Shoot culture	Increasing secondary metabolites and 0.155% increasing alkaloid contents
Bio Reactor	Increasing alkaloid content than suspension culture

8.5.3 Research Work in Alkaloid Enhancement

The following table is based upon the reviews on enhancement techniques, that is, synthetic methods of increasing secondary metabolites contents (Table 1).

9 Conclusion

Various culture techniques show that the alkaloid content of plants have increased. And such techniques are cell culture, tissue culture, and root culture. Along with that the analysis is made on extracted parts from the living plants. All of the papers deal with increasing alkaloid production by synthetic growing methods. Further deep research is required in the direction of the natural method, that is, how can we naturally increase alkaloid contents in living plants.

References

1. Lassen LM (2014) Redirecting photosynthetic electron flow into light driven synthesis of alternative products including high value bio active natural compounds. ACS 3:1–12
2. Seger C, Sturm S (2007) Analytical aspects of plant metabolite profiling platforms: current standings and future aims. J Proteome Res 6:480–497
3. Tom et al (2019) Deciphering Metabolic heterogeneity by single cell analysis
4. Terada BY (2014) Identification of indole alkaloid structural units important for stimulus selective TRPM8 inhibition: SAR study of naturally occurring Iboga derivatives. Am Chem Soc
5. Kingston DGI (2019) Tubulin- interactive natural products as anticancer agents. Am Chem Soc 72:507–515
6. Poompavai S, Gowri Sree V, Raji S (2018) Effect of pulsed electric field on alkaloids of vinca rosea root. J Pharm 1:4–8
7. Sathya S (2008) Antibiogram of catharanthus roseus extracts. Glob J Mol Sci 3:1–7
8. Nakabayashi R (2017) Top down metabolomic approaches for nitrogen containing metabolites. ACS
9. Kutney JK (1981) Alkaloid production in catharanthus roseus cell cultures V. Alkaloids from the 176G, 299Y, 340Y and 951G cell lines. J Nat Prod 44(5):536–540

10. Svoboda GH, Neuss N, Gorman M (1959) Alkaloids of vinca Rosea linn: preparation and characterization of alkaloids. J Am Pharm Assoc 48(11):659–666
11. Kutny JP (1996) Plant cell culture combined with chemistry-a powerful route to complex natural products. Pure Appl Chem 68(11):2073–2080
12. Hussain SM, Fareed S, Ansari S (2012) Current approaches toward production of secondary metabolites. J Pharm Bioallied Sci 4:10–20
13. Kagri F, Ganapathi B, Maricic K (1990) Indole alkaloid formation by catharanthus roseus cells in a biofilm reactor. Biotechnol Prog 6(4)
14. Duflous A, Kruczynski A, Barret J-M (2002) Novel aspects of natural and modified vinca alkaloids. Curr Med Chem Anticancer Agents 2:55–70
15. Sasaki Y (2010) Asymmetric total synthesis of vindorosine, vindoline and key vinblastine analogues. J Am Chem Soc 132:13533–13544
16. Azwanndia NN (2010) A review on the extraction methods use in medicinal plants, principle, strength and limitation. Med Aromat Plants 4:1–6
17. Archambault J, Volesky B, Kurz WGW (1990) Production of indole alkaloids by surface Immobilized C. Roseus Cell. Biotechnol Bioeng 35:660–667
18. Atta-ur-Rahman, Ali I, Bashir M (1983) Isolation and structural studies on the alkaloids in catharanthus roseus. J Nat Prod 47(3):554
19. Huffman BJ, Shenvi RA (2019) Natural products in the "marketplace": interfacing synthesis and biology. J Am Chem Soc 141:3332–3346
20. Balsevich J, Hogge R (1998) Analysis of indole alkaloids from leaves of catharanthus roseus by means of supercritical fluid chromatography/mass spectrometry. J Nat Prod 51(6):1173–1177
21. Canopy MS (2016) Production of indole alkaloids in hairy root cultures of Catharanthus roseus L. and their antimicrobial activity. South Afr J Botany 105:9–18
22. Ciau-Uitz R, Mirandha-Ham ML, Coello J (1994) Indole alkaloid production by transformed and non-transformed root cultures of catharanthus roseus. In Vitro Cell Dev Biol Plant 84–88
23. Mant C, Alisaraie I (2014) Metabolites of vinca alkaloid vinblastine: tubulin binding and activation of Nausea–associated receptors. Am Chem Soc 1–12
24. Barnett CJ (1987) Structure activity relationships of dimeric catharanthus alkaloids 1. Deacetylvinblastine Amide (vinblastine) sulphate. J Med Chem 21(1):88–96
25. Ferreres F (2008) New phenolic compounds and anti-oxidant potential of catharanthus roseus. Am Chem Soc 56:9967–9974
26. Hemalatha RG, Pradeep T (2013) Understanding the molecular signatures in leaves and flowers by desorption electrospray ionization mass spectrometry (DESI MS) imaging. J Agric Food Chem 61:7477–7487
27. Flores HE (1987) Use of plants and organ culture in the production of biological chemicals. Am Chem Soc 66–86
28. Roepke J (2010) Vinca drug components accumulate exclusively in leaf exudates of Madagascar periwinkle. PNAS 107(34):15287–15292
29. Shams KA (2009) Isolation and characterisation of antineoplastic alkaloids from catharanthus roseus L. Don cultivated in Egypt. Research Gate 6(2): 118–122
30. Toivonen L (1993) Utilisation of Hairy root cultures for production of secondary metabolites. Biotechnol Prog 9:12–20
31. Lindsey K, Yeoman MM (1983) The Relationship between Growth rate, Differentiation and alkaloid accumulation in cell cultures. J Exp Botany 34(145):1055–1065
32. Jamshed M (2011) Simultaneous determination of secondary metabolites from vinca rosea plant extractives by reverse phase high performance liquid chromatography. Pharmacognosy Mag 7:92–96
33. Koul M (2013) Catharanthus Roseus and prospects of its endophytes: a new avenue for production of bioactive metabolites. Int J Pharm Sci Res 4:2705–2716
34. Mishra MR (2018) Enhanced alkaloid production from cell culture system of catharanthus roseus in combined effect of nutrient salts, sucrose and plant growth regulators. J Biomed Biochem Sci 1:14

35. Toso RJ et al (1993) Kinetic stabilization of Microtubule Dynamic instability in vitro by vinblastine. Biochemistry 32:1285–1293
36. Silvestri R (2013) New prospects for vinblastine analogues as anticancer agents. Am Chem Soc 56:625–627
37. Akhtar S, Jha SR, Mughees M (2017) Simultaneous effect on metabolic profiling of terpene indole alkaloid pathway in leaves and roots of catharanthus Roseus under drought stress. Int J Pharm Biol Sci 7:10–20
38. Sadowska A, Racka M, Rek J (1983) Alkaloids of catharanthus roseus/vinca rosea/and possible ways of increasing their production. Acta Hortic 132:285–289
39. Ahmad S, Garg M, Tamboli ET, Abdin MZ (2013), In vitro production of alkaloids: factors, approaches, challenges and prospects. Pharmocognosy Rev 7 (2013).
40. Miyazaki T (2007) Synthesis of vinblastine and its analogues. 9(23):4737–4740
41. de Luca V, Cutler AJ (1987) Sub cellular localization of enzymes involved in indole alkaloid bio synthesis in catharanthus roseus. Plant Physiol 85:1099–1102
42. de Luca V (1998) Developmental regulation of enzymes of indole alkaloid bio synthesis in catharanthus roseus. Plant Physiol 86:447–450
43. Verpoorte R, van der Heijden R, Schripsma J (1993) Plant cell biotechnology for the production of alkaloids: present status and prospects. J Nat Prod 565(2):186–207
44. Chen X Metabolic engineering strategies for sustainable terpenoid flavor and fragrance synthesis. J Agric Food Chem
45. Zhu X, Zeng X, Sun C (2014) Biosynthetic pathway of terpenoid indole alkaloids in catharanthus roseus. Front Med 8:285–293

Chapter 37
Machine Learning Based Early Prediction of Rainfall Induced Landslide – A Detailed Review

V. Aarthi and V. Vijayarangan

1 Introduction

Landslides are one of the biggest calamities commonly recurring in many hilly regions around the world. Landslide incidence, being a ramification of spatial climatic changes, geological factors and precipitation, is a highly uncertain phenomenon. Among all the factors inducing landslides, rainfall holds major importance. Landslides cause high destruction, damage and disorder during the monsoon season, incurring human fatalities, considerable damage to surroundings and natural resources often shattering the livelihood of local people [1].

And nowadays the most widely used method for the landslide prediction-related work is machine learning. The machine learning method will be a way of identifying patterns in data and this data can be used for automatically making predictions or decisions in the future [2]. A wide range of methods and models are utilized in the different papers consistent with the dataset which was planned to be used. Some of them had comparisons between the various algorithms for finding out the best possible option for their model. But their results were found very contrasting to each other.

V. Aarthi (✉) · V. Vijayarangan
Department of Electronics and Communication Engineering, Kovilpatti, India
e-mail: 1958001@nec.edu.in

N. Marriwala et al. (eds.), *Soft Computing for Intelligent Systems*, Algorithms for Intelligent Systems, https://doi.org/10.1007/978-981-16-1048-6_37

2 Neural Network

2.1 Artificial Neural Network (ANN)

Feedforward artificial neural network prediction based on the input data selection [3]. Input parameters selected for landslide prediction such as rainfall, displacement, temperature, landslide area, slope inclination and depth. During this study it had been also displayed that the repetition of data within training datasets has a positive effect on the output hh (Fig. 1).

Ghorbanzadeh et al. [4] presented Artificial Neural Network (ANN) to solve the complex problems by discovering patterns [5]. During each cycle, the weights are randomly selected and it is updated using the backward propagation process to reduce the error. The 〖ANN〗 _5 and 〖ANN〗 _8 methods result in the true positive rate of 499.83 and 445.9; false positive rate of 152.03 and 459.81 and then false negative rate of 340.12 and 220.04.

Artificial neural network approach is presented to predict landslide susceptibility and also to quantify the best hazard map of the landslide area [6, 7]. Input parameters selected for analysis are SPOT 5 Satellite Pan Sharpened images, slope, aspect, curvature, drainage, lithology type and Normalized Different Vegetation Index (NDVI). Among all these input parameters, slope has a highest weightage than twice those of the other parameters (Fig. 2).

Backpropagation Neural Network (BPNN)
Backpropagation neural network is a conjunction of an input layer, output layer and at least one hidden layer [1]. The number and structure of hidden layers determine the ability of a network to generate more accurate results. The MSE values of three input datasets are 0.0212, 0.0148 and 0.0136. The RMSE values of these three datasets are 117.6350, 95.5339 and 90.4859.

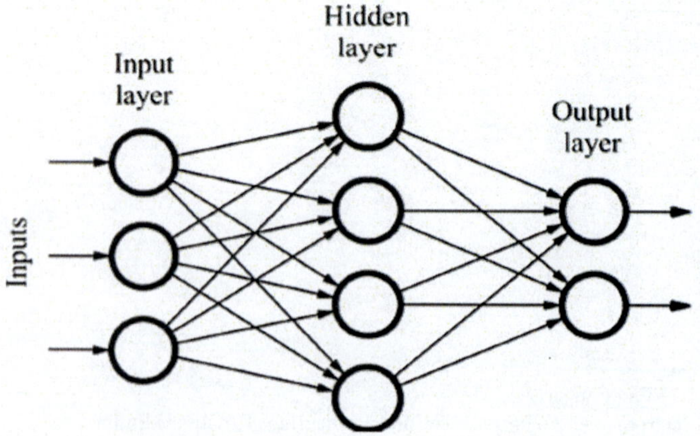

Fig. 1 Architecture of feedforward neural network

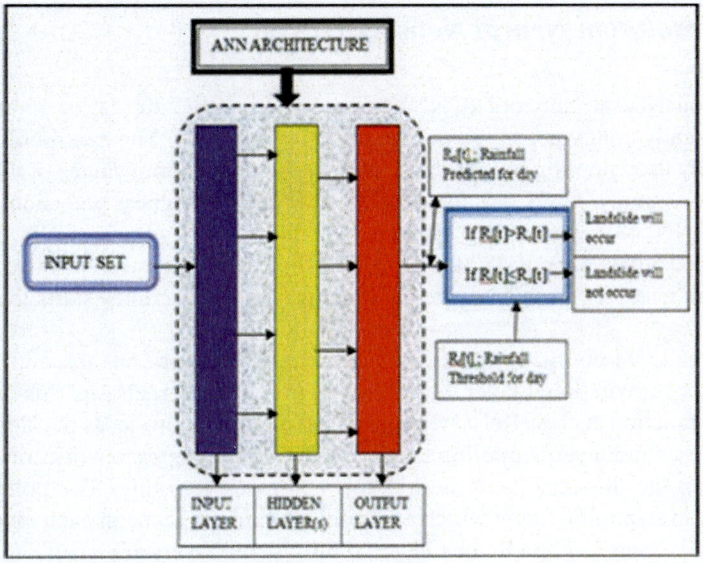

Fig. 2 ANN methodology

Backpropagation model is the most popular precedent in ANN used widely in 90% applications [8]. This method maps the input to output through forward propagation of input and backpropagation of errors. Here, BPNN consists of one input layer, one or more hidden layer and one output layer. Four different models C1, C2, C3 and C4 each with one (C1A, C2A, C3A and C4A) and two hidden layers (C1B, C2B, C3B and C4B) are developed. Among all the model C3B produced better results. Hence, it shows that two hidden layer models performed better than one hidden layer model.

Backpropagation neural network is used to train the Multilayer Perceptron (MLP), which is an artificial neural network that has been widely used for landslide prediction [9]. The accuracy of MLP is 96.85%. The True Positive Rate (TPR) and False Positive Rate (FPR) of MLP are 0.01 and 0.01, respectively. The AUC value and sensitivity are about 0.79 and 1.99, respectively.

Radial Basis Function Neural Network (ANN-RBF)

The author Pham et al. [10] describes Radial Basis Function Neural Network (ANN-RBF) is a hidden layer neural network which uses radial basis function to analyse the spatial relationship between a set of landslide-affecting factor and landslide occurrence. The Area Under the Curve (AUC) value and Root Mean Square Error (RMSE) value of this approach are about 0.936 and 0.31, respectively.

2.2 Convolution Neural Network (CNN)

Convolution Neural Network (CNN) has a specific structure called hidden layer which typically contains convolution and pooling layer [4]. The max-pooling is the most widely used pooling layer, which keeps only the maximal values of the feature maps. In this study, CNN_16, 5 methods achieved the highest precision value of 83.31%, CNN_48, 5 results in high false negative value of 106.56, CNN_32, 5 has high false positive value of 380.03 and CNN_22, 5 with high true positive value of 325.24. But in case we use larger input window size, the accuracy starts to decrease (Fig. 3).

Bhatt et al. [11] propose that convolution neural network has three layers. The first layer is convolutional layer which consists of many kernels and this layer uses activation function such as ReLU (Rectilinear Unit) which produces non-linearity of the model, second layer is pooling layer which reduces the framework computation by reducing the dimensionality and also it controls overfitting. The polling layer used is the max-pooling layer which takes only maximum value in each block. Then complex reasoning is done by last layer. Totally seven layers are used, namely, two convolutional layers, two pooling layers, one flattening layer and two fully connected layers. On convolution stage 32 activation map is produced using 3 × 3 feature detector with input image of 64 × 64 pixel. The convolution neural network has better performance than other ML algorithms, resulting in an accuracy of 96.90%. This model can be used for the early landslide prediction purpose.

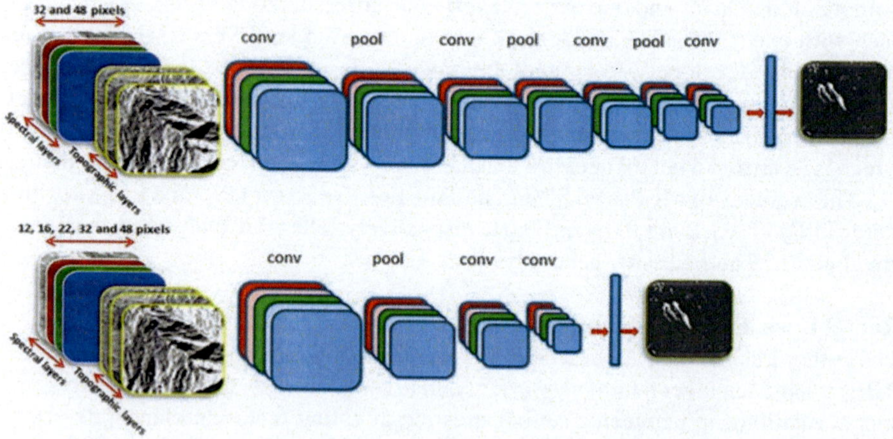

Fig. 3 The CNN architectures with **a** seven layers (D-CNN) and **b** four layers

2.3 Deep Belief Network (DBN)

Huang et al. [12] developed an unsupervised learning algorithm of Deep Belief Network (DBN) to extract the intrinsic characteristics of landslide factors. Deep Belief Network (DBN) model is performed by SoftMax classifier and a dropout mechanism for rainfall-induced early warning of precipitation landslide. The SoftMax classifier included to the top layer of DBN, which combine the supervised SoftMax regression model with the unsupervised parts of DBN, to make better prediction performance. Dropout mechanism is introduced in the DBN training process to overcome the overfitting problems. This model resulted in better performance than former PSO-BP algorithm. The factors considered for these studies are geomorphological types, slope, seismic intensity, elevation and road type, hydrological type, average annual rainfall, daily rainfall and cumulative precipitation in previous 7 days.

2.4 Long Short-Term Memory

Srivastava et al. [1] presented Long Short-Term Memory (LSTM), which is an extension of the Recurrent Neural Network (RNN). For long duration of time, long-term dependencies, time-series data and continuous selectively remembering patterns are learned by this method [13–15]. LSTM network characteristic is confined to the hidden layer structure containing memory blocks instead of simple artificial neurons. Similar to RNN each and every memory cell has a feedback loop for allowing the neural of inputs. The LSTM contains input gate, forget gate and output gate. These three input gates are fed by the current input at a time and the output is generated from the last hidden layer at a time. The forget gate is responsible for discarding the unnecessary information from the cell. The MSE values of three input datasets are 0.0203, 0.0480 and 0.0660. The RMSE values are 111.1900, 102.4577 and 94.7686.

3 Support Vector Machine (SVM)

Support Vector Machine (SVM) proves very good performance in the case of structured data, which the SVM uses to train itself [2]. The factors considered are amount of rainfall (mm), maximum precipitation, cloud percentage, average wind speed, average temperature, pressure, humidity, number of days of consecutive rainfall, slope of curvature and the impact of the landslide on a scale of 1–3. The maximum accuracy the SVM obtained is 70.1% and it can be further improved by increasing the size of the dataset.

Ghorbanzadeh et al. [4] deal with the optimal hyperplane developed for separating the dataset features. SVM method is mostly used for data classification, regression and analysis in various domains. SVM_5 and SVM_8 methods obtain true positive rate of 385.9 and 403.07, false positive rate of 318.28 and 395.65 and then the false negative rate of 50.67 and 58.47 and also SVM_5 and SVM_8 achieve precision value of 54.8 and 50.51%.

Support vector machine is a powerful method for data classification process [16]. Here, SVM-based model is also created on the concept of maximum margin classifiers, which is less sensitive to noise. The True Positive Rate (TPR) and True Negative Rate (TNR) obtained by SVM model are 0.972 and 0.889 and the False Positive Rate (FPR) and False Negative Rate (FNR) are about 0.028 and 0.111, respectively. The precision value obtained by SVM model is 0.897.

Pham et al. [6, 7] present support vector machine for the spatial landslide prediction purpose. The performance of this method is based on the choices of the kernel function. Radial Basis Function (RBF) is used here for landslide models among kernel functions. The very high landslide density of SVM model is 4.772. And also, SVM achieves the Area Under Curve (AUC) value of about 0.881.

Support vector machine is based on statistic learning and the structural risk minimization principle [9]. In this model, the author chose the polynomial kernel function, since it outperformed than other kernel [17]. The accuracy obtained by SVM is 96.28%. The True Positive Rate (TPR) and False Positive Rate (FPR) of SVM are 0.001 and 0.001. And the AUC and sensitivity values are about 0.50 and 0.00, respectively.

Support vector machine algorithm is developed to predict landslide occurrence by using some causative factors such as slope, land use, elevation [18], geology, distance from road, distance from the river, soil textures and rainfall. SVM performs the classification task of constructing hyperplanes in a multi-dimensional space that separate cases of various class labels.

The author Pham et al. [10] deals with support vector machine which uses the statical learning theory and structural risk minimization principal to segregate two classes with the linear hyperplane. The Area Under Curve (AUC) value and Root Mean Square Error (RMSE) value of AODE model are about 0.948 and 0.29, respectively. The accuracy achieved by this method is 87.92%.

3.1 Support Vector Regression (SVR)

Support Vector Regression (SVR) aims to deduce an estimation function that has at the most ε deviation from the actual resulted outputs for all the training data, keeping it as flat as possible [1, 19]. The MSE values of SVR for three datasets are 0.0896, 0.0627 and 0.0591. The RMSE values of these datasets are 123.1122, 100.1048 and 95.2466.

3.2 Support Vector Machine with Bagging (SVM Bagging)

SVM bagging is a combination of support vector machine and bagging ensemble algorithm [20]. The data considered for landslide susceptibility prediction are slope, aspect, elevation, plan curvature, profile curvature, drainage density, curvature, distance to drainage network, Topographic Position Index (TPI), land use, rainfall and lithological unit. The False Positive Rate (FPR), precision and sensitivity value of SVM-Bagging (testing data) model are 0.172, 0.834 and 0.829, respectively. And the Root Mean Square Error (RMSE), Mean Absolute Error (MAE) and Area Under the Curve (AUC) values are about 0.351, 0.248 and 0.899, respectively.

3.3 Support Vector Machine with Boosting (SVM Boosting)

SVM boosting approach is a hybrid model of Support Vector Machine (SVM) and boosting ensemble algorithms [20]. The boosting algorithm includes two steps: (i) initialize distribution of classifiers and (ii) evaluate the error until reaching an end condition. The data considered for analysis are elevation, plan curvature, profile curvature, slope, aspects, drainage density, curvature, distance to drainage network, Topographic Position Index (TPI), land use, rainfall and lithological unit. The False Positive Rate (FPR), precision and sensitivity value of SVM-boosting model (testing results) are about 0.165, 0.841 and 0.835, respectively. The RMSE, MAE and AUC values are 0.344, 0.248 and 0.911, respectively.

3.4 Support Vector Machine with Stacking (SVM Stacking)

SVM-stacking algorithm is the combination of Support Vector Machine (SVM) and stacking ensemble algorithm [20]. The stacking algorithm works by assessing raw classifiers of the poor performance with independent or bootstrapped reference data. The parameters considered are slope, plan curvature, profile curvature, elevation, aspect, distance to drainage network, topographic position index, land use, rainfall and lithological unit. The False Positive Rate (FPR), precision and sensitivity value of SVM stacking (testing results) are about 0.317, 0.685 and 0.683, respectively. The RMSE, MAE and AUC values are about 0.45, 0.401 and 0.743, respectively (Fig. 4).

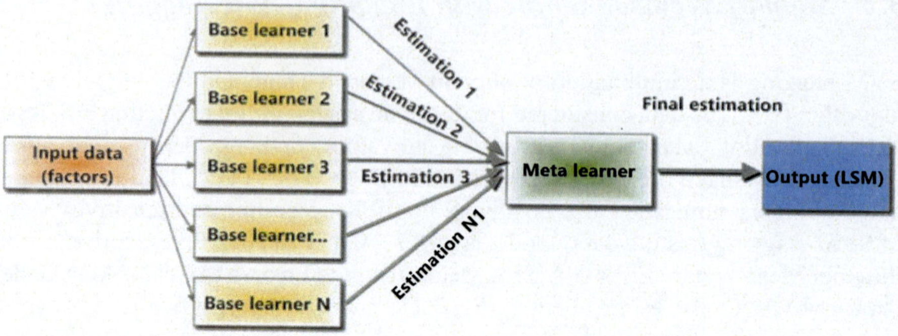

Fig. 4 Simple stacking visualization

3.5 *Sequential Minimal Optimization-Based Support Vector Machine (SMOSVM)*

Pham et al. [21] deal with Sequential Minimal Optimization-based Support Vector Machine (SMOSVM) algorithm to landslide susceptibility using GIS. Since SVM is the most effective method for classification with high accuracy, a hybrid model of SMO and SVM is developed for landslide prediction [22–25]. The factors affecting landslide such as slope angle, elevation, curvature, lithology, soil, land cover, slope aspects, distance to roads, distance to rivers, distance to lineament and rainfall were considered for landslide analysis. The land density and sensitivity value of SMOSVM are about 5.42 and 82.14%, respectively. The accuracy obtained by SMOSVM is 78.87%.

3.6 *Least Square Support Vector Machine (LSSVM)*

Least Square Support Vector Machine (LSSVM) is discussed to predict landslide occurrence [7, 21]. This method is a modification of support vector machine using 'least square' version for function estimation. The factors affecting landslide which are used for landslide susceptibility assessments are slope, aspect, elevation, curvature, lithology, distance to roads, distance to faults, land use and distance to rivers and rainfall. The Area Under Curve (AUC) value is about 0.803.

4 Random Forest (RF)

Kuraduseng et al. [26] presented a Random Forest algorithm to predict a landslide earlier by reducing prediction error and False Negative Rate (FNR). Random Forest

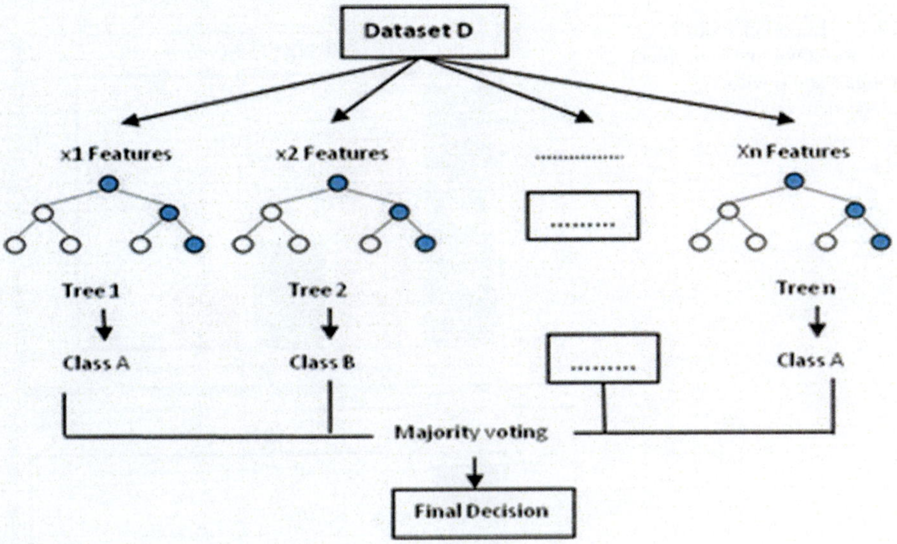

Fig. 5 Random forest with decision tree

(RF) constructs several decision trees from the datasets, as the number of trees increases the accuracy also increase. Factors considered for landslide prediction are rainfall, antecedent rainfall, slope, soil type, soil depth and land cover. Random Forest has a performance rate of 84.61% (without antecedent rainfall), 95.19% (with antecedent rainfall) and prediction error or False Negative Rate (FNR) of 4.80% (Fig. 5).

Random Forest yields good results with satellite images [4]. Therefore, Random Forest is considered as one of the useful non-parametric ensembles learning method in image analysis. The methods RF_5 and RF_8 have a true positive rates of 393.9 and 380.2, false positive rates of 86.71 and 89.6 and then false negative rates of 85.32 and 98.31, and also RF_5 and RF_8 achieve precision values of 81.95% and 80.95%, respectively (Fig. 6).

Random Forest model is associated with various decision trees for landslides to spatially predict for a landslide area [27]. Totally 100 trees are used with a single randomly split variable to grow the trees. The Area Under the Curve (AUC) obtained by the Random Forest (RF) approach is 78.7% and RMSE value is about 0.428, respectively.

Dang et al. [16] deal with Random Forest Classifier (RFC) which is a powerful tool to develop an ensemble of classification trees. This model results in True Positive Rate (TPR) and True Negative Rate (TNR) of 0.983 and 0.901 and then False Positive Rate (FPR) and False Negative Rate (FNR) of about 0.017 and 0.100, respectively. The precision value achieved by RFC is 0.908.

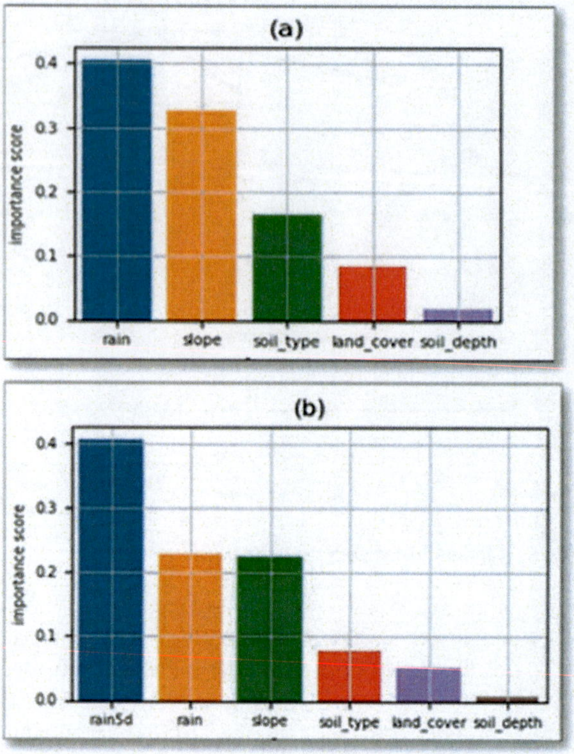

Fig. 6 Importance of RF features **a** without antecedent rainfall and **b** with antecedent rainfall

Random Forest model exploits random binary trees that use the subset of attributes through the bootstrapping method [9, 28]. Random Forest (RF) model achieves accuracy of 97.28% and True Positive Rate (TPR) and False Positive Rate (FPR) of 0.58 and 0.01 respectively. The AUC and sensitivity value of RF model is 0.90 and 2.53 respectively.

The Random Forest algorithm is discussed to predict landslide susceptibility earlier [29]. For that some conditioning factors are to be considered such as slope aspects, soil type, soil thickness, land use, landform, Sediment Transport Index (STI), Topographic Witness Index (TWI), geology, hydrology, Stream Power Index (SPI) and rainfall. The preliminary results for maximum information gain ratio are about the soil thickness (0.1568), geology (0.1456), landform (0.101), land use (0.025) and soil type (0.0035), respectively.

Random Forest (RF) produces landslide susceptibility maps using this data mining model [30]. Some landslide conditioning factors were prepared, namely, slope, aspect, distance from faults, rainfall, altitude, lithology, plan curvature, profile curvature, distance from streams, distance from roads, slope angle and land use. The Area Under the Curve (AUC) value obtained by this approach is 78.3%. The prediction rate of RF is about 81.2%.

Dou et al. [20, 31] present a Random Forest algorithm to model the massive rainfall-induced landslide occurrence. The factors selected for landslide analysis are altitude, slope angle, slope, aspect, total curvature, stream power index, distance to drainage network, compound topographic index, plan curvature, drainage density, distance to geological boundaries, lithology and cumulative rainfall. The False Positive Rate (FPR) obtained or two sets of training results are 51.010 and 51.880, respectively. The precision values of 0.932, 0.931 and accuracies of two sets of training data are 94.577 and 94.441, respectively.

4.1 Random Forest Machine (RFM)

Dang et al. [16] presented a new model named as Random Forest Machine (RFM), which is an integration of Support Vector Machine (SVM) and Random Forest (RF) and the F_1 score result by RFM is 0.96 which is greater than SVM (0.933) and RF (0.944), respectively. The True Positive Rate (TPR) and True Negative Rate (TNR) attained by RFM are 0.947 and 0.988 and False Positive Rate (FPR) and False Negative Rate (FNR) are about 0.012 and 0.053, respectively. The precision value obtained by the RFM is 0.989.

4.2 Distributed Landslide Displacement Prediction with Random Forest (DLDP-RF)

Distributed Landslide Displacement Prediction with Random Forest (DLDP-RF) algorithm is developed for landslide displacement monitoring [32]. The input parameters considered for landslide prediction are rainfall, humidity, weather and reservoir water level. The Mean Square Error (MSE) and Mean Absolute Error (MAE) of DLDP-RF are about 94.4642 and 73.7350, respectively.

5 Logistic Regression (LR)

Logistic regression is most widely used in the landslide susceptibility assessment, as it produces probability of occurring landslide hazards [26]. Factors considered for landslide prediction are rainfall, slope, soil type, antecedent rainfall, soil depth and land cover. The ROC-AUC value of LR is 90.38% (without antecedent rainfall), 96.15% (with antecedent) and produces False Negative Rate (FNR) of 3.84%. It results in better performance than Random Forest (Fig. 7).

Satinelli et al. [33] estimate logistic regression, which calculates the probability that the predicted output belongs to a particular category or class. If the probability

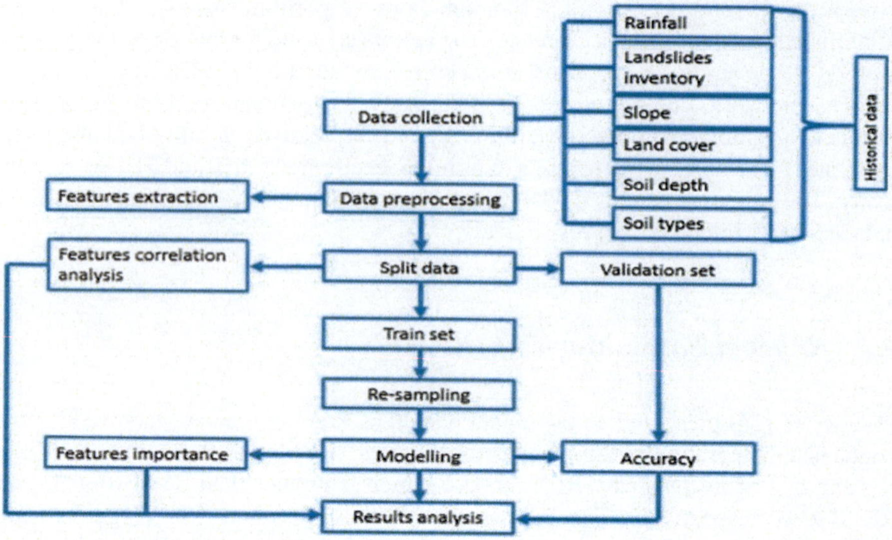

Fig. 7 Landslide prediction flow chart

of occurrence is greater than 50% the model classifies the output as 1 (landslide), otherwise 0 (non-landslide). This technique achieves the result of AUC greater than 0.88.

The authors [9, 34] presented a logistic regression model in landslide detection as it provides a probability of landslide occurrence against every data point. It achieves better accuracy of about 96.25%. The True Positive Rate (TPR) and False Positive Rate (FPR) of this model obtained are about 0.02 and 0.001 and then Area Under Curve (AUC) value and sensitivity of LR model are 0.79 and 1.67, respectively.

The author discussed Logistic Regression (LR) with the following causative factors such as rainfall, seismic intensity, landform, distance to river, distance to road, elevation, plan curvature, slope and aspects [35]. The accuracy of training data and precision value of training data of this model are 0.716 and 0.714, respectively. The Area Under the Curve (AUC) value of LR is 0.785.

Logistic Regression (LR) method is discussed to quantify the best hazard map in the landslide area of assessment using this method [36]. The factors selected for analysis with SPOT 5 Satellite Pan Sharpened images are slope, aspect, rainfall, curvature, drainage, lithology type, lineament, soil type and Normalized Different Vegetation Index (NDVI). Logistic regression is a multivariate analysis model, which is useful for predicting the presence or absence of outcome based on values of a set of predictor variables.

Pham et al. [10] show logistic regression approach is discussed to proper landslide prediction. The AUC value and RMSE value of this approach are about 0.928 and 0.32, respectively. The accuracy obtained by this approach is 86.20%.

A logistic regression algorithm is developed for landslide susceptibility assessment using GIS [21]. Here binary model logistic regression is used for spatial landslide susceptibility prediction [37, 38]. The factors considered are angle, slope, aspects, elevation, lithology, curvature, distance to roads, distance to rivers, land cover and rainfall. The land density and sensitivity value are about 4.01 and 73.66%, respectively. The accuracy achieved by LR model is about 73.64%.

5.1 Logistic Model Tree (LMT)

The author [27] presents Logistic Model Tree (LMT) which is a hybrid model of linear logistic regression and decision tree in order to improve their advantages. This model is most probably used for susceptibility mapping. The Area Under Curve (AUC) value obtained by the LM tree model is 80.9% and RMSE value is 0.364, respectively.

Logistic model tree uses formation gain to split and logistic boost algorithm to produce a logistic regression model at every tree node [35]. Here some conditioning factors are considered for landslide prediction such as rainfall, seismic intensity, landform, distance to the river, distance to road, elevation, plan curvature, slope and aspects. The accuracy and precision value of the LMT are 0.727 and 0.713 (training data), respectively. The Area Under the Curve (AUC) value is about 0.805.

5.2 Bagging Ensemble and Logistic Model Tree (BE-LM Tree)

The author [39] presented a new modelling approach that is a hybrid intelligence of BE-LM tree for landslide susceptibility mapping. Construction and verification of BE-LM Tree were done by GIS database. The quality of the final BE-LM tree model was verified using confusion matrices and several model measures. The accuracy and prediction capability obtained by BE-LM tree are 93.81% and 83.4%, respectively. Therefore, the integration BE-LM tree has significantly improved the quality and accuracy of the result in better performance.

5.3 Random Subspace and Logistic Modes Tree (RSLMT)

The author [35] deals with Random Subspace and Logistic Model Tree (RSLMT), which is a combination of Random Subspace (RS) and Logistic Model Tree (LMT). This model was developed for predicting landslide susceptibility map of the study area. Landslide causative factors considered for predictions are elevation, rainfall, profile curvature, slope, plan curvature, distance to faults, aspects, distance to the river, landform, distance to road and seismic intensity. The accuracy and precision values obtained with this model are 0.738 and 0.715, respectively. The Area Under Curve (AUC) value of RSLMT model is 0.815.

6 Naive Bayes (NB)

The author [35] used Naive Bayes algorithm to predict landslide susceptibility using some input factors such as rainfall, slope, landform, distance to the river, distance to road, elevation, plan curvature, seismic intensity and aspects. The accuracy and precision values obtained by NB are 0.703 and 0.682, respectively. The Area Under the Curve (AUC) value is 0.771.

The Naive Bayes algorithm is discussed, which is a simpler form of Bayesian network and it is popularly known for data mining applications [10]. The AUC and RMSE values of NB are 0.928 and 0.33, respectively. The accuracy of NB is 86.20%.

6.1 Naive Bayes Classifier

The Naive Bayes Classifier algorithm is one of the easiest soft computing methods based on maximum posterior hypothesis and Bayesian theory [18]. This model possesses a strong capability to predict landslides by causative factors such as slope, land use, soil textures, distance from road, distance from the river, elevation, geology and rainfall. These factors are considered to predict landslide occurrence more accurately.

6.2 Naive Bayes Tree Model (NBT)

The Naive Bayes Tree is considered as a hybrid algorithm on each tree leaf node [27]. NBT consists of decision tree classifier and Naive Bayes. This model produces better accuracy and precision value of about 0.799 and 0.745, respectively. NBT achieves the Area Under Curve (AUC) value of 82.4% and RMSE value of 0.301.

The author Pham et al. [6, 7] deals with Naive Bayes Tree (NBT) which is based on Bayesian theorem that considers all attributes are independent to increase the posterior probability in determining the classified class [40, 41]. The very high landslide density of NBT model is 5.205 and the Area Under the Curve (AUC) value is about 0.832, respectively.

7 Decision Tree (DT)

Decision Tree is a hierarchical model which consists of decision rules [9]. The main advantages of the Decision Tree model are that they can handle categorical as well as numeric variables and can incorporate them without strict assumptions on data. The accuracy obtained by the decision tree method (C4.5 model) is 97.17%. The True Positive rate (TPR) and False Positive Rate (FPR) are 0.44 and 0.01, respectively. The AUC and sensitivity value are about 0.77 and 2.18, respectively.

The Decision Tree algorithm is discussed to predict rainfall-triggered landslide occurrence [20, 31]. The factors considered are slope angle, slope, aspects, altitude, plan curvature, distance to drainage network, total curvature, compound topographic index, stream power index, drainage density, distance to geological boundaries, lithology and cumulative rainfall. The FPR values of two sets of training data are 62.430 and 60.570, respectively. The precision values of two set of training data are 0.918 and 0.920 and accuracies are about 93.422 and 93.610, respectively.

7.1 ID3 Decision Tree Classification

ID3 Decision Tree Classification method is constructed using the concept of Shannon entropy [42]. The rainfall parameters and soil properties with some other geological factors improve the prediction performance.

7.2 Classification and Regression Tree (CART)

Classification and Regression Tree (CART) approach is presented to develop landslide susceptibility maps using this type of data mining methods [30]. The factors selected include slope, aspect, distance from faults, altitude, lithology, rainfall, distance from streams, distance from roads, plan curvature, progfile curvature, slope angle and land use. The accuracy and prediction rate achieved by this method are 82.1% and 76.9%, respectively.

8 Linear Regression

Srivastava et al. [1] presented linear regression which is most commonly used for the correlation between the dependent and independent variables [43, 44]. The main principle behind this model is to obtain the optimal curve that fits the observed data. It uses only one input variables which is known as simple linear regression. When considering multiple independent variables the approach is termed as multiple linear regression.

9 Stacked Autoencoder (StAE)

The author [45] deals with the new technique called Stacked Autoencoder (StAE) which is a special type of multiplayer perception in artificial neural network, and it is unsupervised machine learning algorithm with an asymmetric structure in which the middle layer denotes the encoding of the input data in the bottleneck layer [46]. Here the author hybrids the Stacked Autoencoder (StAE) with Random Forest (RF). It produces better performance when StAE and RF works individually.

10 Sparse Autoencoder (SpAE)

Sparse Autoencoder (SpAE) is composed of an input layer, hidden layer and an output layer [45]. This method can effectively prevent overfitting and gradient disappearance [47]. The author combines the SpAE with RF to improve their performance. Thus, the result shows better performance when combining SpAE with RF than individual results.

11 SAR Interferometry

For a successful generation of an interferogram, proper data pair selection is important [48]. The main challenge in preliminary identification is spatial and temporal variability state of atmosphere which superimposes another signal. Synthetic Aperture Radar (SAR) considered input data, such as rainfall and slope data, and the performance is measured by confusion matrix which results in better accuracy (Fig. 8).

12 Novel Ensemble Classifier Model

The author [18] developed a new approach named as a novel ensemble classifier model, which is a combination of Naive Bayes classifier and SVM. The conditioning factors selected for landslide prediction are slope, land use, rainfall, elevation, geology, soil texture, distance from the road and distance from the river. This model has the best degree of fit to landslide susceptibility assessment compared to other models.

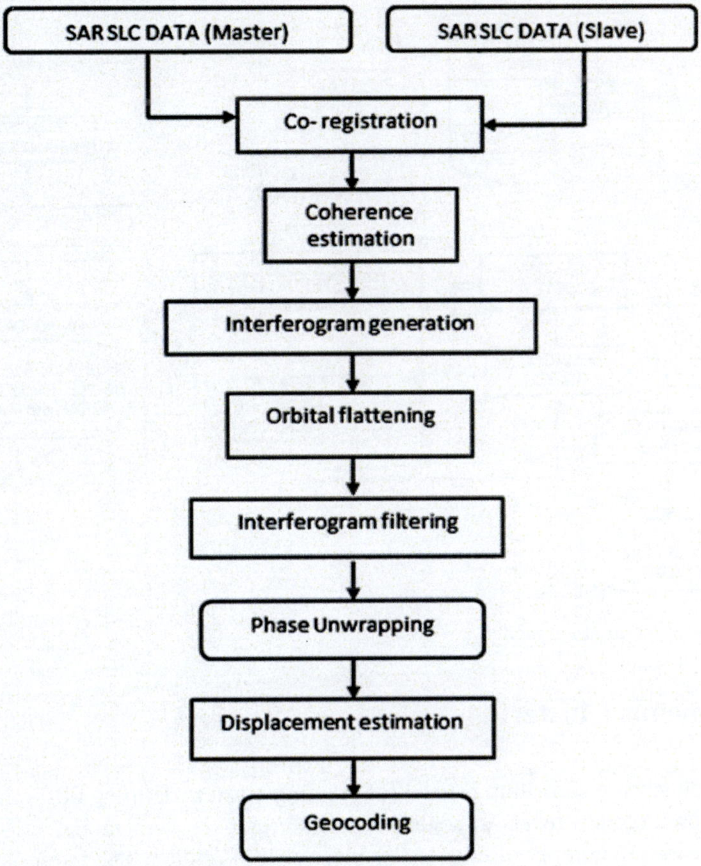

Fig. 8 Flow diagram depicting the two-pass InSAR technique

13 Distributed Landslide Displacement Prediction with Gradient Boosted Trees (DLDP–GBTs)

Zhao et al. [32] present Distributed Landslide Displacement Prediction with Gradient Boosted Trees (DLDP–GBTs) algorithm to predict the horizontal displacement of landslide monitoring. The factors considered for landslide prediction are rainfall, weather, reservoir water level and humidity. The Mean Square Error (MSE) and Mean Absolute Error (MAE) of DLDP–GBTs are 88.127 and 7.2071, respectively.

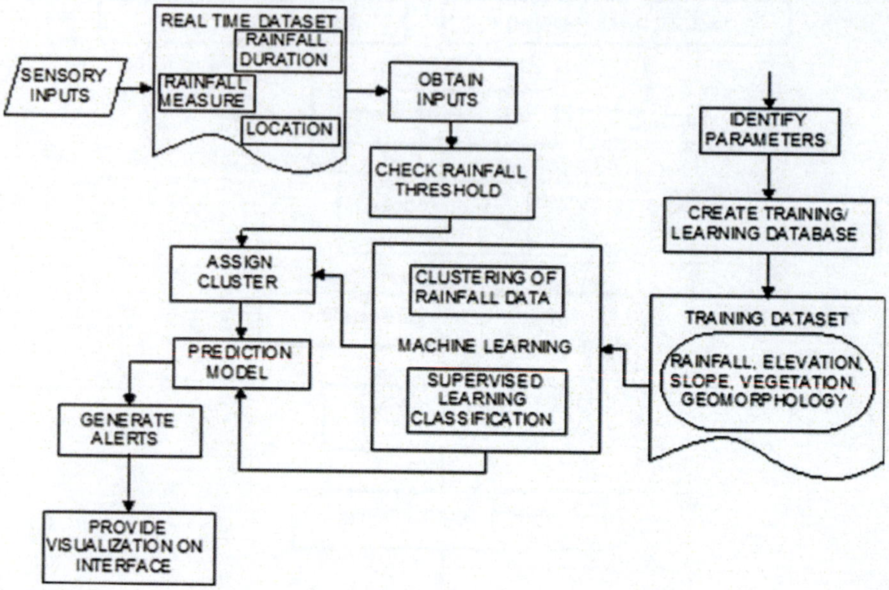

Fig. 9 Flow diagram of the model

14 K-means Clustering

K-means clustering algorithm is used for creating clusters defining different rainfall values to predict alert levels in a susceptibility area [42]. This model separates the whole data in a *K* number of clusters by mean values of data. The input parameters considered for triggering a landslide are soil properties along with other geological actors and rainfall (Fig. 9).

15 Boosted Regression Tree (BRT)

Boosted Regression Tree (BRT) is developed to produce landslide susceptibility maps [30]. The factors considered for these studies are slope, aspect, altitude, distance from faults, plan curvature, lithology, profile curvature, rainfall, distance from streams, distance from roads, slope angle and land use. The accuracy and prediction rate obtained from this model are about 95.8% and 86.2%, respectively.

16 Frequency Ratio

Pradhan et al. [36] deal with frequency ratio method to predict landslide susceptibility using GIS and remote sensing data. The conditioning factors considered with SPOT 5 Satellite Pan Sharpened images are slope, aspects, curvature, drainage, lithology type, lineaments, rainfall soil type and Normalized Different Vegetation Index (NDVI). A frequency ratio method is based on observed relationship between the distribution of landslide and each landslide-related factor to reveal the communication between landslide spot and the factors in the landslide zone.

17 Aggregating One Dependence Estimators (AODE)

The author Pham et al. [10] presented a new method named as Aggregating One Dependence Estimators (AODE) classifier which has not been used earlier for landslide prediction. AODE belongs to the larger class of probability classification learning technique that detects the probability distribution over a set of classes using Bayes theorem with the strong dependence assumption amid the input variables [49]. The Area Under the Curve (AUC) value obtained is 0.968 and Root Mean Square Error (RMSE) value is 0.27, respectively. The accuracy achieved by AODE is about 90.25%.

18 Vote Feature Intervals (VFI)

Vote feature intervals classification algorithm is based on attribute discretization [21]. It is a non-incremental approach using a set of feature intervals in representing a range of feature values. The factors selected for landslide analysis are slope angle, slope aspects, curvature, lithology, soil, land cover, distance to roads, elevation, distance to rivers, distance to lineament and rainfall. The land density and sensitivity value are about 4.68 and 76.74%, respectively. The accuracy of this model is about 77.85%.

19 Multiclass Alternating Decision Tree (MADT)

Multiclass alternating decision tree algorithm is an extension of alternating decision tree by multiclass assessment by separating the issues into several two class sub-issues using Logit Boost techniques [6, 7]. The factors considered for analysis are slope, curvature, lithology, land use, distance to road, distance to faults, aspect, elevation, distance to the rivers and rainfall. The Area Under Curve (AUC) value obtained is 0.853.

20 Conclusion

The natural hazards are beyond the control of human beings, but their destruction is often reduced if prediction mechanisms are carried out in advance. In this paper, a detailed survey and description of various machine learning algorithms for landslide prediction were discussed. This review shows the possibility of improving the accuracy of hybrid algorithms than other conventional methods. Further improvement in coupling of machine learning algorithms for landslide hazard field still needs to develop in different cases with various landslide-affecting factors.

References

1. Srivastava S, Anand N, Sharma S, Dhar S, Sinha LK (2020) Monthly rainfall prediction using various machine learning algorithms for early warning of landslide occurrence. In: International conference for emerging technology (INCET), pp 1–7
2. Uma K, Rameshkumar C, Saravanan TR, Khaja MB (2019) Prediction of rainfall induced landslides using machine learning algorithms. Int J Innov Technol Explor Eng (IJITEE) 9:4274–4278
3. Logar J, Turk G, Marsden P, Ambrozic T (2017) Prediction of rainfall induced landslide movements by artificial neural networks. Natl Hazards Earth Syst 1–18
4. Ghorbanzadeh O, Blaschke T, Gholamnia K, Meena SR, Tiede D, Aryal J (2019) Evaluation of different machine learning methods and deep-learning convolutional neural networks for landslide detection. MDPI J 1–21
5. Goldaragm YJ, Mohammadzadeh A, Ardakani A (2016) Fire risk assessment using neural network and logisticregression. J. Indian Soc. Remote Sens. 44:885–894
6. Pham BT, Bui DT, Dholakia MB, Prakash I, Pham HV (2016) A comparative study of least square support vector machines and multiclass alternating decision trees for spatial prediction of rainfall-induced landslides in a tropical cyclones area. Geotech Geol Eng
7. Pham BT, Bui DT, Prakash I, Dholakia MB (2016) Evaluation of predictive ability of support vector machines and Naive Bayes trees methods for spatial prediction of landslides in Uttarakhand state (India) using GIS. J Geomatics 10:71–79
8. Renuga Devi S, Agarwal P, Venkatesh C, Arulmozhivarman P (2014) Daily rainfall forecasting using artificial neural networks for early warning of landslides. In: International conference on advances in computing, communications and informatics (ICACCI), pp 2218–2224
9. Agrawal K, Baweja Y, Dwivedi D, Saha R, Prasad P, Agrawal S, Kapoor S, Chaturvedi P, Mali N, Kala VU, Dutt V (2017) A comparison of class imbalance techniques for real-world landslide predictions. In: International conference on machine learning and data science, pp 1–8
10. Pham BT, Prakash I, Jaafari A, Bui DT (2018) Spatial prediction of rainfall-induced landslides using aggregating one-dependence estimators classifier. J Indian Soc Remote Sens
11. Bhatt J, Gangwar A, Nijhawan R, Gangodkar D (2019) A research on deep learning advance for landslide classification using convolutional neural networks. Int J Innov Technol Explor Eng (IJITEE) 8:903–906
12. Huang L, Xiang L-Y (2018) Method for meteorological early warning of precipitation-induced landslides based on deep neural network. Neural Process Lett
13. Hu C, Wu Q, Li H, Jian S, Li N and Lou Z (2018) Deep learning with a long short-term memory networks approach for rainfall-runoff simulation. Water 10. https://doi.org/10.3390/w10111543

14. Le XH, Ho HV, Lee G, Jung S (2019) Application of long short term memory (LSTM) neural network for flood forecasting. Water. https://doi.org/10.3390/w11071387
15. Swapna M, Sudhakar N (2018) A hybrid model of rainfall prediction using both parameterized and time series models. Int J Pure Appl Math 119(14):1549–1556
16. Dang V-H, Hoang N-D, Nguyen L-M-D, Bui DT (2020) A novel GIS-based random forest machine algorithm for the spatial prediction of shallow landslide susceptibility. MDPI J 1–20
17. Bui T, Pradhan B, Lofman O, Revhaug I (2012) Landslide susceptibility assessment in vietnam using support vector machines, decision tree, and Naïve Bayes models. Math Probl Eng 2012:1–26
18. Madawala CN, Kumara BTGS, Indrathilaka L (2019) Novel machine learning ensemble approach for landslide prediction. Smart Comput Syst Eng 78–84
19. Smola AJ, Schölkopf B (2004) A tutorial on support vector regression. Stat Comput 14(3):199–222. https://doi.org/10.1023/B:STCO.0000035301.49549.88
20. Dou J, Yunus AP, Bui DT, Merghadi A, Sahana M, Zhu Z, Chen C-W, Khosravi K, Yang Y, Pham BT (2019) Assessment of advanced random forest and decision tree algorithms for modeling rainfall-induced landslide susceptibility in the Izu-Oshima Volcanic Island, Japan. Sci Total Environ 332–346
21. Pham BT, Bui DT, Prakash I, Nguyen LH, Dholakia MB (2017) A comparative study of sequential minimal optimization-based support vector machines, vote feature intervals, and logistic regression in landslide susceptibility assessment using GIS. Environ Earth Sci
22. Kavzoglu T, Sahin EK, Colkesen I (2014) Landslide susceptibility mapping using GIS-based multi-criteria decision analysis, support vector machines, and logistic regression. Landslides 11:425–439
23. Peng L, Niu R, Huang B, Wu X, Zhao Y, Ye R (2014) Landslide susceptibility mapping based on rough set theory and support vector machines: a case of the three Gorges area, China. Geomorphology 204:287–301
24. Pourghasemi HR, Jirandeh AG, Pradhan B, Xu C, Gokceoglu C (2013) Landslide susceptibility mapping using support vector machine and GIS at the Golestan Province, Iran. J Earth Syst Sci 2:349–369
25. Yilmaz I (2010) Comparison of landslide susceptibility mapping methodologies for Koyulhisar, Turkey: conditional probability, logistic regression, artificial neural networks, and support vector machine. Environ Earth Sci 61:821–836
26. Kuradusenge M, Kumaran S, Zennaro M (2020) Rainfall-induced landslide prediction using machine learning models: the case of Ngororero district, Rwanda. Int J Environ Res Public Health 1–20
27. Nsengiyumva JP, Valentino R (2020) Predicting landslide susceptibility and risks using GIS-based machine learning simulations, case of upper Nyabarongo catchment. Geomatics Nat Hazards Risk 11:1250–1277
28. Breiman L (2001) Random forests. Mach Learn 45(1):5–32
29. Liyanage LC, Weerakoon OS, Palliyaguru ST, Wimalaratne GDSP (2019) Towards prediction of landslide susceptibility using random forest for Kalutara district, Sri Lanka. In: IEEE region 10 humanitarian technology conference, pp 216–221
30. Youssef AM, Pourghasemi HR, Pourtaghi ZS, Al-Katheeri MM (2018) Landslide susceptibility mapping using random forest, boosted regression tree, classification and regression tree, and general linear models and comparison of their performance at Wadi Tayyah Basin, Asir region, Saudi Arabia. Geocarto Int 33(9)
31. Dou J, Yunus AP, Bui DT, Merghadi A, Sahana M, Zhu Z, Chen C-W, Han Z, Pham BT (2019) Improved landslide assessment using support vector machine with bagging, boosting, and stacking ensemble machine learning framework in a mountainous
32. Zhao J, Liu Y, Ming Hu (2018) Optimisation algorithm for decision trees and the prediction of horizon displacement of landslides monitoring. Asian Conf Artif Intell Technol 2018:1698–1703
33. Santinelli G, Herrera M (2019) A framework for predicting rainfall-induced landslides using machine learning methods. In: European conference soil mechanics and geotechnical engineering (ECSMGE), pp 1–8

34. Brenning A (2005) Spatial prediction models for landslide hazards: review, comparison and Evaluation. Nat Hazards Earth Syst Sci 5(6):853–862
35. Luo X, Lin F, Chen Y, Zhu S, Xu Z, Huo Z, Yu M, Peng J (2019) Coupling logistic model tree and random subspace to predict the landslide susceptibility areas with considering the uncertainty of environmental features. Sci Rep
36. Pradhan B, Lee S (2010) Delineation of landslide hazard areas on Penang Island, Malaysia, by using frequency ratio, logistic regression, and artificial neural network models. Environ Earth Sci 1037–1054
37. Lee S (2005) Application of logistic regression model and its validation for landslide susceptibility mapping using GIS and remote sensing data. Int J Remote Sens 26:1477–1491
38. Ohlmacher GC, Davis JC (2003) Using multiple logistic regression and GIS technology to predict landslide hazard in northeast Kansas, USA. Eng Geol 69:331–343
39. Truong XL, Mitamura M, Kono Y, Raghavan V, Yonezawa G, Truong XQ, Do TH, Bui DT, Lee S (2018) Enhancing prediction performance of landslide susceptibility model using hybrid machine learning approach of bagging ensemble and logistic model tree. MDPI J 1–22
40. Pham BT, Bui DT, Pourghasemi HR, Indra P, Dholakia MB (2015) Landslide susceptibility assesssment in the Uttarakhand area (India) using GIS: a comparison study of prediction capability of Naïve Bayes, multilayer perceptron neural networks and functional trees methods. Theor Appl Climatol 1–19
41. Soni J, Ansari U, Sharma D, Soni S (2011) Predictive data mining for medical diagnosis: an overview of heart disease prediction. Int J Comput Appl 17(8):43–48
42. Kainthura P, Singh V, Gupta S (2015) GIS based model for monitoring and prediction of landslide susceptibility. In: International conference on next generation computing technologies (NGCT), pp 584–587
43. Navid M, Niloy NH (2018) Multiple linear regressions for predicting rainfall for Bangladesh. Communications 6(1), 1–4. Available https://doi.org/10.11648/j.com.20180601.11
44. Prabakaran S, Kumar PN, Tarun PSM (2017) Rainfall prediction using modified linear regression. ARPN J Eng Appl Sci 12(12)
45. Nam K, Wang F (2020) An extreme rainfall-induced landslide susceptibility assessment using autoencoder combined with random forest in Shimane Prefecture, Japan. Geoenviron Disasters 1–16.
46. Yu S, Príncipe JC (2019) Understanding autoencoders with information theoretic concepts. Neural Netw 117:104–123
47. Huang F, Zhang J, Zhou C, Wang Y, Huang J, Zhu L (2019) A deep learning algorithm using a fully connected sparse autoencoder neural network for landslide susceptibility prediction. Landslides 17:217–229
48. Kalaranjini VS, Ramakrishnan SS (2020) Landslide investigation using SAR Interferometry on selected regions of Idukki district, Kerala, India. Indian J Geo Marine Sci 49:882–888
49. Htike ZZ, Win SL (2013) Classification of eukaryotic splicejunction genetic sequences using averaged one-dependence estimators with subsumption resolution. Procedia Comput Sci 23:36–43

Chapter 38
Fabrication of Serpentine-Structured Flexible Strain Sensor of Graphene and Their Potential Applications in Robotics

Karamvir Singh, Monish Gupta, and Chandra Charu Tripathi

1 Introduction

From last decade, the demand for flexible electronics is increasing day by day. Strain sensor is the basic component of flexible electronics. It finds applications in various fields like robotics, aerospace, human motion detection, etc. [1, 2]. Sensitive and conductive functional materials, flexible substrates, simple and economical fabrication methods, suitable design and structures are some important requirements of a flexible strain sensor [3]. The strain sensors detect the change in resistance or capacitance (depending on the type of sensor used) resulted from the deformation created by applied force/strain [4].

The proper selection of functional materials improves the sensitivity and durability of the device. Conventionally used materials were rigid in nature and were not suitable for flexible applications. But the development of some advanced nanomaterials increases the growth in the fields of flexible electronics. Metal nanowires, liquid metals, conducting polymers and carbon-based material (CNT, graphene) are currently used materials for the fabrication of strain sensors [5–9]. Liquid metals are attracting deformable candidate but their toxic nature and sensitivity in small strain applications limit their use [10]. The metal nanowires are good conductor and their compatibility in composite formation with other solvents is excellent. But high cost and non-linear behaviour in large-scale applications are not preferred in sensing applications. Metal nanowires also produce poor adhesion on polymer substrates [11]. The conductive polymers faces the issue of poor conductivity and complex material structure [12]. The carbon nanomaterials, especially graphene, have gained a huge attention in the area of flexible electronics due to its excellent electrical, mechanical and optical properties [13–15].

K. Singh (✉) · M. Gupta · C. C. Tripathi
University Institute of Engineering and Technology, Kurukshetra University, Kurukshetra 136119, India
e-mail: karamvir2018@kuk.ac.in

N. Marriwala et al. (eds.), *Soft Computing for Intelligent Systems*, Algorithms for Intelligent Systems, https://doi.org/10.1007/978-981-16-1048-6_38

In recent times, the demand of electronics devices is increasing in robotics applications. For the accurate performances of any robot, some suitable sensitive devices are required. Flexible strain and pressure sensors are commonly used in bending and stress measurements. The strain sensors are used to control the movement of a flexible part at a particular angle. The strain sensor observes the deformation created by bending movement of a robotic part by detecting the change in resistance [16, 17]. In a bending movement, some breaks in conductive paths take place that results in increasing resistance. Higher sensitive device produces large resistance variations [18].

In this paper, a flexible strain sensor is fabricated on a flexible polymer substrate. Graphene is used as conductive functional nanomaterials. The sensitivity of a specific structured device is higher in comparison to wide area thin film. So here we demonstrated serpentine-structured strain sensor on a PET sheet. A standard and simple fabrication method is used for the development of thin layer and multi-layer of graphene. With the complete process of fabrication, the bending measurements of the device are highlighted for the potential robotic applications.

2 Materials and Methods

Ugray^{tm} few layer graphene (3–6 layer) with a purity of 96–99% purchased from United Nanotech Innovations Pvt. Ltd India was mixed with NMP (N-Methyl-2-pyrrolidone) (1:1) and sonicated for 5 h. Then 10 drops of ethylene glycol were added in the solution and magnetic stirring was given for half an hour to achieve the suitable dispersions. Thin film of graphene was achieved by using Langmuir–Blodgett (LB) technique. In LB technique, the base tank is filled with DI water and the prepared graphene solution was spread over the DI water using a syringe. In the aqueous solution, the amphiphilic substance molecules started to float onto the air–water interface. The two heads mounted above the tank were moved from corner to the centre of the tank to compress the surface layer and a monomolecular surface layer is produced at the water–air interface. Then a PET (polyethylene terephthalate) substrate is immersed into the solution and a thin layer is formed on both sides of the substrate during extraction as shown in Fig. 1. The performance of flexible sensing devices like sensitivity and strain capabilities can be enhanced by developing ingenious structures. In this paper, we prepared a serpentine structure on flexible substrate and tested for the strain-sensing applications. After thin film deposition, a brown tape was used for the patterning of thin film. A serpentine-structured tape is prepared and pasted onto the thin film. Then the deposited film was etched with isopropyl alcohol and serpentine-structured deposition was achieved. Silver paste is used for contact formation.

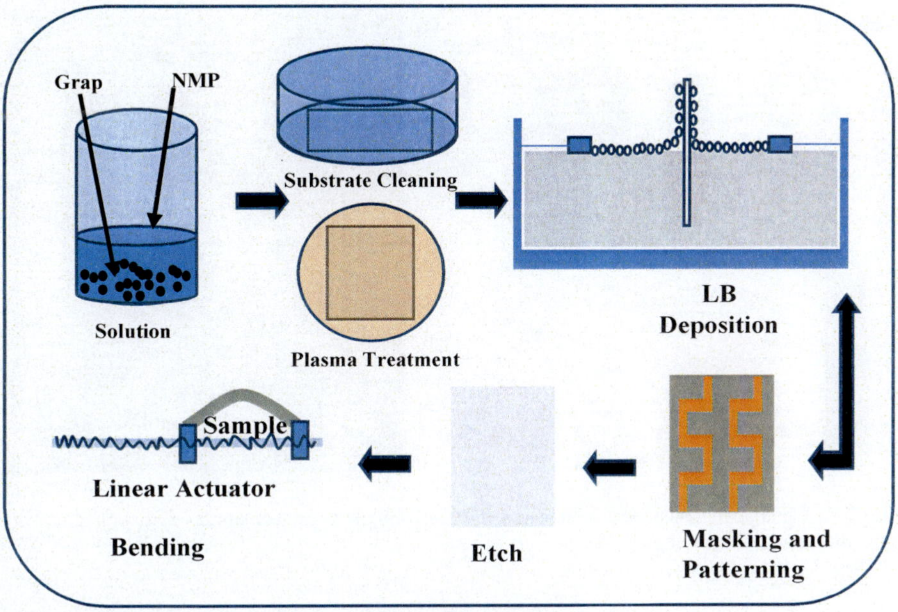

Fig. 1 Serpentine-structured strain sensor fabrication process

3 Result and Discussion

The fabricated device was tested for bending measurements by using a linear actuator which has one fixed plate and one movable plate. The substrate was attached to the top surface of the plates. The variable bending was provided by varying the chord length with the help of moving plate. The corresponding change in resistance with respect to change in chord length was measured and plotted as shown in Fig. 2a. The chord length varied from 45 to 10 mm. As the chord length decreases, a bending in the layer surface takes place. This bending creates gap in the graphene particles and this gap breaks the conductive path in the deposited layer and results in an increment of resistance. More the bending, more will be the resistance increment. The serpentine structures providing higher sensitivity as disturbances in particle spacing in the conducting paths may take place at multiple places during bending applications and lesser number of conductive paths are available in this shape in comparison to wide area thin films.

The direct deposition of graphene layer on PET substrate is difficult due to poor adhesion. So before the deposition process, substrate was cleaned and treated with plasma oxidation. The plasma oxidation improves surface functionalization and wettability which enhances the adhesion capability of the substrate [19, 20]. The plasma was achieved in a chamber by providing a suitable vacuum and continuous flow of oxygen gas. The time of plasma also plays important role in the surface treatment. A 5 min plasma provided better adhesion in comparison to 10, 15 and 20 min

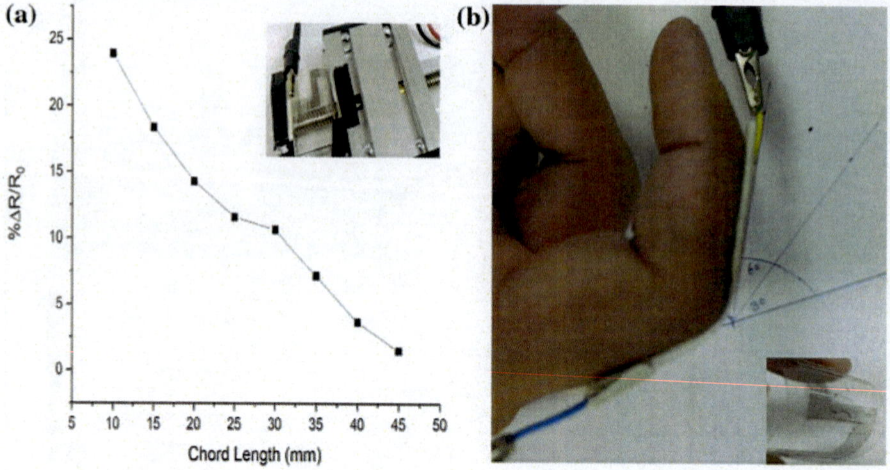

Fig. 2 **a** Change in resistance with variable chord length **b** Sensor attached to finger and its movements at different angles

exposure. The adhesion and the conductivity of the deposited film also depend on the number of dipping/immersing as shown in Table 1. The conductivity of thin film increased with the increase in number of dipping as more layers stacked onto the substrate. The immersion of the substrate is precisely controlled and perfectly maintained so that a thin and uniform layer can be deposited. The thickness of the film increases with the number of dipping and with increasing number of stacked layers onto the substrate the adhesion and uniformity of the film decrease.

The fabricated device was implemented for strain-sensing applications. Various sensors are required in any robotic device for smooth movement of individual part. The controlling of these parts significantly depends on the performance of the sensor used. A suitable strain sensor can measure the movement of a robotic device precisely and can control its direction of movement. In this paper, a flexible strain sensor is presented which is developed by LB technique.

The sensitivity of the strain sensor was observed by using a linear actuator and a linear variation can be seen as the chord length varies from 45 to 10 mm in Fig. 2a. To be used in a robotic application the device must respond at variable angles. So we attached the fabricated strain sensor onto a finger and various movements were implemented. A significant resistance variation was observed during bending of the finger at different angles. The bending of finger with attached strain sensor on PET

Table 1 Resistance variation with plasma oxidation and number of dipping

	Plasma oxidation (minutes)			Number of dipping		
	5	10	15	1	2	3
Resistance	67 KΩ	200 KΩ	1.2 MΩ	67 KΩ	22 KΩ	6 KΩ

substrate is shown in Fig. 2b. The PET substrate is suitable for bending applications but the stretchability of this substrate is almost nil. So this is not suitable for stretching applications. The device was tested for limited number of cycles, but during repeated strain applications, the surface layer may not be able to regain its original state due to non-stretchability of PET substrate.

4 Conclusion

One key path in future of robotics is to develop conductive and flexible devices with suitable sensing capabilities. Here we presented a serpentine-structured stain sensor which can be implemented as a sensing part of any robotic device in the detection of various movements. The sensor is fabricated by using LB technique which is a standard method of thin film deposition. The impact of plasma oxidation on adhesion of graphene solution is shown. The conductivity of the deposited film can be controlled by speed of immersion and number of dipping. To achieve suitable sensitivity, the thin film is patterned into serpentine structure by using a suitable mask. The strain sensing is observed by using linear actuator and the device has shown a significant variation during variable bending. A disturbance in conductive path takes place as the chord length of the substrate decreases from 45 to 10 mm. To fit for robotic applications, the device was attached to the finger and significant sensing performances were observed in finger movements at different angles.

Acknowledgements This work has been carried out at Kurukshetra University Kurukshetra-136119. We are thankful to TEQIP (World Bank project) and NPIU (National Project Implementation Unit) India for providing research assistantship.

References

1. Li Y, Samad YA, Taha T, Cai G, Fu S-Y, Liao K (2016) Highly flexible strain sensor from tissue paper for wearable electronics. ACS Sustain Chem Eng 4(8):4288–4295
2. Ju H, Jeong J, Kwak P, Kwon M, Lee J (2018) Robotic flexible electronics with self-bendable films. Soft Robot 5(6):710–717
3. Yan Z et al (2019) Stretchable micromotion sensor with enhanced sensitivity using serpentine layout. ACS Appl Mater Interfaces 11(13):12261–12271
4. Sun Y, Thompson SE, Nishida T (2009) Strain effect in semiconductors: theory and device applications. Springer Science & Business Media
5. Kim T, Kim D, Lee BJ, Lee J (2019) Soft and deformable sensors based on liquid metals. Sensors 19(19):4250
6. Chen C, Ye C (2018) Metal nanowires. Transparent Conduct Mater Mater Synth Charact Appl 105–131
7. Borghetti M, Serpelloni M, Sardini E, Pandini S (2016) Mechanical behavior of strain sensors based on PEDOT: PSS and silver nanoparticles inks deposited on polymer substrate by inkjet printing. Sens Actuators A 243:71–80

8. Yu Y et al (2017) Flexible and transparent strain sensors based on super-aligned carbon nanotube films. Nanoscale 9(20):6716–6723
9. Nie M, Xia Y, Yang H (2019) A flexible and highly sensitive graphene-based strain sensor for structural health monitoring. Cluster Comput 22(4):8217–8224
10. Ren Y, Sun X, Liu J (2020) Advances in liquid metal-enabled flexible and wearable sensors. Micromachines 11(2):200
11. Guo X et al (2017) Highly stretchable strain sensor based on SWCNTs/CB synergistic conductive network for wearable human-activity monitoring and recognition. Smart Mater Struct 26(9):95017
12. Le T-H, Kim Y, Yoon H (2017) Electrical and electrochemical properties of conducting polymers. Polymers (Basel) 9(4):150
13. Anzar N, Hasan R, Tyagi M, Yadav N, Narang J (2020) Carbon nanotube-a review on synthesis, properties and plethora of applications in the field of biomedical science. Sens Int 1:100003
14. Mehmood A et al (2020) Graphene based nanomaterials for strain sensor application—a review. J Environ Chem Eng 8(3):103743
15. Singh K, Sharma S, Shriwastava S, Singla P, Gupta M, Tripathi CC (2020) Significance of nano-materials, designs consideration and fabrication techniques on performances of strain sensors-a review. Mater Sci Semicond Process 105581
16. Girão PS, Ramos PMP, Postolache O, Pereira JMD (2013) Tactile sensors for robotic applications. Measurement 46(3):1257–1271
17. Ramalingame R, Bouhamed A, Rajendran D, da Veiga Torres R, Hu Z, Kanoun O (2020) Highly sensitive polymer/multiwalled carbon nanotubes based pressure and strain sensors for robotic applications. In: New trends in robot control. Springer, Singapore, pp 371–382
18. Duan L, D'hooge DR, Cardon L (2019) Recent progress on flexible and stretchable piezoresistive strain sensors: from design to application. Prog Mater Sci 100617
19. Mandolfino C (2019) Polypropylene surface modification by low pressure plasma to increase adhesive bonding: effect of process parameters. Surf Coat Technol 366:331–337
20. Martínez-Landeros VH, Vargas-Islas SY, Cruz-González CE, Barrera S, Mourtazov K, Ramírez-Bon R (2019) Studies on the influence of surface treatment type, in the effectiveness of structural adhesive bonding, for carbon fiber reinforced composites. J Manuf Process 39:160–166

Chapter 39
Conformal Patch Antenna Array for ISM Band

Monish Gupta, Nikhil Marriwala, Vikas Mittal, Karamvir Singh, Parveen Singla, and Sandeep Sharma

1 Introduction

The structure of microstrip patch antenna includes a radiating patch, ground plane, substrate and a feeding line. Substrate is sandwiched between ground and patch. Signal to be radiated is applied between patch and ground using microstrip line. Low-profile microstrip patch antenna structure is shown in Fig. 1.

The frequency and efficiency of radiation for patch antenna depend on the dimensions, structure, and material of both patch and substrate. Radiation frequency and efficiency of a patch antenna are dependent on the length of patch antenna and width of patch antenna, respectively. The optimized expressions proposed for design of patch antenna are [1–4] as follows.

For radiating a signal of frequency (f), required patch antenna width (w) is:

$$w = \frac{c_0}{2f\sqrt{\frac{\varepsilon+1}{2}}} \tag{1}$$

where ε is the permittivity of the substrate.

M. Gupta (✉) · K. Singh · S. Sharma
University Institute of Engineering and Technology, Kurukshetra University, Kurukshetra 136119, India
e-mail: monish_gupta1976@kuk.ac.in

V. Mittal
National Institute of Technology, Kurukshetra 136119, India

P. Singla
Chandigar Engineering College, Landran, Mohali, Punjab 140307, India

N. Marriwala
Department of Electronics and Communication Engineering, Kurukshetra University, Kurukshetra, Haryana, India

N. Marriwala et al. (eds.), *Soft Computing for Intelligent Systems*, Algorithms for Intelligent Systems, https://doi.org/10.1007/978-981-16-1048-6_39

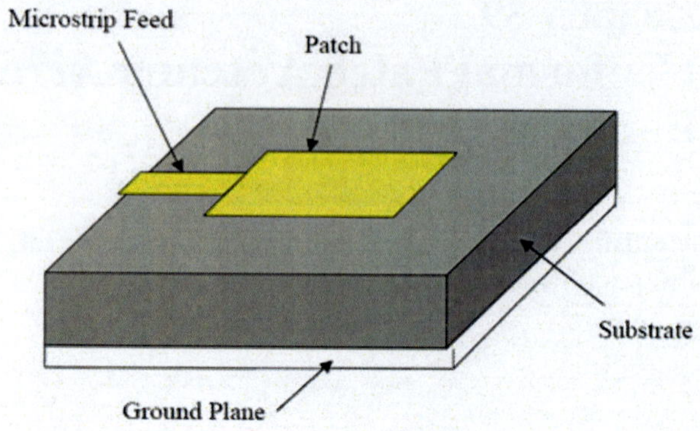

Fig. 1 Structure of microstrip patch antenna

And $c_0 = 10^8$ m/s is velocity of light.

The effective optimized permittivity of medium is:

$$\varepsilon(\text{eff}) = \frac{\varepsilon + 1}{2} + \frac{\varepsilon - 1}{2}\sqrt{1 + 12\frac{h}{w}} \tag{2}$$

where h is substrate.

Optimized required patch antenna length *(L)* is

$$L = \frac{C}{2f\sqrt{\varepsilon}} - 2\Delta l \tag{3}$$

where Δl is extension in length of patch due to available fields.

$$\frac{\Delta l}{h} = 0.412\frac{[\varepsilon(\text{eff}) + 0.3][\frac{w}{h} + 0.264]}{[\varepsilon(\text{eff}) - 0.258][\frac{w}{h} + 0.8]} \tag{4}$$

Patch antenna elements can be connected in series (Fig. 2) or in corporate manner (Fig. 3) to improve the gain of antenna. In series-fed microstrip antenna array, one antenna element is used to feed the successive element in array. This is the simplest possible architecture to obtain antenna array. In this type of feeding methodology, it is difficult to account for mutual coupling between the elements and hence it is very difficult to account for the radiation frequency of the structure. Further the strength of feeding signal goes on reducing as signal travels from one antenna element

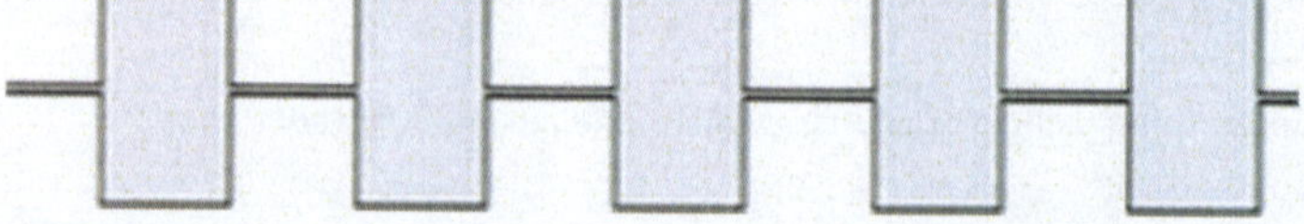

Fig. 2 Series-fed patch array

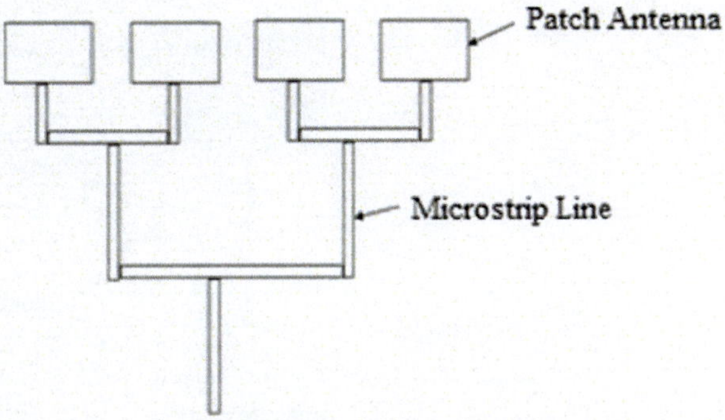

Fig. 3 Corporate-fed patch array

to successive antenna element. This results in reduction of overall gain of array. Due to these drawbacks this method of feeding the antenna elements is rarely used. Generally, antenna array is obtained using corporate-fed system.

In corporate-fed system, all elements in the array are fed simultaneously. Phase and amplitude of excitation signal to all the elements remain the same. Mutual coupling between the elements in corporate-fed system is corporately less than series method which can further be reduced by using filter between the elements. Further the impedance-matching techniques of corporate-fed method are better than that of series-fed method. Various feeding methods like Wilkinson power divider, Balun, and feeding using N node microstrip line can be used to divide power and impedance-matching technique in corporate-fed patch array.

In this study, a corporate-fed microstrip patch antenna array is designed and simulated. As per the maximum power transfer theorem, maximum power will be transferred from source to antenna elements if impedance of source is equal to impedance of antenna elements. Impedance matching between the elements and source is achieved using microstrip lines. It is possible to achieve the desired characteristics impedance of microstrip line by varying its width [5–7]. As microstrip lines are required to be connected in parallel with source in order to feed the antenna elements, required characteristic impedance of microstrip line between source and patch element should be equal to n times the impedance of the source. Here n is the number of twigs emanating from the source. If available source impedance is Z_s then for n twigs from the source required characteristics impedance of microstrip line is $Z_l = n\, Z_s$.

For the structure shown in Fig. 4, impedance of microstrip line required is $Z_l = Z_s$ (source impedance). For the structure shown in Fig. 5 required $Z_l = 2Z_s$.

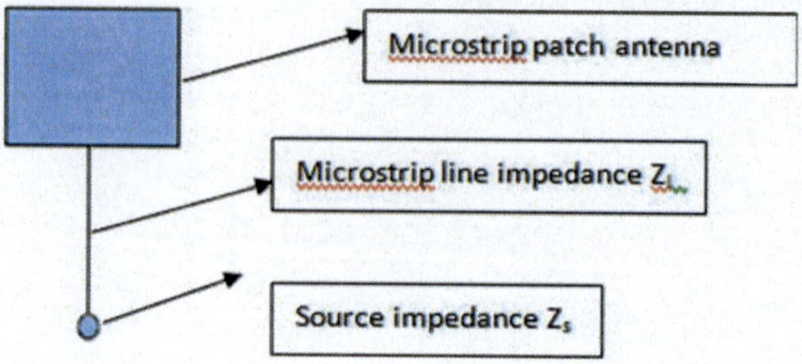

Fig. 4 Patch antenna fed with source

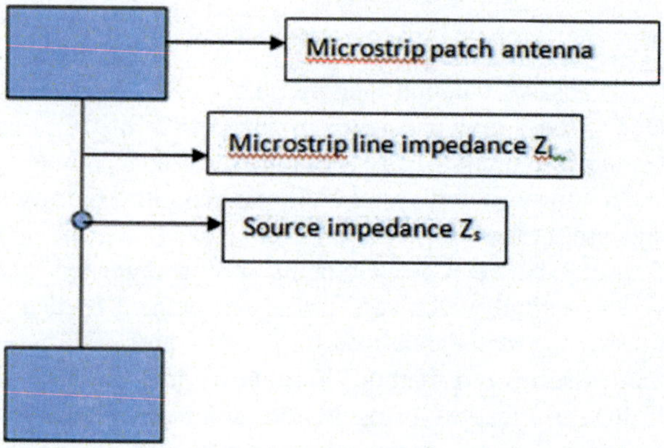

Fig. 5 Patch antenna array fed with source

2 Results and Discussion

The optimized patch antenna dimensions for radiating signal of frequency 2.4 GHz in this study are presented in Table 1.

The optimized microstrip line dimensions used for 50 Ω characteristics impedance at 2.4 GHz are presented in Table 2.

Microstrip patch antenna designed using HFSS is shown in Fig. 6.

The polar radiation pattern of gain for this structure is shown in Fig. 7.

The radiation pattern for this antenna is plotted in Fig. 8.

The maximum directive gain achieved for designed patch antenna is 6.9 dB.

A two-element patch antenna array designed using HFSS is drawn in Fig. 9.

Table 1 Optimized patch antenna dimensions

Substrate	Rogers RT/duroid 58,809 (tm)
Substrate permittivity	2.2
Substrate height	0.887 mm
Patch length	31.7 mm
Patch width	37.93 mm
Patch thickness	0.05 mm

Table 2 Optimized microstrip line dimensions for 50 Ω

Substrate material	Rogers RT having permittivity 2.2 (mm)
Substrate height	0.787
Microstrip line length	6.5
Microstrip line width	3.46
Microstrip thickness	0.05

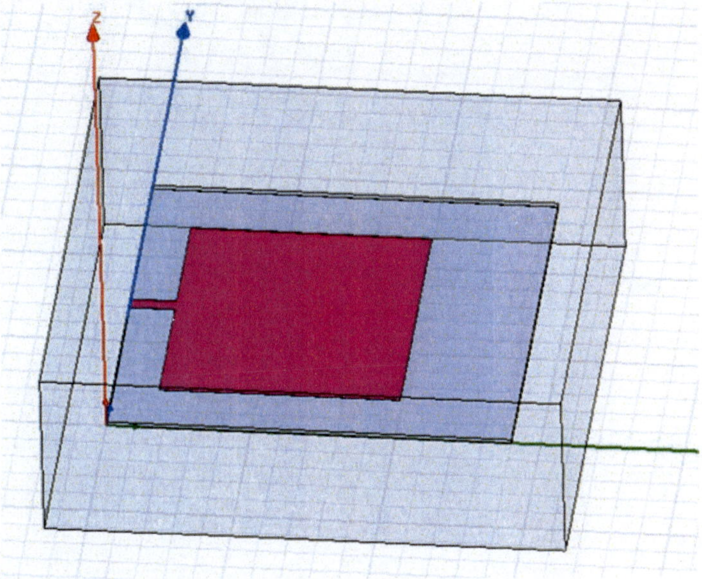

Fig. 6 Microstrip patch antenna using HFSS

Fig. 7 Polar radiation pattern

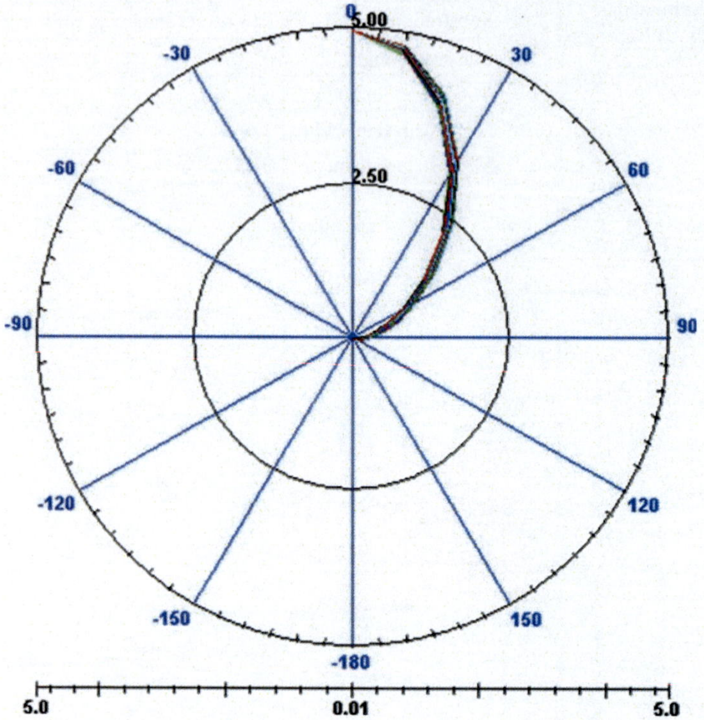

Fig. 8 Radiation pattern of patch antenna

Dimensions of microstrip line for characteristics impedance of 100 Ω at 2.4 GHz are given in Table 3.

The scattering parameters obtained for this structure are shown in Fig. 10.

The polar radiation pattern of gain for this structure is shown in Fig. 11.

The radiation pattern for this antenna is plotted in Fig. 12.

From the plotted radiation pattern, it is evident that maximum directive gain achieved for this antenna array is 9.02 dB.

3 Conclusion

A two-element microstrip patch antenna array is successfully designed using microstrip line. The gain of array had been improved by a factor of 2.1 dB. Gain of the antenna system can further be improved by increasing the patch elements in the array.

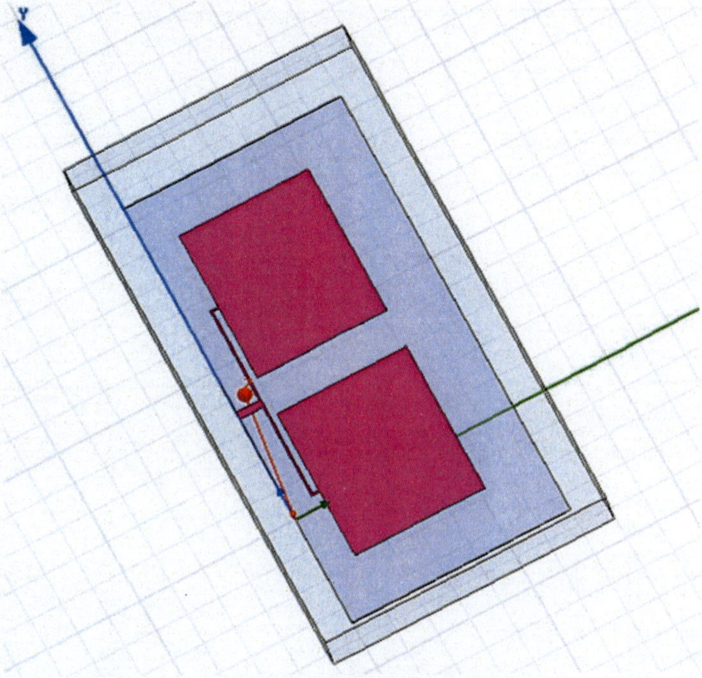

Fig. 9 Microstrip patch antenna array designed using HFSS

Table 3 Dimensions of microstrip line for characteristics impedance of 100 Ω

Substrate material	Rogers RT having permittivity 2.2 (mm)
Substrate height	0.787
Microstrip line length	13.5
Microstrip line width	3.46
Microstrip thickness	0.05

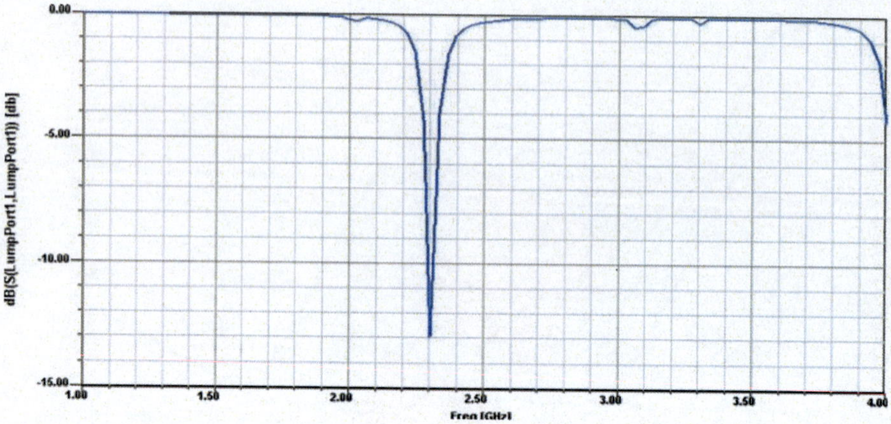

Fig. 10 S-parameters of microstrip patch antenna array

Fig. 11 Polar radiation pattern

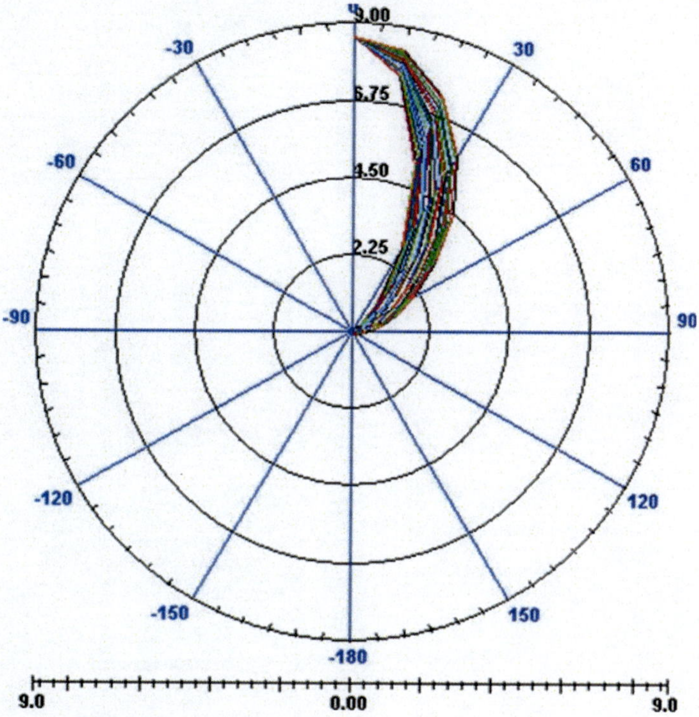

Fig. 12 Radiation pattern of patch antenna array

References

1. Gupta M, Saxena J, Vohra A (2009) Highly directive patch antenna using planar metamaterial. Int J Appl Eng Res (IJAER) 4(3):89–96
2. Balanis CA (1997) Antenna theory: analysis and design, 2nd edn. Wiley, New York
3. Pozar DM (1998) Microwave engineering. Wiley
4. Hammerstad E, Bekkadal FA (1975) Microstrip handbook, ELAB report, STF 44 A74169. University of Trondheim, Norway
5. Hammerstad E (1980) Accurate models for microstrip computer aided design. In: Symposium on microwave theory and techniques, pp 407–409
6. Hammerstard EO (1975) Equations for microstrip circuit design. In: Proceedings of the European microwave conference. Germany, pp 268–272
7. Hong JS, Lancaster MJ (2001) Microstrip filters for RF/microwave applications. Wiley, New York, NY, USA

Chapter 40
Techno-Economic Analysis of a Microgrid for a Small Community

Vijay Kumar Garg and Sudhir Sharma

1 Introduction

Worldwide, the electricity demand is increasing day by day and still there is a gap between demand and availability. There is a huge population who does not have access to electricity or it is very less as compared to the demand for electricity. The energy demand is increasing at a rate of 1.3% every year [1]. As per the Paris agreement (UNFCCC, 2015), to limit the global temperature rise, more and more use of renewable energy sources is essential. The curtailment in greenhouse emission is not as per the desire. Renewable energy use is increasing but the rate of increment in the use of renewable energy sources is not up to the mark. Solar and wind power can play an important role to meet the demand for electricity worldwide. Cost reduction and digital technology can play a vital role to raise generation from solar and wind energy sources.

Electricity demand is made by various energy resources like thermal, hydro, nuclear and renewable. Energy share of renewable energy sources mainly solar PV and wind are growing. Since 2000, approximately 700 million people got electricity access whereas LPG connection has been distributed to about 80 million people for clean cooking [2]. The Government of India is providing secure, affordable and sustainable energy while meeting the target to set up new renewable energy plants to reduce local air pollution. The total installed capacity as on 30.04.2020 is 370347.52 MW having contribution of thermal as 62.68%, renewable as 23.06%, hydro as 12.40% and nuclear power as 1.85% as shown in Fig. 1. As per a report of the Ministry of New and Renewable Energy (MNRE) on 30.04.2020, out of total

V. K. Garg (✉)
(Research Scholar) IKG Punjab Technical University, Jalandhar, India

Department of Electrical Engineering, UIET, Kurukshetra University, Kurukshetra, India

S. Sharma
Department of Electrical Engineering, DAV Institute of Engineering and Technology, Jalandhar, India

N. Marriwala et al. (eds.), *Soft Computing for Intelligent Systems*, Algorithms for Intelligent Systems, https://doi.org/10.1007/978-981-16-1048-6_40

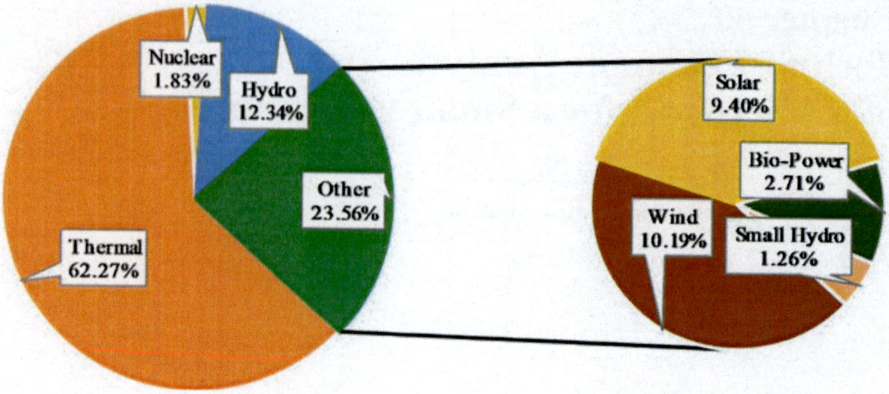

Fig. 1 Total installed capacity of India

renewable energy of 84,399.90 MW, wind power is 37278.68 MW, solar power is 32527.55 MW, bio power is 9946.11 MW and small hydropower is 4647.56 MW [3].

Renewable Energy Resources (RERs) can fulfill the growing energy demand but their intermittent nature is a major obstacle towards their sustainable development. By operating renewable sources in hybrid mode, this problem can be overcome. HRES can meet the growing energy demand and gives a feasible and cost-effective solution.

As from literature, it appears that various researchers have performed hybrid optimization studies using HOMER, MATLAB and other software tools. As per the availability of resources at the study site, researchers have performed various studies with different combinations of RERs such as PV, wind, biomass, biogas, gasifier, biodiesel, micro-hydro, etc. with various inverter and battery systems in a grid and off-grid mode. Analysis of various HRES has been performed to meet the electricity demand at various study sites worldwide. As per the conveniences of resources at the study site, researchers have implemented various studies with different configurations of RERs such as photovoltaic (PV), wind, biomass, biogas, gasifier, biodiesel, micro-hydro, etc. with various inverter and Energy Storage System (ESS) in a grid and off-grid mode. Analysis of various HRES has been performed to meet the electricity demand at various study sites worldwide. Krishan et al. [4] propose an optimum design of HRES based on techno-economic analysis to meet domestic and agriculture needs of a small community at Yamuna nagar, India. The results show that PV/wind/battery hybrid system is the most economical and sustainable solution at the study site. Size optimization of a hybrid system has been performed for off-grid and grid-connected microgrid for various locations in Bangladesh [5]. Sensitivity analysis on the optimum combination shows its feasibility with a variation of different input parameters. Techno-economic analysis on an off-grid hybrid system consisting of PV/wind/diesel/battery system has been designed for a village in Jordan [6]. The result shows that a 50% renewable faction is technically and economically feasible to meet the load demand. Oueslati et al. [7] analyse the hybrid energy system on a technical and economic basis at Tunisia and emphasizes the integration of renewable energy sources with electric grid to cope with increasing electricity charges and future energy need. Techno-economic analysis for the two cities of Iran was conducted, and it is found that Zanjan is feasible for solar and wind power plant. At the same time, Zehadan is economical for wind power plants [8]. Therefore, to achieve sustainable development, a proper feasibility study of sources should be performed. Bendaoud et al. [9] investigated the benefit of installing a solar system on public buildings at Algeria. The results show that maximum production has been achieved at an optimum angle of inclination of 32° at the study site. A study conducted on the remote villages at Karnataka, India, shows that a grid-independent hybrid system consisting of PV/MHP/battery is a feasible solution for un-electrified villages [10]. The energy cost reduces by 35% with usage of Li–ion battery instead of lead–acid battery. Similarly, a techno-economic analysis has been conducted at sub-tropical residents at Aligarh, India [11]. The result shows that during a grid outage, PV/diesel/battery system is the best-suited system with lesser overall operating costs and reduced environmental hazards. The use of renewable energy sources reduces the cost and improves the overall health of the environment. From the surveyed literature, it was observed that RERs vary as per the climate and geographical location of the study site. The utility-grid-connected hybrid system is more reliable and efficient. The excess electricity generated in the off-grid period can be sold back to the utility grid. Keeping all in mind, the contribution of this research is to determine the optimum sizes of HRES combinations (PV/wind/battery/utility grid) along with economic parameters by considering load demand and availability of RERs at the study site.

The paper is arranged as follows. Section 2 described about system modelling; Sect. 3 gives a brief overview about software tool considered in this study; Sect. 4 depicts load and resource assessment at study site and various cost parameters related to the HRES; Sect. 5 presents the simulation results; Sect. 6 described about various sensitivity analyses carried out on optimum system; and Sect. 7 concludes the papers with future research directions.

2 System Modelling

Techno-economic analysis and optimum design of HRESs have been performed by using HOMER Pro (version 3.13.8). HOMER can perform optimization, feasibility and sensitivity analysis. The software is capable to analyse hourly energy balance as well as a feasibility study by imposing several constraints on the island and grid-connected systems for remote, autonomous and distributed generation applications. Energy supply systems are evaluated based on technical, economic and environmental aspects. The optimum system configuration is found out by minimizing an objective function like NPC with system constraints like a balance of power, demand management and penetration of renewable sources. Before performing the optimization, various assessments are to be done as Pre-HOMER analysis. In the first step, annual electricity load demand has been estimated based on similar types of work by other researchers and personal judgement. After that the resource data is being collected from different references. Based on HRES components, the micro-grid system model has been prepared in the software along with cost data with technical specifications. The pre-HOMER assessments are given to the software for techno-economic analysis.

The study site is a small location near village Tandwal, having a population of 50 which are residing in 10 households located at 30° 14.2′N and 76° 58.4′E, Ambala, Haryana, India.

2.1 Load Profile

The energy requirement of the residential community is not as par with the requirement of urban community. The detailed information of load is given in Table 1. The electricity load demand estimation is done carefully for two sessions by considering the appliances requirement and duration of usage for residential and agriculture load. The demand of electricity is high in early morning and evening time, when most of the family members are at home. In day time, they are either busy in field job or the children at school so load will be on lower side.

2.2 Solar Power

The solar cell is a basic module of electricity generation from solar energy. The P_{PV} output power of the PV array in kW can be calculated as [12, 13]

$$P_{PV} = Y_{PV} f_{PV} \frac{G(\beta, \alpha)}{G_{STC}} [1 + \gamma (T_C - T_{STC})] \tag{1}$$

where Y_{PV} denotes the rated capacity in kW, f_{PV} represents the PV derating factor, $G(\beta, \alpha)$ characterizes net irradiance in kWm^{-2}, G_{STC} is incident irradiance in Wm^{-2}, γ represents temperature coefficient, T_C is cell temperature in °C and T_{STC} is cell temperature under STC in °C. The cost of PV system is considered as 800 $/kW, 750 $/kW, 10 $/kW/year for capital, replacement, and operating and maintenance cost, respectively. The cost parameters considered are variable and extracted from available literature, local Indian manufacturer and distributor.

The bar and curve charts in Fig. 2 represent the monthly average daily solar radiation and clearness index. At the study site, July and August being monsoon season, so due to cloud, the value of clearness index is low in this duration of the

Table 1 Estimated load demand for a small community

Load	No. in use	Power (Watts)	Summer (April–October)		Winter (November-March)	
			Hrs/day	Wh/day	Hrs/day	Wh/day
Domestic load						
LEDs	2	15	5	150	6	180
LEDs	2	12	6	144	7	168
Ceiling fan	1	60	10	600	0	0
Table fan	1	75	10	750	0	0
LED TV	1	60	3	180	4	240
Refrigerators	1*0.5	500	10	2500	8	2000
Total				4324		2588
No. of houses	10	43,240			25,880	
Community load						
Temples (LEDs)	2	15	8	240	8	240
Street light (LED)s	5	24	8	960	10	1200
Grass cutting machine	2	750	0.5	750	0.5	750
Total	1950				2190	
	45,190				28,070	

* 50% of total houses have a refrigerator

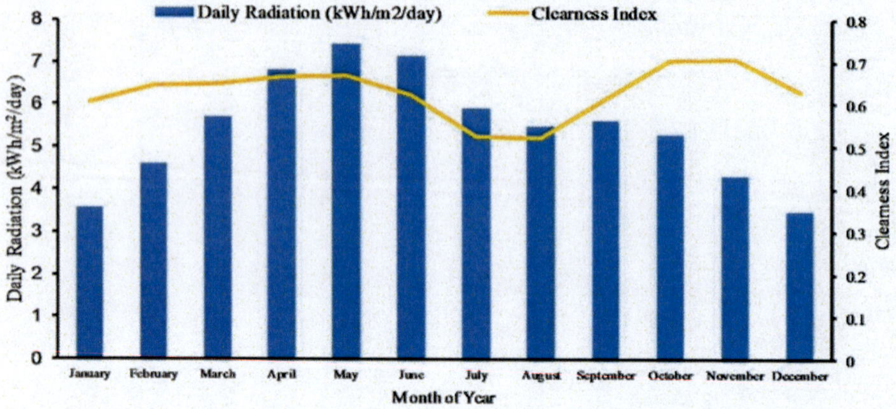

Fig. 2 Monthly solar radiation and clearness index at the study site

year. Solar radiation intensity will be high in summer and low in the winter season. The annual average value of solar radiation at the study site is 5.44 kWh/m^2/day.

2.3 *Wind Turbine Generation*

Wind generator output power, P_W depends on wind velocity V and is given as follows [14, 15]:

$$P_W = \left\{ \begin{array}{ccc} P_R & \text{if} & V_R < V < V_{C_0} \\ P_R\left(\frac{V^3 - V_{C_1}^3}{V_R^3 - V_{C_1}^3}\right) & \text{if} & V_{C_1} < V < V_R \\ 0 & \text{otherwise} & \end{array} \right\} \tag{2}$$

where P_R, V_R, V_{C_0} and V_{C_1} are rated power, rated speed, cut-off and cut-in speeds of the wind turbine, respectively. If η and A_W represent the generator efficiency and the total swept area, respectively, then the actual power of the wind turbine, P_A will be

$$P_A = P_w A_w \eta \tag{3}$$

Each wind turbine considered for the system is of 1 kW rating, having capital and replacement cost of 11,539$/unit, O&M cost of 230$/unit/year, the hub height of 25 m and lifetime is set as 20 years.

Month-wise average wind speed data for the study site is collected from NASA surface meteorology and solar energy database at 50-m anemometer height. The

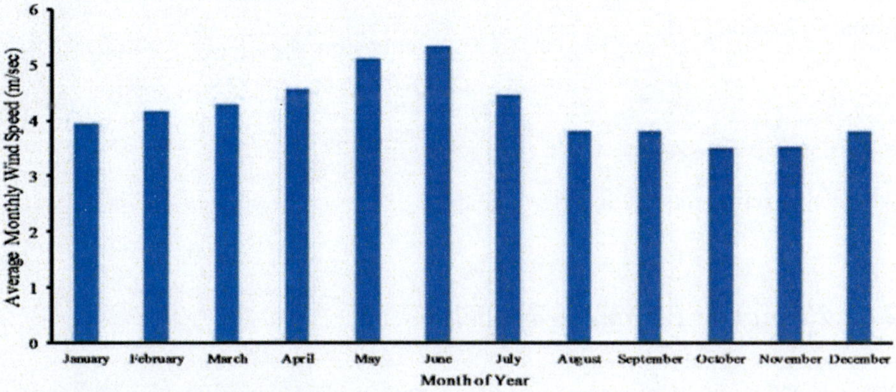

Fig. 3 Month-wise wind speed at study site

wind speed at the study site varies from 3.5 to 5.34 m/s with an average value of 4.19 m/s as shown in Fig. 3.

2.4 *Energy Storage System*

Intermittent nature of renewable energy sources can be overcome by energy storage systems and they are helpful to improve the reliability as well as penetration level of RERs. It is like virtual inertia of the generator, which helps in absorbing the temporary transients. The storage capacity of the BESS can be calculated as [16]

$$P_{\mathrm{BAT}}^{\mathrm{MAX}} = \frac{N_{\mathrm{BAT}} V_{\mathrm{BAT}} I_{\mathrm{BAT}}^{\mathrm{MAX}}}{1000} \tag{4}$$

where N_{BAT}, V_{BAT} and I_{BAT} represent the number of BESS, voltage rating of the battery unit and charging current rating in amperes, respectively. A lead–acid Surrette S-260 battery has been considered, which has a maximum capacity of 260Ah, the capital cost of 300$/unit, the replacement cost of 250$/unit, O&M cost of 5$/unit/year, round trip efficiency of 80% and a lifetime of 10 years. The maximum and minimum SOCs have taken 90% and 30%, respectively.

2.5 *Power Converter*

To link AC and DC buses with other micro-grid components, there is a requirement of a power converter, which can act as a rectifier or inverter. The converter size depends on peak demand (PLM). The inverter power rating PINV can be calculated by the given expression [14]:

$$P_{\mathrm{INV}} = \frac{P_L^M}{\eta_{\mathrm{INV}}} \tag{5}$$

where η_{INV} represents the inverter efficiency.

2.6 Economic Parameter Modelling

Various economic parameters associated with renewable power production are described in this section.

Net Present Cost (NPC) includes all the cost and revenue that occur in project lifetime. It can be calculated by the given expression:

$$\mathrm{NPC} = \frac{\text{Net Annual Cost}}{\mathrm{CRF}(i, N)} \tag{6}$$

where Net Annual Cost is annual cost minus revenue, i describe the annual real interest rate in %, N represents project duration and Capital Recovery Factor (CRF) which converts the present value in a sequence of equivalent annual cash flow over a project period.

Levelized Cost of Energy (LCOE) represented the average cost for generating electricity per kWh and is represented as

$$\mathrm{LCOE} = \frac{\text{Net Annual Cost}}{\text{Useful Electric Energy Produced}} \tag{7}$$

2.7 Grid Parameters

The village is connected with the utility grid but the availability of the supply depends on various factors. The agriculture supply is governed by time-to-time review of electricity administration. The per-unit tariff of electricity varies from 0.06 $/kWh to 0.05 $/kWh [17]. The electricity tariff has been considered as 0.08 $/kWh for the case study. The net metering facility is also considered for the study site to sell back the excess electricity to the utility grid. The sell back price has been considered as 0.05$/kWh. It will be helpful from the economic feasibility point of view and improves sustainability and payback period.

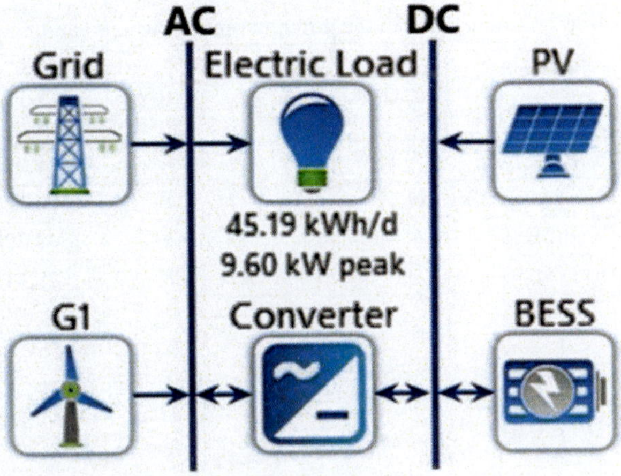

Fig. 4 Configuration and various components of HRES

3 Hybrid Renewable Energy System Design

The schematic diagram of the microgrid consisting of a photovoltaic system, wind generator, battery energy storage system, converter, utility grid and various types of interconnected loads of the consumers are shown in Fig. 4. The bidirectional power converter with the control system maintains the flow of electricity between the buses. The battery bank will maintain the supply in the event of a shortage of power supply and charge during the excess availability of electricity.

The feasibility of the proposed system can be analysed by its economic analysis. The specification related to technical and economic parameters has to be investigated and fed into the proposed system.

4 Microgrid Simulation Result and Discussion

Based on various input parameters, an analysis has been performed in HOMER with various input parameters as shown in Fig. 4. The study has been done by taking a realistic load profile of the study site. The economic feasibility of the system is evaluated mainly on the basis of NPC and COE. NPC is the sum of an initial cost of RERs and their components. COE depends on renewable fraction, number of battery units, utility grid connection and dispatch strategies. In this simulation, a cycle charging strategy is used. In this strategy, the primary load is supplied by the generating unit on a priority basis and in case of excess power, it goes to battery charging or sold back to the utility grid.

The optimal results of AC microgrid are shown in Figs. 5, 6, 7, 8, 9 and 10. The optimum solutions are tabulated based on NPC and COE as shown in Table 2. The best configuration is PV–BESS grid, which consists of 11 kW PV, 11 units of BESS

Table 2 Comparison of the different optimized configurations

Configuration	Solar (KW)	Wind (kW)	BESS	Converter (kW)	NPC ($)	LCOE ($/kWh)	Operating cost ($/kWh)	Initial cost ($/kWh)
PV/BESS/grid	11	–	11	8	18,995	0.0559	389.57	13,700
PV/wind/BESS/grid	11	1	11	8	24,330	0.0711	487.80	17,700
Wind/BESS/grid	–	1	19	11	33,398	0.149	1582	11,900
Grid /BESS	–	–	21	8	28,443	0.127	1511	7900

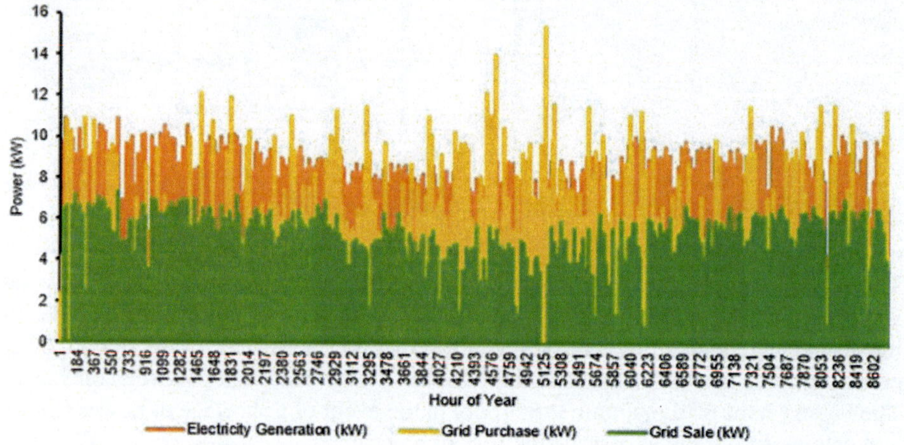

Fig. 5 Variation of electricity generated, purchase and sale on an hourly basis for a year

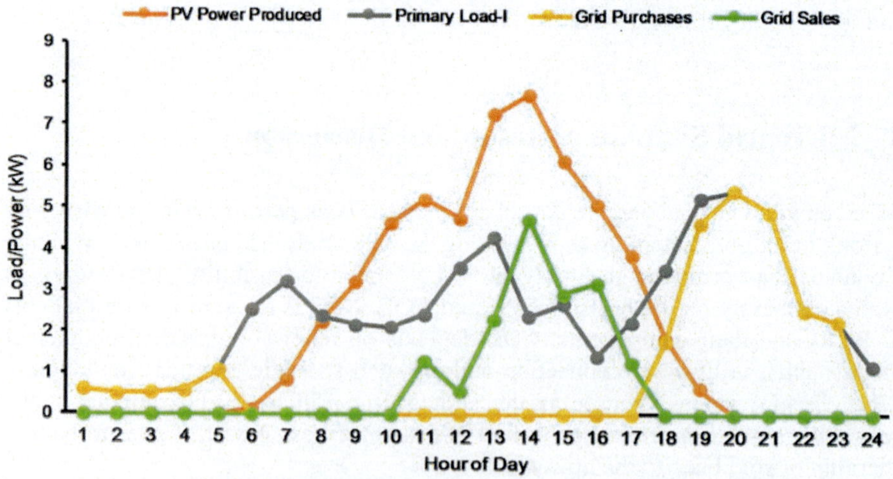

Fig. 6 Hour-wise variation of different output parameters in a day

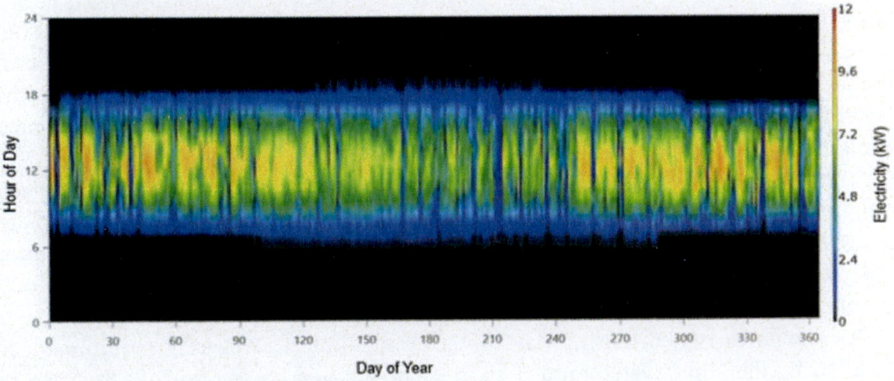

Fig. 7 Electricity produced from solar power

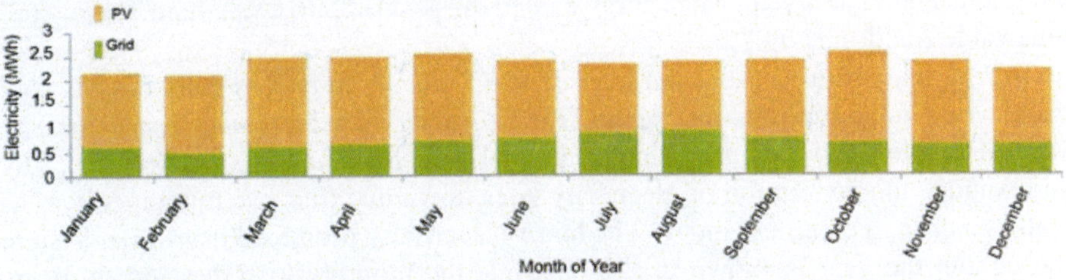

Fig. 8 Monthly average electric production by proposed HRES

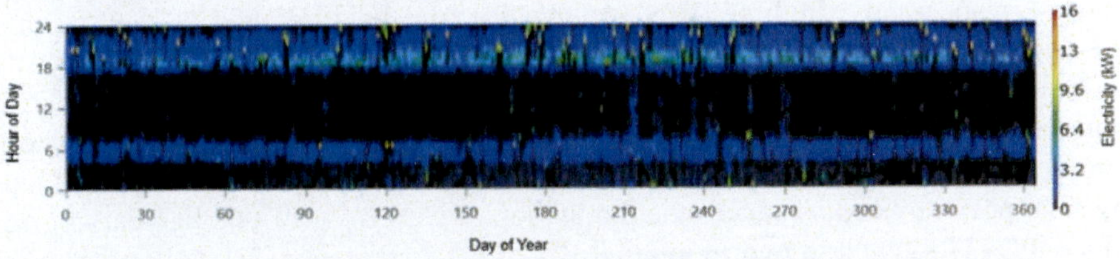

Fig. 9 Hourly pattern of energy purchased from the grid throughout the year

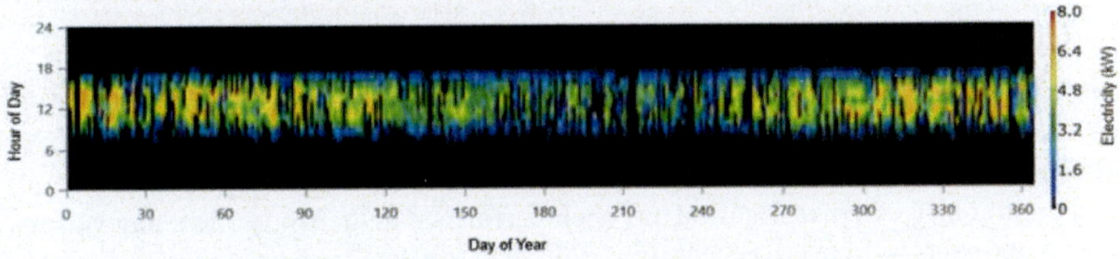

Fig. 10 Hourly pattern of energy sold to grid throughout the year

and 8 kW of the converter with grid integration. The NPC of the system comes out to be 18,995 $ and COE as 0.0559 $/kWh. The COE, NPC, initial and O&M cost for the best four combinations are depicted in Table 2. The COE, NPC and O&M cost of PV–battery–utility grid system is lowest. The initial cost is 13700 $, which is high due to initial investment in PV, battery and converter system.

The solar system generates 19,746 kWh/year by operating 4374 h/year. The mean output of the solar system is 54.1 kWh/day, with the maximum output of the plant as 10.9 kW in a day. The capacity factor of the solar system is 20.5%. The converter operates for 5116 h/year, with a capacity factor of 24.9%. The hour-wise electricity generated from solar energy, grid purchase and grid sale is shown in Fig. 5.

As shown in Fig. 6, during day time, the maximum electricity requirement is fulfilled by the solar system and if short then supplied by the grid. In the case of excess electricity production, it is sold back to the utility grid through net metering facility, which reduces the overall system cost. The electric energy purchased from the grid is 8603 kWh/year. The energy sale to the grid is 8536 kWh/year. The overall renewable fraction is 69.7%.

During the winter season, the load demand is less, so excess electricity is sold back, while in the summer, the demand for electricity increases, so generated energy is consumed locally, and only a small fraction is sold back. In the month of June, July and August, the generation of electricity goes down due to a rise in temperature as well as a drop in clearness index. The hourly electricity produced from solar system throughout the year is shown in Fig. 7, while the monthly load demand fulfill by solar and grid electricity has been shown in Fig. 8.

The hour-wise electricity purchased and sold back to the grid is shown in Figs. 9 and 10, respectively. In June, July and August, the load demand is high so the electricity sold back to grid is less as compared to other months.

Generation of electricity from fossil fuels produces different types of harmful pollutants like CO_2, carbon monoxide, sulphur dioxide, nitrogen oxides and unburned hydrocarbon. In proposed system, most of the electricity requirement are fulfilled by solar power and so it reduces the production of CO_2 from 11,068 kg/year in grid mode to 5437 kg/year in hybrid microgrid.

5 Conclusion

In this paper, a techno-economic analysis has been done by using HOMER on the proposed grid-integrated HRES. It is feasible to meet the electricity load of a small rural community. An investigation has been performed at the study site using various cost components used in HRES configuration. Among the four best-optimized configurations, i.e. PV–BESS–grid, Wind–PV–BESS–grid, wind–BESS– grid and grid–BESS, PV–BESS–grid system consisting of 11 kW PV, 11 units of 1 kW battery and 8 kW converter comes to be the most economical configuration having COE of 0.0559 $/kWh and NPC of 18,995 $. The proposed system may not be feasible for all sites. Therefore, a comprehensive study is required and this paper proposed an approach that can be applicable for doing study for any site or geographical region.

References

1. International Energy Agency (2019) World energy outlook (Online). Available: https://www.iea.org. Accessed 15 Mar 2020
2. International Energy Agency (2019) India-countries and regions (Online). Available: https://www.iea.org/countries/india. Accessed 15 Mar 2020
3. Central Electricity Authority (CEA) (2020) Executive summary on power sector (Online). Available: https://www.cea.nic.in. Accessed 15 May 2020
4. Krishan O, Suhag S (2019) Techno-economic analysis of a hybrid renewable energy system for an energy poor rural community. J. Energy Storage 23:305–319
5. Nurunnabi M, Roy NK, Hossain E, Pota HR (2019) Size optimization and sensitivity analysis of hybrid wind/PV micro-grids-a case study for Bangladesh. IEEE Access 7:150120–150140
6. Almashakbeh AS, Arfoa AA, Hrayshat ES (2019) Techno-economic evaluation of an off-grid hybrid PV-wind-diesel-battery system with various scenarios of system's renewable energy fraction. Energy Sour Part A Recover Util Environ Eff 1–24
7. Oueslati H, Ben Mabrouk S (2019) Techno-economic analysis of an on-grid PV/Wind/Battery hybrid power system used for electrifying building. Energy Sour Part A Recover Util Environ Eff 1–14
8. Razmjoo A, Shirmohammadi R, Davarpanah A, Pourfayaz F, Aslani A (2019) Stand-alone hybrid energy systems for remote area power generation. Energy Rep 5:231–241
9. Bendaoud B, Malek A, Loukarfi L, Maammeur H (2020) Conceptual study of photovoltaic power plant connected to the urban electrical network in northern Algeria. Energy Sour Part A Recover Util Environ Eff 1–20
10. Ramesh M, Saini RP (2020) Effect of different batteries and diesel generator on the performance of a stand-alone hybrid renewable energy system. Energy Sour Part A Recover Util Environ Eff 1–23
11. Bin Arif MS, Mustafa U, Prabaharan N, Ayob SBM, Ahmad J (2020) Performance evaluation of a hybrid solar PV system with reduced emission designed for residential load in subtropical region. Energy Sour Part A Recover Util Environ Eff 1–23
12. Phurailatpam C, Rajpurohit BS, Wang L (2018) Planning and optimization of autonomous DC microgrids for rural and urban applications in India. Renew Sustain Energy Rev 82:194–204
13. Bhakta S, Mukherjee V (2017) Performance indices evaluation and techno economic analysis of photovoltaic power plant for the application of isolated India's island. Sustain Energy Technol Assess 20:9–24
14. Das BK, Hoque N, Mandal S, Pal TK, Raihan MA (2017) A techno-economic feasibility of a stand-alone hybrid power generation for remote area application in Bangladesh. Energy 134:775–788
15. Dhundhara S, Verma YP, Williams A (2018) Techno-economic analysis of the lithium-ion and lead-acid battery in microgrid systems. Energy Convers Manage 177:122–142
16. Kaur M, Dhundhara S, Verma YP, Chauhan S (2020) Techno-economic analysis of photovoltaic-biomass-based microgrid system for reliable rural electrification. Int Trans Electr Energy Syst 1–20
17. UHBVN (2019) Schedule of tariff for supply of energy (Online). Available: https://www.uhbvn.org.in. Accessed 31 Mar 2020

Chapter 41
Smart Street Lighting System: An Approach Towards Effective Power Utilization

R. P. Ram Kumar, A. V. N. S. Sumanth, R. Sai Sumanth, Anandita Thakur, Asha Sree Chintala, and Banothu Geethanjali

1 Introduction

Electricity has a great impact in our daily life. Though, electricity has a great importance many of them fail to think about how life could be without electricity? In our day-to-day life, we use electricity for many purposes like lighting to charging electronics. In this way, most of the electricity are being wasted. Without even bothering about the availability of resources from which electricity is generated, people are simply wasting it. If we look at scenarios where electricity is mostly wasted, in most of the cases the lights are kept on continuously without proper use. The lights especially in our streets are continuously kept on even in the day time. The street lights are essential elements of our routine life. Without street lights, we cannot travel at night. But, if we see that the street lights are kept on even when there is visibility, it ultimately leads to the wastage of the electricity and even the lifetime of the bulbs will be reduced. From this, we understand the wastage of many resources like fossil fuels, coal and water from which electricity is produced. To evade the wear and tear of electricity due to street lights in a minimum level, SSLS is suggested. By the conversion of present conventional lighting system into a SSLS and by replacing bulbs with the LED lights, we can preserve the electricity as LEDs consume less power. In the conventional street light system, during night, even in the absence of vehicles the lights will be on in streets throughout night. But, in SSLS, the sensor will first detect whether there is any vehicle crossing that way. After detecting, the corresponding lights will be turned on. Figure 1 shows the traditional street lighting systems.

R. P. Ram Kumar (✉) · A. V. N. S. Sumanth · R. Sai Sumanth · A. Thakur · A. S. Chintala · B. Geethanjali
Department of Computer Science and Engineering, Malla Reddy Engineering College (Autonomous), Secunderabad, Telangana, India

N. Marriwala et al. (eds.), *Soft Computing for Intelligent Systems*, Algorithms for Intelligent Systems, https://doi.org/10.1007/978-981-16-1048-6_41

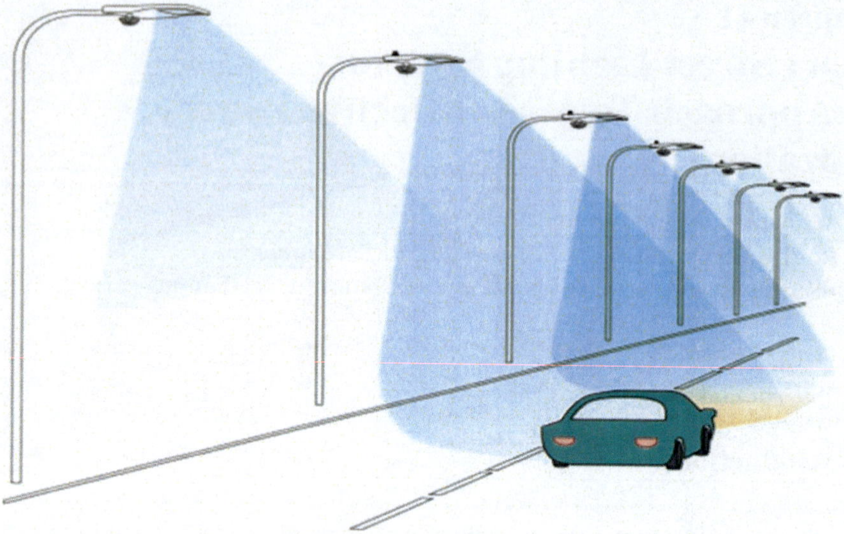

Fig. 1 Traditional street lighting system. *Courtesy* Source [1]

2 Literature Review

The following section deals with the summary of few existing smart lighting approaches.

Kokilavani and Malathi [2] designed smart lighting system in streets with the trending Internet of Things technology that would reduce the energy consumption by automatically switching off street lights via sensor when vehicle movement is less. Rajput et al. [3] devised street lighting system using GSM module where the architecture included light sensor, smoke sensor, gas sensor and sound sensor. Depending on the traffic intensity and climatic changes, the lighting system saves energy and eases maintenance of the system. Jagdeesh et al. [4] designed intelligent street lights to save the electricity for future. The design read sensor data from street and depending on the data value the intensity of light was dimmed or used fully. The data was communicated to successive unit and the process was repeated iteratively thus saving energy consumption. Yusaki et al. [5] designed lighting system that saves energy by using sensor. The smart system design consisted of lamp unit, sensor unit and access point. The motion sensor in units is communicated to other units if motion is detected. While the lamp unit dims the light if motion is not detected, the sensor unit communicates about the motion status to succeeding units fitted in various places. Access points were used when the distance between the sensor units exceeds a certain limit.

Vinoth et al. [6] designed light controller system using GPS-linked security system as the back bone that ensures streetlight and human safety. Apart from energy-saving feature, the system ensured theft-free feature due to fingerprint scanner and helpline calling facility like ambulance, police during emergencies as well as linked the system with AADHAR to avoid misuse except the pre-approved individuals by the government. Parkash et al. [7] devised an IoT-supported SSLS that allows accessing real-time information ON/OFF of the street lights. The smart embedded system used in the design detected the obstacle in the street and depending on the movement of the object detected the sensor switches the street light ON. Additionally, the street light switch can be controlled and accessed through Internet thus minimizing the power wastage and eliminating manual operation of street lights. Sakshi Anand and Neelu Jain [8] reviewed wireless transmission-based street lighting system. The survey of the system revealed intelligent street light system detected vehicle movement and switched the LED street light ON/OFF depending on the light intensity detected by sensors. Moreover, fault detection was possible when the lighting system was monitored through wireless technology. Archana et al. [9] developed an intelligent street light system that comprises sensors to detect vehicle movement, controls the switch ON/OFF through LDR sensor, controls the intensity of light using PIR sensor, and an in-built camera is mounted for safety and fault detection system to report the authorities. The system also provides the facility of manual operation under crucial circumstances.

2.1 Drawbacks of the Existing Street Light System

- The main disadvantage of the existing street light system is huge wastage of electricity.
- Difficulty in finding out that a certain street light is not working and need to be replaced.
- High electricity bills are generated due to this conventional system.
- The lifetime of the lights used is reduced due to unnecessarily switching on the lights at day.

2.2 Summary of the Findings from the Existing Methods

- Nowadays, humans are engaged irrespective of the place, home or streets, and they could not switch off the lights in time.
- In many cases, the street lights are turned on early in nights even before the dusk and they will be turned off after the dawn or sometimes noon.
- So, the street lights are the major area from which electricity is mostly being wasted.
- The main concept of our project is to save electricity using Light-Dependent Resistors (LDR) in the street light system.
- The circuit facilitates to turn on the street lights only during the vehicle traffic exclusively when it is dark.

3 Proposed Smart Street Lighting System

The proposed method helps in controlling the street light system efficiently. According to the necessity, the street lights will be put on and off without human intervention. In this way, we are trying to stop the unnecessary wastage of electricity and using electricity efficiently as required.

3.1 *Working*

- There is 9 V battery supply to the circuit of smart street light, which contains LDR sensors, LEDs, resistors (1 k range), transistors (2222A), battery with 9v capacity and required wires.
- The circuit allows LEDs to glow if and only if a movement is detected and the LDR sensor is covered completely dark.
- Here the main motive of the sensor is entrenched to glow the LEDs one after the other in specific time interval based on vehicle movement.
- In general, this application provides a single sensor to a particular set of lights to save electricity and to decrease the capital.
- So, when a motion is being detected the street lights glow.

3.2 *Components Required*

Table 1 shows the required components.

Table 1 Required components

S. No.	Name of the component	Capacity/range	Quantity required
1	Light-emitting diode (LED)	–	4
2	Light-dependent resistor (LDR)	5 mm	4
3	Resistor	1 K	8
4	Transistor	2222 A	8
5	Battery	9 V	2
6	Connecting wires	–	As per requirement

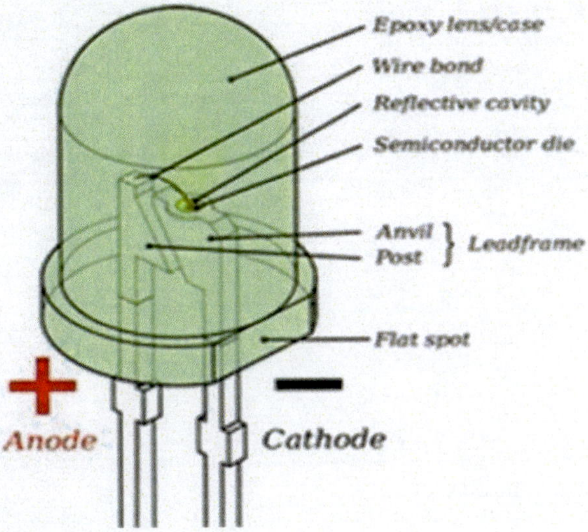

Fig. 2 LED. *Courtesy* Source [10]

3.3 Light-Emitting Diode (LED)

Light-emitting diode is a tiny bulb which emits light while electrical current is passed. These bulbs are used in our project. These LEDs produce less heat and consume less power compared to other bulbs. Figure 2 shows the simple LED used in the proposed method.

3.4 Light-Dependent Resistor (LDR)

These light-dependent resistors, when surrounded by darkness, turns ON an electric light or turns OFF the electric light when they are not surrounded by darkness. It gives a lesser amount of resistance in high light intensity and vice versa. To reduce the wastage of power due to unusual circumstances, we use LDRs in SSLS. Figure 3 shows the simple LDR used in the proposed method.

3.5 Resistor

Resistors can be seen in many electrical circuits. Resistor is a component that opposes the flow of current when a voltage is applied across it in a circuit. We measure resistance in ohms. These are mainly used for controlling currents in circuits. Figure 4 shows the simple resistor used in the proposed method.

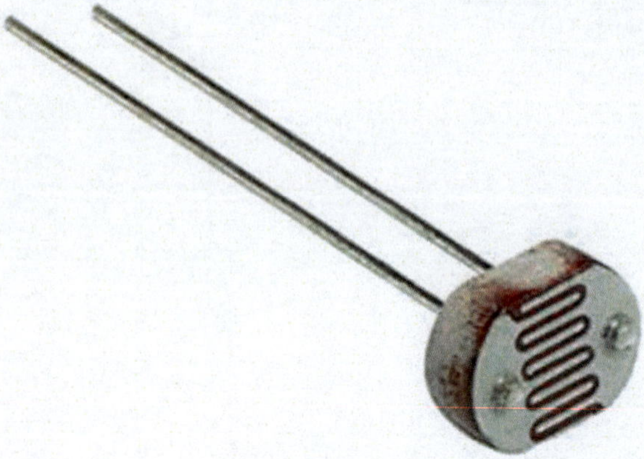

Fig. 3 LDR. *Courtesy* Source [11]

Fig. 4 Resistor. *Courtesy* Source [12]

3.6 Transistor

A transistor is made up of semiconductor material which is capable of carrying electric current and it can work as a switch in electrical signals. A change in current in the inner semiconductor layer can lead to a large change in current through the component. Here, the transistor is used as a two-way switch. If the voltage is greater than specific range, it is sent through the emitter or otherwise collector is used. Figure 5 shows the simple transistor used in the proposed method.

3.7 Battery

It is an electrochemical cell that provides power or releases electrical energy when needed. Generally, a battery contains an anode, cathode and an electrolyte. Here, we use two 9 V batteries for power supply. Figure 6 shows the simple battery used in the proposed method.

Fig. 5 Transistor. *Courtesy* Source [13]

Fig. 6 Battery. *Courtesy* Source [14]

3.8 Circuit Diagram

In Fig. 7, the circuit for connecting the components regarding the proposed method of a SSLS is shown.

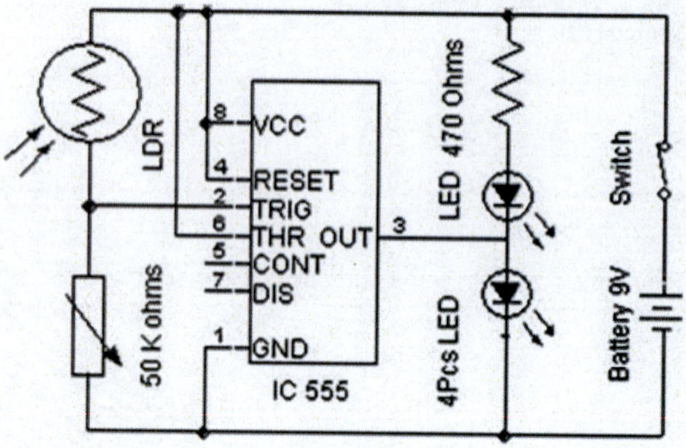

Fig. 7 Circuit diagram of the SSLS. *Courtesy* Source [15]

3.9 *Circuit Connection Details*

The following section exhibits the working principle of SSLS.

- Circuit connections in smart street light start with considering two transistors and then one of the emitters and collectors of two transistors is connected with each other.
- Next, resistor is connected to the collector of the transistor.
- Further, one more resistor is connected to the base of one transistor followed by connecting one to another base of the first transistor and another end of the wire to the collector of second transistor.
- Later, a LED is connected to the negative end to emitter of the first transistor and positive to the two resistors.
- Subsequently, LDR is connected to resistors.
- Later, the pins of IC 555 are connected to the corresponding components as shown in Fig. 7
- Finally, when the LDR is covered with darkness, the LED glows.

3.10 *Flow Chart*

The following section unfolds the flow chart of the proposed SSLS. Figure 8 shows the flow chart of the proposed method. Figures 9 and 10 represent the model diagram and working model of the proposed method, respectively.

4 Results and Discussion

The major motive of the project is to reduce the problems faced by the prevailing lighting system, and to identify a way out to conserve electricity. The initial phase

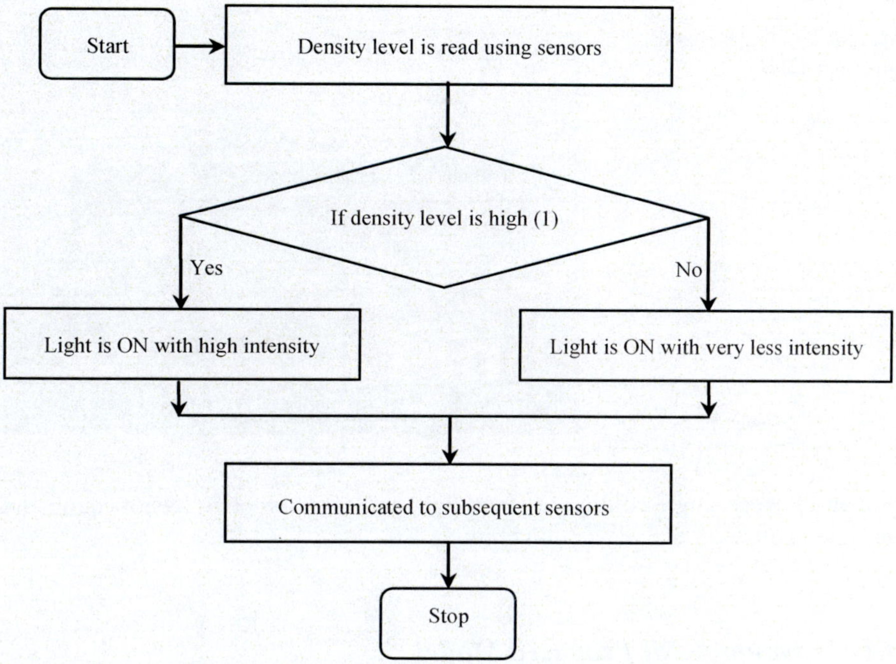

Fig. 8 Flow chart for the proposed SSLS

Fig. 9 Model diagram of proposed SSLS

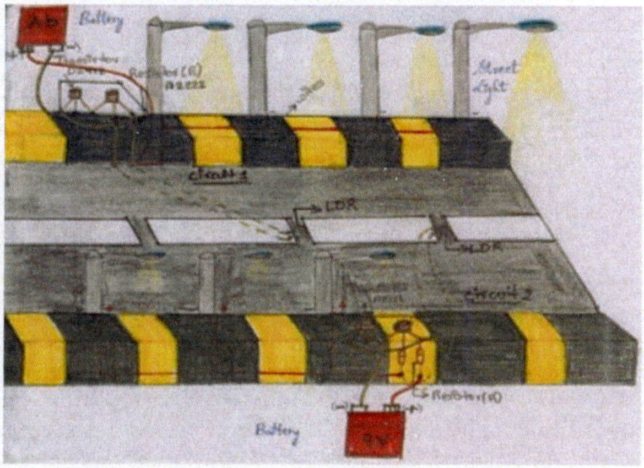

of the project is to set up the connections for the inputs and outputs of the system to handle the lighting system in streets. The prototype of this project shows the implementation of the project and will prove to be useful for humanity by less power consumption and adhering the rules and requirements of the government thus making the project futuristic. The micro-technology used in this system allows a single person

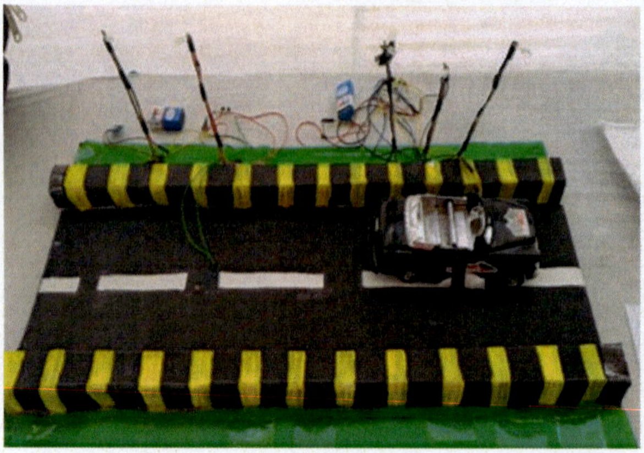

Fig. 10 Working model of proposed SSLS

to control every single column of street light and eases problem identification and detection with effective power utilization.

4.1 *Advantages of Proposed Model*

- Electrical energy will be saved.
- The cost of maintenance will be reduced.
- Energy efficiency will be increased.
- Eliminates manpower.
- Easy to detect and resolve the problems if any.

5 Conclusion

This paper gives us an idea about the design and construction of an SSLS and its functioning vis-à-vis put on and off the lights depending on the necessity. This solves the problem of wastage of electrical energy and thereby decreases the power consumption. It also increases the road safety and efficiency in switching on and off street lights automatically.

References

1. In Norvegia Lampioni Intelligenti Che Si Illuminano Quando Passano Auto E Persone - GreenMe.it, GreenMe.it. (Online). Available: https://www.greenme.it/abitare/risparmio-energetico/lampioni-intelligenti-led-norvegia/. Accessed 15 May 2020

2. Kokilavani MM, Malathi A (2017) Smart street lighting system using IoT. Int J Adv Res Appl Sci Technol 3(11):8–11
3. Rajput PKY, Gargeyee K, Monica P, Priyanka Y (2013) Intelligent street lighting system—J236069.pdf. Int J Eng Sci Invent. 2(3):60–69
4. Jagadeesh YM, Akilesh S, Karthik S, Prasanth (2015) Intelligent street lights. Procedia Technol 21:547–551. https://doi.org/10.1016/j.protcy.2015.10.050
5. Fujii Y, Yoshiura N, Takita A, Ohta N (2013) Smart street light system with energy saving function based on the sensor network. In: Proceedings of the 4th international conference on future energy systems, pp 271–272. https://doi.org/10.1145/2487166.2487202. (e-Energy 2013)
6. Vinoth Kumar N, Subhash VK, Viswanathan T, Surya Prakash S (2018) Intelligent street light controller with security system. Int J Innov Technol Explor Eng 8(2):296–299
7. Parkash PV, Rajendra D (2016) Internet of Things based intelligent street lighting system for smart city. Int J Innov Res Sci Eng Technol 5(5):7684–7691
8. Anand S, Jain N (2015) Intelligent street light system using wireless transmission: a review. Int J Adv Res Electron Commun Eng 4(2):271–275
9. Archana G, Aishwarya N, Anitha J, Kumar V (2015) Intelligent street light system. Int J Recent Adv Eng Technol 3(4):16–18
10. Light-emitting Diodes Circuit, Working Principle and Application, ElProCus—Electronic Projects For Engineering Students. (Online). Available: https://www.elprocus.com/light-emitting-diode-led-working-application/. Accessed: 15 May 2020
11. "LDR 5mm Light Dependent Resistor Photo Resistor.photocell, Calcutta Electronics. (Online). Available: https://calcuttaelectronics.com/product/ldr-5-mm-light-dependent-resistance-photocell/. Accessed 15 May 2020
12. "Motogadget Load Resister (82 Ohms, 5 Watt) For Fuel Reserve Indicators. (Online). Available: https://www.mvecycles.com/Motogadget-Load-resister-82-Ohms-5-Watt-for-fuel-reserve-indicator. Accessed 15 May 2020
13. Wikipedia, "BC548," Wikipedia, 15 May 2020. (Online). Available: https://en.wikipedia.org/wiki/BC548. Accessed 15 May 2020
14. "HW Battery 9V with Connector for Arduino/ Robotics. (Online). Available: https://www.flyrobo.in/hw-battery-9v-with-connector-for-arduino-robotics. Accessed 15 May 2020
15. Automatic Street Light | Electronics Project and Circuits, Circuiteasy. (Online). Available: https://www.circuiteasy.com/automatic-street-light. Accessed 15 May 2020

Chapter 42
An Improved Application-Oriented Teaching Style by Integrating Design Thinking and Project-Based Learning

R. P. Ram Kumar, Sanjeeva Polepaka, and S. Udaya Bhaskar

1 Introduction

The traditional method of Project-Based Learning (PBL) imparts the learning approach by doing activities. PBL is a systematic teaching method with innovative and sophisticated questions to understand the procedures. The significant features of PBL are summarized from [1], which includes the following:

(1) Requires the ability/interest to learn something new.
(2) Integrates skills and essential academic content.
(3) Organized around the open-ended questions.
(4) Includes repeated revisions based on the feedback.
(5) Mainly depends on the critical thinking, collaboration with proper communication.
(6) Finally, the project is built according to the problem definition.

The advantages of PBL approach is summarized from [2, 3].

(1) PBL promotes an interdisciplinary approach.
(2) Promotes the collaboration of more participants to reach the goal(s).
(3) Allows the students to investigate and explore more through queries.
(4) Aids in improving the self-directed learning.
(5) Develops motivation towards attaining the goal.
(6) Provides diverse learning styles.
(7) Improves recall and retention abilities.
(8) Encourages interpersonal and teamwork skills.

R. P. Ram Kumar (✉) · S. Polepaka
Department of Computer Science and Engineering, Malla Reddy Engineering College (Autonomous), Secunderabad, Telangana, India

S. Udaya Bhaskar
Department of Mechanical Engineering, Malla Reddy Engineering College (Autonomous), Secunderabad, Telangana, India

N. Marriwala et al. (eds.), *Soft Computing for Intelligent Systems*, Algorithms for Intelligent Systems, https://doi.org/10.1007/978-981-16-1048-6_42

The pitfalls of PBL are abridged from [4] as follows:

(1) The absence of being encouraged to accomplish the goal.
(2) Lack of interest makes a distraction to the students.
(3) Becomes more overhead to the teachers, when preparing the steps to get back to the students on track.
(4) Sometimes, rarely leads to access of inappropriate content.
(5) Occasionally, inappropriate for disabled students.

Figure 1 shows various modules in the traditional method of learning. They are (1) lecture, (2) activity, (3) quiz, (4) review, (5) exam and (6) project. Based on the complications of the problem, the lecture, activity and quiz procedures can be repeated for the better understanding of the problem. On the other hand, the PBL initiates with the rubrics and creates the environment for 'Need to Know' activities in the initial stage. Later stages proceed with the diverse exploring approaches such as activities, workshop, homework, simulations, discussions, modelling, reading, interviews, building, preparing, drafting, quiz and feedback. Finally, the project is developed based on the problem statement.

The creative and contemporary approach to solve the problem is the Design Thinking approach. The DT is a solution-based iterative procedure to discover the user requirements. Dym and Little [6] represented the engineering design as a thoughtful development to perform the necessary functions within limits. The Engineering Project in Community Services (EPICS) put forth the human-centred design process for the 'Thoughtful development' to meet the goals. Figure 2 shows the various phases in the EPICS design process, namely,

(1) Project identification phase to recognize the specific needs.
(2) Specification development to identify the stakeholder requirements, project requirements and pitfalls of current project with the evaluation criteria.
(3) Conceptual design phase to determine the best approach from the possible solutions.

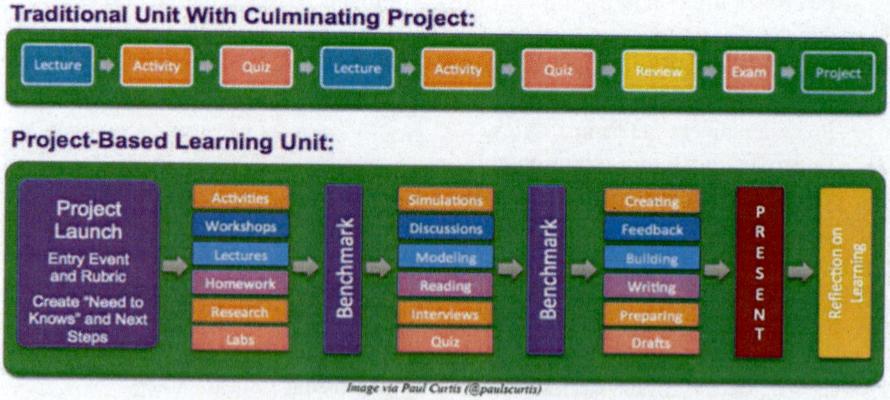

Fig. 1 Traditional versus PBL method. *Source* [5]

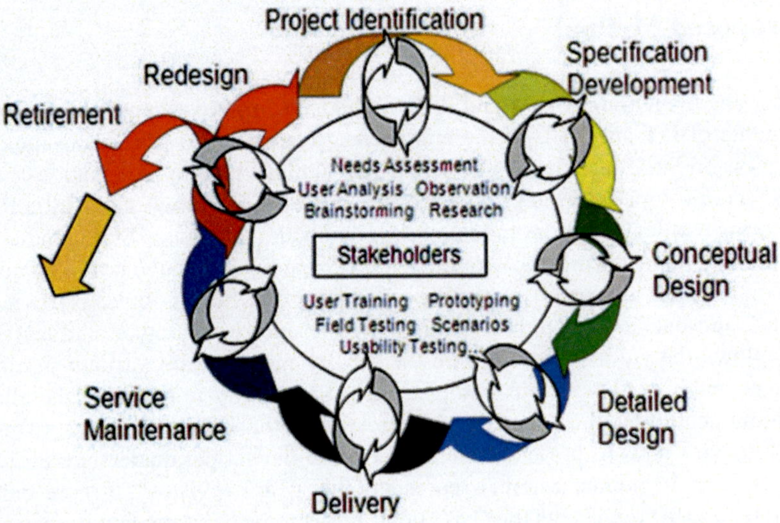

Fig. 2 EPICS design process. *Source* [7]

(4) Detailed design phase deals with the construction of a working prototype to fulfill the functional requirements of testing procedures.
(5) Delivery phase deals with the production of the product through the detailed design, necessary training material and user manual.
(6) Service/maintenance phase ensures the project performance.
(7) Finally, the retirement or the redesign process is carried out based on the evaluation of the project in the service phase.

The DT is defined as the process of solving the problem creatively [8]. The significant features of DT are as follows:

(1) Diminishes risk correlated with new ideas.
(2) Helps to learn faster.
(3) Follows innovative approach and not the incremental approach.
(4) Ensures that the delivered project meets the client's objectives and requirements.

The benefits of DT process are summarized from [9] as follows:

(1) DT always focuses on end users.
(2) DT leverages multi-disciplinary expertise.
(3) DT employs better understanding.
(4) DT follows the iterated and retest procedures until the requirements are met.
(5) DT provides solutions to the real-world problems.

2 Proposed Method

By analysing the features of DT and PBL methods, the proposed approach integrates the benefits of DT and PBL in the engineering education to present an improved application-oriented teaching style. Here, the 'Online Faculty Evaluation System (OFES)' is developed by integrating the DT and PBL approaches at Malla Reddy Engineering College (Autonomous), Secunderabad, Telangana State, India. The initial study deals with the exploration of OFES modules through complex queries (based on the PBL approach). Later, the refinement of OFES modules is carried out using DT approaches (through empathize, define, ideate, prototype and test). The OFES allows the students to evaluate the faculty members on various significant parameters such as (1) syllabus completion, (2) regularity in handling the classes, (3) communication skills, (4) creating interest while handling the topics, (5) usage of teaching aids and (6) presentation methods. All the six parameters are evaluated on the scale of 10 points, where 1 represents the 'needs revision', 5 represents the 'moderate' and 10 represents the 'Excellent' performances of the faculty members.

After identifying the evaluating parameters, the OFES implementation process is initiated by integrating the features of DT and PBL methods. Henceforth, the significant characteristics of OFES are determined using query sessions, various activities, discussions, interviews and drafts, followed by specification development, conceptual design, detailed design and feedback with repeated iterations to meet the stakeholder requirements. There are three modules in the proposed OFES, namely, (1) Login, (2) Submit Feedback and (3) Admin. The login module provides access to OFES for the authorized users.

The feedback module allows the students to provide/submit the feedback about the faculty members for the subjects being taught in the current semester. The admin module provides the consolidated reports based on the requirements. For example, class-wise, year-wise, subject-wise and faculty-wise reports provide provision for adding/removing the Subjects/Faculty members/Students in the respective department(s). Once the reports are generated, they are stored in the form of Portable Document Format (PDF) for documentation purpose. Figures 3, 4 and 5 show the login page, feedback form with evaluating parameters and admin privileges, respectively, in the OFES. The OFES is implemented using the front-end tools such as HTML and CSS along with the JavaScript. Java Servlet and Java Server Pages (JSP) are used for server-side scripting tools. MySQL is the database used to store the feedback details along with the details of faculty members and students in the OFES. Currently, the limitation of the OFES is that it is tested only on the intranet, which is executed within the organization.

Fig. 3 Login page

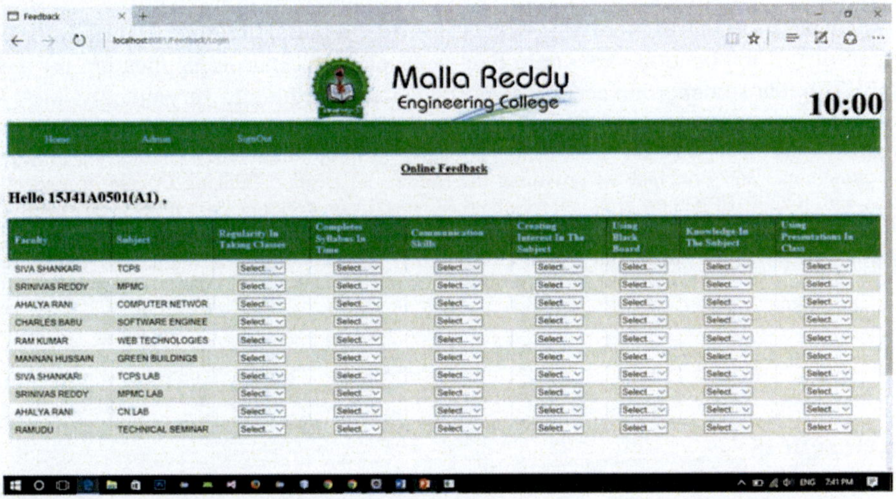

Fig. 4 Evaluating parameters in the feedback form

3 Conclusion

In connection with the traditional teaching method, the understanding and fulfilling of the stakeholder requirements become difficult. In the PBL approach, the user requirements are identified using numerous queries, whereas the DT strategy follows the innovative and creative problem-solving approach. The integration of DT and PBL makes the stakeholders understand the complex problems efficiently and leads to the development of effective 'Online Faculty Evaluation System' in our college. The implementation of OFES not only reduces the overhead in the manual feedback

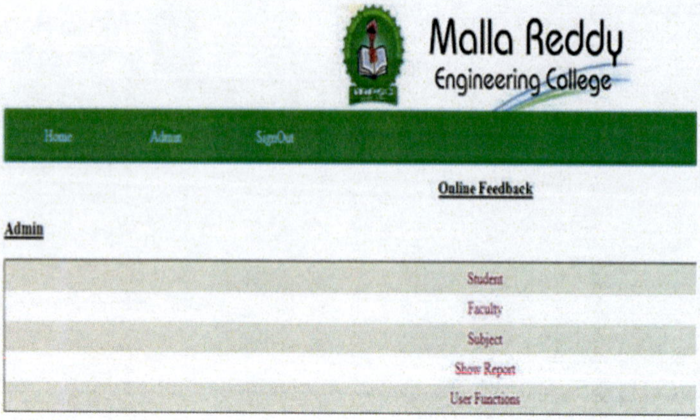

Fig. 5 Admin privileges

system but it also saves more time with customized reports. In future, as the advancement of this application, we would like to develop an Android application for the OFES so that students can access the portal using smartphones very quickly.

Acknowledgements I would like to thank our management of Malla Reddy Engineering College (Autonomous) and Principal for providing the training in Design Thinking Course in association with IUCEE-EPICS Programme, which encouraged us to work towards OFES in an effective manner. I thank our faculty members for their support. Finally, I would like to represent my gratitude to Mr. Vamsi Krishna Kalakuntla (Register Number: 15J41A0557) and Mr. Nadikoppu Venkatesh (Register Number: 15J41A0538), B. Tech., CSE students for their complete participation in OFES implementation process.

References

1. Noodle. (Online). Available: https://www.noodle.com/articles/what-is-project-based-learning. Accessed 06 Dec 2020
2. "Six Benefits to Project Based Learning and Technology Integration—TechnoKids Blog, TechnoKids Blog. (Online). Available: https://www.technokids.com/blog/technology-integration/benefits-project-based-learning/. Accessed 06 Dec 2020
3. Advantages (Online). Available: https://facdent.hku.hk/pbl/pbl10/advantage.htm. Accessed 06 Dec 2020
4. Advantages and Disadvantages of Project Based Learning—Lessons—Blendspace, Blendspace (Online) Available: https://www.tes.com/lessons/y16eZLeH9fPfrA/advantages-and-disadvantages-of-project-based-learning. Accessed 06 Dec 2020
5. The Difference between Projects and Project-Based Learning, TeachThought, 18 Oct 2019. (Online). Available: https://www.teachthought.com/project-based-learning/difference-between-doing-projects-and-project-based-learning/. Accessed 06 Dec 2020
6. Dym CL, Little P (2004) Engineering design: a project-based introduction. Wiley, New York

7. Teacher Tool Box—The EPICS Design Process and Design Cycle, Online: https://engineering.purdue.edu/EPICS/k12/resources/1.6%20Teacher%20Toolbox%20EPICS%20High%20Design%20Process%20and%20Cycle.pdf
8. What Is Design Thinking? IDEO U. (Online). Available: https://www.ideou.com/blogs/inspiration/what-is-design-thinking. Accessed 06 Dec 2020
9. Benefits of Desing Thinking (Online). Available: https://www.drewlepp.com/blog/5-big-benefits-design-thinking/. Accessed 06 Dec 2020

Chapter 43
Predicting and Estimating the Major Nutrients of Soil Using Machine Learning Techniques

Supreet Kaur and Kamal Malik

1 Introduction

India is an agrarian country and its primary resource is soil. Soil plays an imperative position in the financial system of the country, as the agriculture sector plays a lead in a large fraction of the Indian economy [1]. Agriculture is the science with the skill of cultivating vegetation as well as livestock and soil is the most important component in the agriculture field for efficient crop yielding.

The introduction of machine learning in the agricultural field has made huge remunerations in the research field using the analysis techniques [2, 3]. The machine learning analysis techniques play an important role in any field of science in establishing the fundamental variations by finding the eminent relationship present among the objects and concepts [1, 4]. It furthermore provides necessary information in the course of which research can be processed by following an orderly approach.

Using machine learning the classification of soil according to the quality will be easy and efficient. The machine learning applications in soil properties prediction include strategies of developing a decision support system to predict soil quality [5], irrigation requirement in arid and semi-arid regions [6], and improving the smart farming practices [7].

2 Materials and Methods

In our approach, simple linear regression was used to study the relationship between nitrogen (N), phosphorus (P), and potassium (K) content of the soil. We have also

S. Kaur (✉)
CT University, Ludhiana, India

K. Malik
Department of Computer Science and Engineering, CT University, Ludhiana, India

N. Marriwala et al. (eds.), *Soft Computing for Intelligent Systems*, Algorithms for Intelligent Systems, https://doi.org/10.1007/978-981-16-1048-6_43

Table 1 Attributes description

Soil nutrient symbol	Description
Ca	Calcium (kg/ha)
Mg	Magnesium (kg/ha)
K	Potassium (kg/ha)
S	Sulfur (kg/ha)
N	Nitrogen (kg/ha)
Lime	Used to increases soil pH (kg/ha)
C	Carbon (kg/ha)
P	Phosphorus (kg/ha)
Moisture	Water content

used the multiple linear regression model to study the impact of N over carbon (C), sulfur (S), lime, moisture, calcium (Ca), magnesium (mg), P, and K.

2.1 Data Collection

The dataset used in this study was taken from Kaggle [8]. The dataset was named soil nutrients data. It consists of 9 attributes and 1557 observations [9]. The dataset consists of nine soil nutrients which are explained in Table 1.

The data were pre-processed using the R programming language. The summary () function was used to study the dataset. The data was free from any redundant value and was ready to use. The linear regression was applied to the dataset using the function lm which means linear model. It was a very easy and fast approach to study the relation based on simple linear regression and multiple linear regression.

2.2 Automated System

In soil sciences, statistical analysis, artificial intelligence, and machine learning methods have been widely applied. The use of statistical methods to predict soil properties dates back to the early twentieth century [10]. The machine learning applications in the soil properties prediction include strategies of developing a system to predict soil quality, irrigation requirements in arid and semi-arid regions, and improving the smart farming practices [11, 12].

An autonomous model was proposed to detect the relationship between the major soil components N, P, and K, also known as primary nutrients of the soil. In our approach, we have used simple linear regression and multiple linear regression. The simple linear regression was used to make three models where each model individually studied the impact of N over P, N over K, and P over K. The fourth model, the multiple regression model, was made to study the impact of the response variable which was N over the other attributes of the dataset. The results obtained from these models can be used for further comparative study with logistic regression and other classification techniques of machine learning.

3 Results and Discussion

3.1 Simple Linear Regression Models

To discuss the N–P–K relation in a soil sample and their significance over each other, simple linear regression was used to create three prediction models.

Model 1 [Analysis of N–P relation]
The lm function also known as the linear model function was used to apply simple linear regression to the given dataset. The results obtained from model 1 are described in Table 2.

Table 2 is showing the required results. From the above results, we can conclude that the N content and the P content in a particular soil sample are highly relatable. The results showed us the estimated value, standard error value, t-value, and probability value. The probability value was used to predict that the N variable is significantly affecting the P variable value. This can be justified by looking at the significance code shown in the above figure. The "***" means that both the variables are 99.9% significantly related. In other words, we can say that the N content has a 99.9% significant impact over the P content in the soil samples of particular data. The multiple R-squared = 0.01574, which meant that nearly 1.57% of variance was being explained by this small model. The value of r^2 can be improved by increasing the number of variables. The result can be better explained by looking at the scatterplot of these two variables shown in Fig. 1.

In the scatterplot shown in Fig. 1, each point is represented using the small circles and they do have a kind of linear relationship. The blue color line is known as the

Table 2 Results of simple regression model 1 for N–P

	Estimate	Standard error	*t*-value	*p*-value(>|t|)	Significance code
(Intercept)	194.40771	2.73558	71.066	< 2e-16	***
P	0.10912	0.02189	4.984	6.91e-07	***
Residual standard error: 75.79 on 1554 degrees of freedom					
Multiple R-squared = 0.01574, **Adjusted R-squared** = 0.0151					
***F*-statistic** = 24.84 on 1 and 1554 DF, ***p*-value**: 6.91e-07					

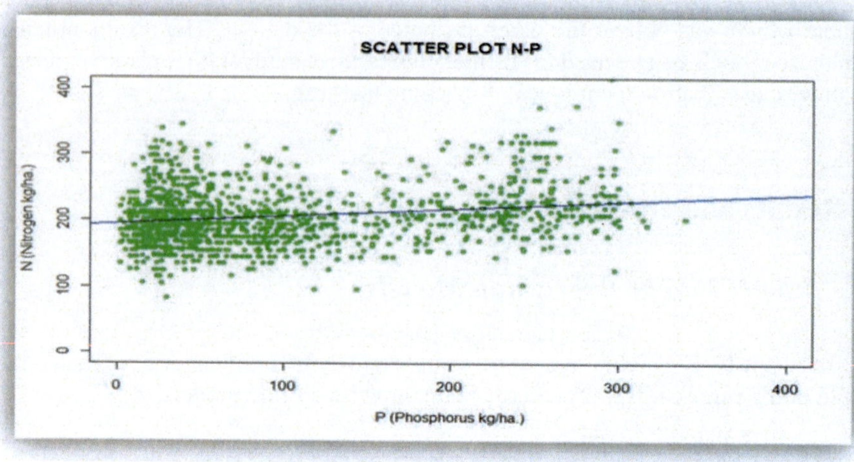

Fig. 1 Scatterplot for the N–P relationship

predicted line which we have predicted using the simple linear regression model. This line is approximately predicting the values of N.

Model 2 [Analysis of N–K relation]

The second model based on simple linear regression was designed using the lm function. This model was designed to study the significant impact of N nutrient content over K in the given soil samples. The results obtained are discussed in Table 3.

Table 3 is showing the required results. From the above results, we can conclude that the N content which is the dependent variable, and the K content being the independent variable in a particular soil sample is highly relatable. This can be justified by looking at the significance code shown in the above figure. The "**" means that both the variables are 95% significantly related. In other words, we can say that the N content has 95% of significant impact over the K content in the soil samples of particular data. The result can be better explained by looking at the scatterplot of these two variables shown in Fig. 2.

Table 3 Results of simple regression model 2 for N-K

	Estimate	Standard error	*t*-value	*p*-value(>|*t*|)	Significance code
(Intercept)	215.52946	4.36079	49.42	<2e-16	***
K	−0.03497	0.01198	−2.92	0.00355	**
Residual standard error: 76.18 on 1554 degrees of freedom					
Multiple R-squared = 0.005456, **Adjusted R-squared** = 0.004816					
***F*-statistic** = 8.525 on 1 and 1554 DF, ***p*-value**: 0.003553					

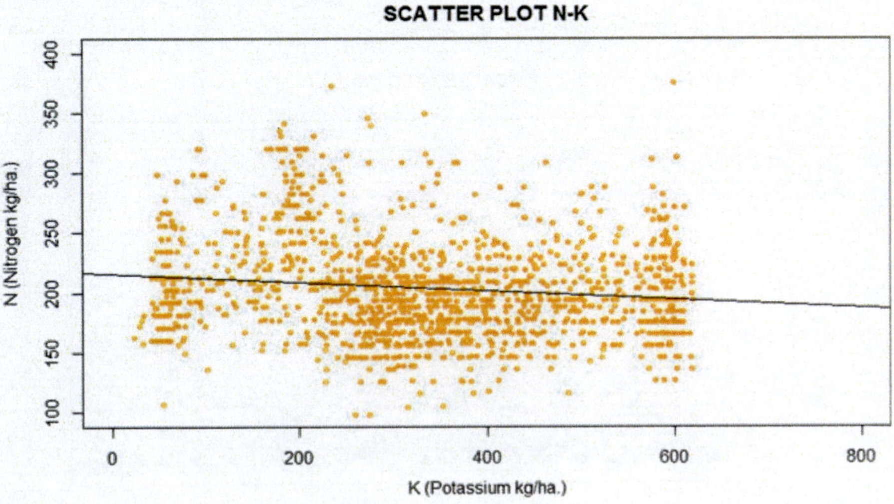

Fig. 2 Scatterplot for the N–K relationship

In the scatterplot shown in Fig. 2, each point is represented using the small circles and they do have a kind of linear relationship. The black color line is known as the predicted line which we have predicted using the simple linear regression model. This line is approximately predicting the values of N.

Model 3 [Analysis of P–K relation]
The third model was designed to predict and estimate the P content of the soil concerning the K content of the soil samples in the datasets. The results obtained from model 3 are described in Table 4.

Table 4 is showing the required results. From the above results, we can conclude that the P content and the K content in a particular soil sample are highly relatable. This can be justified by looking at the significance code shown in the above figure. The "***" means that both the variables are 99.9% significantly related. In other words, we can say that the P content has a 99.9% significant impact over the K content in the soil samples of particular data. The result can be better explained by looking at the scatterplot of these two variables shown in Fig. 3.

Table 4 Results of simple regression model 3 for P-K

	Estimate	Standard error	*t*-value	*p*-value(>|*t*|)	Significance code
(Intercept)	72.91435	5.00626	14.565	<2e-16	***
K	0.04911	0.01375	3.572	0.000365	***
Residual standard error :87.46 on 1554 degrees of freedom					
Multiple R-squared = 0.008144, **Adjusted R-squared** = 0.007505					
***F*-statistic** = 12.76 on 1 and 1554 DF, ***p*-value**: 0.0003651					

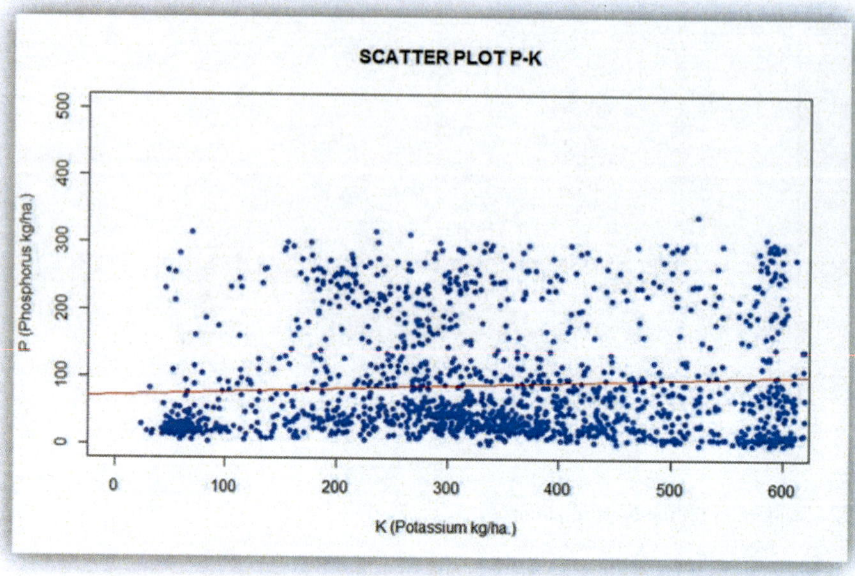

Fig. 3 Scatterplot for the P–K relationship

In the scatterplot shown in Fig. 3, each point is represented using the small circles and they do have a kind of linear relationship. The red color line is known as the predicted line which we have predicted using the simple linear regression model. This line is approximately predicting the values of P.

3.2 *Multiple Linear Regression Model*

The multiple linear regression, a well-known regression technique of machine learning, was used to discuss the significance of the presence of N over the other important variables present in the dataset. The other important variables present in the dataset were Ca, Mg, S, lime, moisture, C, P, and K. The R studio interface was used to obtain the required analysis. The lm function was used to apply multiple linear regression to the given dataset. The response variable was N and the model was used to predict the effect of other variables over the response variable. The results obtained are shown in Table 5.

Table 5 is showing the results obtained from the multiple linear regression model. The result included the estimate values (beta values), standard error, *t*-values, and *p*-values of all the variables toward the response variable. These "***" are telling us the level of significance of beta being non-zero. For example, 0.001 means that we are 99.9% confident that the beta is non-zero. From Fig. 3, we have obtained 5 variables with a good level of significance. Among the 5 variables, we have 4 variables about

Table 5 Results of multiple regression model

	Estimate	Standard error	*t*-value	*p*-value(>|*t*|)	Significance code
(Intercept)	182.664343	11.420132	15.995	< 2e-16	***
Ca	0.833224	0.458765	1.816	0.069528	
Mg	−0.209187	0.175136	−1.194	0.232496	
K	−0.030700	0.012852	−2.389	0.017024	*
S	0.262066	0.072196	3.630	0.000293	***
Lime	−0.006422	0.101142	−0.063	0.949379	
C	29.271722	4.853761	6.031	2.04e-09	***
P	0.087151	0.021903	3.979	7.24e-05	***
Moisture	−4.298094	13.323108	−0.323	0.747038	
Residual standard error: 74 on 1547 degrees of freedom					
Multiple R-squared = 0.9632, **Adjusted R-squared** = 0.9311					
***F*-statistic** = 89.9 on 8 and 1547 DF, ***p*-value**: < 2.2e-16					

which are 99.9% sure that they are related to the N variable, and about the 1 variable, we are 95% sure that it is related to the N variable. For other variables, although they have non-zero beta values, we are not sure that they are impacting the response variable or not.

For the whole model, the residual standard error is coming out to be 74 on 1547 degrees of freedom. The r^2 value is coming out to be 0.9623 which means that 96% of the variance in the soil dataset is being explained by this model. The adjusted r^2 takes into account the number of variables so it is coming less as compared to the r^2 value. The adjusted r^2 value is coming out to be 0.9311. The adjusted r^2 is the better parameter to report in the case of the multiple linear regression model. The F-statistic value is coming out to be 84.34 with a p-value very small. So we can confidently say that these variables mentioned in the figure are affecting the response variable N.

4 Conclusion

From the above findings, it can be concluded that the N content of the soil significantly affects the P content of the soil by 99.9%. Whereas the N content is having a less significant impact of 95% over K. The P–K is also strongly related as the simple regression model and showed that the response variable P had 99.9% of significant impact over variable K. So from these statements it can be concluded that the primary nutrients of soil known as N, P, and K are strongly related and do have a very good significant impact on each other value in the given soil samples in the soil datasets. The multiple regression model output results made it clear that the model worked very well with the adjusted r^2 value 0.9311 and the F-statistics value 84.34. The multiple regression models have shown 93.11% accuracy. The results obtained from the multiple linear regression model concluded that the C, P, K, and S nutrients of the soil do have a very good impact on the response variable, i.e., N. The variables are having a positive impact on the response variable. So it is established that the soil content of N does have a strong relationship with the soil content of N, P, K, C, and S. The results may vary for the different datasets and also using different machine learning algorithms over the same dataset.

References

1. Masri D, Woon WL, Aung Z (2020) Soil property prediction: an extreme learning machine approach. Neural Inf Process Lect Notes Comput Sci 18–27
2. Morellos A, Pantazi XE, Moshou D, Alexandridis T, Whetton R, Tziotzios G, Wiebensohn J, Bill R, Mouazen AM (2019) Machine learning based prediction of soil total nitrogen, organic carbon and moisture content by using VIS-NIR spectroscopy. Biosys Eng 15(2):104–116
3. Oladipupo T (2020) Types of machine learning algorithms. New Adv Mach Learn 10(7):5772–9385
4. Lisbin G (2019) Crop yield and rainfall prediction in Tumakuru district using machine learning. J Environ Sci Toxicol Food Technol 2(1):32–42
5. Abbal P, Sablayrolles JM, Matzner-Lober É, Boursiquot JM, Baudrit C, Carbonneau A (2020) A decision support system for vine growers based on a Bayesian network. J Agric Biol Environ Stat 21(1):131–151
6. Ali M, Deo RC, Downs NJ, Maraseni T (2020) Multi-stage hybridized online sequential extreme learning machine integrated with Markov Chain Monte Carlo Copula-Bat algorithm for rainfall forecasting. Atmos Res 21(3):450–464
7. Pantazi XE, Moshou D, Tamouridou AA (2019) Automated leaf disease detection in different crop species through image features analysis and one class classifiers. Comput Electron Agric 15:96–104
8. Kaggle Homepage. https://www.kaggle.com/datasets. Last Accessed 07 Nov 2020
9. Soil Nutrient Dataset, Retrieved from: https://www.kaggle.com/search?q=Soil. Last Accessed 7 Nov 2020
10. Mohapatra S (2020) A novel approach to analyze and predict the crop yield productivity using machine learning algorithms. J Adv Res Dyn Control Syst 12(3):21–26
11. Balducci F, Impedovo D, Pirlo G (2020) Machine learning applications on agricultural datasets for smart farm enhancement. Machines 6(3):38–49
12. Rossel J, Williams J, Daily G, Noble A, Matthews N, Gordon L, Wetterstrand H, DeClerck F, Shah M, Steduto P, de Fraiture C (2019) Sustainable intensification of agriculture for human prosperity and global sustainability. Ambio 46(1):4–17

Chapter 44
Evaluation of Spectral Efficiency for 5G Waveform Contenders

Sumina Sidiq, Farhana Mustafa, Javaid A. Sheikh, and Bilal A. Malik

1 Introduction

There is a dramatic increase in data rates over the past few years and hence 4G wireless and virtual communication systems are rolled out significantly [1]. The main functioning values of 5G wireless transmission systems that must be improved which include the highest information rates, spectral plus energy efficiency, latency, power consumption, etc., and also flexibility essentially. IMT2020 [2] concept characterizes the 5G wireless communication to proclaim a period of immersive expertise. 5G performance services are alienated into three key categories which include enriched mobile broadband (eMBB), Machine-Type Communication (mMTC), and Ultra-Reliable Low-Latency Communications (URLLC) individually [3]. To go through these constraints, novel acquiescent 5G waveform contenders are intended. In this paper, the principal purpose is to provide an analysis of 5G waveforms based on spectral efficacy.

2 Fundamental Waveforms of 5G

Numerous wireless gatherings are working globally to characterize the needs of 5G technology. A standard 5G waveform ought to fulfill the requirements of wireless communication as well as networks like increased spectral efficacy (SE), high information rates, low dormancy, power utilization, and out-of-band emissions (OOB). Also, to attain robustness against the wireless fading channels, various challenges are addressed by the 5G wireless communication systems which include massive

S. Sidiq (✉) · F. Mustafa · J. A. Sheikh
Department of Electronics and IT, University of Kashmir, Srinagar, India

B. A. Malik
Institute of Engineering and Technology, University of Kashmir, Srinagar, India

N. Marriwala et al. (eds.), *Soft Computing for Intelligent Systems*, Algorithms for Intelligent Systems, https://doi.org/10.1007/978-981-16-1048-6_44

connectivity, increased traffic, and essentially expanding the wide scope of wireless applications having different wireless channel attributes.

- **UFMC**

It is one of the MCM techniques used for 5G networks in which sub-carriers are filtered in groups in the frequency domain to diminish the OOB emission. It reduces the filter length and time to perform modulation is also reduced. UFMC uses the QAM type of modulation. CP is not required in UFMC. The absolute block length is proportional to CP-OFDM. It is increasingly delicate to small-time misalignment than CP-OFDM. In UFMC, all the allocated spectrum is utilized efficiently as there is no repetition of similar bits. The interference on adjacent sub-carrier decreases in UFMC as side lobes decrease. UFMC has numerous points of interest while keeping backward compatibility with legacy OFDM.

- **GFDM**

It is an adaptable MCM scheme [1] proposed for wireless communication for upcoming wireless network systems (5G). The flexible nature of GFDM makes it pertinent for CP-OFDM and single-carrier frequency domain equalization (SC-FDE) [4]. It depends on the inflection of individual blocks, where each block comprises various sub-carriers and sub-symbols. In GFDM, sub-carriers are filtered with a model filter that is shifted circularly in frequency and time domain. The OOB emissions are reduced by this process [5].

- **FBMC/OQAM**

It is the improvement of OFDM [6] which emphasis on shortcomings of various issues while providing higher throughput data. It is an MCM technique that is being studied to be applied in remote and cellular networks [7]. FBMC is considered one of the best MCM techniques for 5G because of the following reasons.

3 Comparison of the Spectral Efficiency for 5G Waveforms

In this, section the functioning of 5G waveforms based on spectral efficacy (SE) is studied.

Initially, 5G waveforms are evaluated based on spectral efficacy. The Long-Term Evolution (LTE) framework channel transfer speed of 12 MHz is considered in this paper plus two users are considered in simulation results for the non-concurrent mode multiuser access method. Various multicarrier modulation techniques like OFDM, GFDM, SC-FDMA, FBMC, and UFMC are taken into consideration, the SE is one of the functions of FFT length, modulation parameters plus the sequence of intonation but does not rely upon the burst time. Also, SC-OFDM and OFDM are having a similar spectral efficacy which is expressed as

Table 1 Simulation parameters

Size of FFT	K_{FFT}	1024
Bit per symbol	N	2
Size of resource block	K_{RB}	12
No of RBs	K_{Re}^{1} K_{Re}^{2}	3 for U_1 9 for U_2
Sampling frequency	F_{c}	15.63 MHz

$$\eta_{\text{OFDM}} = \eta_{S_1C-\text{OFDM}} = M \times \frac{K_{\text{FFT}}}{(K_{\text{FFT}} + K_{\text{cp}})} \tag{1}$$

Here, M is the order of modulation.

The SE of UFMC decreases because of the fleeting circumstances of the shaping filter. The SE for UFMC can be written as follows:

$$\eta_{\text{UFMC}} = M \times \frac{K_{\text{FFT}}}{(K_{\text{FFT}} + L - 1)} \tag{2}$$

Now to get a similar SE for UFMC and OFDM, we take $L = K_{cp} + 1$. In GFDM, the cyclic prefix is inserted per symbol; hence, the SE is given by

$$\eta_{\text{FBMC}} = \frac{M \times s \times K_{\text{FFT}}}{\frac{(2s-1)K_{\text{FFT}}}{2} + K_{\text{FFT}}z} = \frac{Ms}{s + z - \frac{1}{2}} \tag{3}$$

The SE of GFDM is dependent on the size of FFT. K represents the length of each prototype filter. The spectral efficacy of FBMC depends upon the burst length. If the duration of a burst is more than 3 ms, then GFDM exhibits improved SE in comparison to OFDM, FBMC. Also, for UFMC the length of the filter is $L = 72$, and stopband attenuation is 40 dB. The overall parameters of the simulation are given in Table 1.

However, on comparing the spectral efficiencies of GFDM and OFDM, we have.

I. If the amount of data carriers, as well as the size of FFT, is the same then the SE of OFDM is less than that of the GFDM. When there is no CP insertion the magnitude of the OFDM sub-symbol is the same as that of the sub-symbols of GFDM. Every GFDM symbol has more complex modulated samples. GFDM uses only one CP for N sub-symbols and hence that is the reason that GFDM has increased SE.

II. The OFDM SE is similar to that of the SE of GFDM, let us take into account an information block of constant magnitude (i.e., $K_{\text{S}} \times N = K_{\text{S}}^{\text{OFDM}}$ and $ZN = K_{\text{FFT}}$, which signifies that frequency grid is divided by N plus is also modified). Hence, the size of the symbols of GFDM is equal as OFDM; also, the frequency spacing is more by N times. As the frequency grid is constant, the SE of GFDM is greater than the SE of OFDM.

Moreover, the spectral efficacy of FBMC is determined by the frame length. In FBMC, spectral efficacy decreases because of the fleeting state of the comprehensive shaping of a filter. Hence, no persistent deficit per symbol arises in comparison to supplementary waveforms plus the spectral efficacy enhances through the length of the burst. Spectral efficacy of FBMC is given as

$$\eta_{\mathrm{FBMC}} = \frac{M \times s \times K_{\mathrm{FFT}}}{\frac{(2s-1)K_{\mathrm{FFT}}}{2} + K_{\mathrm{FFT}}z} = \frac{Ms}{s + z - \frac{1}{2}} \quad (4)$$

Here, the number of symbols is denoted by S and Z represents the spreading factor.

4 Simulation Results

In this section, the Spectral efficiency in comparison to the burst duration is evaluated for different waveforms of 5G. The simulation outcomes are given in Fig. 1, where modulation $M = 2$. From the figure, it is seen that the spectral efficiency for OFDM and UFMC is the same. Also, the loss of SE for GFDM is very less because the CP in GFDM is included merely once for each symbol which implies that there are N times more CP for OFDM in comparison to GFDM.

Plus, SE of FBMC relies on the time durations, besides is effective than that of the OFDM and UFMC, if the time of burst is more than 3.5 ms ($z = 4$, $M = 2$).

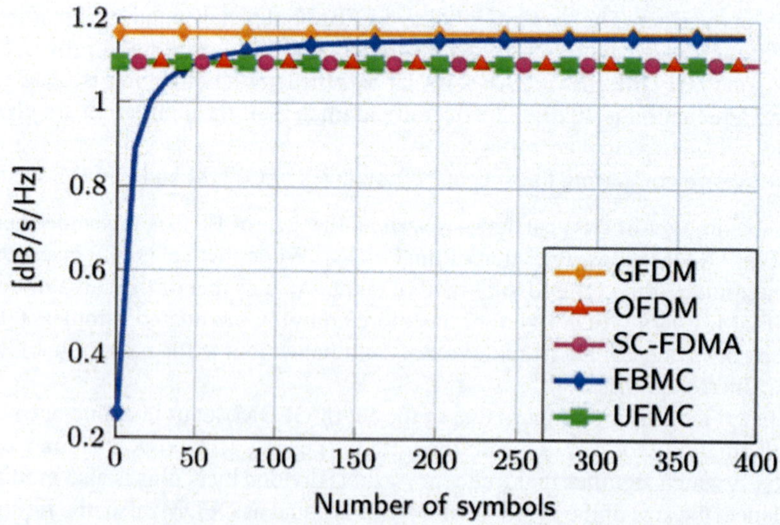

Fig. 1 Spectral efficiency of different waveforms of 5G

Table 2 Modulation techniques

Modulations techniques		
Waveforms	Advantages	Disadvantages
OFDM	• Trouble-free FDE • Simple MIMO assimilation • Supple frequency allocation • Small execution intricacy	• Eminent OOBE plus PAPR • Precise management prerequisite • Reduced functioning for elevated mobility uses
GFDM	• Flexible design • Good frequency localization • ReducedPAPR	• Sophisticateddormancy because to block processing • Demanding MIMO incorporation plus the design of a pilot • High implementation complexity
UFMC	• Effective localization frequency • Concise length of filter in comparison to subcarrier-wise processes (i.e., GFDMplus OQAM-FBMC) • Having compatibility to MIMO	• Because of the absence of CP no resistance to ISI • Because an increase in the size of the FFT receiver has complexity
FBMC	• Smallest OOBE • Effective spectral efficacy • Suitable for high-mobility functions • Suitable for asynchronous broadcast	• Difficult MIMO incorporation and pilot draw up plans • Because of the absence of CP, no resistance to ISI • High implementation complexity • Intensified consumption of power because of OQAM

The advantages and disadvantages of different multicarrier modulation techniques for 5G are discussed in Table 2.

5 Conclusion

In this paper, all 5G modulation techniques are studied and it is found that the 5G waveforms offer lesser OOBE in comparison to the SC-OFDM and OFDM techniques. In FBMC, the subcarriers are filtered individually and hence result in excellent frequency localization between the 5G waveform contenders. However, GFDM is also one of the waveforms with individual subcarrier filtering. The OOBE is increased in GFDM plus rapid evolutions are caused because of the quadrilateral window form in the time domain. The particular carrier methods are ideal in energy-constrained usage instances alongside the utilization of guard intervals which help in presenting better spectral efficacy. The spectral efficacy is another basic plan that is profoundly influenced by the window/channel design, the filter shape plus additional overheads.

References

1. Krishna P (2019) Spectral efficiency analysis of multicarrier scheme for 5G communications. Int J Recent Technol Eng (IJRTE) 7(6):1682–1686
2. ITU-R (2015) IMT Vision—Framework and overall objectives of the future development of IMT for 2020 and beyond. Technical report M. 2083–0
3. Zhang X, Chen L, Qiu J, Abdoli J (2016) On the waveform for 5G. IEEE Commun Mag 54(11):74–80
4. Sidiq S, Mustafa F. Sheikh J, Malik B (2019) FBMC and UFMC: the modulation techniques for 5G. In: 2019 international conference on power electronics, control and automation (ICPECA). New Delhi. https://doi.org/10.1109/ICPECA47973.2019.8975581
5. Bhasker AM, Kushwaha R (2018) Modulation schemes for future 5G cellular networks, IRACST. Int J Comput Netw Wireless Commun (IJCNWC) 8(1)
6. Farhang-Boroujeny B, Moradi H (2016) OFDM inspired waveforms for 5G. IEEE Commun Surv Tutorials 18(4):2474–2492
7. Krishna P (2019) BER performance evaluation of FBMC/OQAM and OFDM multicarrier system. Int J Innov Technol Exploring Eng (IJITEE) 8(10):1564–1568

Chapter 45
The Nuts and Bolts of the India-Abusive Fake Government of Telangana: Cyberpolicing Against Online Sedition

B. Malathi, K. Pavan Johar, N. Santhoshi, N. Srihari Rao, and K. Chandra Sekharaiah

1 Introduction

Business Intelligence systems are considered to be those which handle extremely large databases (ELDBs) which are based on datasets of the order of terabytes of data, whereas the higher order predictive analytics systems are the Big Data systems that handle data of the order of petabytes or Exabyte characterized basically by volume, velocity, variety and veracity (Four Vs).

The rapid development of the Internet and the technological advances of data systems of use and abuse led to a volley of concerns of cybersecurity, cybercrimes and other cyber threats such as cyber terrorism and cyber warfare so much so that such concerns are raised spanning the ordinary user of the Internet to the governments of all statures which adopted the e-governance paradigm, for example, from the State Governments in India to the Union Government of India [1–5]. Fake data is also on the rise.

The rest of the paper is organized as follows. Section 2 includes why the case study is treated as a big data crime and Sect. 3 explains the four crimes. What action by government and JNTUH is the need of hour is mentioned in Sects. 4 and 5, respectively, and Sect. 6 concludes.

B. Malathi (✉) · N. Santhoshi · K. Chandra Sekharaiah
JNTUH, Hyderabad, Telangana, India

K. Pavan Johar
PAIRS Foundation, Hyderabad, Telangana, India

N. Srihari Rao
JNTUA, Anantapuramu, Andhra Pradesh, India

N. Marriwala et al. (eds.), *Soft Computing for Intelligent Systems*, Algorithms for Intelligent Systems, https://doi.org/10.1007/978-981-16-1048-6_45

2 How It is a Big Data Crime?

The case study involves four crimes—State Emblem of India (Prevention of Improper Use) Act 2005 violation, Identity Theft under Section 66-C IT Act violation, Sedition Law violation and Cheating the Nation [4, 5]. Two cybercriminal organizations JNTUHJAC and FGoT (Fake Government of Telangana) are involved. The enormity of the offence is very high. This is because the fake identity that has surfaced is not something like a fake degree or a fake university, etc., but a Fake Government altogether that was blatantly publicly abusive of the Indian constitution and the Indian nation. More than 2000 online registrations were obtained as part of the organizational activities. Thus, 4*2*2000 = 16,000 units of crime are involved. This is the degree of cybercrime [3, 6–9]. As the data is pertinent to crime and is abusive of the Indian nation and not just that of any single individual or a small group or a small organizational setting, the enormity of the offence is considered high and the crime is considered Big Data Crime. Apart from the figure of 16,000, what is covered as data obtained by the website organizers/owners/developers under each registration online in the cybercriminal website https://jntuhjac.com matters too. This adds data value to the case under study. Another dimension is that the crime is abusive of the academic environment of a premier university [10], the first technological university of India, that is, JNTU Hyderabad. The four Vs may not matter for the arithmetic figures noted in this case if the case is something other than a crime case. Since this is a crime case, aptly this is a big data crime case. We do not have a pinpointing line as to what arithmetic figures could qualify a crime case as a big data crime case. But, we say that the enormity of the details involved in this case makes it a big data crime case. Our argument is that the big data crime definition should not be at par with the big data definition. The four Vs that characterize a data system as a big data system should be made applicable even for low-value definitions w.r.t. the four Vs when the data system is that of a crime case. The academic character of JNTUH is at stake on account of the lack of any proper checks against cybercriminal organizations [11, 12]. Neither the Cyberabad Police nabbed the culprits for conviction nor are the police practices followed in this regard of any considerable quality.

3 Two Crimes or Four Crimes?

3.1 About Crime 1

Crime no.0053/2014 was registered in KPHB Police Station, Hyderabad, in January 2014 for the police complaint against the abuse of the Indian National Emblem (under State Emblem of India (Prevention of Improper Use) Act 2005) in the JNTUHJAC website.

The police failed to get the accused convicted. Charges were laid against A1 and A2 in the charge sheet. In the charge sheet against A1, the police mentioned that A2 was absconding and that a separate charge sheet would be filed w.r.t. A2. A1 was arrested and trialed. The Cyberabad Police submitted a confession statement (bearing the signatures of two witnesses) of A1. When examined by the court, the witnesses revealed that the police got their signatures on white papers and that they were novices to the details of the confession statement. Owing to the hostile witnesses, A1 was acquitted. In the response to our RTI application, the police revealed that the cased ended in acquittal. On the Cyberabad website, the case status information indicates "court disposal" owing to "acquittal of the accused" as shown in Fig. 1. Figure 2 shows the court judgment is dt.14 October 2015, whereas the Court Disposal dt.12 December 2014 due to acquittal mentioned in Fig. 1 is wrong. The police failed to

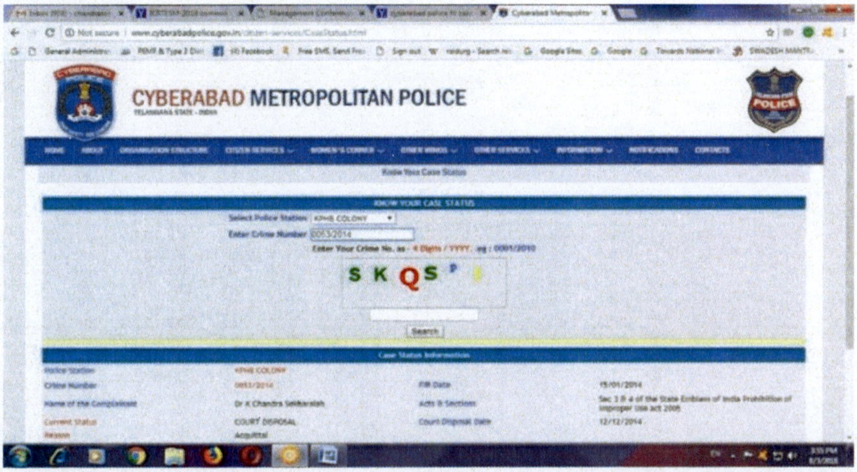

Case Status Information			
Police Station	KPHB COLONY		
Crime Number	0053/2014	FIR Date	15/01/2014
Name of the Complainant	Dr K Chandra Sekharaiah	Acts & Sections	Sec 3 & 4 of the State Emblem of India Prohibition of Improper Use act 2005
Current Status	COURT DISPOSAL	Court Disposal Date	12/12/2014
Reason	Acquittal		

Fig. 1 FIR 0053/2014: The Police mentioned "Court Disposal" as the Current Status of the Case owing to Acquittal of A1, whereas they failed to produce A2 in the court by a separate charge sheet as said in the charge sheet against A1?

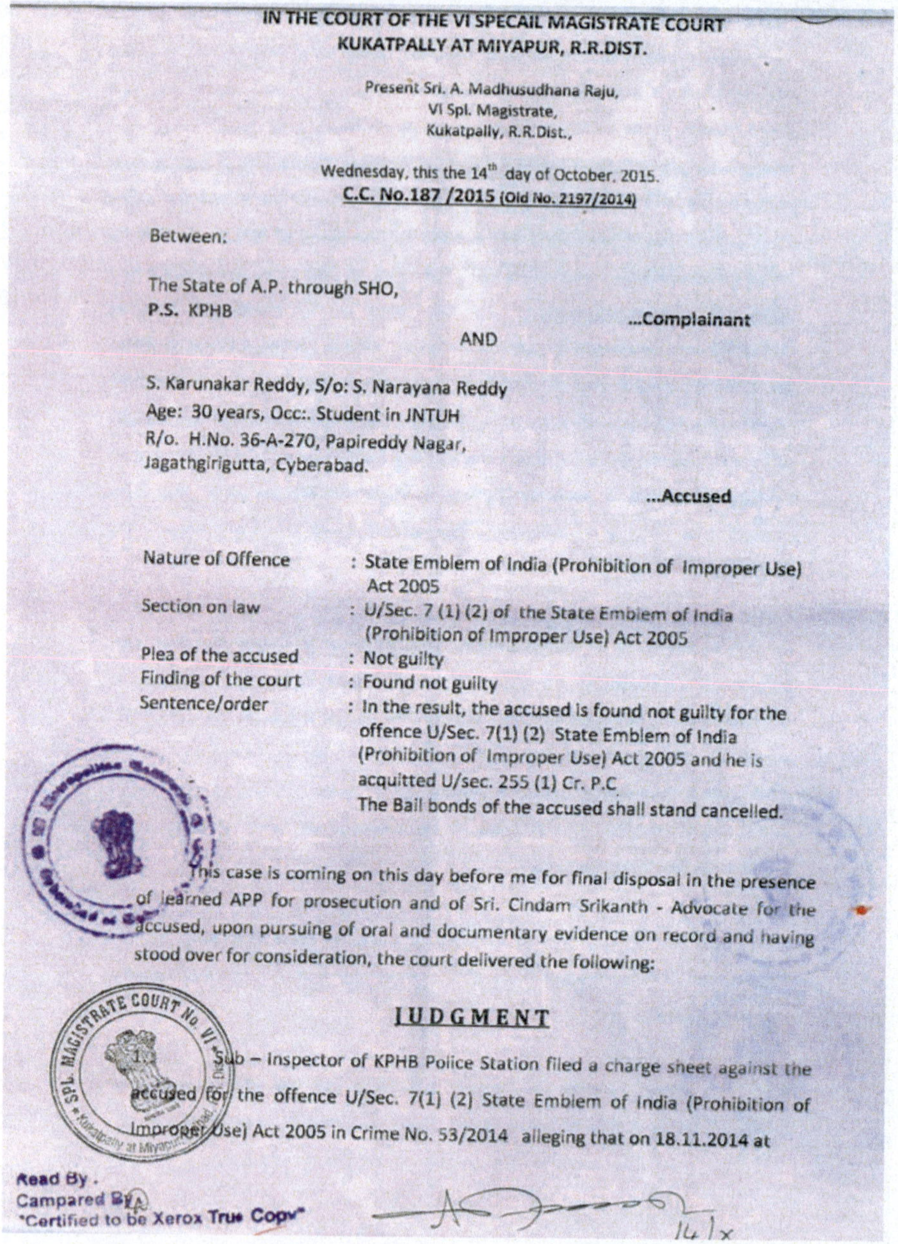

IN THE COURT OF THE VI SPECAIL MAGISTRATE COURT
KUKATPALLY AT MIYAPUR, R.R.DIST.

Present Sri. A. Madhusudhana Raju,
VI Spl. Magistrate,
Kukatpally, R.R.Dist.,

Wednesday, this the 14th day of October, 2015.
C.C. No.187 /2015 (Old No. 2197/2014)

Between:

The State of A.P. through SHO,
P.S. KPHB ...Complainant

AND

S. Karunakar Reddy, S/o: S. Narayana Reddy
Age: 30 years, Occ:. Student in JNTUH
R/o. H.No. 36-A-270, Papireddy Nagar,
Jagathgirigutta, Cyberabad.

...Accused

Nature of Offence	: State Emblem of India (Prohibition of Improper Use) Act 2005
Section on law	: U/Sec. 7 (1) (2) of the State Emblem of India (Prohibition of Improper Use) Act 2005
Plea of the accused	: Not guilty
Finding of the court	: Found not guilty
Sentence/order	: In the result, the accused is found not guilty for the offence U/Sec. 7(1) (2) State Emblem of India (Prohibition of Improper Use) Act 2005 and he is acquitted U/sec. 255 (1) Cr. P.C The Bail bonds of the accused shall stand cancelled.

This case is coming on this day before me for final disposal in the presence of learned APP for prosecution and of Sri. Cindam Srikanth - Advocate for the accused, upon pursuing of oral and documentary evidence on record and having stood over for consideration, the court delivered the following:

JUDGMENT

Sub – Inspector of KPHB Police Station filed a charge sheet against the accused for the offence U/Sec. 7(1) (2) State Emblem of India (Prohibition of Improper Use) Act 2005 in Crime No. 53/2014 alleging that on 18.11.2014 at

Read By .
Campared By
"Certified to be Xerox True Copy"

Fig. 2 Judgement Page 1: The court judgment is dt.14Oct2015. Court Disposal dt. 12/12/2014 due to acquittal mentioned in Fig. 1 is wrong. The police failed to produce A2 in the court by a separate charge sheet as said in the Chargesheet against A1 and mentioned wrong details as "Court Disposal" due to acquittal of A1?

produce A2 in the court by a separate charge sheet as said in the charge sheet against A1 and mentioned wrong details as "Court Disposal" due to acquittal of A1. Thus, the whole exercise of handling cybercrime indicated ineffective policing.

3.2 About Crime 2

Crime no.0006/2018 was registered in Cybercrimes Police Station@Gachibowli, Hyderabad, in January 2018 for the police complaint against the Fake Government of Telangana used in the JNTUHJAC website [3, 6–9]. The complaint was against the crime of Sedition Law Violation, but the police registered the crime under Section 66-C IT Act 2008 which covers Identity Theft as shown in Fig. 3. The police response is considered unbecoming as it is generally felt that Sedition is involved in the case.

Is it an Identity Theft crime or the crime of Sedition?
We opine that sedition crime is applicable in the case. "Identity theft" is meaningful when an authoritative identity is already there and when that identity is used unofficially unauthorizedly falsely. But the FGoT is not a false organization in the presence of an existing Government of Telangana. It was used as a false organization in the absence of an existing Government of Telangana. When something exists, it can be thieved. When something does not exist, it cannot be thieved. Thus, aptly, it is viewed that "sedition" crime is applicable in the case rather than "identity theft"

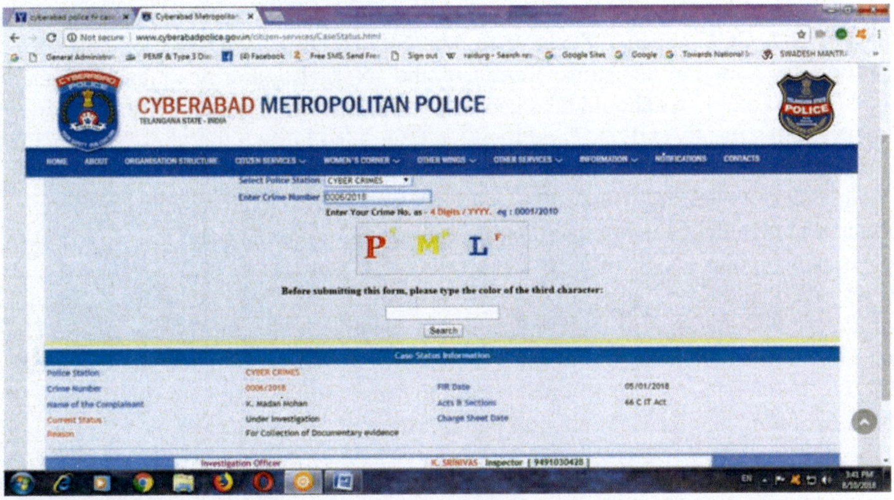

Fig. 3 FIR 0006/2018 is against violation of "identity theft" under Section 66C-IT Act whereas the complaint aptly claimed to register under "Sedition Law Violation" against the Fake Government of Telangana (GoT2011)

crime. The police proceedings, as well as the judicial proceedings, should rather be in such regard.

3.3 About Crime 3

The crime covers Cheating the Nation. This is reported in [10]. But, as such, the police did not yet register the case under this crime.

3.4 About Crime 4

The crime covers Sedition Law Violation. This is reported in the complaint for which Crime no.0006/2018 was registered in Cybercrimes Police Station@Gachibowli, Hyderabad, in January 2018 [2, 6, 7]. But, as such, the police did not yet register the case under this crime. This is mooted.

4 Governmental Action is the Need of the Hour

Both the GoI and the GoT2014 have to take measures w.r.t. handling the cybercriminal academic environment that prevailed in the JNTUH academic environment. The MHRD, UGC and AICTE have to take cognizance of the situation and adopt appropriate, remedial measures. Similarly, the JNTUH authorities and the Cyberabad Police authorities should take necessary actions such as issuing the prohibition orders against the JNTUHJAC and the FGoT(GoT2011) such that these cybercriminal organizations can not undertake any activity overtly further [6–9]. The police should intensify efforts to nab the criminals and for judicial accounts of conviction and punishment such that similar crimes do not recur. GoI and the GoT should release white papers on the issue of cybercriminal organizations in JNTUH academic environment and for policing for effective policing in the future. JNTUH authorities should take cognizance of the cybercriminal organizations that thrived in JNTUH for 3 years or so and take appropriate measures. Thus, it is upon the MHRD, UGC, AICTE, JNTUH, Cyberabad Police, GoI, GoT and GoAP to uphold the sanctity of the constitution and the nation by taking appropriate measures (Fig. 4).

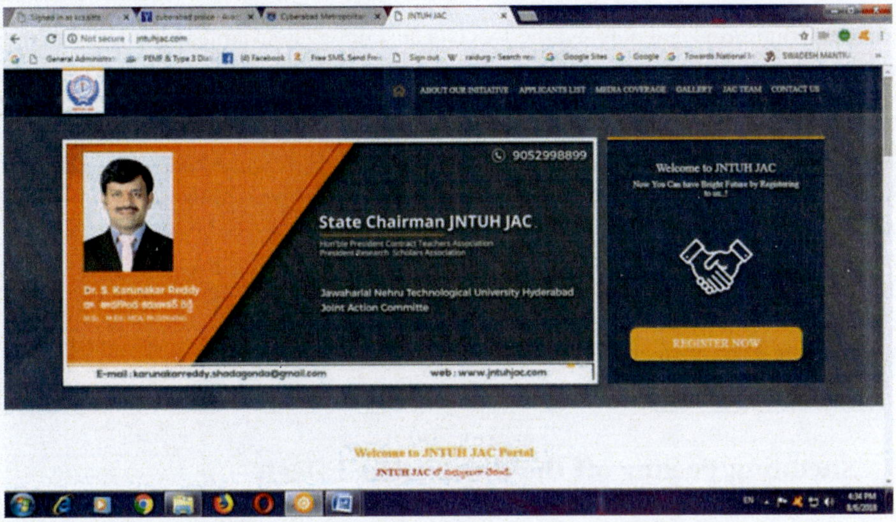

Fig. 4 Is the new JNTUHJAC organization whose home page dt. Aug2018 is captured in this figure not a new version in the continuum of the earlier India-abusive Cybercriminal Organizations such as JNTUHJAC & GoT2011 against which the Police failed to issue Prohibition Orders or to ensure Conviction?

4.1 Analysing the Cybercrimes from Individualistic Versus Organizational Perspectives: What Better JNTU Hyderabad, Cyberabad Police, the Government of Telangana (GoT2014) and the GoI Could Do

Certain cybercrimes are by an individual or two or three individuals. In the case of a crime by an individual, the police and judicial processes will be against the individual who perpetrated the crime. The police or judicial action will be against the individual only. But, in the case of an organized crime, the crime has an organizational setting. Many individuals form an organizational setting and the crime involved accordingly is in an organized sense. In the case of a crime by a single individual, the single individual is convicted and undergoes punishment. But, in the case of an organized crime such as the present case study, one or more organizations rather than merely one or more individuals are to be the subjects for penal action. The police and the judiciary should take action against the organizations as well as the individuals responsible for the same [8, 9].

In the case of the twin criminal organizations, JNTUHJAC and the Fake Government of Telangana, seemingly, neither the police nor the judiciary took action such as the issuance of prohibition orders against these organizations [6, 7]. In the absence of any police action against the twin organizations, the organizations tend to thrive covertly if not overtly like the Dark Web. Until and unless the Government declares or releases a white paper about it that the criminal organizations are defused or that the criminal conditions that plagued the JNTUH academic environment are defused, the impression is that the Fake Government of Telangana exists as the Dark Government of Telangana(DGoT) just as the Dark Web has significance.

Cybercriminal DGoT has prevailed in Telangana such that the police evidence, in the court of law was hostile towards the police [8, 9]. Accused A1 was acquitted. Accused A2 was absconding. The case was left into darkness of inattention without conviction. Thus, in the light of the police proceedings against the registered cybercrimes in JNTUH academic environment, it is high time that the police, judiciary and the Government rise to the occasion and respond to the prevailing circumstances that are plagued by it. The police, judiciary and the Government should issue prohibition orders against the twin cybercriminal organizations.

5 Shedding/Peeling off the Deleterious Effects of the Maladaptive Dimension of the Organizational Development: Telangana State Organization Versus JNTUH University Organization

India-abusive Cybercrimes in the Telangana movement [10] need a check by the GoI and the central agencies such as AICTE, UGC, DIT, MeitY, MHRD through the JNTUH authorities and this issue does not seem to have been taken up. This means that the national development through Telangana requires a review in the sense that the GoI should take measures to review what steps the GoT2014 has taken to recover the losses of national consciousness and sentiments due to the Telangana movement. Now that the Telangana state formation is a reality, it is high time that the GoT2014 take tangible measures of its commitment to Mother India to recover the losses caused in the process.

Loss of national consciousness in a state gives rise to the tardy development of the nation because national development really means the sense of the nation concept held by its people. National development is not merely the numerical figures of GDP, etc., that are indicative of economic development.

The sense of the nation concept means the national consciousness and respect for the nation and the national insignia and constitution of the nation which alone serve as an index for the national commitment held by people [8, 9]. Now the responsibility in this regard lies with the GoT2014 first. GoT2014 should make special endeavours for recovering and restoring the national consciousness of the people. For example, it should spare special endeavours, funds and provisions for special training of the students in the state by special NCC training. In the immediate future, for its true organizational development, JNTUH should take remedial measures by releasing a white paper w.r.t. cybercrimes that prevailed in its academic environment [6, 7]. It should identify and take action against the responsible authorities.

6 Conclusions

Is the Fake Government of Telangana (GoT2011) activity online organizing through online registrations thus streaming the inmates of the JNTU Hyderabad which is an institution of academic merit? Fig. 5 shows the clipping of cybercriminal JNTUH-JAC activity organizing Telangana youth on the day of "Pragathi Nivedana Sabha" that was organized by the TRS Government on 2 September 2018. If the educated elite of the university fall prey to the evil designs that are technologically abusive of the nation, who is to save the nation? The educated elite of the university are to be the torchbearers of knowledge and of good motivation and guidance to the general masses in the society. The Cyberabad Police action of ignoring the seditious dimension of the case is mooted [6, 7]. After repeated reviews, it is generally felt that the registration of the crime by the Cyberabad Police should have clearly squarely covered the sedition dimension of the crime case.

It is not enough for a cybercrimes researcher to merely study cybercrimes. One should get into the field of the police station, lodge police complaints against cybercrimes, track lifespan/life cycle of the running cybercrime case such as attending to the court procedures related to the crime case handling. It is dry research and

Fig. 5 "Andhra Jyothi" Telugu Newspaper Clipping dt.3Sep2019 showing the "Cybercriminal JNTUHJAC vide FIR 53/2014@KPHB Police Station, Cyberabad, Hyderabad" Actively Organizing Telangana Youth

teaching if one merely handles the subject only within the four walls of the classroom or the lab. There are many allied issues such as cyber policing ethics, cyber ethics, cyber laws that go hand in hand with the subject of cybercrimes. One becomes a better researcher by studying cybercrimes from a 360-degrees perspective such as the applicability of the big data dimension for a cybercrime, for example, as explored in this paper. The paper leads to insights in the field of ethics in academic governance and establishing the same in the cybercrime-affected JNTUH. The selfish inefficacy of the university administration has set aside the issue of inculcation, among the students, of national affection, national fervour, national consciousness, national spirit and love and commitment to the nation. Thus, the university has failed to fulfil the national goals in academics.

References

1. Usha Gayatri P, Chandra Sekharaiah K (2013) Encasing the baneful side of internet. In: National conference on computer science and security (COCSS 2013), 5–6 Apr 2013, Sardar Vallabhbhai Patel Institute of Technology, Vasad, Gujarat, India
2. Tirupathi Kumar B, Chandra Sekharaiah K, Mounitha P (2015) A case study of web content mining in handling cybercrime. Int J Adv Res Sci Eng (IJARSE) 04(1):665–668
3. Santhoshi N, Chandra Sekharaiah K, Madan Mohan K, Ravi Kumar S, Malathi B (2018) Cyber intelligence alternatives to offset online sedition by in-website image analysis through webcrawler cyber forensics. In: Proceedings of international conference on soft computing and signal processing (ICSCSP 2018)
4. Pavana Johar K, Malathi B, Ravi Kumar S, Srihari Rao N, Madan Mohan K, Chandra Sekharaiah K (2018) India-abusive Government-of-Telangana (GoT2011): a constitutional IT (an SMI) solution. In: Proceeding of international conference on science, technology and management (ICSTM2018) Indian Council of Social Science Research, North West Regional Center, Punjab University Campus, Chandigarh, India on 12th August 2018. ISBN: 978-93-87433-34-2; Int J Res Electron Comput Eng (IJRECE) 6(3) (July–September 2018). ISSN: 2393-9028 (PRINT) | ISSN: 2348-2281 (ONLINE), pp. 1118–1124
5. Ramesh Babu J, Chandra Sekharaiah K (2018) Adaptive management of cybercriminal, maladaptive organizations, in the offing, that imperil the nation. ICDMAI, Pune, 19–21 Jan 2018
6. Pavana Johar K, Gouri Sankar M, Ravi Kumar S, Madan Mohan K, Punitha P, Santhoshi N, Malathi B, Ramesh Babu J, Srihari Rao N, Chandra Sekharaiah K (2018) Cyberpolicing the multifaceted cybercriminal, fake Government of Telangana : what is sauce for the goose is sauce for the gander. In: Proceedings of 3rd international conference on research trends in engineering, applied science and management (ICRTESM-2018), Osmania University Centre for International Programmes, 4th November 2018, pp 321–330. Univ Rev 7(11), November 2018, pp 256–264
7. Srihari Rao N, Chandra Sekharaiah K, Ananda Rao A (2019) An RTI Act implementation analysis: a solution approach to defuse cyber criminally seditious Government of Telangana. Int J Emerg Technol Innov Res (JETIR) 6(3):13–21
8. Malathi B, Chandra Sekharaiah K (2018) Web intelligent information systems: a PGF-mediated social media evaluation perspective. In: Book of abstracts of andhra pradesh science congress on the theme integrating science and technology for global sustainability (APSC-2018), 9–11 Nov 2018, pp 420–421. ICRTESM-2018, 4 Nov 2018. ISBN: 978-93-87433-44-1. IJMTE, 8(11), NOVEMBER/2018, pp 684–690. ISSN NO: 2249–7455

9. Madan Mohan K, Chandra Sekharaiah K, Premchand P, Gopal Upakaram Pullaiah, Malathi B (2018) Approving psycho-neuro-computer systems to prevent (systemic Vs individualistic perspective) cybercrimes in information highway. In: 2018 IEEE 3rd international conference on computing, communication and security (ICCCS), Kathmandu (Nepal), 25–27 Oct 2018, pp 205–209
10. https://sites.google.com/site/chandraksekharaiah/india-against-corruption-jntu
11. https://sites.google.com/site/sekharaiahk/apeoples-governanceforumwebpage
12. Srihari Rao N, Chandra Sekharaiah K, Ananda Rao A (2018) An IT-theoretic defusion of OOFGoT. In: The book of abstracts of Andhra Pradesh science congress on the theme integrating science and technology for global sustainability (APSC-2018), 9–11 Nov 2018, pp 423–424. ICRTESM2018, 16 Dec 2018, pp 42–50. ISBN: 978-93-87433-48-9. IJMTAE 8(12), December 2018, pp 4050–4057

Chapter 46
A Survey of Ship Detection and Classification Techniques

D. Princy and V. R. S. Mani

1 Introduction

Synthetic Aperture Radar (SAR), which has an all-day imaging and reconnaissance capability, is an indispensable and important monitoring tool in the field of remote sensing. This technology plays an irreplaceable role in many civil and miltary applications, like marine fishery management, ocean environment protection, handling abnormal sea situations, maritime navigation monitoring and control, and key target reconnaissance and surveillance in water. The strong application demand has greatly promoted the event and progress of related technologies, and SAR ship detection in remote sensing is now receiving increasing attention. Although many achievements are published within the past few decades, there are still many challenges to affect, such as ship detection in complex scenarios (e.g., inshore, offshore, and inland river environments) and detection of ships of various sizes, densely arranged ships, and defocused SAR ships. In recent years, especially the last 5 years, the CNN has shown a robust ability for object detection and image feature representation. Region-based approaches, CNN techniques like Faster-RCNN series [1], and regression-based methods, like the YOLO series [2] and SSD [3, 4], have achieved state-of-the-art performance in both accuracy and efficiency in object detection. Due to their excellent detection performance, these data-driven algorithms and their variants are being introduced to object detection in remote sensing images, like optical remote sensing images and SAR images, to reinforce their accuracy, efficiency, and robustness. However, the foremost popular and best performing CNN-based detectors always contain many of learnable parameters, which means that training such a

D. Princy (✉) · V. R. S. Mani
Department of Electronics and Communication Engineering, National Engineering College, Kovilpatti, India
e-mail: princybalan1998@gmail.com

V. R. S. Mani
e-mail: vrsmece@nec.edu.in

N. Marriwala et al. (eds.), *Soft Computing for Intelligent Systems*, Algorithms for Intelligent Systems, https://doi.org/10.1007/978-981-16-1048-6_46

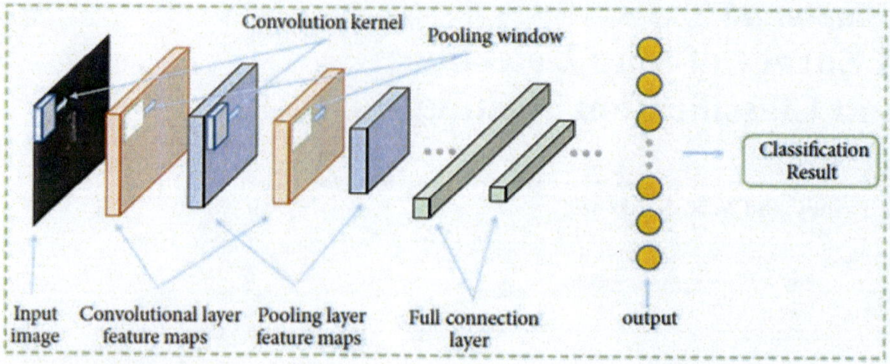

Fig. 1 Architecture of CNN

detector demands large numbers of bounding box-level annotated images. To realize good performance without large number of annotated images, most existing SAR ship detectors are developed supported transfer learning; they create use of pre-trained backbones to initialize the parameters of the networks then fine-tune the detectors with SAR datasets. There are three benefits to transfer learning-based SAR ship detectors: (1) fast convergence, (2) accurate and stable detection performance, and (3) the tiny demand for annotated SAR images. This paper is organized as follows. It begins with an introduction in Sect. 1, Followed by CNN in Sect. 2, HyperLi-Net in Sect. 3, A Lightweight Network in Sect. 4, Attention Receptive Pyramid Network in Sect. 5, RESNET in Sect. 6 and CFAR Ship detection in Sect. 7, (Hierarchical self-diffusion saliency (HSDS) in Sect. 8, T-NET in Sect. 9, Single Shot Detector (SSD) in Sect. 10, Coarse-to-fine network in Sect. 11, RetinaNet in Sect. 12, Balance scene learning mechanism in Sect. 13, Attention mechanism in Sect. 14, High-Resolution Ship Detector (HRSD) Network in Sect. 15, Recurrent Attention convolutional neural network in Sect. 16, Feature Fusion pyramid network and deep reinforcement learning method in Sect. 17, Feature Balancing and Reinforcement Network in Sect. 18, Squeeze excitation skip-connection path network in Sect. 19, Neural network in Sect. 20, Hybrid network of CNN and NN in Sect. 21, finally, the survey ends with Conclusion and Discussion (Fig. 1).

2 CNN

A CNN consists of several sorts of layers. They are convolutional layer, pooling layer, fully connected input layer, fully connected layer, and fully connected output layer. Convolution layer creates a feature map to predict the category probabilities for every feature by applying a filter that scans whole image. Pooling layer-scales down the amount of data that convolutional layer generated for every feature and maintains the foremost essensial information. Fully connected input layer-flattens the output generated by previous layers to show them into single vector which will be used as input for subsequent layer. Fully connected layer applies weight over the input generated by the feature analysis to predict an accurate label. Fully connected output layer-generates the ultimate probabilities to work out a category for the image

2.1 Various Modified Structures of CNN Used for Ship Detection

Zhu et al. [5] noticed that Speckle noise during a SAR image classification is additionally a severe problem to discriminate the target preciously. This approach extracts the luminance information of SAR image and creates the target's profile with the luminance level, which reduces the speckle noise significantly. Construct a convolutional neural network to coach these luminance outline images rather than the raw SAR images. The experimental results on the MSTAR dataset indicate above approach that achieves a better accuracy rate than the opposite state-of-the-art methods, which is on the brink of 100% in each category.

Ucar (2020) developed a Deep Convolutional Neural Network (DCNN) model for detecting the ships using the satellite images as inputs. This model has acquired an adequate accuracy value by just employing a pre-processed satellite image with a deep learning model built from scratch. The designed CNN model is formed with a transparent and easy to implement form especially to the popular satellite image set visual and graphical results to show that the proposed CNN model provides an efficient detection process with an accuracy of 99.60%.

Long et al. [6] designed a completely unique SAR ship detector from scratch is designed. PCB-MSK (parallel convolutional block of multi-size kernels) consists of two groups of convolutions, each group is made of four convolutional layers. In the Convolutional Module with Features Reused (CMFR), the output and input feature maps of the previous block are connected for current layer to scale back information loss during forward propagation to strengthen the supervision for shallow layers during parameter optimization. For each source prediction layer, the binary classification is conducted to alleviate the positive/negative imbalance; deconvolution and have fusion are utilized to reinforce the feature representation. Then, perform fine detection. Experiments on RDISD-SAR and SSDD achieve a state-of-the-art accuracy and competitive speed, the Average Precision (AP) reaches 88.70 and 90.57% for RDISD-SAR and SSDD.Aps is 10.43 and 4.23% higher than the DSOD and 6.64 and 1.70% higher than scratchDet. Decision speed is 58.24FPS on a GTX 1080 Ti GPU the number of parameters is 18.19 M and amount of computation is 21.33G.

Shen et al. [7] developed a framework, in which deconvolution is used to upsample the PolSAR images, PReLU is added to maintain the numerical properties. A complex structure block is additionally designed to accommodate the PolSAR arrangement. In addition, prior information on the low-resolution image itself is used to reduce the artifacts. Due to the complexity of PolSAR imaging systems, there may be some differences in the data acquired in different flight angles at different times, so it is difficult for training samples to contain all the types of features. Inspired by the downscaling of imagery in the geosciences [8, 9], a simple and effective complex structure-based residual compensation (RC) strategy is proposed for the post-processing of super resolution results. Shena et al. (2019) classified the PolSAR images using SVM supervised classification in PolSARpro v5.0. The AIRSAR Flevoland data [10] was used for the classification. This method shows a superior performance when compared to the traditional methods in both the quantitative evaluation and visual assessment. This method improved the spatial resolution significantly, especially in terms of detail information retention, and it improves the mean PSNR by more than 12% when compared to the traditional methods. By analyzing the phase statistics and polarimetric response, this method has a good polarimetric information retention ability and can obtain a higher classification accuracy.

Zhao et al. [11] developed a coupled CNN for small and densely clustered SAR ship detection. This method mainly consists of two sub-networks. They are an Exhaustive Ship Proposal Network (ESPN) for ship-like region generation from multiple layers with multiple receptive fields and an Accurate Ship Discrimination Network (ASDN) for warning elimination by pertaining to the context information of every proposal generated by ESPN. Experiments are evaluated on two data sets. One is collected from 60 wide-swath Sentinel-1 images and therefore the other is from 20 GaoFen-3 (GF-3) images. Both data sets contain many ships that are small and densely clustered. The quantitative comparison results demonstrate the clear improvements of this method in terms of Average Precision (AP) and F1 score by 0.4028 and 0.3045 for the Sentinel-1 data set compared with the multi-step constant false alarm rate (CFAR-MS) method. The values are verified as 0.2033 and 0.1522 for the GF-3 data set. In addition, the new method is demonstrated to be more efficient than

CFAR-MS

Deng et al. (2018) presented an effective method to learn deep ship detector from scratch. A condensed backbone network, which consists of several dense blocks. Earlier layers can receive additional supervision from the objective function through the dense connections, which makes it easy to train. Then feature reuse strategy is adopted to make it highly parameter efficient. Therefore, the backbone network could be freely designed and effectively trained from scratch without using a large amount of annotated samples and improve the cross-entropy loss to address the foreground–background imbalance and predict multi-scale ship proposals from several intermediate layers to improve the recall rate. Then, position-sensitive score maps are adopted to encode position information into each ship proposal for discrimination. State-of-the-art deep CNNs-based object detection methods can be divided into two groups: (1) Region proposal-based methods and (2) Regression-based methods. Region proposal-based methods, such as R-CNN, fast RCNN, faster R-CNN, PVANET, MSCNN, and R-FCN, divide the framework of detection in two stages. The first stage generates a sparse set of candidate proposals and the second stage classifies the proposals into foreground classes/background. The highest accuracy object detectors to date are region proposal-based methods. In contrast, regression-based methods, such as YOLO and SSD, use a single feedforward convolutional network to directly predict classes and bounding boxes. Results on the Sentinel-1 data set show that learning ship detector from scratch achieved better performance than ImageNet pre-trained model-based detectors and the above method is more effective than existing algorithms for detecting the small and densely clustered ships.

2.2 *Umbrella Network*

Xie et al. (2019) developed a novel deep convolutional neural network architecture named umbrella. Its framework consists of two alternate CNN-layer blocks. One block is a fusion of six 3-layer paths, that is used to extract diverse level features from different convolution layers. The other block is made of convolution layers and pooling layers are mainly utilized to reduce dimensions and extract hierarchical feature information. The combination of the two blocks could extract rich features from different spatial scale and simultaneously alleviate overfitting. The performance of the umbrella model was validated by the Moving and Stationary Target Acquisition and Recognition (MSTAR) benchmark data set (Fig. 2).

The performance of our umbrella network is also compared with some recent results from advanced SAR identification methods, including transfer learning-based

Fig. 2 Architecture of umbrella network

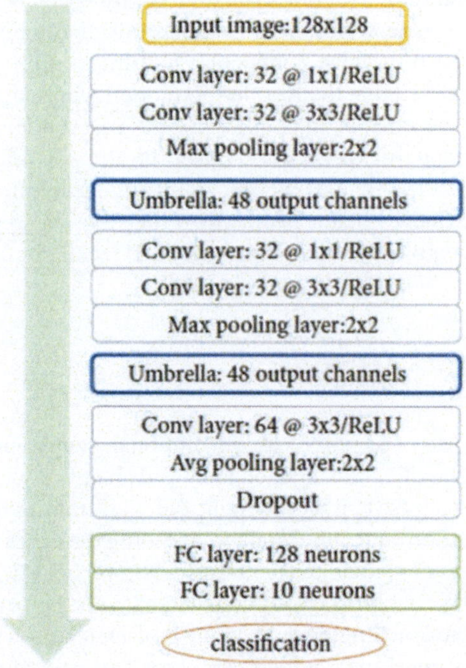

method (CNN-TL-bypass), sparse representation of monogenic signal (MSRC), tritask joint sparse representation (TJSR), A-ConvNet [12], and DCHUN. The accuracy rate of umbrella method is improved by 0.41 to 0.45% with respect to these methods based on deep learning.

This architecture could achieve higher than 99% accuracy for the classification of 10-class targets and higher than 96% accuracy for the classification of 8 variants of the T72 tank, even in the case of diverse positions located by targets. The accuracy of our umbrella is superior to the current networks applied in the classification of MSTAR. The result shows that the umbrella architecture possesses a very robust generalization capability and will be potential for SAR-ART.

2.3 Gabor-CNN

Hu et al. [13] describes the use of Gabor-CNN for object detection based on a small number of samples. A feature extraction convolution kernel library made of multi-shape Gabor and color. Gabor is made and the optimal Gabor convolution kernel group is obtained by means of coaching and screening that is convolved with the input image to obtain feature information of objects with strong auxiliary function. Then, the k-means clustering algorithm is adopted to construct several different sizes of anchor boxes, which improves the standard of the regional proposals). Deeply-Utilized Feature Pyramid Network (DU-FPN) method is to strengthen the feature expression of objects within the image. A bottom-up and a top-down feature pyramid is made in ResNet-50 and has information of objects that are deeply utilized through the transverse connection and integration of features at various scales. Ren et al. proved that the ResNet-50 model, when selected for pre-training, can achieve better performance than other pre-training models, such as VGG and PRELU-net through enough experiments. Based on the performance, the ResNet-50 model is selected as backbone. Experimental results show that the tactic developed during this paper achieves better results than the state-of-art contrast models on data sets with small samples in terms of accuracy and recall rate and thus has a strong application prospect.

2.4 MR-SSD

(Ma et al. 2018) a convolutional neutral network (CNN) model for marine target classification at patch level and an entire scheme for marine target detection in large-scale SAR images. Eight types of marine targets in GF-3 SAR images are labeled based on feature analysis, building the data sets for further experiments, classification task at patch level, a novel CNN model with six convolutional layers, three pooling layers, and two fully connected layers has been designed. With respect to the detection part, a Single Shot Multi-box Detector with a multi-resolution input (MR-SSD) is developed, which can extract more features at different resolution versions. For

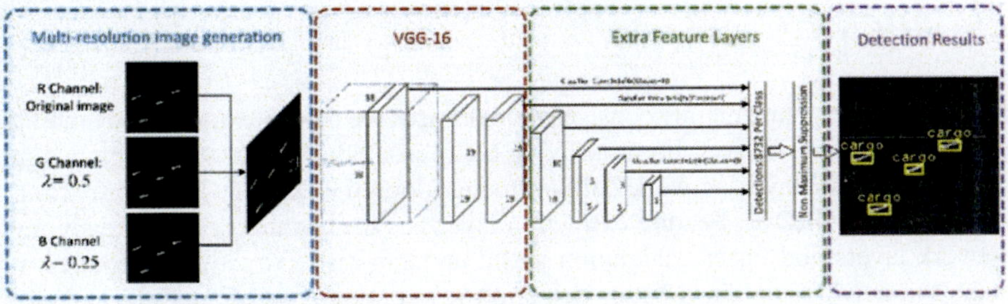

Fig. 3 Structure of MR-SSD

detecting different targets in large-scale SAR images, a whole workflow including sea-land segmentation, cropping with overlapping, detection with MR-SSD model, coordinates mapping, and predicted boxes consolidation is developed (Fig. 3).

Multi-Resolution Single Shot Multi-box Detector (MR-SSD), which has three sections: The first section is multi-resolution image generation, the second section is a standard CNN architecture used for image classification, and the last section is the auxiliary structure containing multi-scale feature maps, convolutional predictors, and default boxes with different aspect ratios. The MR-SSD is capable of extracting features from different resolution images at the same time, which helps to increase the detection precision. The second section of the MR-SSD is a standard CNN architecture, i.e., VGG-16, including five groups of convolutional layers combined with ReLU and pooling layers. The extra feature layers allow the detection at multiple-scales. In this section, the corresponding parameters used in SSD which proves to be effective in object detection challenges. Experiments based on the GF-3 dataset demonstrate the merits of this methods for marine target classification and detection.

Future work

Focused on eliminating false alarms in SAR imageries by image processing methods (Fig. 4).

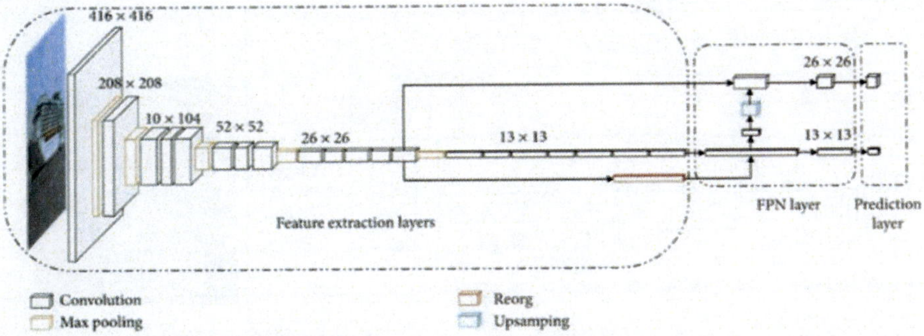

Fig. 4 Architecture of RDCNN By promoting YOLOv2 and v3

2.5 RDCNN

This network structure mainly consists of three parts: the feature extraction layer, FPN layer, and prediction layer. For improving the presented network with a lightweight feature extraction layer, network Zhijian Huang adopts the Darknet-19 feature extraction layer of YOLOv2. Feature extraction layer has the advantage of relatively few network layers and faster calculation speed and can also extract deep features, to make the network obtain a better detection result. Huang et al. [14] developed a network that promotes the multi-scale prediction idea of YOLOv3: Redmon and Farhadi [2] to design a new FPN layer. The clustering algorithm is also used, the prediction method of YOLOv2: Redmon and Farhadi [15] is adopted in the prediction layer. Improved RDCNN network has surpassed YOLOv2/v3 in the AIOU detection of positioning accuracy. It's 0.0153 higher than that of YOLOv2 and 0.0096 higher than that of YOLOv3. The improved network is 0.0044 higher than YOLOv2 in the mAP index. RDCNN method is also compared with Fast R–CNN, Faster R–CNN, SSD, YOLOv2. etc. under different datasets and hardware configurations. Results show that the tactic has advantage in precision and speed, and it also can satisfy the video scene (Fig. 5).

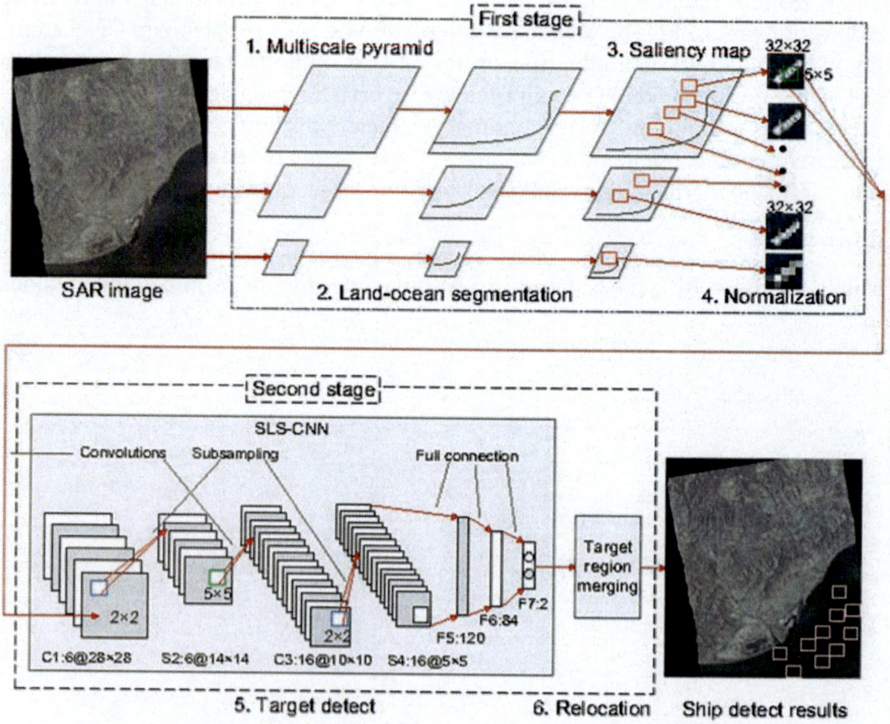

Fig. 5 Architecture of SLS-CNN

2.6 SLS-CNN

Liu et al. 16 developed a framework of Sea-Land Segmentation-based Convolutional Neural Network (SLS-CNN) for ship detection that attempts to combine the SLS-CNN detector, saliency computation, and corner features. For this, sea-land segmentation based on the heat map of SR saliency and probability distribution of the corner is applied, which is followed by SLS-CNN detector, and a final merged minimum bounding rectangles. The framework has been tested and assessed on ALOS PALSAR and TerraSAR-X imagery. Experimental results on representative SAR images of different kinds of ships demonstrate the efficiency and robustness of our proposed SLS-CNN detector (Fig. 6).

2.7 G-CNN

(A) **Overall framework (B) B-CNN (C) D-CNN**

Zhang et al. [17] presented a novel approach for high-speed ship detection in SAR images based on a grid convolutional neural network (G-CNN). This method improves the detection speed by meshing the input image, inspired by the basic thought of you only look once (YOLO), and using depth-wise separable convolution. G-CNN is mainly composed of a backbone convolutional neural network (B-CNN) and a detection convolutional neural network (D-CNN). SAR images to be detected are divided into grid cells and each grid cell is responsible for the detection of specific ships. Then, the whole image is input into B-CNN to extract features. Ship detection is completed in D-CNN under three scales. Experimental results show that the detection speed of the above method is faster than the existing other methods, such as faster-regions convolutional neural network (Faster R-CNN), single shot multi-box detector (SSD), and YOLO, under the same hardware environment with NVIDIA GTX1080 graphics processing unit (GPU) and the detection accuracy is kept within an acceptable range. Our proposed G-CNN ship detection system has great application values in real-time maritime disaster rescue and emergency military strategy formulation

Future work

This G-CNN SAR ship detection system has a slightly lower performance for small and dense ships. Therefore, further improvement is needed to solve this problem (Fig. 7).

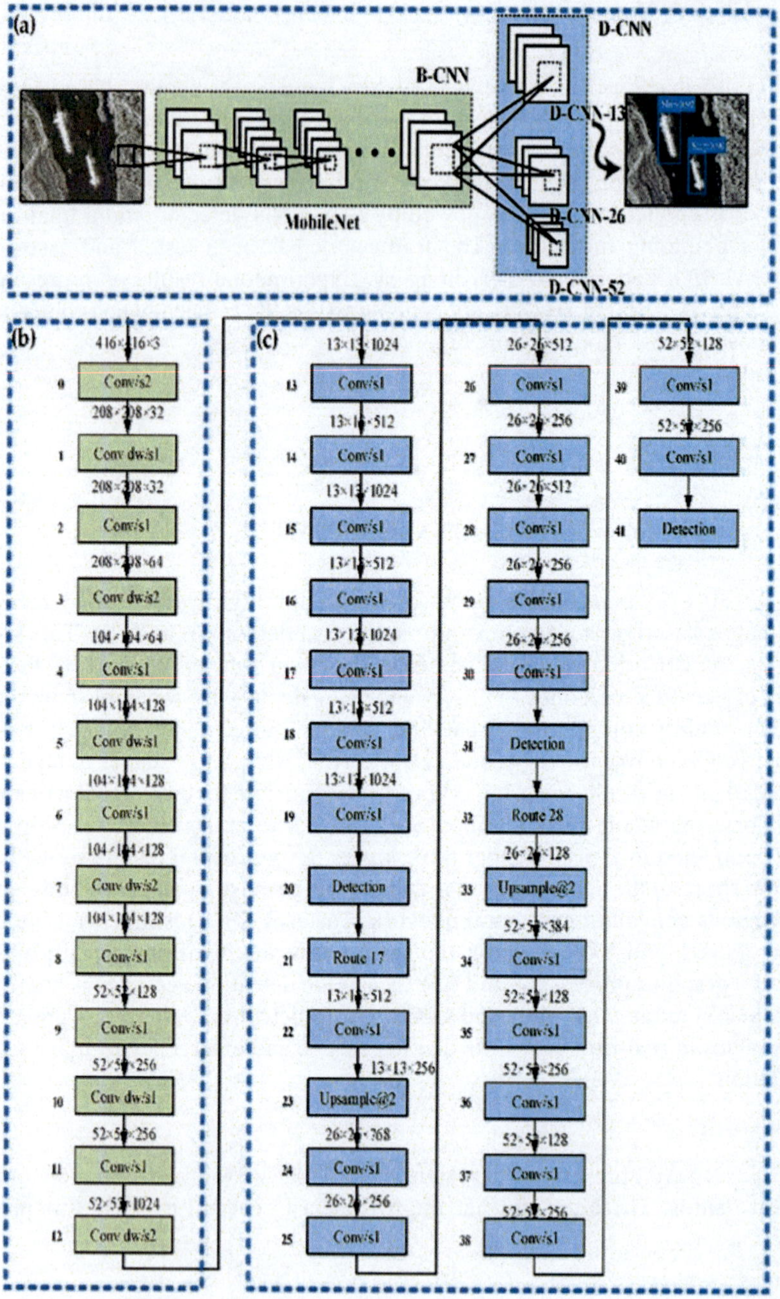

Fig. 6 G-CNN ship detection system

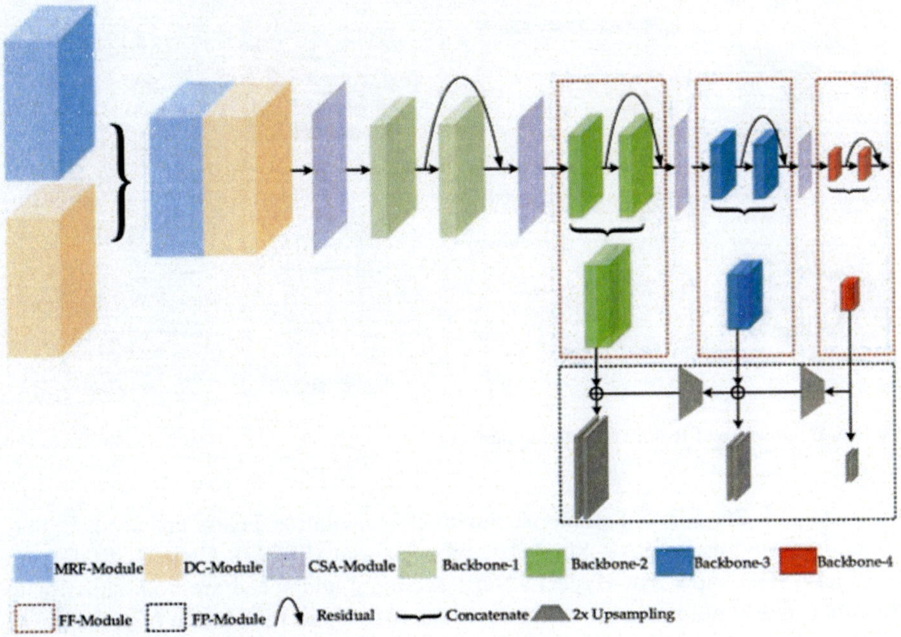

Fig. 7 HyperLi-network architecture

3 HyperLi-Network

HyperLi-Network consists of five external modules to achieve high-accuracy. They are MRF-Module, DC-Module, CSA-Module, FF-Module, and FP-Module. Five internal mechanisms are used to achieve high speed. They are RF-Model, S-Kernel, N-Channel, Separa-Convand BN Fusion. MRF-Module is at the input-end of HyperLi-Net.DC-Module to expand MRF-Module receptive field without increasing too much computation cost. DC-Module is also at the input-end of HyperLi-Net that is parallel to MRF-Module.

Zhang et al. [12] designed 4backbones and each backbone reduces the feature maps size by half(from $L/4$ to $L/32$) in order to learn deep semantic features. CSA-Module can improve the representativeness of important features and suppress inessential features. FF-Module can realize shallow and deep features fusion, making the final features used for detection more robust and contextual. In HyperLi-Net, three detection scales are designed which is big-scale ($L/32$), medium-scale ($L/16$), and small-scale ($L/8$), which jointly constitute a FP-Module. Inspired by Inception structure [18], MRF-Module is designed, Inspired by Multi-Scale Context Aggregation [19], DC-Module is designed, Inspired by YOLOv3 [2], 4backbones is designed. Inspired by CBAM [20], CSA-Module is designed. Inspired by DenseNet [21], FF-Module is designed, Inspired by FPN [22], FP-Module is designed, Inspired by YOLO [23, 24, 2], adopted the idea of RF-Model to design HyperLi-Net, Inspired

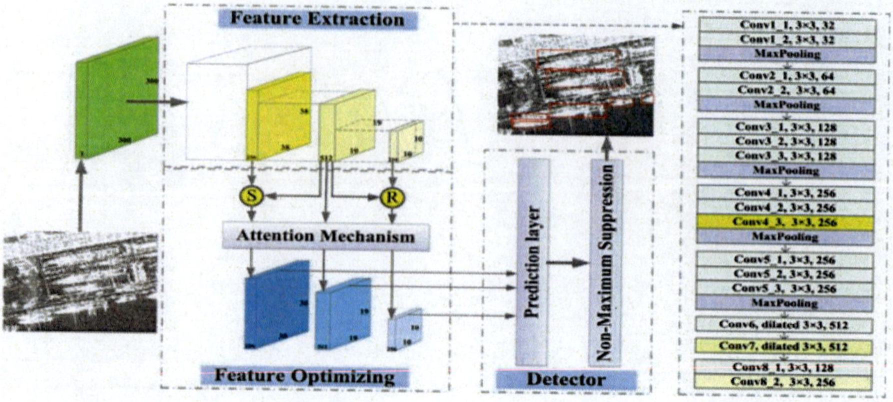

Fig. 8 Architecture of feature optimizing network

by VGG [25, 26, 27], S kernels to construct convolution layers are used. Experimental results on the SAR Ship Detection Dataset (SSDD), Gaofen-SSDD, and Sentinel-SSDD show that HyperLi-Net's accuracy and speed are both superior to the other nine state-of-the-art methods. It can also avoid the trouble of pre-training on ImageNet enhancing training efficiency. For its light model, it can be efficiently trained on CPUs, reducing hardware cost caused by GPUs (Fig. 8).

4 A Lightweight Network

Lightweight model named as LSSD is designed according to the structure of VGG16-based SSD, with one-channel SAR images as inputs to the network. The count of channels of each convolutional layer in original VGG16 is reduced by half. Then the fully connected layers of VGG16 are removed and 3 convolution blocks are added named as conv6, conv7, and conv8, respectively. Similar to SSD model, dilated kernels are used in conv6 and conv7 for expanding the receptive fields. The number of parameters to be trained in LSSD is less than one-fourth of that in SSD, which can improve the training and testing speed significantly.

Zhang et al. [17], a simpler structure called lightweight single shot detector (LSSD) is designed and can be trained from scratch which reduce the training and testing time without affecting accuracy. A new bi-directional feature fusion module including one semantic aggregation block and one feature reuse block is proposed to improve the performance of multi-scale targets detection by enhancing the features of both low feature layers and high feature layers. Then the features are further optimized by leveraging attention mechanism, which is beneficial to catch the silent information more efficiently. This feature fusion method is compared with classical FPN model and TDM model, which are used in many target detection tasks. This method has significant advantages in both speed and accuracy, and outperforms other state-of-art methods. Additionally, a test on GF-3 satellite SAR data with multiple modes verifies the generalization performance of this model.

Long et al. [10] paper proposed a lightweight convolutional neural network, LiraNet, which combines the idea of dense connections, residual connections, and group convolution, including stem blocks and extractor modules. The designed stem block uses a series of small convolutions to extract the input image features, and the extractor network adopts the designed two-way dense connection module, which further reduces the network operation complexity. Mounting LiraNet on the object detection framework Darknet, this paper develops Lira-you only look once (Lira-YOLO), a lightweight model for ship detection in radar images, which can easily be deployed on the mobile devices. Lira-YOLO's prediction module uses a two-layer YOLO prediction layer and adds a residual module for better feature delivery. For verifying the performance of the model, mini-RD, a lightweight distance Doppler domain radar images dataset, is constructed. The SSDD dataset is the ship object dataset of the SAR images proposed in [28], which contains 1260 pictures, including only a class of ships, and the sizes of the radar images are between 300×200 and 550×450. Experiments show that the network complexity of Lira-YOLO is low, being only 2.980 Bflops, and the parameter quantity is smaller, which is only 4.3 MB. The mean average precision (mAP) indicators on the mini-RD and SAR ship detection dataset (SSDD) reach 83.21 and 85.46%, respectively, which is comparable to the tiny-YOLOv3. Lira-YOLO has achieved a good detection accuracy with less memory.

Li et al. (2020) developed a lightweight ship detection model (LSDM) in which the backbone network is improved by using dense connection inspired from DenseNet, and the feature pyramid networks are improved by using spatial separation convolution to replace normal convolution. Two improvements reduce parameters and optimize the network structure effectively.

YOLOv3 is an end-to-end object detection model, and its network structure includes a backbone network and a detection network. The residual units are designed to avoid the vanishing-gradient problem inspired from the Resnet.

DenseNet solves the vanishing-gradient problem by connecting each layer to every other layer in a feed-forward fashion. DenseNet is a narrow network, in which each layer accepts inputs from all previous layers and passes the feature maps to all subsequent layers. Each layer reuses the global features and adds only a few new features remaining other features unchanged. This mechanism of feature reuse makes the DenseNet has fewer parameters than traditional convolutional neural networks. In addition, as each layer can access directly the gradient from the loss function, the DenseNet optimizes the information flow and gradient throughout network, which make it easy to be trained and has low overfitting risk on small training data set.

Shao et al. (2018) developed, a lightweight CNN model for target recognition in SAR image. Based on visual attention mechanism, the channel attention by-pass and spatial attention by-pass enhance the feature extraction ability. Then, the depthwise separable convolution is used to replace the standard convolution to reduce the computation cost and heighten the recognition efficiency. Finally, a new weighted distance measure loss function is introduced to weaken the adverse effect of data

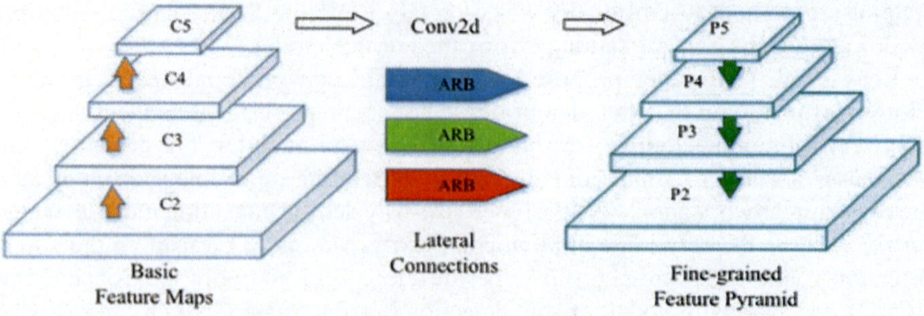

Fig. 9 Architecture of attention receptive pyramid network

imbalance on the recognition accuracy of minority class. A series of recognition experiments based on two open data sets of MSTAR and Open SAR Ship are implemented. Testing the performance of the above network in SAR image recognition, conduct a classification experiment based on the MSTAR dataset, and select four CNN models with good performance in SARA-ATR or CV field, namely, Network-1, developed by Wilmanski et al. (2019). Network-2, A-ConvNets developed by Chen et al. in literature, ResNet18, and SE-ResNet50. Experimental results show that compared with four advanced networks, above network can greatly diminish the model size and iteration time while guaranteeing the recognition accuracy, and it can effectively alleviate the adverse effects of data imbalance on recognition results.

Limitations

1. The method used in this paper belongs to supervised learning in machine learning field. The deep network needs a large number of data to train the parameters adequately, which restricts its application to a certain extent
2. This network needs the same size images as input, if the size of the input images is quite different, the recognition result will be affected. This problem can be solved by introducing Space Pyramid Pooling (SPP) [29], which will be our future research direction.
3. This experimental network is sensitive to noise (Fig. 9).

5 Attention Receptive Pyramid Network

Zhao et al. [30] developed a novel network, called Attention Receptive Pyramid Network (ARPN). ARPN is a two-stage detector and designed to improve the performance of detecting multi-scale ships in SAR images by enhancing the relationships among nonlocal features and refining information at different feature maps. Receptive Fields Block (RFB) and Convolutional Block Attention Module (CBAM) are employed and combined reasonably in attention receptive block to build a top-down fine-grained feature pyramid. RFB, composed of several branches of convolutional layers with specifically asymmetric kernel sizes and various dilation rates, is used for

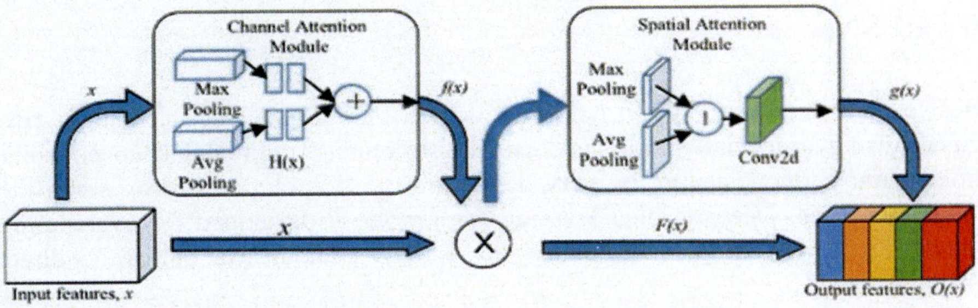

Fig. 10 Convolutional attention block module

grabbing features of ships with large aspect ratios and enhancing local features with their global dependences. CBAM, which consists of channel and spatial attention mechanisms, is utilized to boost significant information and suppress interference caused by surroundings. Evaluating the effectiveness of ARPN, experiments are conducted on SAR ship detection dataset and two large-scene SAR images (Fig. 10).

The detection results illustrate that competitive performance has been achieved by the above method in comparison with several CNN-based algorithms, such as Faster-RCNN, RetinaNet, feature pyramid network, YOLOv3, Dense Attention Pyramid Network, Depth-wise Separable Convolutional Neural Network, High-Resolution Ship Detection Network, and Squeeze-and-Excitation Rank Faster-RCNN.

Future work

Concentrate on combining backscattering properties of ships in SAR images with convolutional design of networks and introducing a strong restriction, e.g., mask, to further improve the detection accuracy as well as detection speed (Fig. 11).

Fig. 11 Architecture of RESNET block

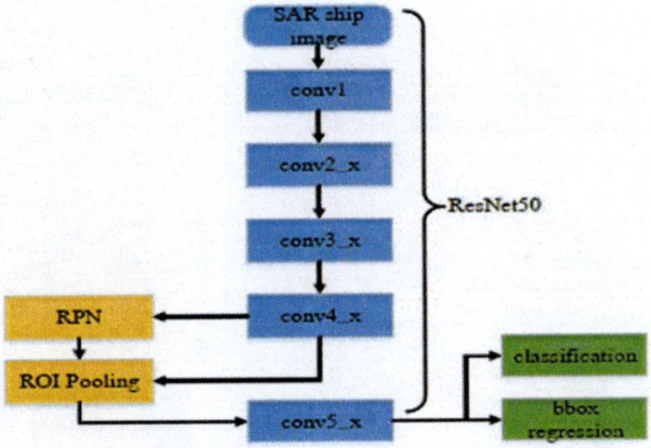

6 RESNet

ResNet is a deep convolution neural network which can achieve very high efficiency using a specially designed residual structure. Traditional deep convolution neural network cannot be very deep, and the accuracy will drop when the depth increases. Deep residual learning framework is developed for overcoming the traditional network, and the network learns the residual instead of learning direct mapping.

Li et al. [31] developed a very deep network ResNet with higher accuracy and faster training speed is applied to train the SAR ship detection model. Transfer learning is applied to combat the small dataset.

Verifying the effectiveness of the above method, CFAR-based method, YOLOv2 (44), and Faster-RCNN (the base network is VGG16) experiments are conducted in the same dataset to compare with our method and achieve higher accuracy and faster training speed, which verifies the effectiveness of our method

Dong et al. (2019) presented a ship classification framework based on deep residual network for high-resolution SAR images. In basic, networks with more layers have higher classification accuracy. The training accuracy degradation and the limited dataset are major problems in the training process. To build deeper networks, residual modules are constructed and batch normalization is to keep the activation function output. Different fine-tuning methods are used to select the best training scheme. To take advantage of the proposed framework, a dataset including 835 ship slices is augmented by different multiples and then used to evaluate above method and other Convolutional Neural Network (CNN) models (Fig. 12).

In the above framework, a SAR image ship dataset should be established and divided into a training set and a test set. The training set has been augmented by different multiples and it is used to train the ResNet-based classification model. In

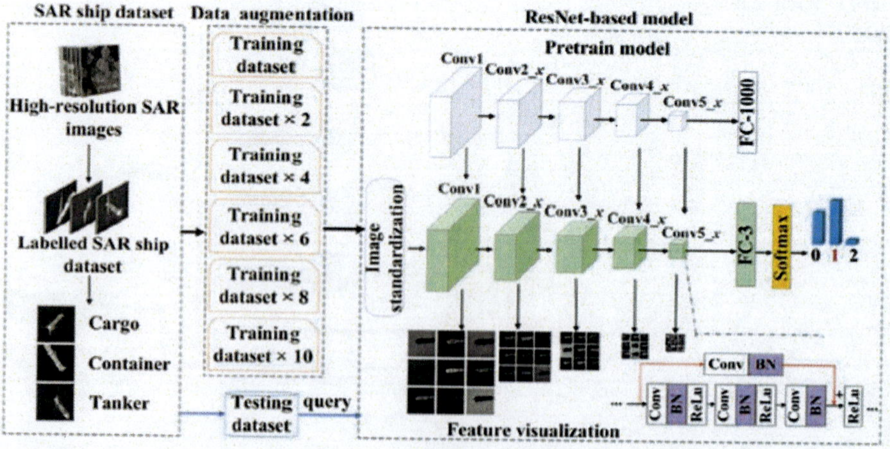

Fig. 12 Framework for RESNET-based ship detection in high-resolution SAR images

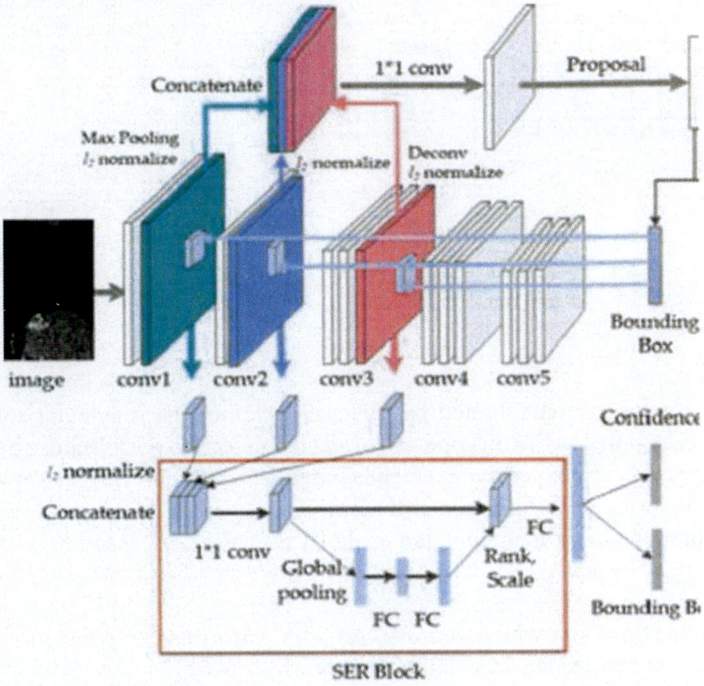

Fig. 13 Architecture of SER RCNN

particular, to observe the operating mechanism of the feature map, feature visualization is added to the training process. The test set is used to validate the models overall accuracy of ResNet50 and ResNet-101 can reach 99.37%. Although ResNet-101 has twice the number of network layers as ResNet-50, it can maintain the accuracy without degradation. This shows the benefits of ResNet. ResNet is a large network, and training on small data sets is likely to cause overfitting, so the amount of training data should be large. Above Framework can achieve a 99% overall accuracy on the augmented dataset under the optimal fine-tuning technique, 3% higher than that in other models, which demonstrates the effectiveness of this method (Fig. 13).

6.1 Faster R-CNN

Lin et al. [11] presented faster R-CNN method to improve the detection performance by using squeeze-and-excitation mechanism. The feature maps are extracted and concatenated to obtain multi-scale feature maps with ImageNet pre-trained VGG network. After region of interest pooling, an encoding scale vector which has values between 0 and 1 is generated from sub-feature maps. The scale vector is ranked, and only top K values will be preserved. Other values will be set to 0. Then, the

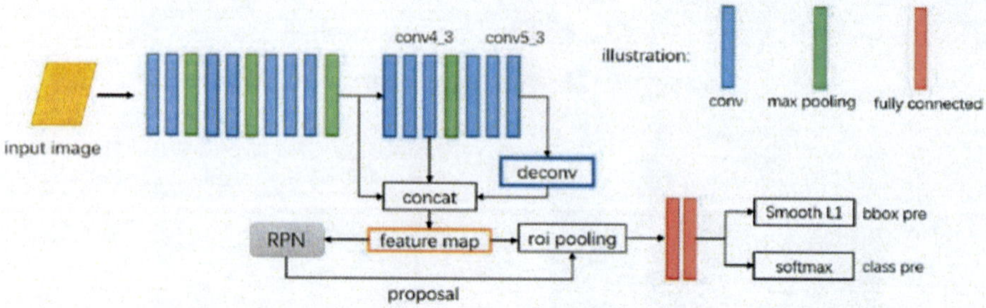

Fig. 14 Modified faster-RCNN structure

sub-feature maps are recalibrated by this scale vector. The redundant sub-feature maps will be suppressed by this operation, and the detection performance of detector can be improved. The experimental results based on Sentinel-1 images show that the detection performance of the proposed method achieves 0.836 which is 9.7% better than the state-of-the-art method when using F1 as matric and executes 14% faster.

Future work

Investigate the third way which improves network performance by designing suitable architecture for the second-stage classification (Fig. 14).

Zhang et al. [32] presented a fast regional-based convolutional neural network (R-CNN) method to detect ships from high-resolution remote sensing imagery. Choose GaoFen-2 optical remote sensing images with a resolution of 1 m and preprocess the images with a Support Vector Machine (SVM) to divide the large detection area into small Regions Of Interest (ROI) that may contain ships. Then, apply ship detection algorithms based on a region-based convolutional neural network (R-CNN) on ROI images. For improving the detection result of small and gathering ships, adopt an effective target detection framework, Faster-RCNN, and improve the structure of its original convolutional neural network (CNN), VGG16, by using multi-resolution convolutional features and performing ROI pooling on a larger feature map in a Region Proposal Network (RPN) and compare the most effective classic ship detection method, the Deformable Part Model (DPM), another two widely used target detection frameworks, the single shot multi-box detector (SSD) and YOLOv2(45), the original VGG16-based Faster-RCNN, and our improved Faster-RCNN. Experimental results show that the above improved Faster-RCNN method achieves a higher recall and accuracy for small ships and gathering ships. Hence it provides a very effective method for offshore and inland river ship detection based on high-resolution remote sensing imagery.

Future work

Try more traditional methods in the preprocessing stage to increase the recall of ROI, such as the variation of LBP, Gaussian Local Descriptors, SML, and PCA classifier. It will be worthwhile to conduct further research on more sophisticated CNN and even RNN methods based on optical remote sensing imagery.

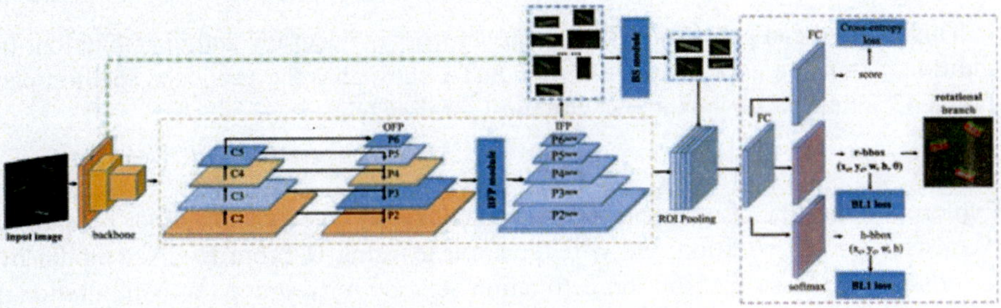

Fig. 15 Structure of libra R-CNN

Li et al. [33, 34, 33] developed a new dataset and four strategies to improve the standard Faster-RCNN algorithm. The dataset contains ships in various environments, such as image resolution, ship size, sea condition, and sensor type, it can be a benchmark for researchers to evaluate their algorithms. The strategies include feature fusion, transfer learning, hard negative mining, and other implementation details.

Proposed method obtains better accuracy (Fig. 15).

6.2 R-CNN

Guo et al. (2020) presented a rotational Libra R-Convolutional Neural Network (CNN) method. Our idea is to balance the three levels of neural networks for predicting the location of ships with rotational angle information, which refers to the feature level, sample level, and objective level. T extract a discriminative feature and improve its robustness against the impact of different sizes of ships, the concept of balanced feature pyramid is introduced. Generate reliable proposals for feature pyramid and efficiently mine hard negative samples, we employ intersection over union(IoU)-balanced sampling. Finally, to eliminate the redundant background and detect densely distributed ships, we bring in a rotational region detection branch with balanced L1 loss. In basics, develop the balanced learning with rotational region detection to achieve consistent improvement on accuracy and visualization. Zhang et al. [17] proposed that ship detection based on CNN (S-CNN) has specially designed techniques to improve accuracy. Tanget et al. (2018) used the compressed domain to achieve fast ship candidate region extraction and employed high-resolution features for classification with the Deep Neural Network (DNN). Kang et al (2018) applied the Faster R-CNNto generate a guide window for the Constant False-AlarmRate (CFAR) algorithm to improve the detection for small ships. Wang et al. [9] employed RetinaNet to address ship detection in Synthetic Aperture Radar (SAR) images. All of the above methods were based on horizontal region detection. Yang et al. (2017) further proposed the R2CNN based on Faster R-CNN to predict a rotational box.

This method can gain consistent improvements in accuracy and visualization. In addition, extensive experiments on the DOTA show that the proposed method can gain 3.43% than R2CNN and 4.09% than LibraR-CNN.

Future work

Explore more efficient strategies for extremely small objects, e.g., super resolution reconstruction. In addition, we will focus on learning R-Libra R-CNN model by introducing a sparse distribution estimation scheme to improve the computational efficiency.

7 CFAR Ship Detection

Liu et al. (2019) developed a systemic analytical framework for CFAR algorithms based on PWF or multi-look PWF (MPWF). This framework covers the entire log-cumulants space in terms of the textural distributions in the product model, including the constant, gamma, inverse gamma, Fisher, beta, inverse beta, and generalized gamma distributions (G_Ds). Derive the analytical forms of the PDF for each of the textural distributions and the probability of false alarm (PFA). Then, the threshold is derived by fixing the False Alarm Rate (FAR). Experimental results using both the simulated and real data demonstrate that the derived expressions and CFAR algorithms are valid and robust (Fig. 16).

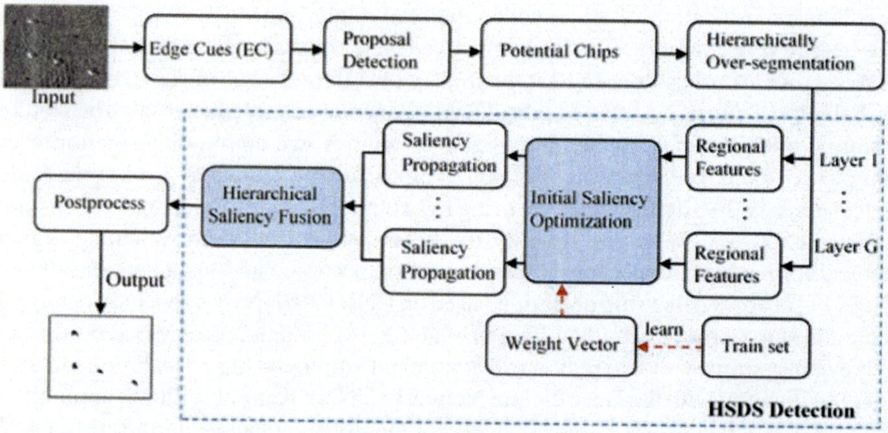

Fig. 16 Architecture of HSDS

8 HSDS

Ma et al. (2019) developed a novel Hierarchical Self-Diffusion Saliency (HSDS) method for detecting vehicle targets in large-scale SAR images. Reducing the influence of cluttered returns on saliency analysis, weight vector from the training set to capture optimal initial saliency of the super-pixels during saliency diffusion. By accounting for the multiple sizes of background objects, the saliency analysis is implemented in multi-scale space, and a saliency fusion strategy employed to integrate the multi-scale saliency maps. A hierarchical self-diffusion saliency detection method is designed to accurately detect the targets from the proposal chips. The graph diffusion is a common saliency detection algorithm, examples including conditional random fields (50), quadratic energy models (51), random walks (52). and manifold ranking (53). In these methods, the images are first partitioned into graphs. Saliency information then propagates throughout the graphs. In this section, promote the performance of the saliency propagation by improving the initial node saliency optimization strategy and the saliency map fusion rule.

Simulation experiments demonstrate that above-developed method can produce a more accurate and stable detection performance, with decreased false alarms, compared to benchmark approaches.

Future work

Attempt to reduce the algorithms reliance on training samples and also plan to integrate task-related information in saliency calculations in order to further enhance the accuracy of target detection (Fig. 17).

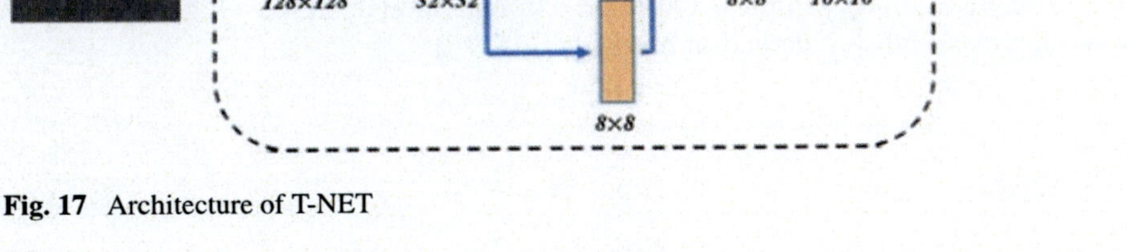

Fig. 17 Architecture of T-NET

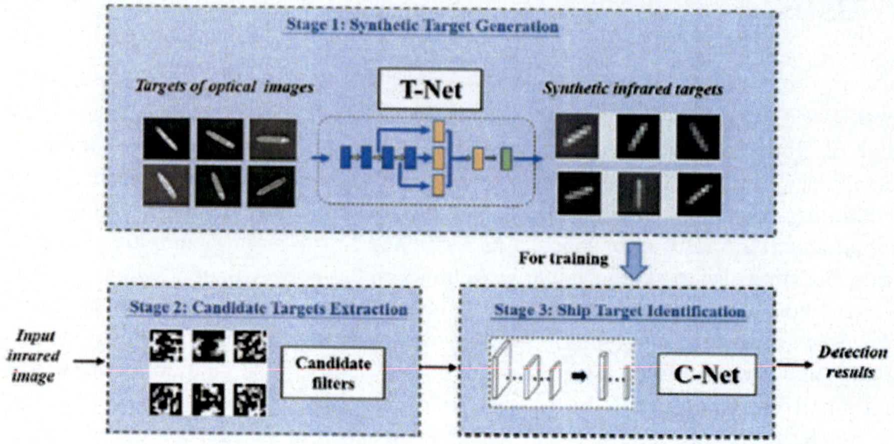

Fig. 18 Overview of the method

9 T-NET

Jiang et al. (2018) developed a new infrared ship detection method based on Convolutional Neural Networks (CNN)) which is trained only with synthetic targets. For the problem of limited infrared training data, Design a Transfer Network (T-Net) to generate large amount of synthetic infrared-style ship targets from Google Earth images. The experiments are conducted on a near infrared band image (0.845–0.885 μm), a short wavelength infrared band image(1.560–1.66 μm), and a long wavelength infrared band image (2.1–2.3 μm) of Landsat-8 satellite (Fig. 18).

There are three parts in the above method: (1) Generating synthetic targets using T-Net; (2) Candidate targets extraction using template filtering method; (3) Identifying candidates using CNet. Experiments are conducted on a near infrared band image (0.845–0.885 μm), a short wavelength infrared band image (1.560–1.66 μm), and a long wavelength infrared band image (2.1–2.3 μm) of Landsat-8 satellite. The results demonstrate the effectiveness of the target generation ability of T-Net. With only synthetic training samples, above detection method achieves a higher accuracy than other classical ship detection methods (Fig. 19).

10 SSD

Wang et al. [9] developed a method that is based on the original SSD (Single Shot Detector), using a rotatable bounding box. This method can learn and predict the category, location, and angle information of ships using only one forward computation. The generated oriented bounding box is much tighter than the normal bounding box and is strong to background disturbances. A developed semantic aggregation method

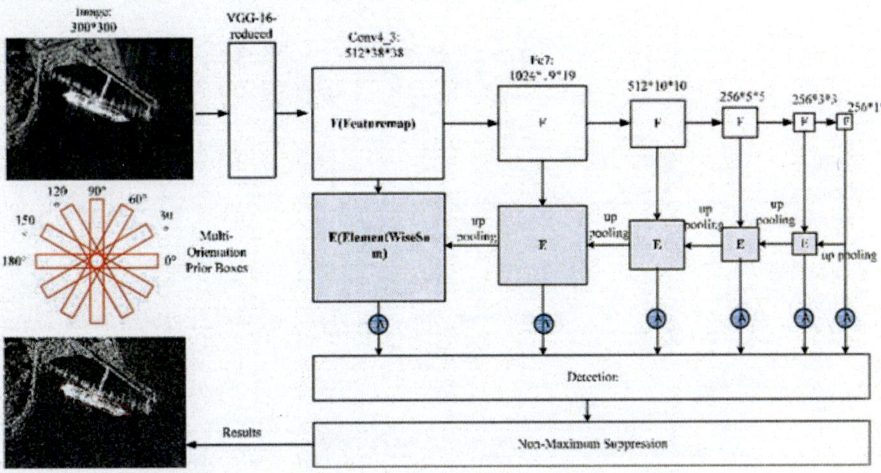

Fig. 19 Structure of SSD

which fuses features during a top-down way. This method can provide abundant location and semantic information, which is useful for classification and site and adopt the eye module for the six prediction layers. It can adaptively select meaningful features and neglect weak ones. This is helpful for detecting small ships. Multi-orientation anchors are designed with different sizes, aspect ratios, and orientations. These can consider both speed and accuracy. Angular regression is embedded into the existing bounding box regression module, and the angle prediction is output with the position and score, without requiring too many extra computations. The loss function with angular regression is employed for optimizing the model. Average Angle Precision (AAP) is used for evaluating the performance. Inspired by [35, 36], we adopt the rotatable bounding box to detect ships and estimate their angles. The rotatable bounding box is tighter, which is robust to the disturbance of the background pixels. This is also helpful for improving the result of detection and angle estimation [37]. Attention was first used in SENet [31, 38]. The squeeze-and-excitation block can adaptively recalibrate channel-wise feature responses by explicitly modeling interdependencies between channels. This is helpful for classification. GRP-DSOD (gated recurrent pyramid-deeply supervised object detector) [27] uses a gate in the prediction layer to adaptively enhance or attenuate supervision at different scales based on the input object size.

The above method can detect ships and estimate their angles with a high-accuracy and speed. It can detect ships not only in open sea areas but also near the shore.

Chen et al. (2019) designed single shot detection framework combined with attention mechanism, which roughly locates the regions of interest via an automatically learned attentional map. This lays the foundation of accurate positioning of extremely small objects since the background interference can be effectively suppressed. Furthermore, a multi-level feature fusion module integrated in top-down and bottom-up manner is adopted to adequately aggregate features from not only

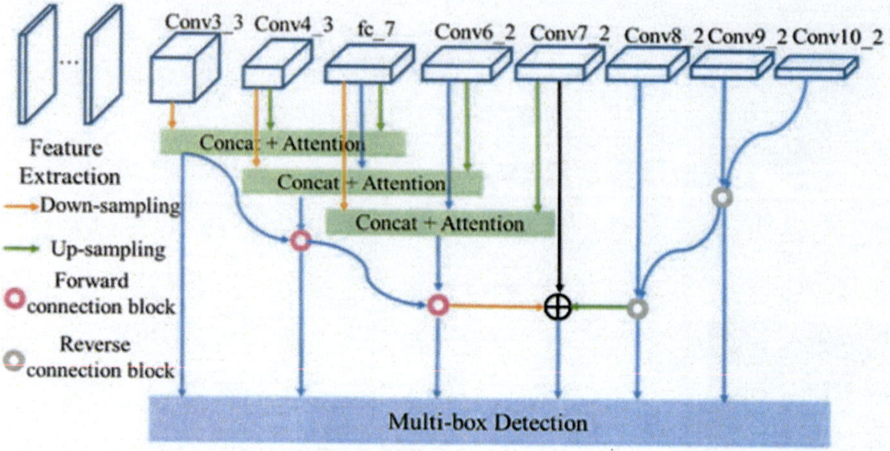

Fig. 20 Architecture of SSD with regional attention

adjacent but also distant layers. This strengthens local details and merge strong semantic information, enabling the generation of higher qualified anchors for the efficient detection of multi-scale and multi-orientated objects (Fig. 20).

Experiments on SAR ship dataset have achieved a promising result, surpassing current state-of-the-art method

11 Coarse-to-Fine Network

Wu et al. [39] developed a sea-land separation algorithm that combines gradient information and gray information is applied to avoid false alarms on land, the Feature Pyramid Network (FPN) is used to achieve small ship detection, and a multi-scale detection strategy is to achieve ship detection with different degrees of refinement. Then the feature extraction structure is adopted to fuse different hierarchical features to improve the representation ability of features. Then a coarse-to-fine ship detection network (CF-SDN) that directly achieves an end-to-end mapping from image pixels to bounding boxes with confidences. A coarse-to-fine detection strategy is applied to improve the classification ability of the network.

A coarse-to-fine ship detection network (CF-SDN), including the feature extraction structure, the distribution of anchor, the coarse-to-fine detection strategy, the details of training and testing. The sea-land separation algorithm [29] used in this paper considers the gradient information and the gray information of the optical remote sensing image comprehensively, combines some typical image morphology algorithms, and generates a binary image (Fig. 21).

In the above structure, 13 convolutional layers and the 4 max pooling layers of VGG-16 which is pre-trained with ImageNet dataset (57) as the basic network, and

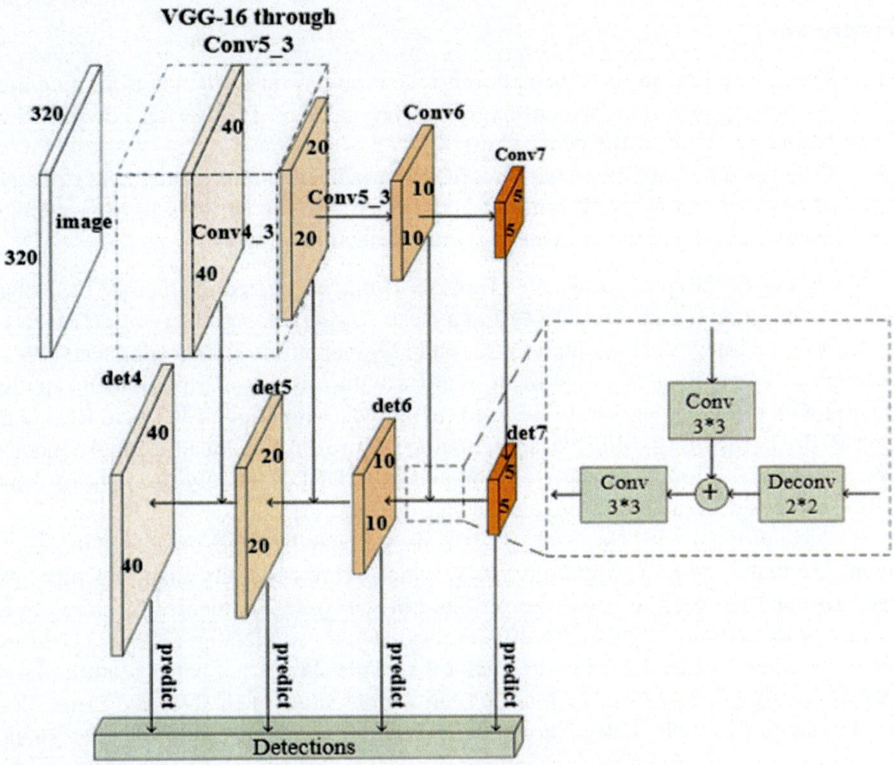

Fig. 21 Feature extraction structure of coarse-to-fine ship detection network

add 2 convolutional layers (conv6andconv7) at the end of the network. The two convolutional layers (conv6 andconv7) reduce the resolution of the feature map to half in sequence. With the deepening of the network, the features are continuously sampled by the max pooling layer, and the resolution of the output feature map gets smaller, but the semantic information is more abundant. This is similar to the bottom-up processing FPN networks (A deep convnet computes an inherent multi-scale and pyramidal shape feature hierarchy). We select four different resolution feature maps that output from conv4_3, conv5_3, conv6, and conv7. The strides of the selected feature maps are 8, 16, 32, and 64. The input size of this network is 320 × 320 pixels and the resolutions of the selected feature map are 40 × 40 (conv4_3), 20 × 20 (conv5_3), 10 × 10 (conv6), and 5 × 5 (conv7). Experimental results on optical remote sensing data set show that the proposed method outperforms other excellent detection algorithms and achieves good detection performance on the dataset including some small-sized ships.

Future work

(1) The orientation angle information is taken into account when determining the position of the ship, which can effectively reduce the overlap between the bounding boxes of the dense ships.
(2) Combined with the characteristics of remote sensing images, the select strategy of positive and negative samples is considered in the network to improve the classification and location ability of the detection network.

Chen et al. (2020) presented a novel coarse-to-fine ship detection method based on Discrete Wavelet Transform (DWT) and a Deep Residual Dense Network (DRDN). multi-spectral images are adopted for sea-land segmentation, and an enhanced DWT is employed to extract ship candidate regions with missing alarms. Panchromatic images with clear spatial details are used for ship classification. The Local Residual Dense Block (LRDB) to fully extract semantic feature through local residual connection and densely connected convolutional layers. DRDN mainly consists of four LRDBs and is designed to remove false alarms (Fig. 22).

ResNet, proposed by He et al. (2020), has obtained record-breaking improvements on many visual recognition tasks, which allows identity mapping through deep residual learning to solve the problem of vanishing gradients. To strengthen feature propagation, Huang et al. (2019) introduced by connecting all layers (within the same dense block) directly with each other. This shows that better feature reuse leads to better performance. Inspired by the architectures of ResNet and DenseNet a novel Deep Residual Dense Network (DRDN) for ship classification is designed.

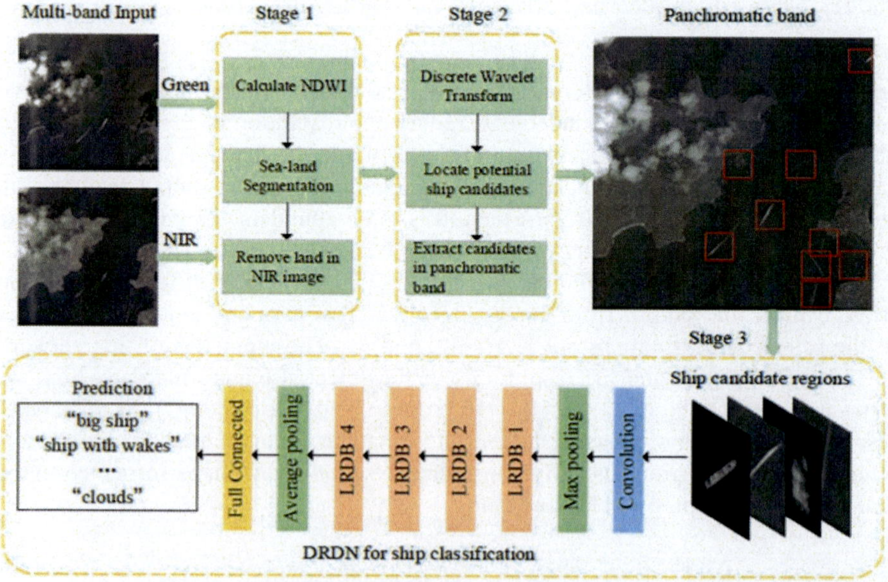

Fig. 22 Overall framework for DRDN Ship classification

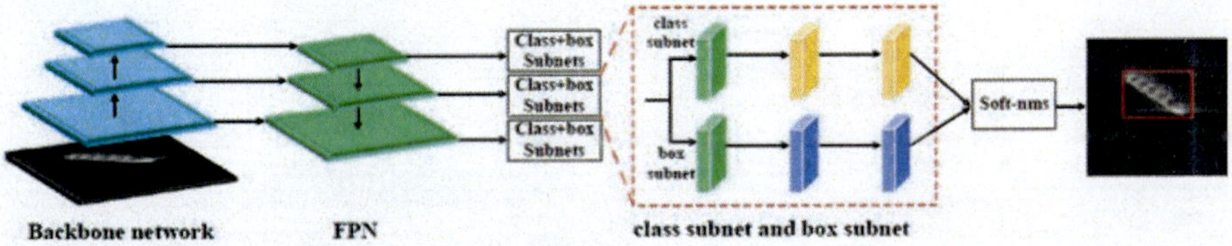

Fig. 23 Architecture of retina Net-Plus method

Extensive experiments demonstrate that the proposed method has high robustness in complex image backgrounds and achieves higher detection accuracy than other state-of-the-art methods.

Future work

Improving the detection performance of inshore ships, further reducing false alarms, and better exploiting multi-spectral features.

12 Retina Net

Su and Wei et al. (2019) introduced a RetinaNet-Plus method for ship detection in high-resolution SAR imagery based on RetinaNet network. In this technique, instead of setting the score for neighboring region proposals to zero as in Non-Maximum Suppression (NMS), Soft-NMS decreases the detection scores as an increasing function of overlap to avoid loss of precision (Fig. 23).

This framework is based on RetinaNet, which has three components: a backbone network for feature extraction and two sub-networks (one for classification and the other for box regression).

Replace Non-Maximum Suppression (NMS) by Soft-NMS, which performs as a post-processing stage to obtain the final set of detections. This method is more accurate than the existing algorithms and is effective for ship detection of high-resolution SAR imagery.

Future work

Technique to optimize anchor boxes for ship detection

13 Balance Scene Learning Mechanism

Zhang et al. [12] introduced a Balance Scene Learning Mechanism (BSLM) for offshore and inshore ship detection in SAR images. BSLM consists of three steps:

(1) Based on unsupervised representation learning, a Generative Adversarial Network (GAN) is used to extract the scene features of SAR images; (2) using these

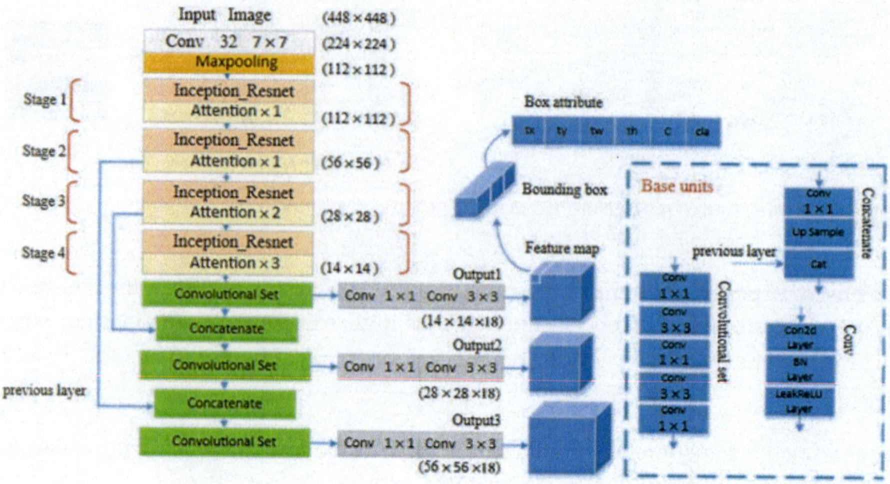

Fig. 24 Structure of target network integrating attention mechanism

features, a scene binary cluster (offshore/inshore) is conducted by K-means; (3) finally, the small cluster's samples (inshore) are augmented via replication, rotation transformation, or noise addition to balance another big cluster (offshore), to eliminate scene learning bias, and to obtain balanced learning representation ability that can enhance learning benefits and improve detection accuracy. This letter applies BSLM to four widely used and open-sourced deep learning detectors, i.e., Faster Region Convolutional Neural Network (Faster R-CNN), Cascade R-CNN, Single Shot multi-box Detector (SSD), and RetinaNet, to verify its effectiveness.

Experimental result on the open SAR Ship Detection Dataset (SSDD) reveals that BSTM can greatly improve detection accuracy, especially for more complex inshore scenes (Fig. 24).

14 Attention Mechanism

In the target network integrating algorithm, the dimension of the input image is adjusted by 7 * 7 convolution layer, and then downsampling is done by Max pooling layer. The feature extraction network consists of four stages. Each stage uses Inception-ResNet as the basic unit to construct the feature pyramid, which can enhance the ability to acquire upper information. In each stage, salient feature maps of different depths are obtained by concatenating several attention models in series, and the features of different depths are fused to highlight the advantages of location.

Chen et al. (2019) introduced an object detection network which combines attention mechanism to enhance the network's ability to accurately locate targets in complex background. Dealing with the diversity of ship target scales, this paper

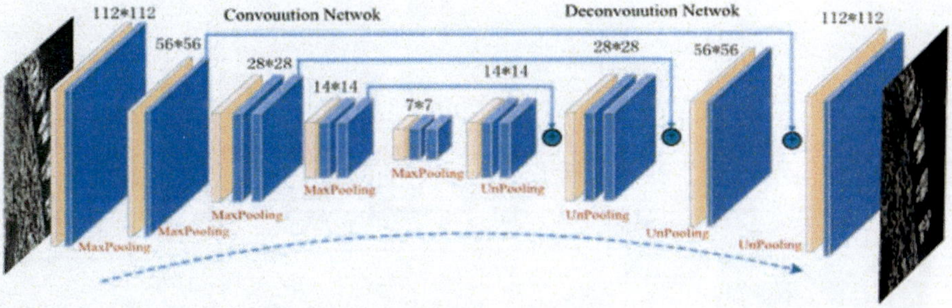

Fig. 25 Structure of convolution and deconvolution block

describes a loss function that incorporates Generalized Intersection over Union (GIoU) loss to reduce the sensitivity of the algorithm to scale (Fig. 25).

This method achieves good results for ship target detection in complex backgrounds based on the extended SAR Ship Detection Dataset (SSDD), while maintaining a fast detection speed.

15 HRSD Network

Wei and Su et al. (2020) introduced a novel ship detection method based on a high-resolution ship detection network (HR-SDNet) for high-resolution SAR imagery. The HR-SDNet adopts a novel high-resolution feature pyramid network (HRFPN) to take full advantage of the feature maps of high-resolution and low-resolution convolutions for SAR image ship detection. In this technique, the HRFPN connects high-to-low resolution sub-networks in parallel and can maintain high resolution. The Soft Non-Maximum Suppression (Soft-NMS) is used to improve the performance of the NMS, thereby improving the detection performance of the dense ships. Then, introduce the Microsoft Common Objects in Context (COCO) evaluation metrics, which provides not only the higher quality evaluation metrics Average Precision (AP) for more accurate bounding box regression, but also the evaluation metrics for small, medium, and large targets, so as to precisely evaluate the detection performance of above method (Fig. 26).

The High-Resolution Ship Detection Network (HR-SDNet) has four components: a High-Resolution Feature Pyramid Networks (HRFPN) as the backbone for feature extraction to build a multi-level representation; an Region Proposal Network (RPN) for generating candidate object bounding box proposals; three cascades Fast RCNN with thresholds $U = \{0.5,0.6,0.7\}$ for bounding box regression and classification; the Soft-NMS is executed as a post-processing step to obtain fine detection results. Experimental results on SSDD dataset and TerraSAR-X high-resolution images: (1) This approach based on HRFPN has superior detection performance for both inshore and offshore scenes of the high-resolution SAR imagery, which achieves

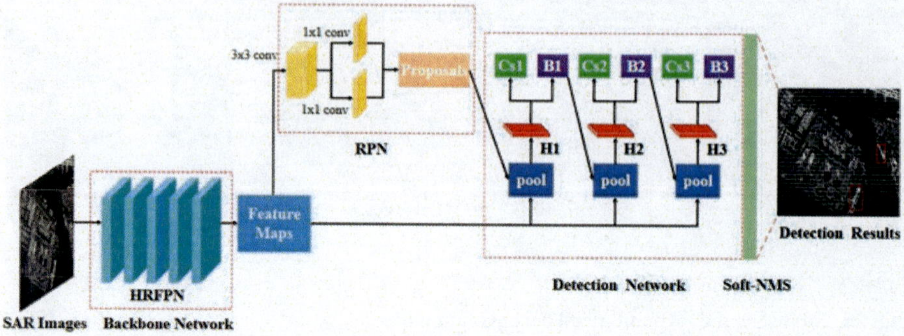

Fig. 26 Structure of HRSD network

nearly 4.3% performance gains compared to FPN in inshore scenes; thus, proving its effectiveness; (2) Compared with the existing algorithms, this approach is more accurate and robust for ship detection of high-resolution SAR imagery, especially inshore and offshore scenes; (3) With the Soft-NMS algorithm, our network performs better, which achieves nearly 1% performance gains in terms of AP; (4) The COCO evaluation metrics is effective for SAR image ship detection; (5) The displayed thresholds within a certain range have a significant impact on the robustness of ship detectors.

Future work

Focus on ship instance segmentation for high-resolution SAR imagery.

16 Recurrent Attention Convolutional Neural Network

Xu et al. (2020) introduced an improved recurrent attention convolutional neural network. This network has a multi-scale architecture and consists of three cascading sub-networks, each with a VGG19 network for image feature extraction and an attention developed network for locating feature area. A scale-dependent pooling algorithm is designed to select an appropriate convolution in the VGG19 network for classification, and a multi-feature mechanism is introduced in attention developed network to describe the feature regions. The VGG19 and attention network are cross-trained to accelerate convergence and to improve detection accuracy (Fig. 27).

This method is trained and validated on a self-built ship database and effectively improves the detection accuracy to 86.7% outperforming the baseline VGG19 and recurrent attention convolutional neural network methods.

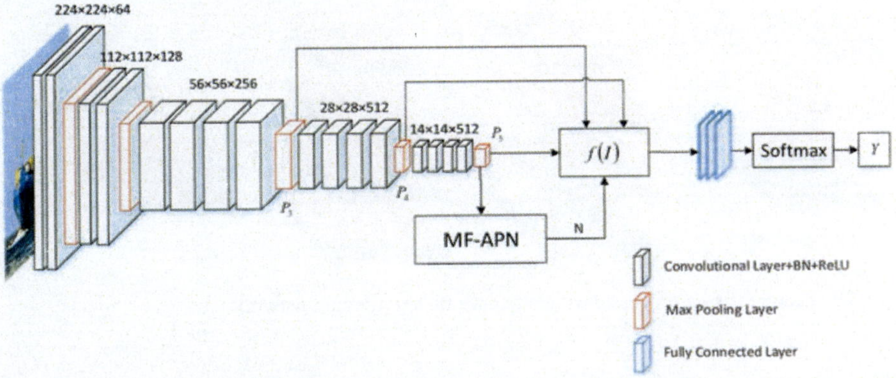

Fig. 27 Architecture of VGG-SDP network

17 Feature Fusion Pyramid Network and Deep Reinforcement Learning Method

Fu et al. (2018) introduced a ship rotation detection model based on a Feature Fusion Pyramid Network and deep reinforcement learning (FFPN-RL). In this paper, the detection network can efficiently generate the inclined rectangular box for ship. Develope the Feature Fusion Pyramid Network (FFPN) that strengthens the reuse of various scales features, and FFPN can extract the low-level location and high-level semantic information that has a crucial impact on multi-scale ship detection and precise location of dense parking ships. In order to get accurate ship angle information, apply deep reinforcement learning to the inclined ship detection task for the first time. In addition, suggests forward prior policy guidance and a long-term training method to train an angle prediction agent constructed through a dueling structure Q network, which is in position to iteratively and accurately obtain the ship angle. In addition, design soft rotation non-maximum suppression to scale back the missed ship detection while suppressing the redundant detection boxes. Detailed experiments on the remote sensing ship image dataset are administered (Figs. 28 and

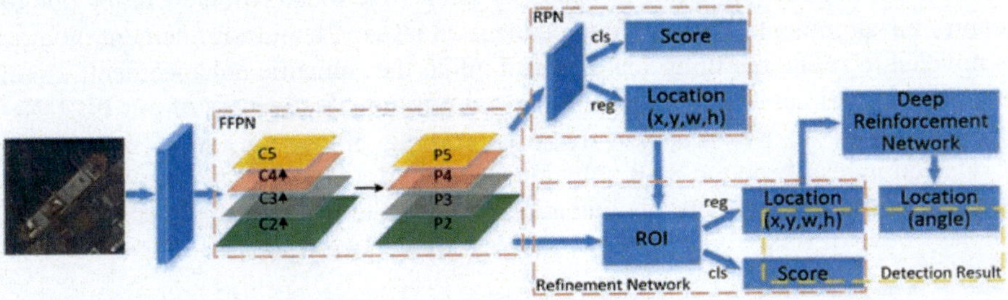

Fig. 28 Architecture of feature fusion pyramid network and deep reinforcement learning techniques

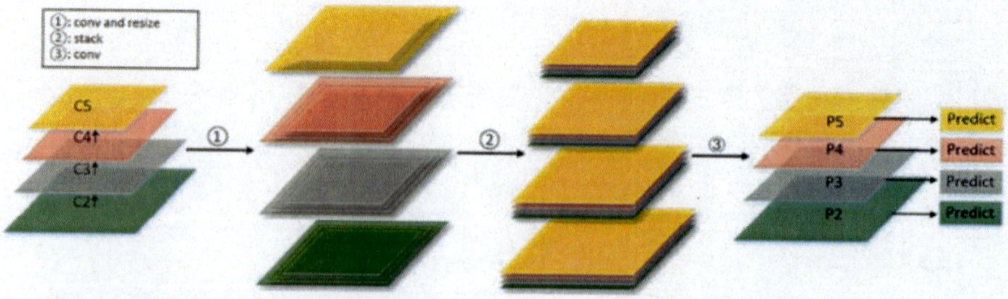

Fig. 29 Feature fusion process during a feature fusion pyramid network

29).

In FFPN take ResNet50 as the network backbone. The outputs of each stage last residual block in ResNet50 are used as original feature maps, which are $C2$, $C3$, $C4$, and $C5$. Through convolution, resizing, and stacking operations, the original feature maps are converted to fused feature maps. Defined fused feature maps as $\{P2, P3, P4, P5, P6\}$. Experiments validate that the above FFPN-RL ship detection model has efficient detection performance.

Future work

Also explore extending our ship detection models to other objects.

18 Feature Balancing and Reinforcement Network

Jiamei Fu (2) introduced a novel detection method named feature balancing and refinement network (FBR-Net) is proposed. This method eliminates the effect of anchors by adopting a general anchor-free strategy that directly learns the encoded bounding boxes. Leverage the developed attention-guided balanced pyramid to balance semantically the multiple features across different levels. It can help the detector learn more information about the small-scale ships in complex scenes. Considering the SAR imaging mechanism, the interference near the ship boundary with the similar scattering power probably affects the localization accuracy due to feature misalignment. To tackle the localization issue, a feature-refinement module is noveled to refine the thing features and guide the semantic enhancement. Then, extensive experiments are conducted to point out the effectiveness of our FBR-Net compared with the overall anchor-free baseline (Fig. 30).

Detection results on the SAR ship detection dataset (SSDD) and AIR-SARShip-1.0 dataset illustrate that our method achieves the state-of-the-art performance.

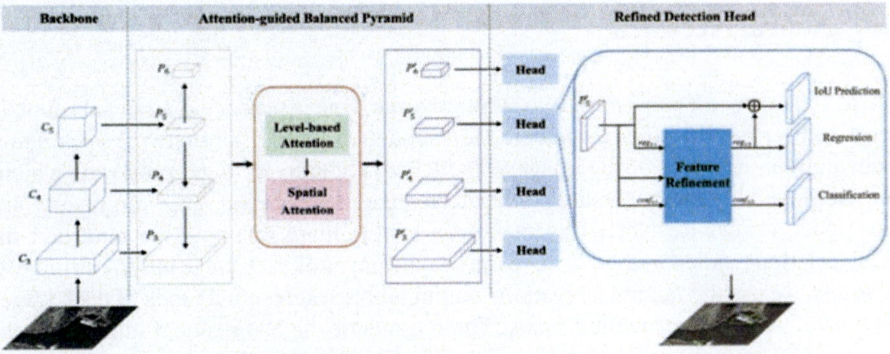

Fig. 30 Architecture of FBR-NET

19 Squeeze Excitation Skip-Connection Path Network

Huang et al. [21] introduced novel neural network architecture named squeeze excitation skip-connection path networks (SESPNets). A bottom-up path is added to feature pyramid network to improve feature extraction capability, and path-level skip-connection structure is firstly proposed to enhance in-formation flow and reduce parameter redundancy. Also, squeeze excitation module is adopted, which may adaptively recalibrate channel-wise feature responses by adding an additional branch after each shortcut path connection block. The multi-scale fused Region of Interest (ROI) align is then proposed to get more accurate and multi-scale proposals. Then, soft-non-maximum suppression is used to beat the matter of Non-Maximum Suppression (NMS) in ship detection (Fig. 31).

Experiments show that SESPNets model has achieved the state-of-the-art performance with the recall, precision, and F1 of 0.841, 0.831, and 0.836.

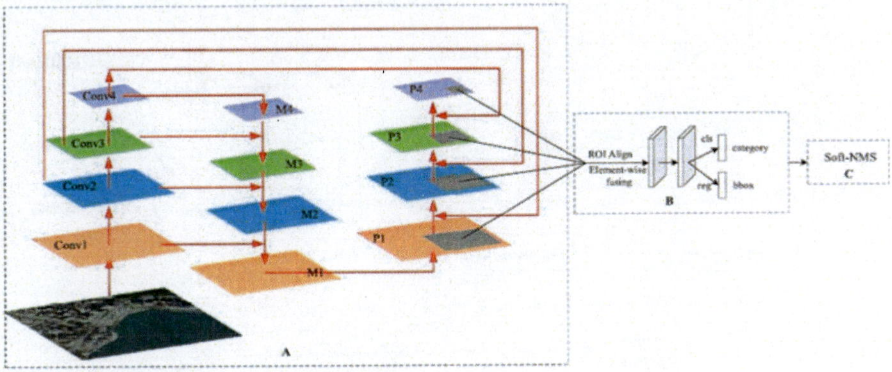

Fig. 31 Architecture of squeeze excitation skip-connection path network

20 Neural Network

DCMSNN network consists of two sub-networks. One is the region proposal subnetwork (RPN) and the other is the detection subnetwork. Two sub-networks share the convolutional features of the images. The RPN is employed to get proposals which are employed by detection subnetwork to realize more refined detection results. In this paper, we use ResNet101 as backbone, feature maps that have the same size in ResNet101 are called a stage. Due to the deepest layer of each stage has the strongest features, we use the feature activations output by the last residual block of each stage as our reference set of feature maps. The outputs of the last residual block in each stage conv2, conv3, conv4, conv5 as $C2; C3; C4; C5$ (Fig. 32).

Zhang et al. [12] presented a densely connected multi-scale neural network based on faster-RCNN framework to solve multi-scale and multiscene SAR ship detection. Instead of employing a single feature map to get proposals, Yue Zhang densely connects one feature map to each other feature maps from top to down and generates proposals from each fused feature map. In addition, he developed a training strategy to scale back the load of easy examples within the loss function, so that the training process more specialize in the hard examples to reduce false alarm.

The public SSDD dataset that has a similar procedure as PASCAL VOC is provided by Li et al. [17]. It includes SAR images of resolution from 1 to 15 m which are collected from RadarSat-2, TerraSAR-X, and Sentinel-1. Li et al. [17] utilize

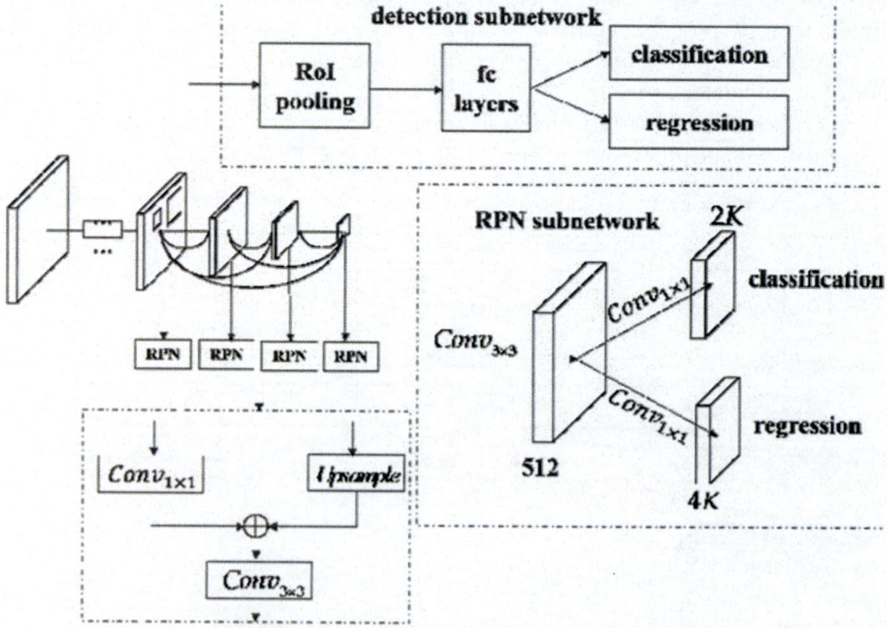

Fig. 32 Architecture of DCMSNN

feature fusion, transfer learning, hard negative mining, and other implementation details to improve the AP from 70.1 to 78.8% compared to Faster in SSDD. Experiments on expanded public SAR ship detection dataset, verify the above method can achieve an excellent performance on multi-scale SAR ship detection in multiscene.

Liu et al. [16] presented a multi-scale full convolutional network (MS-FCN) based sea-land segmentation method and applies a rotatable bounding box-based object detection method (DRBox) to unravel the inshore ship detection problem. The ocean region and land region are separated by MS-FCN then DRBox is applied on sea region. The above method combines global information and native information of SAR image to realize high accuracy. The networks are trained with ChineseGaofen-3 satellite images. Experiments show that the proposed method performs well on this task by applying multi-scale information in segmentation and angle information in detection.

21 Hybrid Network of CNN and NN

Lei (2020) presented a target detection method with multi-features in SAR imagery. It is composed of two parallel sub-channels. DL features and hand-crafted features are extracted in these channels. For capture the DL features of original SAR images, CNN is used. Deep Neural network (NN) is used to further analyze hand-crafted features. These two sub-channels are combined in the main channel. Further processing of several layers of network, fused deep features are extracted. Softmax classifier is used to discriminate ship target, validate the detection ability of the above method, and collect the sentinel-1 satellite SAR data which form test and training test.

Detection performance is improved by fusing multi-features. Hybrid network achieves higher performance result than the other two methods 2.59 and 6.26%, respectively.

22 Discussion and Conclusion

From the above-detailed survey it becomes clear that all the works that were carried out in ship detection in SAR images techniques are aimed at achieving, high accuracy, and high detection speed. For ship detection in SAR images both high-resolution images and low-resolution images were used by different authors. The SSDD is so far the first SAR images on public for researchers to evaluate the performance of their detectors. Experimental results using SSDD, Gaofen-SSDD, and Sentinel-SSD in HyperLi-Net Architecture shows that detection accuracy and speed are both superior compared to the other ship detection methods. HyperLi-Net's detection speed on SSDD is 222 FPS on SAR image with size of 160 $\times$ 160 pixels and need 523.16 ms for detecting all 116SAR images in the test set. The accuracy on Gaofen-SSDD is inferior to that on SSDD, because (1) images in Gaofen-SSDD have more complex scenarios and these images possessing many inshore ships account for a large proportion of the total test set, inevitably bringing great obstacles to detect; (2) images from low-resolution Gaofen-3 are frequently accompanied with severe speckle noise, declining the detection accuracy. HyperLi-Net's detection speed on Gaofen-SSDD is 247FPS. It takes only 4.05 ms to accomplish ship detection tasks in a 160 × 160 SAR image. The detection accuracy on Sentinel-SSDD is slightly superior to that on Gaofen-SSDD, because SAR images from high-resolution Sentinel-1 satellite have better quality than that from low-resolution Gaofen-3 satellite, which contains severe speckle noise. Small ships with dense distribution account for smaller proportion among the total test set of Sentinel-SSDD than Gaofen-SSDD's. The detection speed on Sentinel-SSDD of HyperLi-Net is 248 FPS that is similar to Gaofen-SSDD. It takes only 4.03 ms to accomplish ship detection tasks in a SAR image with 160 × 160 size. This paper also encourages researchers to develop novel algorithms for Ship Detection in SAR images using deep learning networks, alternative to the conventional detection networks. Also there is still more scope for research in dense and small ship detection.

References

1. Girshick R, Donahue J, Darrell T, Malik J (2016) Region-based convolutional networks for accurate object detection and segmentation. IEEE Trans Pattern Anal Mach Intell 38(1):142–158
2. Redmon J, Farhadi A (2018) YOLOv3: an incremental improvement. arXiv:1804.02767 [Online] Available http://arxiv.org/abs/1804.02767
3. Liu W, Anguelov D, Erhan D, Szegedy C, Reed S, Fu C-Y, Berg AC (2016) SSD: single shot multi box detector. In: Proceedings European conference computer vision Amsterdam, The Netherlands: Springer, pp 21–37
4. Liu W, Anguelov D, Erhan D, Szegedy C, Reed S, Fu C-Y, Berg AC (2016) Ssd: Single shot multibox detector. In: Proceedings of the European conference on computer vision, Amsterdam, The Netherlands, 8–16 October 2016, pp 21–23
5. Zhu H, Wang W, Leung R (2020) SAR target classification based on radar image luminance analysis by deep learning. IEEE Sens. Council 4(3)
6. Long Z, Suyuan W, Zhongm C, Jiaqi F, Xiaoting Y, We D (2020) Lira-YOLO: a lightweight model for ship detection in radar images. J Syst Eng Electron 99:1–7
7. Shen H, Lin L, Li J, Qiang Yuan Q, Zhao L (2020) A residual convolutional neural network for polarimetric SAR image super-resolution. 161:90–108
8. Wang Q, Shi W, Atkinson PM, Zhao Y (2015) Downscaling MODIS images with area to point regression kriging. Remote Sens Environ 166:191–204
9. Wang Q, Atkinson PM (2018) Spatio-temporal fusion for daily Sentinel2 images. Remote Sens Environ 204:31–42
10. Zhou F, Fan W, Sheng Q, Tao M (2018) Ship detection based on deep convolutional neural networks for PolSAR images. In: IEEE International geoscience and remote sensing symposium, pp 681–684
11. Zhao J, Guo W, Zhang Z, Yu W (2019) A coupled convolutional neural network for small anddensely clustered ship detection in SAR images. Springer 62
12. Zhang T, Zhang X, Shi J, Wei S (2020) HyperLi-Net: a hyper-light deep learning network for high-accurate and high-speed ship detection from synthetic aperture radar imagery. Elsevier 167:123–153

13. Hu X, Wang X, Meng F, Hu X, Yan Y, Lib Y, Huang J, Jiang X (2020) Gabor-CNN for object detection based on small samples, KeAi 16:1116–1129
14. Huang Z, Sui B, Wen J, Jiang G (2020) An intelligent ship image/video detection and classification method with improved regressive deep convolutional neural network. Hindawi Complexity
15. Redmon J, Farhadi A (2017) YOLO9000: better faster stronger. In: Proceedings of the IEEE conference on computer vision and pattern recognition, IEEE, Hon-olulu, HI, USA, July 2017, pp 6517–6525
16. Liu Y, Zhang M-H, Xu P, Guo Z-W (2020) SAR ship detection using sea-land segmentation-based convolutional neural network
17. Zhang X, Wang H, Xu C, Lv Y, Fu C, Xiao HC, He Y (2019) A lightweight feature optimizing network for ship detection in SAR image. IEEE Access 7
18. Szegedy C, Liu W, Jia Y et al (2014) Going deeper with convolutions. 1–9. https://arxiv.org/abs/1409.484
19. Yu F, Koltun V (2015) Multi-scale context aggregation by dilated convolutions
20. Woo S, Park J, Lee J-Y, Kweon (2018) In So. CBAM: Convolutional block attention module. arXiv preprint. arXiv:1807.06521. https://arxiv.org/abs/1807.06521
21. Huang G, Zhuang L, Laurens van der M, Weinberger KQ (2018) Densely connected convolutional networks. arXiv preprint. arXiv:1608.06993. https://arxiv.org/abs/1608.06993
22. Lin Z, Ji K, Kang M, Leng X, Zou H (2017) Deep convolutional highway unit network for SAR target classification with limited labeled training data. IEEE Geosci. Remote Sens. Lett. 14(7):1091–1095
23. Redmon J, Divvala S, Girshick R, Farhadi A (2015) You only look once: unified, real-time object detection. arXiv preprint. arXiv:1506.02640. https://arxiv.org/abs/1506.02640
24. Redmon J, Divvala S, Girshick R, Farhadi A (2016) You only look once: unified, real-time object detection. In: Proceedings IEEE conference computer vision pattern recognition (CVPR), June 2016, pp 779–788
25. Simonyan K, Zisserman A (2014) Very deep convolutional networks for large-scale image recognition. arXiv preprint. arXiv:1409.1556. https://arxiv.org/abs/1409.1556
26. Simonyan K, Zisserman A (2014) Very deep convolutional networks for large-scale image recognition. arXiv 2014
27. Simonyan K, Zisserman A (2014) Very deep convolutional networks for large-scale image recognition. Comput Sci
28. Li JW, Qu CW, Peng SJ et al (2018) Ship detection in SAR im-ages based on convolutional neural network. Syst Eng Electron 40(9):1953–1959
29. Ding Z, Yu Y, Wang B, Zhang L (2012) An approach for visual attention based on biquaternion and its application for ship detection in multispectral imagery. Neurocomputing 76:9–17
30. Zhao Y, Zhao L, Xiong B, Kuang G (2020) Attention receptive pyramid network for ship detection in SAR images. J Appl Earth Observ Remote Sens 6
31. Li Y, Ding Z, Zhang C, Wang Y, Chen J (2019) SAR Ship detection based on resnet and transfer learning. IGARSS
32. Zhang S, Wu R, Xu K, Wang* J, Sun W (2019) R-CNN-Based ship detection from high resolution remote sensing imagery. Remote Sens vpl.11
33. Li J, Qu C, Shao J (2017) Ship detection in SAR images based on an improved faster R-CNN. In: Proceedings BIGSARDATA, Beijing, China, November 2017, pp 1–6
34. Li J, Qu C, Shao J (2017) Ship detection in SAR images based on an improved faster R-CNN. In: Proceedings SAR Big Data Era Models Methods Applications (BIGSARDATA), Nov. 2017, pp 1–6
35. Qin X, Zhou SL, Zou HX et al (2013) A CFAR detection algorithm for generalized gamma distributed background in high-resolution SAR images. IEEE Geosci Remote Sens Lett 10:806–810
36. Zheng T, Wang J, Li P (2020) Deep learning based target detection method with multi = features in SAR imagery. IEEE Acceess
37. Copernicus open access hub. https://scihub.copernicus.eu/

38. Khan HM, Yunze C (2018) Ship detection in SAR Image using YOLOv2. In: Proceedings of the 37th Chinese control conference
39. Ma F, Gao F, Wang J, Hussain A, Zhou H (2020) A novel biologically-inspired target detection method based on saliency analysis for synthetic apertureradar(SAR)imagery. Elsevier 402:66–79
40. Han L, Ran D, Ye W, Yang W, Wu X (2020) Multi-size convolution and learning deep network for sar ship detection from scratch. IEEE Access 8

Chapter 47
General Solution and Fundamental Solution in Anisotropic Micropolar Thermoelastic Media with Mass Diffusion

Vijay Chawla and Sanjeev Ahuja

1 Introduction

Fundamental solution plays an important role in the study of applied mathematics. It plays a crucial role in the solution of numerous problems in the mechanics and physics of solids. Fundamental solution frequently used to derive many analytical solutions of practical problems in case when boundary conditions are imposed. Fundamental solution play a key role in an integral equation representation of a boundary value problem more easily solved by analytical methods in comparison to a differential equation with specified initial and boundary conditions. This type of solution technique (numerical methods technique) makes the subject more attractive mainly for those researchers whose area of interest is numerical methods. Fundamental Solution also provides a wonderful platform to overcome the main drawbacks in the boundary element method which also uses the fundamental solution to satisfy the governing equation. Consequently in a single line we can say that with the latest technological demand, no boundary element method can be made more advanced without further developments in the area of fundamental solutions or in other words we can say that fundamental solution is a basic building block of many further works. Fundamental solution plays an important role in the solution of numerous problems in the mechanics and physics of solids. They are a basic building block of many further works. For example, fundamental solutions can be used to construct many analytical solutions of practical problems when boundary conditions are imposed. They are essential in the boundary element method as well as the study of cracks, defects and inclusions.

V. Chawla (✉)
Department of Mathematics, Maharaja Agrasen Mahavidyalya, Jagadhri, Yamuna Nagar, India

S. Ahuja
Department of Mathematics, University Institute of Engineering and Technology, Kurukshetra University, Kurukshetra, India

N. Marriwala et al. (eds.), *Soft Computing for Intelligent Systems*, Algorithms for Intelligent Systems, https://doi.org/10.1007/978-981-16-1048-6_47

Dunn and Wienecke [1] established the half space Green's functions for transversely isotropic piezoelectric solid and also derived closed-form expressions. Pan and Tanon [2] constructed the Green's functions for three-dimensional problem in anisotropic piezoelectric solids and also discussed its applications. Chen [3] derived general solution for transverse isotropic thermo-piezo-elastic media in dynamic as well as in static case and derived an exact solution for a penny-shaped cracked subjected to uniform temperature load. Chen et al. [4] presented three-dimensional exact solution for a penny-shaped crack in an infinite piezoelectric medium subjected to an arbitrarily point temperature load by using potential theory method for both impermeable and permeable cracks.

Sharma [5] derived the fundamental solution in transversely isotropic thermoelastic material in an integral form. Ciarletta et al. [6] derived the fundamental solution in micropolar isotropic thermoelastic material with voids by the potential method. Hou et al. [7] constructed Green's function for three-dimensional problem in transversely isotropic biomaterials by using operator theory. Hou et al. [8] studied the Green's functions for two-dimensional problem in a semi-infinite orthotropic thermoelastic media by introducing new harmonic functions. Xiong et al. [9] discussed the Green's functions for two-dimensional problems in orthotropic piezo thermoelastic material by trial and error method. Hou et al. [10] constructed the general solution and fundamental solution of the two-dimensional problem in orthotropic thermoelastic material. Seremet [11] constructed an exact Green's function and integral formula for a boundary value problem (BVP) for a thermoelastic wedge in terms of elementary functions. Seremet [12] derived New Green's function and a New Green-type integral formula for a boundary value problem (BVP) in thermoelastic quadrant. Kumar and Kansal [13] studied the plane wave propagation and fundamental solution in the generalized theory of thermoelastic diffusion.

Kumar and Chawla [14, 15] derived the fundamental solution and Green's function for two-dimensional problem in orthotropic thermoelastic diffusion media by using the operator theory and obtained result also presented graphically. Also Kumar and chawla [16, 17] derived the fundamental solution and Green's function in orthotropic piezothermoelastic diffusion media by trial and error method. Kumar and Chawla [18] discussed the problem of reflection and transmission in thermoelastic media with three-phase-lag model for isotropic case. Kumar and Vandna [19] derived Green function for three-dimensional problem in transversely isotropic thermoelastic biomaterial for concentrated heat source. Kumar and Chawla [20] presented the fundamental solution for two-dimensional problem in orthotropic thermoelastic media with voids by introducing nine newly harmonic functions. Şeremet [21] derived new constructive formulas in thermoelastic Green's functions for boundary value problem of thermoelasticity in steady-state case and also the constructive formulas are expressed in terms of Green's functions for Poisson's equation. Pan et al. [22] derived the general solution and fundamental solution for fluid-saturated, orthotropic, poroelastic materials in case of steady-state problem. Chawla et al. [23] constructed general solution and fundamental solution for two-dimensional problem in micropolar thermoelastic material. Dang et al. [24] investigated a planar crack of arbitrary shape embedded in three-dimensional isotropic hygrothermoelastic media by using Hankel transform technique. Zhao et al. [25] derived the three-dimensional general solution and fundamental solution in hygrothermoelastic media by using operator theory. Tomar et al. [26] studied plane waves in thermo-viscoelastic material with voids under different theories of thermoelasticity. Biswas [27] investigated fundamental solution in steady oscillations equations in nonlocal thermoelastic medium with voids.

However, the important general solution and fundamental solution for two-dimensional problem for a steady point heat source in anisotropic thermoelastic material with mass diffusion and voids have not been discussed so far in the literature.

The fundamental solution for two-dimensional in micropolar transversely isotropic thermoelastic medium is investigated in this paper. Based on the two-dimensional general solution of micropolar transversely isotropic thermoelastic media, the fundamental solutions for a steady point heat source acting on the surface of a micropolar thermoelastic material is obtained by nine newly introduced harmonic functions. From the present investigation, some special cases of interest are also deduced.

2 Basic Equations and General Solution

Following Aoudi [28], the two-dimensional basic equations for homogenous transversely isotropic micropolar thermoelastic diffusion material (for static case) in the absence of body forces, body couples and heat sources and mass diffusion sources by assuming the displacement vector $\vec{u} = (u, 0, w)$, temperature change $T(x, z, t)$, mass concentration $C(x, z, t)$, and microrotation vector $\vec{\Phi} = (0, \phi_2, 0)$ are given by

$$\left[A_{11}\frac{\partial^2}{\partial x^2} + A_{55}\frac{\partial^2}{\partial z^2}\right]u + \left[(A_{13} + A_{56})\frac{\partial^2}{\partial x \partial z}\right]w + \left[K_1\frac{\partial}{\partial z}\right]\phi_2 - \left[\beta_1\frac{\partial}{\partial x}\right]T - \left[\gamma_1\frac{\partial}{\partial x}\right]C = 0, \tag{1}$$

$$\left[(A_{13} + A_{56})\frac{\partial^2}{\partial x \partial z}\right]u + \left[A_{66}\frac{\partial^2}{\partial x^2} + A_{33}\frac{\partial^2}{\partial z^2}\right]w + \left[K_2\frac{\partial}{\partial x}\right]\phi_2 - \left[\beta_3\frac{\partial}{\partial z}\right]T - \left[\gamma_3\frac{\partial}{\partial z}\right]C = 0, \tag{2}$$

$$\left[K_1\frac{\partial}{\partial z}\right]u + \left[K_2\frac{\partial}{\partial x}\right]w + \left[B_{77}\frac{\partial^2}{\partial x^2} + B_{66}\frac{\partial^2}{\partial z^2} - X\right]\phi_2 = 0, \tag{3}$$

$$\left(K_1^*\frac{\partial^2}{\partial x^2} + K_3^*\frac{\partial^2}{\partial z^2}\right)T = 0, \tag{4}$$

$$\frac{\partial}{\partial x}\left[\gamma_1\left(\alpha_1^*\frac{\partial^2}{\partial x^2} + \alpha_3^*\frac{\partial^2}{\partial z^2}\right)\right]u + \frac{\partial}{\partial z}\left[\gamma_3\left(\alpha_1^*\frac{\partial^2}{\partial x^2} + \alpha_3^*\frac{\partial^2}{\partial z^2}\right)\right]w$$

$$+\left[a\left(\alpha_1^*\frac{\partial^2}{\partial x^2}+\alpha_3^*\frac{\partial^2}{\partial z^2}\right)\right]T-\left[\mathrm{d}\left(\alpha_1^*\frac{\partial^2}{\partial x^2}+\alpha_3^*\frac{\partial^2}{\partial z^2}\right)\right]C=0 \tag{5}$$

$$t_{11}=A_{11}\frac{\partial u}{\partial x}+A_{13}\frac{\partial w}{\partial z}-\beta_1 T-\gamma_1 C \tag{6a}$$

$$t_{33}=A_{11}\frac{\partial u}{\partial x}+A_{33}\frac{\partial w}{\partial z}-\beta_3 T-\gamma_3 C \tag{6b}$$

$$t_{31}=A_{65}\frac{\partial w}{\partial x}+A_{55}\frac{\partial u}{\partial z}+K_1\phi_2 \tag{6c}$$

where

$$\beta_1=A_{11}\alpha_1+A_{13}\alpha_3,\quad \beta_3=A_{31}\alpha_1+A_{33}\alpha_3,$$
$$K_1=A_{56}-A_{55},\quad K_2=A_{66}-A_{56},\quad X=K_2-K_1$$

α_1,α_3 are coefficients of linear thermal expansion.

Equations (1–5) can be written as

$$D\{u,w,\phi_2,T,C\}^t=0, \tag{7}$$

where D is the differential operator matrix given by

$$\begin{bmatrix}
A_{11}\frac{\partial^2}{\partial x^2}+A_{55}\frac{\partial^2}{\partial z^2} & (A_{13}+A_{56})\frac{\partial^2}{\partial x\partial z} & K_1\frac{\partial}{\partial z} & -\gamma_1\frac{\partial}{\partial x} & -\beta_1\frac{\partial}{\partial x}\\
(A_{13}+A_{56})\frac{\partial^2}{\partial x\partial z} & A_{66}\frac{\partial^2}{\partial x^2}+A_{33}\frac{\partial^2}{\partial z^2} & K_2\frac{\partial}{\partial x} & -\gamma_3\frac{\partial}{\partial z} & -\beta_3\frac{\partial}{\partial z}\\
K_1\frac{\partial}{\partial z} & K_2\frac{\partial}{\partial x} & \left(B_{77}\frac{\partial^2}{\partial x^2}+B_{66}\frac{\partial^2}{\partial z^2}-X\right) & 0 & 0\\
\frac{\partial}{\partial x}\left[\gamma_1\left(\alpha_1^*\frac{\partial^2}{\partial x^2}+\alpha_3^*\frac{\partial^2}{\partial z^2}\right)\right] & \frac{\partial}{\partial z}\left[\gamma_3\left(\alpha_1^*\frac{\partial^2}{\partial x^2}+\alpha_3^*\frac{\partial^2}{\partial z^2}\right)\right] & 0 & \left[a\left(\alpha_1^*\frac{\partial^2}{\partial x^2}+\alpha_3^*\frac{\partial^2}{\partial z^2}\right)\right]T & -\left[d\left(\alpha_1^*\frac{\partial^2}{\partial x^2}+\alpha_3^*\frac{\partial^2}{\partial z^2}\right)\right]\\
0 & 0 & 0 & 0 & \left(K_1\frac{\partial^2}{\partial x^2}+K_3\frac{\partial^2}{\partial z^2}\right)
\end{bmatrix} \tag{8}$$

Equation (7) is a homogeneous set of differential equations in u, w, ϕ_2, T, C. The general solution by the operator theory as follows

$$u=A_{i1}F+\bar{A}_{i1}G,\quad w=A_{i2}F+\bar{A}_{i2}G,$$
$$\phi_2=\bar{A}_{i3}G,\quad C=\bar{A}_{i4}F+A_{i4}G,\quad T=A_{i3}F+\bar{A}_{i3}G, \tag{9}$$

where A_{ij} are algebraic cofactors of the matrix D, of which the determinant is

$$|D| = \left(a^*\frac{\partial^8}{\partial z^8} + b^*\frac{\partial^8}{\partial x^2\partial z^6} + c^*\frac{\partial^8}{\partial x^4\partial z^4} + \mathrm{d}^*\frac{\partial^8}{\partial x^6\partial z^2} + \mathrm{e}^*\frac{\partial^8}{\partial x^8}\right)$$
$$\times\left(K_1^*\frac{\partial^2}{\partial x^2} + K_3^*\frac{\partial^2}{\partial z^2}\right) + \left(\bar{a}\frac{\partial^6}{\partial z^6} + \bar{b}\frac{\partial^6}{\partial x^2\partial z^4} + \bar{c}\frac{\partial^6}{\partial x^4\partial z^2} + \bar{d}\frac{\partial^6}{\partial z^6}\right)$$
$$\times\left(K_1^*\frac{\partial^2}{\partial x^2} + K_3^*\frac{\partial^2}{\partial z^2}\right)(i = 1, 2, 3, 4, 5) \tag{10}$$

where

$$a^* = \alpha_3^* B_{66}A_{55}\delta_1,\ b^* = A_{55}\delta_2(b_3\gamma_3 - bA_{33}) + \alpha_3^* B_{66}(A_{11}\delta_1 - bA_{66}A_{55}),$$
$$c^* = A_{11}\delta_2(b_3\gamma_3 - bA_{33}) - bA_{66}(\alpha_3^* B_{66}A_{11} + A_{55}\delta_2) + \alpha_1^* B_{77}A_{55}\delta_1,$$
$$d^* = \alpha_1^* B_{77}(A_{11}\delta_1 - bA_{66}A_{55}) - bA_{66}A_{11}\,\delta_2,\ \mathrm{e}^* = -b\alpha_1^* B_{77}A_{66}A_{11}.$$
$$\bar{a} = X\alpha_3^* A_{55}\delta_1, \bar{b} = \alpha_3^*(bA_{55}\delta_3 - XA_{11}\delta_1),$$
$$\bar{c} = \alpha_1^*(bA_{55}\delta_3 - XA_{11}\delta_1) + b\alpha_3^* A_{11}\delta_3$$

and

$$\delta_1 = b_3\gamma_3 - bA_{33},\ \delta_2 = \alpha_1^* B_{66} + \alpha_3^* B_{77},$$
$$\delta_2 = X\,A_{66} + K_2^2\bar{a} = -\delta_4(\delta_1\delta_8 + \varepsilon_1),$$
$$\bar{b} = \delta_8(\delta_2^2 - \delta_4) + \delta_5(\delta_2\varepsilon_1 - \delta_1\varepsilon_2)$$
$$- \delta_3(\delta_1\delta_8 + \varepsilon_1),\ \bar{c} = -\delta_3\delta_8 - \varepsilon_2\delta_5.$$

The functions F and G in Eq. (15) satisfies the following homogeneous equation

$$|D|F = 0 \text{ and } |D|G = 0 \tag{11}$$

It can be seen that if $i = 1, 2, 3, 4$ are taken in Eq. (9), four general solutions are obtained in which $T = 0$. These solutions are identical to those without thermal fact and are not discussed here. Therefore if $i = 5$ should be taken in Eq. (9), the following solution is obtained

$$u = \left(p_1\frac{\partial^6}{\partial x^6} + q_1\frac{\partial^6}{\partial z^2\partial x^4} + r_1\frac{\partial^4}{\partial z^4\partial x^2} + v_1\frac{\partial^6}{\partial z^6}\right)\frac{\partial F}{\partial x}$$
$$+\left(\bar{p}_1\frac{\partial^4}{\partial x^4} + \bar{q}_1\frac{\partial^4}{\partial z^2\partial x^2} + \bar{r}_1\frac{\partial^4}{\partial z^4}\right)\frac{\partial G}{\partial x}, \tag{12a}$$

$$w = \left(p_2\frac{\partial^6}{\partial x^6} + q_2\frac{\partial^6}{\partial z^2\partial x^4} + r_2\frac{\partial^4}{\partial z^4\partial x^2} + v_2\frac{\partial^6}{\partial z^6}\right)\frac{\partial F}{\partial z}$$

$$+\left(\bar{p}_2\frac{\partial^4}{\partial x^4}+\bar{q}_2\frac{\partial^4}{\partial z^2\partial x^2}+\bar{r}_2\frac{\partial^4}{\partial z^4}\right)\frac{\partial G}{\partial z}, \quad (12b)$$

$$\phi_2=\left(\bar{p}_3\frac{\partial^4}{\partial x^4}+\bar{q}_3\frac{\partial^4}{\partial x^2\partial z^2}+\bar{r}_3\frac{\partial^4}{\partial z^4}\right)\frac{\partial^2}{\partial x\partial z}G, \quad (12c)$$

$$C=\left(p_4\frac{\partial^8}{\partial x^8}+q_4\frac{\partial^8}{\partial x^6\partial z^2}+r^4\frac{\partial^8}{\partial x^4\partial z^4}+v_4\frac{\partial^8}{\partial x^2\partial z^6}+w_4\frac{\partial^8}{\partial z^8}\right)F$$
$$+\left(\bar{p}_4\frac{\partial^6}{\partial x^6}+\bar{q}_4\frac{\partial^6}{\partial x^4\partial z^2}+\bar{r}_4\frac{\partial^6}{\partial x^2\partial z^4}+\bar{v}_4\frac{\partial^6}{\partial x^6}\right)G, \quad (12d)$$

$$T=\left(a^*\frac{\partial^8}{\partial z^8}+b^*\frac{\partial^8}{\partial x^2\partial z^6}+c^*\frac{\partial^8}{\partial x^4\partial z^4}+d^*\frac{\partial^8}{\partial x^6\partial z^2}+e^*\frac{\partial^8}{\partial x^8}\right)F$$
$$+\left(\bar{a}\frac{\partial^6}{\partial z^6}+\bar{b}\frac{\partial^6}{\partial z^4\partial x^2}+\bar{c}\frac{\partial^6}{\partial z^2\partial x^4}+\bar{d}\frac{\partial^6}{\partial x^6}\right)G, \quad (12e)$$

where

Equation (11) can be rewritten as

$$\prod_{j=1}^{5}\left(\frac{\partial^2}{\partial x^2}+\frac{\partial^2}{\partial z_j^2}\right)F=0, \quad (13a)$$

$$\prod_{j=1}^{4}\left(\frac{\partial^2}{\partial x^2}+\frac{\partial^2}{\partial z_j^2}\right)G=0, \quad (13b)$$

where

$z_j=s_jz,\ s_5=\sqrt{\frac{K_1^*}{K_3^*}}$ and $s_j(j=1,2,3,4)$ are four roots (with positive real part) of the following algebraic equation

$$a^*s^8-b^*s^6+c^*s^4-d^*s^2+e^*=0, \quad (14)$$

and

$\bar{z}_j=s_jz,\ \bar{s}_4=\sqrt{\frac{K_1^*}{K_3^*}}$ and $s_j(j=1,2,3)$ are three roots (with positive real part) of the following algebraic equation

$$\bar{a}s^6-\bar{b}s^4+\bar{c}s^2-\bar{d}=0. \quad (15)$$

As known from the generalized **Almansi (proved by Ding et al.** [1]) theorem, the function F and G can be expressed, respectively, in terms of five and four harmonic functions

(i) $F = F_1 + F_2 + F_3 + F_4 + F_5$ for distinct $s_j (j = 1, 2, 3, 4, 5)$

$$G = G_1 + G_2 + G_3 + G_4 \quad \text{for} \quad \text{distinct}\, \bar{s}_j (j = 1, 2, 3, 4). \tag{16a}$$

(ii) $F = F_1 + F_2 + F_3 + F_4 + zF_5$ for $s_1 \neq s_2 \neq s_3 \neq s_4 = s_5$.

$$G = G_1 + G_2 + G_3 + zG_4 \quad \text{for} \quad \bar{s}_1 \neq \bar{s}_2 \neq s_3 = s_4. \tag{16b}$$

(iii) $F = F_1 + F_2 + F_3 + zF_4 + z^2 F_5$ for $s_1 \neq s_2 \neq s_3 = s_4 = s_5$.

$$G = G_1 + G_2 + zG_3 + z^2 G_4 \quad \text{for} \quad \bar{s}_1 \neq \bar{s}_2 = s_3 = s_4. \tag{16c}$$

(iv) $F = F_1 + F_2 + zF_3 + z^2 F_4 + z^3 F_5 \quad \text{for} \quad s_1 \neq s_2 = s_3 = s_4 = s_5.$ (16d)

(v) $F = F_1 + zF_2 + z^2 F_3 + z^3 F_4 + z^4 F_5 \quad \text{for} \quad s_1 = s_2 = s_3 = s_4 = s_5.$

$$G = G_1 + zG_2 + z^2 G_3 + z^3 G_4 \quad \bar{s}_1 = \bar{s}_2 = \bar{s}_3 = \bar{s}_4 \tag{16e}$$

where $F_j (j = 1, 2, 3, 4, 5)$ and $G_j (j = 1, 2, 3, 4)$ satisfies the following harmonic equation

$$\left(\frac{\partial^2}{\partial x^2} + \frac{\partial^2}{\partial z_j^2} \right) F_j = 0, \quad (j = 1, 2, 3, 4) \tag{17a}$$

$$\left(\frac{\partial^2}{\partial x^2} + \frac{\partial^2}{\partial z_j^2} \right) G_j = 0, \quad (j = 1, 2, 3). \tag{17b}$$

The general solution for the case of distinct roots can be derived as follows

$$\begin{gathered}
u = \sum_{j=1}^{5} p_{1j} \frac{\partial^7 F_j}{\partial x \partial z_j^6} + \sum_{j=1}^{4} \bar{p}_{1j} \frac{\partial^5 G_j}{\partial x \partial z_j^4}, \\
w = \sum_{j=1}^{5} s_j p_{2j} \frac{\partial^7 F_j}{\partial z_j^7} + \sum_{j=1}^{4} s_j \bar{p}_{2j} \frac{\partial^5 G_j}{\partial z_j^5} \\
\phi_2 = \sum_{j=1}^{4} \bar{p}_{3j} \frac{\partial^6 G_j}{\partial x \partial z_j^5}, \; C = \sum_{j=1}^{5} p_{4j} \frac{\partial^8 F_j}{\partial z_j^8} + \sum_{j=1}^{4} \bar{p}_{4j} \frac{\partial^6 G_j}{\partial z_j^6}, \\
T = p_{55} \frac{\partial^8 F_5}{\partial z_5^8} + \bar{p}_{54} \frac{\partial^6 G_4}{\partial z_4^6},
\end{gathered} \tag{18}$$

where

$$p_{kj} = -p_k + q_k s_j^2 - r_k s_j^4 + v_k s_j^6, \quad (k = 1, 2) \;\&\; j = 1, 2, 3, 4, 5$$
$$p_{4j} = p_4 - q_4 s_j^2 + r_4 s_j^4 - v_4 s_j^6 + w_4 s_j^8,$$
$$p_{55} = a^* s_5^8 - b^* s_5^6 + c^* s_5^4 - d^* s_5^2 - \mathrm{e}^*,$$
$$\bar{p}_{kj} = \bar{p}_k - \bar{q}_k \bar{s}_j^2 + \bar{r}_k \bar{s}_j^4, \quad (k = 1, 2, 3) \;\&\; j = 1, 2, 3, 4$$
$$\bar{p}_{54} = \bar{a}\ \bar{s}_4^6 - \bar{b}\bar{s}_4^4 + \bar{c}\bar{s}_4^2 - \bar{d}$$

In the similar way general solution for the four three cases can be derived. Equation (18) can be further simplified by taking

$$p_{1j} \frac{\partial^6 F_j}{\partial z_j^6} = \psi_j, \tag{19a}$$

and

$$\bar{p}_{1j} \frac{\partial^4 G_j}{\partial \bar{z}_j^4} = \bar{\psi}_j. \tag{19b}$$

$$u = \sum_{j=1}^{5} \frac{\partial \psi_j}{\partial x} + \sum_{j=1}^{4} \frac{\partial \bar{\psi}_j}{\partial x}, w = \sum_{j=1}^{5} s_j P_{1j} \frac{\partial \psi_j}{\partial z_j} + \sum_{j=1}^{4} \bar{s}_j \bar{P}_{1j} \frac{\partial \bar{\psi}_j}{\partial \bar{z}_j}$$
$$\phi_2 = \sum_{j=1}^{4} \bar{s}_j \bar{P}_{2j} \frac{\partial^2 \bar{\psi}_j}{\partial x \partial z_j}, C = \sum_{j=1}^{5} P_{3j} \frac{\partial^2 \psi_j}{\partial z_j^2} + \sum_{j=1}^{4} \bar{P}_{3j} \frac{\partial^2 \bar{\psi}_j}{\partial \bar{z}_j^2},$$
$$T = P_{45} \frac{\partial^2 \psi_5}{\partial z_5^2} + \bar{p}_{44} \frac{\partial^2 \bar{\psi}_4}{\partial \bar{z}_4^2} \tag{20}$$

where

$$P_{1j} = p_{2j}/p_{1j}, \quad P_{2j} = p_{3j}/p_{1j}, \quad P_{3j} = p_{4j}/p_{1j}, \quad P_{45} = p_{55}/p_{15}.$$
$$\bar{P}_{1j} = \bar{p}_{2j}/\bar{p}_{1j}, \quad \bar{P}_{23} = \bar{p}_{33}/\bar{p}_{13}, \quad \bar{P}_{34} = \bar{p}_{44}/\bar{p}_{14}.$$

The functions ψ_j $(j = 1, 2, 3, 4, 5)$ and $\bar{\psi}_j$ $(j = 1, 2, 3, 4)$ satisfies the harmonic equations

$$\left(\frac{\partial^2}{\partial x^2} + \frac{\partial^2}{\partial z_j^2}\right)\psi_j = 0, \quad j = 1, 2, 3, 4, 5 \tag{21a}$$

$$\left(\frac{\partial^2}{\partial x^2} + \frac{\partial^2}{\partial z_j^2}\right)\bar{\psi}_j = 0 \quad j = 1, 2, 3, 4. \tag{21b}$$

Making use of Eq. (20) in Eq. (6a–c), we obtain

$$t_{11} = \sum_{j=1}^{5} (-A_{11} + A_{13}s_j^2 P_{1j} - \beta_1 P_{4j} - \gamma_1 P_{3j}) \frac{\partial^2 \psi_j}{\partial z_j^2}$$

$$+ \sum_{j=1}^{4} (-A_{11} + A_{13}\bar{s}_j^2 \bar{P}_{1j} - \beta_1 \bar{P}_{4j} - \gamma_1 \bar{P}_{3j}) \frac{\partial^2 \bar{\psi}_j}{\partial \bar{z}_j^2}, \quad (22a)$$

$$t_{33} = \sum_{j=1}^{5} (-A_{11} + A_{33}s_j^2 P_{1j} - \beta_3 P_{4j} - \gamma_3 P_{3j}) \frac{\partial^2 \psi_j}{\partial z_j^2}$$

$$+ \sum_{j=1}^{4} (-A_{11} + A_{33}\bar{s}_j^2 \bar{P}_{1j} - \beta_3 \bar{P}_{4j} - \gamma_3 \bar{P}_{3j}) \frac{\partial^2 \bar{\psi}_j}{\partial \bar{z}_j^2}, \quad (22b)$$

$$t_{31} = \sum_{j=1}^{5} (A_{65} P_{1j} + A_{55}) s_j \frac{\partial^2 \psi_j}{\partial x \partial z_j} + \sum_{j=1}^{4} A_{65} \bar{P}_{1j} + A_{55} + K_1 \bar{P}_{2j}) \bar{s}_j \frac{\partial^2 \bar{\psi}_j}{\partial x \partial z_j}, \quad (22c)$$

where

$$P_{41} = P_{42} = P_{43} = P_{44} = 0, \ \bar{P}_{41} = \bar{P}_{42} = \bar{P}_{43}.$$

Substituting the values of t_{11}, t_{33} and t_{31} from (22a–b) in Eq. (1) with the aid of Eq. (21a–b), we get

$$A_{11} - A_{13}s_j^2 P_{1j} + \beta_1 P_{4j} + \gamma_1 P_{3j} = (A_{65} P_{1j} + A_{55}) s_j^2 \ (j = 1, 2, 3, 4, 5) \quad (23a)$$

$$-A_{11} + A_{33}s_j^2 P_{1j} - \beta_3 P_{4j} - \gamma_3 P_{3j} = A_{65} P_{1j} + A_{55} \quad (23b)$$

$$A_{11} - A_{13}\bar{s}_j^2 P_{1j} + \beta_1 \bar{P}_{4j} + \gamma_1 \bar{P}_{3j} = (A_{65} \bar{P}_{1j} + A_{55} + K_1 \bar{P}_{2j}) \bar{s}_j^2 \quad (j = 1, 2, 3, 4) \quad (24a)$$

$$-A_{11} + A_{33}\bar{s}_j^2 \bar{P}_{1j} - \beta_3 \bar{P}_{4j} - \gamma_3 \bar{P}_{3j} = A_{65} \bar{P}_{1j} + A_{55} + K_1 \bar{P}_{2j} \quad (24b)$$

The general solution (22a–b) with the help of (23a–b) and (24a–b) can be simplified as

$$t_{11} = -\sum_{j=1}^{4} s_j^2 w_{1j} \frac{\partial^2 \psi_j}{\partial z_j^2} - \sum_{j=1}^{3} \bar{s}_j^2 \bar{w}_{1j} \frac{\partial^2 \bar{\psi}_j}{\partial \bar{z}_j^2},$$

$$t_{33} = \sum_{j=1}^{4} w_{1j} \frac{\partial^2 \psi_j}{\partial z_j^2} + \sum_{j=1}^{3} \bar{w}_{1j} \frac{\partial^2 \bar{\psi}_j}{\partial \bar{z}_j^2},$$

$$t_{31} = \sum_{j=1}^{4} s_j w_{1j} \frac{\partial^2 \psi_j}{\partial x \partial z_j} + \sum_{j=1}^{4} \bar{s}_j \bar{w}_{1j} \frac{\partial^2 \bar{\psi}_j}{\partial x \partial \bar{z}_j}, \tag{25}$$

where

$$\begin{aligned} w_{1j} &= \frac{A_{11} - A_{13} s_j^2 P_{1j} + \beta_1 P_{4j} + \gamma_1 P_{3j}}{s_j^2} \\ &= A_{65} P_{1j} + A_{55} = -A_{11} + A_{33} s_j^2 P_{1j} - \beta_3 P_{4j} - \gamma_3 P_{3j} \end{aligned} \tag{26a}$$

$$\begin{aligned} \bar{w}_{1j} &= \frac{A_{11} - A_{13} \bar{s}_j^2 P_{1j} + \beta_1 \bar{P}_{4j} + \gamma_1 \bar{P}_{3j}}{\bar{s}_j^2} \\ &= A_{65} \bar{P}_{1j} + A_{55} + K_1 \bar{P}_{2j} = -A_{11} + A_{33} \bar{s}_j^2 \bar{P}_{1j} - \beta_3 \bar{P}_{4j} - \gamma_3 \bar{P}_{3j} \end{aligned} \tag{26b}$$

3 Fundamental Solution for a Point Heat Source in a Semi-infinite Transversely Isotropic Micropolar Thermoelastic Diffusion Material

We consider a semi-infinite transversely isotropic micropolar thermoelastic diffusion material with $z \geq 0$. A point heat source H is applied at the origin and the surface $z = 0$ is stress free, equilibrated, concentration and thermally insulated. The complete geometry of the problem is shown in Fig. 1. The general solution given by Eq. (20)

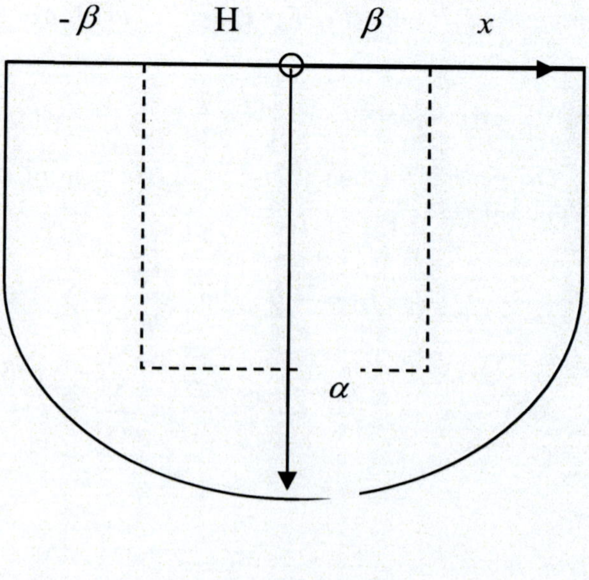

Fig. 1 Geometry of the problem

and (22a–b) is derived in this section.

Introduce the harmonic functions as

$$\psi_j = A_j\left[\frac{1}{2}(z_j^2 - x^2)\left(\log r_j - \frac{3}{2}\right) - xz_j \tan^{-1}\frac{x}{z_j}\right] \quad j = 1, 2, 3, 4, 5 \tag{27}$$

where A_j ($j = 1, 2, 3, 4, 5$) are arbitrary constants to be determined and

$$r_j = \sqrt{x^2 + z_j^2}, \tag{28}$$

and

$$\bar{\psi}_j = \bar{A}_j\left[\frac{1}{2}(\bar{z}_j^2 - x^2)\left(\log \bar{r}_j - \frac{3}{2}\right) - x\bar{z}_j \tan^{-1}\frac{x}{\bar{z}_j}\right], \quad j = 1, 2, 3, 4 \tag{29}$$

where $\bar{A}_j$ ($j = 1, 2, 3, 4$) are arbitrary constants to be determined and

$$\bar{r}_j = \sqrt{x^2 + \bar{z}_j^2}. \tag{30a}$$

Here $\bar{A}_4$ can be written as linear combination of A_3, i.e.,

$$\bar{A}_4 = \eta A_3 \tag{30b}$$

4 Boundary Conditions

For stress-free surface $z = 0$, the boundary conditions are

(i) **Vanishing of the normal stress component, i.e.,**

$$t_{33} = 0 \tag{31a}$$

(ii) **Vanishing of the tangential stress component, i.e.,**

$$t_{31} = 0 \tag{31b}$$

(iii) **Vanishing of the tangential couple stress component, i.e.,**

$$m_{32} = 0 \quad \text{or} \quad \frac{\partial \phi_2}{\partial z} = 0 \tag{31c}$$

(iv) **Vanishing of the temperature gradient field, i.e.,**

$$\frac{\partial T}{\partial z} = 0, \tag{31d}$$

(v) **Vanishing of the Concentration gradient field, i.e.,**

$$\frac{\partial C}{\partial z} = 0. \tag{31e}$$

When the mechanical, tangential couple stress, concentration and thermal condition for a rectangle of $0 \le z \le \alpha$ and $-\beta \le x \le \beta$ $(b > 0)$ are considered (Fig. 1), four equations can be obtained

$$\int_{-\beta}^{\beta} \sigma_{zz}(x, \alpha)\mathrm{d}x + \int_{0}^{\alpha} [\sigma_{zx}(\beta, z) - \sigma_{zx}(-\beta, z)]\mathrm{d}z = 0, \tag{32a}$$

$$\int_{-\beta}^{\beta} \frac{\partial \varphi_2}{\partial z}(x, \alpha)\mathrm{d}x + \int_{0}^{\alpha} \left[\frac{\partial \varphi_2}{\partial x}(\beta, z) - \frac{\partial \varphi_2}{\partial x}(-\beta, z)\right]\mathrm{d}z = 0, \tag{32b}$$

$$\int_{-\beta}^{\beta} \left[\frac{\partial C}{\partial z}(x, \alpha)\right]\mathrm{d}x - \int_{0}^{\alpha} \left[\frac{\partial C}{\partial x}(\beta, z) - \frac{\partial C}{\partial x}(-\beta, z)\right]\mathrm{d}z = 0. \tag{32c}$$

$$-a_3 \int_{-\beta}^{\beta} \left[\frac{\partial T}{\partial z}(x, \alpha)\right]\mathrm{d}x - a_1 \int_{0}^{\alpha} \left[\frac{\partial T}{\partial x}(\beta, z) - \frac{\partial T}{\partial x}(-\beta, z)\right]\mathrm{d}z = H. \tag{32d}$$

Substituting the values of ψ_j and $\bar{\psi}_j$ from Eqs. (27) and (29) in Eq. (20) and (22a, b), we obtain the expressions for components of displacement, temperature change, volume fraction field and stress components as follows:

$$u = -\sum_{j=1}^{5} A_j\left[x(\log r_j - 1) + z_j \tan^{-1}\frac{x}{z_j}\right] - \sum_{j=1}^{4} \bar{A}_j\left[x(\log \bar{r}_j - 1) + \bar{z}_j \tan^{-1}\frac{x}{\bar{z}_j}\right], \tag{33a}$$

$$w = \sum_{j=1}^{5} s_j P_{1j} A_j\left[z_j(\log r_j - 1) - x \tan^{-1}\frac{x}{z_j}\right] + \sum_{j=1}^{4} \bar{s}_j \bar{P}_{1j} \bar{A}_j\left[\bar{z}_j(\log \bar{r}_j - 1) - x \tan^{-1}\frac{x}{\bar{z}_j}\right], \tag{33b}$$

$$\phi_2 = -\sum_{j=1}^{4} \bar{A}_j \bar{P}_{2j} \tan^{-1} \frac{x}{\bar{z}_j}, \tag{33c}$$

$$C = \sum_{j=1}^{5} A_j P_{2j} \log r_j + \sum_{j=1}^{4} \bar{A}_j \bar{P}_{2j} \log \bar{r}_j, \tag{33d}$$

$$T = A_5 P_{45} \log r_5 + \bar{A}_4 \bar{P}_{44} \log \bar{r}_4, \tag{33e}$$

$$\sigma_{xx} = -\sum_{j=1}^{5} s_j^2 w_{1j} A_j \log r_j - \sum_{j=1}^{4} \bar{s}_j^2 \bar{w}_{1j} \bar{A}_j \log \bar{r}_j, \tag{33f}$$

$$\sigma_{zz} = \sum_{j=1}^{5} w_{1j} A_j \log r_j + \sum_{j=1}^{4} \bar{w}_{1j} \bar{A}_j \log \bar{r}_j, \tag{33g}$$

$$\sigma_{zx} = -\sum_{j=1}^{5} s_j w_{1j} A_j \tan^{-1} \frac{x}{z_j} - \sum_{j=1}^{4} \bar{s}_j \bar{w}_{1j} \bar{A}_j \tan^{-1} \frac{x}{\bar{z}_j}. \tag{33h}$$

Making use the values of σ_{zz}, σ_{zx}, C, ϕ_2 and T from Eqs. (33c, 32d, e, g, h) in Eqs. (31a–31e), we obtain

$$\sum_{j=1}^{5} w_{1j} A_j = 0, \tag{34a}$$

$$\sum_{j=1}^{5} \bar{w}_{1j} \bar{A}_j = 0, \tag{34b}$$

$$\sum_{j=1}^{4} s_j w_{1j} A_j = 0, \tag{34c}$$

$$\sum_{j=1}^{4} \bar{s}_j \bar{w}_{1j} \bar{A}_j = 0, \tag{34d}$$

$\frac{\partial C}{\partial z}$, $\frac{\partial T}{\partial z}$ and $\frac{\partial \varphi_2}{\partial z}$ are automatically satisfied at the surface $z = 0$.

Making use the values of σ_{zz} and σ_{zx} from Eqs. (33g, 33h) in Eq. (32a), we obtain

$$\sum_{j=1}^{5} w_{1j} A_j I_3 + \sum_{j=1}^{4} \bar{w}_{1j} \bar{A}_j I_4 = 0, \tag{35}$$

where

$$I_3 = \left[x\left(\log\sqrt{x^2 + s_j^2\alpha^2} - 1\right) + s_j\alpha \tan^{-1}\frac{x}{s_j\alpha} \right]_{x=-\beta}^{x=\beta} - 2\left[z_j \tan^{-1}\frac{\beta}{s_j z} + b\log\sqrt{\beta^2 + s_j^2 z^2} \right]_{z=0}^{z=\alpha} = 2\beta(\log\beta - 1), \tag{36a}$$

and

$$I_4 = \left[x\left(\log\sqrt{x^2 + \bar{s}_j^2\alpha^2} - 1\right) + \bar{s}_j\alpha \tan^{-1}\frac{x}{\bar{s}_j\alpha} \right]_{x=-\beta}^{x=\beta} - 2\left[\bar{z}_j \tan^{-1}\frac{\beta}{\bar{s}_j z} + \beta\log\sqrt{\beta^2 + \bar{s}_j^2 z^2} \right]_{z=0}^{z=\alpha} = 2\beta(\log\beta - 1). \tag{36b}$$

By virtue of the Eqs. (36a, b), the Eqs. (35) degenerate to Eq. (34a, b), i.e., eqs. (34a, b) and (35) are satisfied automatically.

Some useful integrals are given as follows

$$\int \frac{\partial\phi_2}{\partial z} = -\sum_{j=1}^{4} \bar{A}_j \bar{P}_{2j} \int -\frac{s_j x}{x^2 + \bar{s}_j^2\bar{z}^2}\mathrm{d}x = \sum_{j=1}^{5} \bar{A}_j \bar{P}_{2j} \tan^{-1}\frac{x}{\bar{z}_j}, \tag{37a}$$

$$\int \frac{\partial\phi_2}{\partial x}\mathrm{d}z = \sum_{j=1}^{4} \bar{A}_j \bar{P}_{2j} \int \frac{z_j}{x^2 + z_j^2}\mathrm{d}z = -\sum_{j=1}^{4} \bar{A}_j \bar{P}_{2j} \tan^{-1}\frac{x}{\bar{z}_j}. \tag{37b}$$

$$\int \frac{\partial C}{\partial z} = \sum_{j=1}^{5} A_j s_j^2 P_{2j} \int \frac{z}{x^2 + s_j^2 z^2}\mathrm{d}x + \sum_{j=1}^{5} A_j s_j^2 P_{2j} \int \frac{z}{x^2 + s_j^2 z^2}\mathrm{d}x$$

$$= \sum_{j=1}^{5} A_j s_j P_{2j} \tan^{-1}\frac{x}{z_j} + A_j s_j P_{2j} \tan^{-1}\frac{x}{\bar{z}_j}, \tag{37c}$$

$$\int \frac{\partial C}{\partial z}\mathrm{d}z = \sum_{j=1}^{5} A_j P_{2j} \int \frac{x}{x^2 + s_j^2 z^2}\mathrm{d}z + \sum_{j=1}^{4} \bar{A}_j s_j^2 \bar{P}_{2j} \int \frac{x}{x^2 + \bar{s}_j^2\bar{z}^2}\mathrm{d}z$$

$$= -\sum_{j=1}^{5} \frac{A_j}{s_j} P_{2j} \tan^{-1}\frac{x}{z_j} - \sum_{j=1}^{5} \frac{A_j}{s_j} \bar{P}_{2j} \tan^{-1}\frac{x}{\bar{z}_j}, \tag{37d}$$

$$\int \frac{\partial T}{\partial z}\mathrm{d}x = s_5 P_{45} A_5 \int \frac{z_5}{r_j^2}\mathrm{d}x + s_4 P_{44} \bar{A}_4 \int \frac{\bar{z}_4}{\bar{r}_j^2}\mathrm{d}x$$

$$= s_5 P_{45} A_5 \tan^{-1}\frac{x}{z_5} - \bar{s}_5 \bar{P}_{45} \bar{A}_5 \tan^{-1}\frac{x}{\bar{z}_4}, \tag{37e}$$

$$\int \frac{\partial T}{\partial z}\mathrm{dz} = P_{45}A_5 \int \frac{x}{r_5^2}\mathrm{dz} + \bar{P}_{44}\bar{A}_4 \int \frac{x}{\bar{r}_4^2}\mathrm{dz}$$
$$= -\frac{P_{45}}{s_5}A_5 \tan^{-1}\frac{x}{z_5} - \frac{\bar{P}_{44}}{\bar{s}_4}\bar{A}_5 \tan^{-1}\frac{x}{\bar{z}_4}. \tag{37f}$$

Making use of Eq. (33e) in Eq. (32d), with the aid of $s_5 = \sqrt{\frac{K_1}{K_3}} = \bar{s}_4$ and the integrals (37e, 37f), we obtain

$$P_{45}A_5I_5 + \bar{P}_{44}\bar{A}_4I_6 = \frac{H}{\sqrt{K_3/K_1}}, \tag{38}$$

$$I_5 = -\left[\tan^{-1}(\frac{x}{s_5\alpha})\right]_{x=-\beta}^{x=\beta} + \left[\tan^{-1}(\frac{\beta}{s_5 z})\right]_{z=0}^{z=\alpha} = -\pi, \tag{39a}$$

$$I_6 = -\left[\tan^{-1}(\frac{x}{\bar{s}_4\alpha})\right]_{x=-\beta}^{x=\beta} + \left[\tan^{-1}(\frac{\beta}{\bar{s}_4 z})\right]_{z=0}^{z=\alpha} = -\pi. \tag{39b}$$

A_5 can be determined from Eqs. (38) and (39a, 39b), as follows

$$A_5 = -\frac{H}{\pi(P_{45} + \alpha\bar{P}_{44})\sqrt{K_3/K_1}}. \tag{40}$$

Substituting the value of ϕ_2 from Eq. (33c) in Eq. (32b) and with the aid of the integrals (37a, 37b), we obtain

$$\sum_{j=1}^{4} \bar{P}_{2j}\bar{A}_j = 0. \tag{41}$$

Substituting the value of C from Eq. (33d) in Eq. (32c) and with the aid of the integrals (37c, 37d), we obtain

$$\sum_{j=1}^{5} q_j P_{2j} A_j = 0, \tag{42}$$

$$\sum_{j=1}^{4} \bar{q}_j \bar{P}_{2j} \bar{A}_j = 0 \tag{43}$$

where

$$q_j = 2(s_j^2 - 1)\tan^{-1}\left(\frac{\beta}{s_j\alpha}\right) + \pi,$$

$$\bar{q}_j = 2(\bar{s}_j^2 - 1)\tan^{-1}\left(\frac{\beta}{\bar{s}_j\alpha}\right) + \pi.$$

Thus the nine constants $A_j(j = 1, 2, 3, 4, 5)$, $\bar{A}_j(j = 1, 2, 3, 4)$ can be determined by nine equations including Eqs. (34a–34d), (40–43) and by the relation given in Eq. 30b, by Cramer rule.

5 Special Cases

Case I: In the absence of diffusion effect:

In the absence of diffusion effects, (32a–33h) reduces to

$$u = -\sum_{j=1}^{4} A_j\left[x(\log r_j - 1) + z_j \tan^{-1}\frac{x}{z_j}\right]$$
$$-\sum_{j=1}^{3} \bar{A}_j\left[x(\log \bar{r}_j - 1) + \bar{z}_j \tan^{-1}\frac{x}{\bar{z}_j}\right], \tag{44a}$$

$$w = \sum_{j=1}^{4} s_j P_{1j} A_j\left[z_j(\log r_j - 1) - x\tan^{-1}\frac{x}{z_j}\right]$$
$$+\sum_{j=1}^{3} \bar{s}_j \bar{P}_{1j} \bar{A}_j\left[\bar{z}_j(\log \bar{r}_j - 1) - x\tan^{-1}\frac{x}{\bar{z}_j}\right], \tag{44b}$$

$$\phi_2 = -\sum_{j=1}^{3} \bar{A}_j \bar{P}_{2j} \tan^{-1}\frac{x}{\bar{z}_j}, \tag{44c}$$

$$T = A_4 P_{34} \log r_4 + \bar{A}_3 \bar{P}_{33} \log \bar{r}_3, \tag{44d}$$

$$\sigma_{xx} = -\sum_{j=1}^{4} s_j^2 w_{1j} A_j \log r_j - \sum_{j=1}^{3} \bar{s}_j^2 \bar{w}_{1j} \bar{A}_j \log \bar{r}_j, \tag{44e}$$

$$\sigma_{zz} = \sum_{j=1}^{4} w_{1j} A_j \log r_j + \sum_{j=1}^{3} \bar{w}_{1j} \bar{A}_j \log \bar{r}_j, \tag{44f}$$

$$\sigma_{zx} = -\sum_{j=1}^{4} s_j w_{1j} A_j \tan^{-1}\frac{x}{z_j} - \sum_{j=1}^{3} \bar{s}_j \bar{w}_{1j} \bar{A}_j \tan^{-1}\frac{x}{\bar{z}_j}. \tag{44g}$$

The above results are similar as obtained by Kumar and Chawla [12].

Case II: Negligible the micropolar and Diffusion effects:

In the absence of micropolar and diffusion effects, (33a–33h) reduces to

$$u = -\sum_{j=1}^{3} A_j \left[x(\log r_j - 1) + z_j \tan^{-1}\frac{x}{z_j} \right], \tag{45a}$$

$$w = \sum_{j=1}^{3} s_j P_{1j} A_j \left[z_j(\log r_j - 1) - x \tan^{-1}\frac{x}{z_j} \right], \tag{45b}$$

$$T = A_3 P_{23} \log r_3, \tag{45c}$$

$$\sigma_{xx} = -\sum_{j=1}^{3} s_j^2 w_{1j} A_j \log r_j, \tag{45d}$$

$$\sigma_{zz} = \sum_{j=1}^{3} w_{1j} A_j \log r_j, \tag{45e}$$

$$\sigma_{zx} = -\sum_{j=1}^{3} s_j w_{1j} A_j \tan^{-1}\frac{x}{z_j}. \tag{45f}$$

The above results are similar as obtained by Hou et al. [10].

Case III: Negligible the Voids, Thermal, and Diffusion effects:

In the absence of voids, thermal, and diffusion effects, we obtain the corresponding results for transversely isotropic elastic medium.

6 Conclusion

The general solution and fundamental solution for two-dimensional problem in micropolar thermoelastic media with mass diffusion has been established. The two-dimensional general solution in transversely isotropic micropolar thermoelastic media with mass diffusion is derived firstly by using the operator theory. On the basis of obtained general solution, the fundamental solution for a steady point heat source on the surface of a semi-infinite transversely isotropic micropolar thermoelastic material with mass diffusion is derived by nine newly introduced harmonic functions. The components of displacement, stress, temperature change, mass concentration, and couple stress are expressed in terms of elementary functions, so it is convenient to use them From the present investigation, some special cases of interest are also deduced and compared with the previous results obtained.

Applications: fundamental solutions for two-dimensional in anisotropic media are important to the solution of inclusion problems and of the boundary integral equations. This type of solution technique (which, I have used in this research paper) is very useful for finding the general solution and fundamental solution in anisotropic media for different theories that is, micropolar thermoelastic material with voids and mass diffusion, micropolar thermoelastic material with mass diffusion and voids, microstretch thermo elastic material, with mass diffusion, etc. This type of solution technique provides a wonderful platform for new researchers to construct the general solution in thermoelasticity with double porosity and triple porosity. Also this type of solution technique will be very useful to construct fundamental solution for three-dimensional problems and Green's function in different symmetries which will be very useful for solving boundary value problems as well as the study of Cracks, defects, and inclusions.

Conflict of Interest Authors declare that they do not have any conflict of interest.

References

1. Ding HJ, Chen B, Liang J (1996) General solutions for coupled equations in piezoelectric media. Int J Solids Struct 33:2283–2298
2. Pan E, Tanon F (2000) Three dimensional green's functions in anisotropic piezoelectric solids. Int J Solids Struct 37:943–958
3. Chen WQ (2000) On the general solution for piezothermoelasticity for transverse isotropy with applications. ASME J Appl Mech 67:705–711
4. Chen WQ, Lim CW, Ding HJ (2005) Point temperature solution for a penny- shaped crack in an infinite transversely isotropic thermo-piezo-elastic medium. Eng Anal with Bound Elem 29:524–532
5. Sharma B (1958) Thermal stresses in transversely isotropic semi-infinite elastic solids. ASME J Appl Mech 23:86–88
6. Ciarletta M, Scalia A, Svanadze M (2007) Fundamental solution in the theory of micropolar thermoelastic for materials with voids. J Therm Stress 30:213–229
7. Hou PF, Leung AYT, He YJ (2008) Three-dimensional green's functions for transversely isotropic thermoelastic biomaterials. Int J Solids Struct 45:6100–6113
8. Hou PF, Wang L, Yi T (2009) 2D Green's functions for semi-infinite orthotropic thermoelastic plane. Appl Math Model 33:1674–1682
9. Xiong SM, Hou PF, Yang SY (2010) 2D Green's functions for semi-infinite orthotropic piezothermoelastic plane. IEEE Trans Ultrason Ferroelectr Freq Control 57:1003–1010
10. Hou PF, Sha H, Chen CP (2011) 2D general solution and fundamental solution for orthotropic thermoelastic materials. Eng Anal Boundary Elem 35:56–60
11. Seremet V (2011) Deriving exact Green's functions and integral formulas for a thermoelastic wedge. Engng Anal with Bound Elem 35:527–532
12. Seremet V (2012) New closed form Green's function and integral formula for a thermoelastic quadrant. Appl Math Model 36:799–812
13. Kumar R, Kansal T (2012) Plane waves and fundamental solution in the generalized theories of thermoelastic diffusion. Int J Appl Math Mech 8:1–20

14. Kumar R, Chawla V (2011) A study of fundamental solution in orthotropic thermodiffusive elastic media. Int Commun Heat Mass Transf 27:456–462
15. Kumar R, Chawla V (2012) Green's functions in orthotropic thermoelstic diffusion media. Eng Anal Bound Elem 36:1272–1277
16. Kumar R, Chawla V (2012) General steady-state solution and green's function in orthotropic piezothermoelastic diffusion medium. Arch Mech 64:555–579
17. Kumar R, Chawla V (2013) fundamental solution for two-dimensional problem in orthotropic piezothermoelastic diffusion media. Mater Phys Mech 16:159–174
18. Kumar R, Chawla V (2013) Reflection and refraction of plane wave at the interface between elastic and thermoelastic media with three-phase-lag. Int Commun Heat Mass Transf 48:53–60
19. Kumar R, Gupta V (2014) Green's function for transversely isotropic thermoelastic diffusion bimaterials. J Therm Stress 37:1201–1229
20. Kumar R, Chawla V (2015) General solution and fundamental solution for two-dimensional problem in orthotropic thermoelastic media with voids. J Adv Math Appl Amer Sci Publ 3:1–8
21. Şeremet V (2016) A method to derive thermoelastic green's functions for bounded domains (on examples of two-dimensional problems for parallelepipeds). Acta Mech 227:3603–3620
22. Pan LH, Hou PF, Chen JY (2016) 2D steady-state general solution and fundamental solution for fluid-saturated. Z Angew Math Phys ZAMP 67–84
23. Chawla V, Ahuja S, Rani V (2017) Fundamental solution for a two-dimensional problem in transversely isotropic micropolar thermoelastic media". Multidiscipline Model Mater Struct 13:409–423
24. Dang HY, Zhao MH, Fan CY, Chen ZT (2018) Analysis of arbitrarily shaped planar cracks in three-dimensional isotropic hygrothermoelastic media. J Therm Stress 6:1–28
25. Zhao MH, Dang HY, Fan CY, Chen ZT (2018) Three dimensional steady-state general solution for isotropic hygrothermoelastic media. J Therm Stress 41:951–972
26. Tomar T, Goyal N, Szekeres A (2019) Plane waves in thermo-viscoelastic material with voids under different theories of thermoelasticity. Int J Appl Mech Eng 24:691–708
27. Biswas S (2020) Fundamental solution of steady oscillations equations in nonlocal thermoelastic medium with voids. J Therm Stress 43:284–304
28. Aoudi A (2009) The coupled theory of micropolar thermoelastic diffusion. Acta Mech 208:181–203

Chapter 48
Statistical Analysis of Factors Affecting COVID-19

Aditya Kapoor, Nonita Sharma, K. P. Sharma, and Ravi Sharma

1 Introduction

The COVID-19 also known as coronavirus is already considered a global pandemic by WHO as the spread of disease increased rapidly from Wuhan, China, to 193 countries [1, 2]. The incubation period of this contact transmissible disease is 6–14 days. While in the acute stage of the novel coronavirus, the early symptoms of the disease can be listed as cough, high temperature, breathing difficulties and fatigue and in extreme conditions it can even lead to death. Due to the lack of vaccine, currently separating oneself is the measure adopted to counter coronavirus. The transmission of novel coronavirus can occur through various large droplets, bio-aerosols, or contact with secretions which are similar to influenza virus [3]. Many researchers have claimed that the transmission of virus is impacted by diverse geographical factors for instance climatic conditions, for example, humidity and temperature, and also density of population (PD) [4]. Furthermore, it was also observed that the spread of the coronavirus is more severe in mid-latitude countries where there is considerably low temperature as compared to the relatively tropical countries. To access, interpret, and respond to a disease outbreak, essentially in the case of pandemics like coronavirus, good geographical knowledge is essential [5]. The global strategy which was used to combat the coronavirus further limiting the outbreak of the disease and decreasing the stress on the healthcare system is regular hand washing and social

A. Kapoor (✉) · N. Sharma · K. P. Sharma · R. Sharma
Department of Computer Science and Engineering, Dr BR Ambedkar National Institute of Technology, Jalandhar 144011, Punjab, India

N. Sharma
e-mail: nonita@nitj.ac.in

K. P. Sharma
e-mail: sharmakp@nitj.ac.in

R. Sharma
e-mail: ravi.cs.19@nitj.ac.in

N. Marriwala et al. (eds.), *Soft Computing for Intelligent Systems*, Algorithms for Intelligent Systems, https://doi.org/10.1007/978-981-16-1048-6_48

distancing. The precaution measures taken by some countries include imposing limitations on social interactions, including lockdown which leads to forbidding all but the important transport and adjourning economic and trading activities.

In the era of COVID-19, the pandemic has asymmetrically affected the industries. There are certain companies which are suffering from restriction on transport and leisure events, whereas other companies such as goods delivery (like Amazon), respirator industry, industries manufacturing protective equipments, and other OTT platform industries to name a few. In general, a lockdown imposed all over the nation reduced the day to day activities by around 20% from regular levels [6]. Cases of COVID-19 are reaching near to 11.5 million all over the world [1, 2]. In India total number of cases is around 700 k from which 280 k cases are active cases. The total number of COVID positive cases per day is also increasing day by day. Thus, it is the need of the hour to analyze this data and evaluate meaningful insights in data and help flatten the curve.

The manuscript presents an analysis for statistical analysis of data taken for Tamil Nadu state of India. The considered dataset lacks many values and directly implementing algorithms and analysis on dataset will alter the accuracy and the final result and higher probability of biasedness. So, initial step taken was data pre-processing. In this step, the missing values are checked. In addition to this, the dataset had missing values for some particular time interval. After finalizing the target dataset, a thorough analysis is done, and we have finalized our research questions as follows.

1.1 Research Question

1. Estimating the relationship between the Status of Patient and Gender?
2. Estimating the relationship between the Status of Patient and the age of Patient?
3. Estimating the relationship between the Age of Patient and Gender of Patient?

The manuscript targets to answer these research questions. After introducing the research problem in introduction section, Sect. 2 presents the related work. Section 3 presents the results of the data analysis followed by conclusion and future scope.

2 Related Work

The coronavirus has affected all the countries around the globe and millions of people are impacted with significant health encumbrance and thus is an important area of research for finding the vaccine to eradicate the disease or to find the factors that cause the increase in the outbreak and how can the spread of disease be reduced. Gupta et al. [7] worked on a spatial association among the long-term climate, social factors, and topography with the confirmed count of coronavirus cases in the country India. In their study, they found a positive relationship between the amount by which infection spreads with the long-term climatic temperature records, SR, WS, and PD. Also, their study shows a negative relationship between the AET, SH rainfall, and elevation. In addition to this, there are a substantial number of studies worldwide that have already attempted to examine correlations between the coronavirus spread and the current weather or climatic situation [8, 9]. According to study conducted by Chen et al. [1], the changes in one weather factor like temperature or humidity do not alter the results of coronavirus outbreak, whereas several meteorological factors combined together could well correlate the pandemic trend better than single-factor models.

For the statistical estimations, Fu et al. [10] employed Boltzmann function to estimate the total count of confirmed positive cases in each muncipality/province and in China and estimated the growing trend of confirmed positive cases. Danon et al. [11] created a model to evaluate initial disease spread trends and peaks in England & Wales and further calculated the effects of seasonal diversification in disease spread counts. On the basis of accumulated datasets of suspected cases, isolation, reports, deaths, Tang et al. [12] consider that the pestilence pattern predominantly relies upon segregation and suspected cases. Hence, it is very imperative to keep on fortifying isolation and detachment systems and enhance the location rate. Wu et al. [13] accumulated and analyzed medical observation, discharge, cure, critical and total death count and applied the state transfer matrix model to estimate the peak inflection time and patient distribution to improve resources medically.

The initially conducted studies have emphasised on the macroscopic perspective of the coronavirus pandemic using the concept of Baidu migration also confirmed diagnosis content at the provincial or national scale. In the study, [14] a multi-source and open-source data are applied that are easy to get and to perceive and for classifying the spatial situation of epidemic, and it further elaborates the role of different spatial elements on the spread of the pandemic from geographical approach. A system containing indicators for assessing the major infection disease is constructed and further on this basis, a grid map is constructed.

3 Data Analysis

The coronavirus is relatively a new type of virus and thus not much information is available about this virus, thereby it becomes a very difficult task to find a vaccine or medicine as soon as it is required. The data analysis and the implementation of algorithms to extract meaningful insights from the data and therefore flatten the curve of the graph which is currently increasing day by day. To analyze these results, we have downloaded the data of Tamil Nadu state and further performed chi-square test to estimate the relationship between various parameters. When the data is filtered state-wise, we checked the data for the missing value. Out of the total 4956 cases, 1583 cases of age and 1519 gender values were missing. In this particular analysis, we removed the missing values. For removing the missing values, we first checked the data range in which missing values are less. Upon visualization of the data, we

Table 1 Description of age w.r.t. male and female

	Gender			Statistic	Std. error
Age bracket	Male	Mean		36.086	0.5111
		95% confidence interval for mean	Lower bound	35.083	
			Upper bound	37.089	
		5% trimmed mean		35.766	
		Median		35.000	
		Variance		237.451	
		Std. Dev		15.4094	
		Minimum		1.0	
		Maximum		98.0	
		Range		97.0	
		Interquartile range		22.0	
		Skewness		0.407	0.081
		Kurtosis		0.155	0.162
	Female	Mean		35.475	0.7363
		95% confidence interval for mean	Lower bound	34.028	
			Upper bound	36.921	
		5% trimmed mean		35.176	
		Median		34.00	
		Variance		276.486	
		Std. Dev		16.6279	
		Minimum		1.0	
		Maximum		85.0	
		Range		84.0	
		Interquartile range		22.0	
		Skewness		0.320	0.108
	Kurtosis			−0.291	0.216

encountered that from April 22, 2020 to May 03, 2020 there are very few missing values and thus are considered (Table 1).

3.1 *RQ1. Estimating the Relationship Between the Status of Patient and Gender?*

At the initial level, the chi-square test was implemented to correlate the gender variable and current status. Table 2 and Fig. 1 demonstrate the results of the evaluation. The value of correlation factor as demonstrated in Table 3 comes out to be 0.648, greater than 0.05 entailing to reject the null hypothesis.

Table 2 Cross table of gender and current status

		Current Status		
Gender		Hospitalized	Deceased	Total
	Male	906	3	909
	Female	509	1	510
Total		1415	4	1419

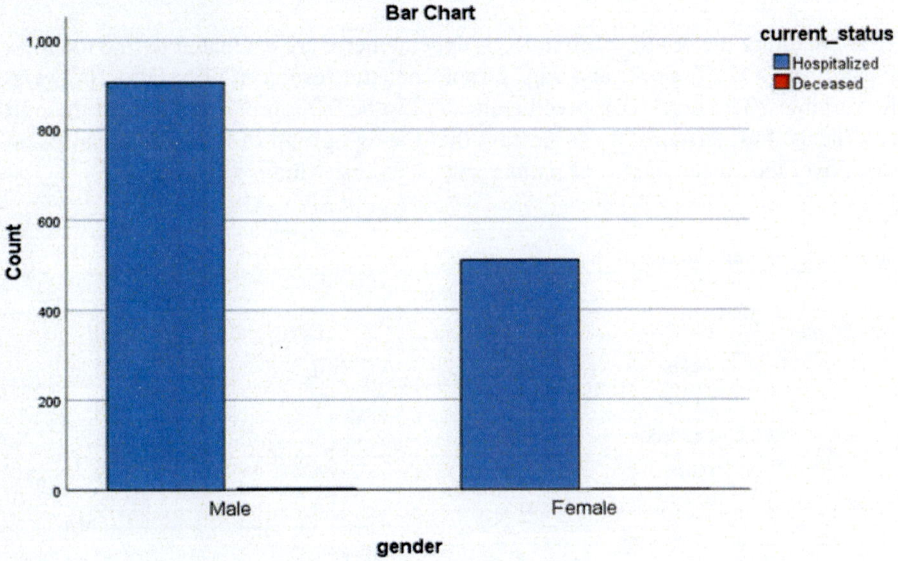

Fig. 1 Bar Chart for gender with respect to current status

Table 3 Chi-square test for gender and current status

	Value	Df	Asymptotic Sig	Exact sig. (2-sided)	Exact Sig. (1-sided)
Pearson Chi—square	0.209	1	0.648		
Continuity correction	0.000	1	1.000		
Likelihood ratio	0.221	1	0.639		
Fisher's exact test				1.000	0.546
Linear by linear association	0.208	1	0.648		
No of valid cases	1419				

To conclude from the value of the test, it can be assumed that gender and current status have no correlation.

3.2 *RQ2. Estimating the Relationship Between the Status of Patient and the Age of Patient?*

After obtaining the relationship in RQ1, dependencies are evaluated to find the relationship in RQ2. Table 4. and Fig. 2 represent the results of correlation between the variables. The results observed comes out to be 0.00 implying to reject the null hypothesis. Furthermore, it can be said that the age group of the patient causes an effect over the current status of patient and vice versa (Table 5).

Table 4 Age group and current status cross table

		Current status		
Age group		Hospitalized	Deceased	Total
	<18	173	0	173
	19–40	739	1	740
	41–65	455	1	456
	>65	48	2	50
Total		1415	4	1419

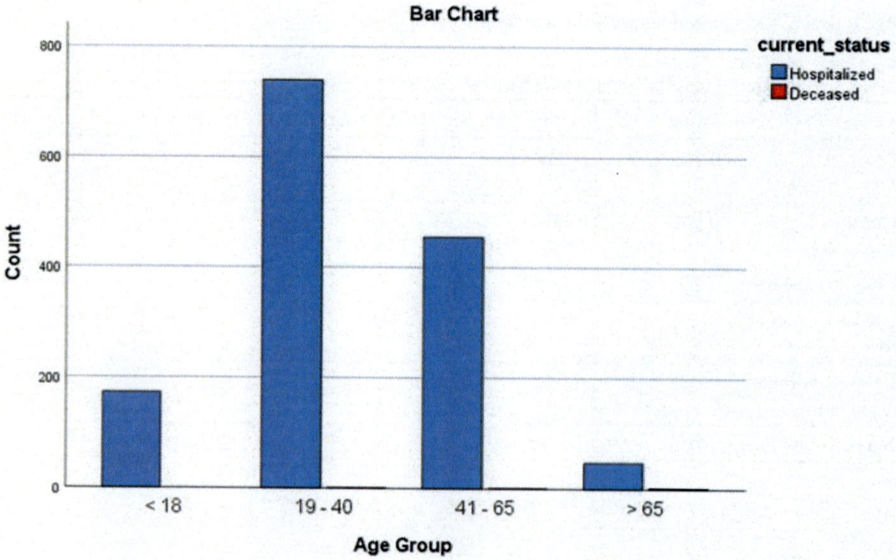

Fig. 2 Bar Chart for age group w.r.t. current status

Table 5 Age group and current status Chi-square test

	Value	*df*	Asymptotic significance (2-sided)
Pearson Chi—square	25.710	3	0.000
Likelihood ratio	80,711	3	0.033
Linear—by—linear association	7.530	1	0.006
No of valid cases	1419		

Table 6 Age group and gender cross table

		Gender		
		Male	Female	Total
Age Group	<18	98	75	173
	19–40	491	249	740
	41–65	291	165	456
	>65	29	21	50
Total		909	510	1419

3.3 RQ3. Estimating the Relationship Between the Age of Patient and Gender of Patient?

After finding the relationship in RQ1 and RQ2, the chi-square test was implemented again to determine the association between age group and gender. The chi-square test evaluates the dependencies of any of the age group with the gender of the patient and vice versa. The crosstabulation results is shown in Table 6 between age group and gender also Fig. 3 represents it graphically. The calculated chi-square value as shown in Table 7 comes to be 0.085 which is greater than 0.05 and thus the null hypothesis is to be rejected. Furthermore, it can be rightly said that the gender of the patient has no effect over the age group and vice versa (Table 7).

4 Conclusion

Coronavirus positive cases are increasing day by day, and it is essential to analyze this data. In the proposed study, Tamil Nadu state's data is estimated and to find the correlation between different variables. Results demonstrate that only Age group and current status show dependencies and the rest of two relations are independent (Table 8).

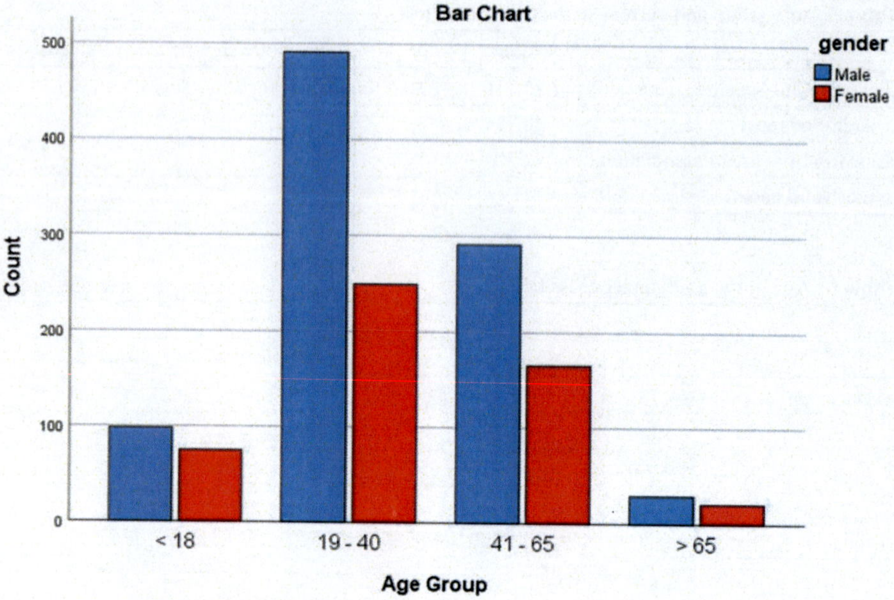

Fig. 3 Bar chart for age group w.r.t. gender of patient

Table 7 Age group and gender of patient Chi-square test

	Value	*df*	Asymptotic significance (2-sided)
Pearson Chi—square	6.626	3	0.085
Likelihood ratio	6.521	3	0.089
Linear—by—linear association	0.191	1	0.662
No of valid cases	1419		

Table 8 Values for chi-square test

	Value
Gender and current status	0.648
Age group and current status	0.000
Age group and gender	0.085

References

1. Chen B, Liang H, Yuan X, Hu Y, Xu M, Zhao Y et al (2020) Roles ofmeteorological conditions in COVID-19 transmission on a worldwide scale. medRxiv
2. Xu Z, Shi L, Wang Y, Zhang J, Huang L, Zhang C, Liu S, Zhao P, Liu H, Zhu L, Tai Y, Bai C, Gao T, Song J, Xia P, Dong J, Zhao J, Wang FS (Apr, 2020) Pathological findings of COVID-19 associated with acute respiratory distress syndrome. Lancet Respir Med 8(4):420–422

3. Li Y, Huang X, Yu IT et al (2005) Role of air distribution in SARS transmission during the largest nosocomial outbreak in Hong Kong. Indoor Air 15:83–95
4. Dalziel BD, Kissler S, Gog JR, Vibourd C, Bjornstad ON, MetcaffJE et al (2018) Urbanization and humidity shape the intensity ofinfluenza epidemics in U.S cities. Science 362:75–79
5. Murugesan B, Karuppannan S, Mengistie AT, Ranganathan M, Gopalakrishnan G (2020) Distribution and trend analysis ofCOVID-19 in India: geospatial approach. J Geogr Stud 4(1):1–9
6. World Health Organization (2020) Coronavirus Disease (COVID-19)Advice for the Public. Accessed: 5 Apr 2020. (Online). Available: https://www.who.int/emergencies/diseases/novel-coronavirus-2019/advice-forpublic
7. Gupta A, Banerjee S, Das S (2020) Significance of geographical factors to the COVID-19 outbreak in India. Model Earth Syst Environ. https://doi.org/10.1007/s40808-020-00838-2
8. Bashir MF, Ma B, Bilal Komal B, Bashir MA, Tan D, Bashir M (2020) Correlation between climate indicators and COVID-19 pandemicin New York, USA. Sci Total Environ 728:138835
9. Sajadi MM, Habibzadeh P, Vintzileos A, Shokouhi S, Miralles-WilhelmF, Amoroso A (2020) Temperature and latitude analysis topredict potential spread and seasonality for COVID-19. Availableat SSRN 3550308
10. Danon L, Brooks-Pollock E, Bailey M, Keeling MJ (2020) A spatialmodel of CoVID-19 transmission in England and Wales: Early spreadand peak timing. medRxiv
11. Tang B, Xia F, Tang S, Bragazzi NL, Li Q, Sun X, Liang J, Xiao Y, Wu J (2020) The evolution of quarantined and suspected cases determinesthe final trend of the 2019-nCoV epidemics based on multi-source data analyses, Technical report
12. Wu K, Zheng J, Chen J (2020) Utilize state transition matrix model topredict the novel corona virus infection peak and patient distribution. Technical Report, 2020
13. Lai S, Bogoch I, Ruktanonchai NW, Watts A, Li Y, Yu J, Lv X, Yang W, Yu H, Khan K (2020) Assessing spread risk of Wuhan novel coronavirus within and beyond China, January–April 2020: a travel network based modelling study. medRxiv, 2020
14. Yang S, He D, Luo J, Chen W, Yang X, Wei M, Kong X, Li Y, Feng X, Zeng Z (2017) 'Simulation of HIV/AIDS distribution using GIS basedcellular automata model.' Biomed Res 28(9):4053–4057

Chapter 49
Performance Analysis of Image Enhancement Techniques on X-Ray Images

Priya and Reeta Devi

1 Introduction

Coronavirus disease (COVID-19) is an infectious disease that emerged in late 2019 in Wuhan China and had spread worldwide. It is caused by coronaviruses which mainly affect the human respiratory system. The common symptoms include fever, cough, fatigue, headache, and breathlessness while the severe cases vary from pneumonia, respiratory failure to death [1]. Clinical, radiographic, and pathological studies of the infected patients reveal the primary infection in lungs [2]. Detection of X-ray images of the lungs of the patients affected by COVID-19 is not easy by the common person but only by the doctors or the experts in that domain. Image processing is an area through which the X-ray images of the lungs of the patients affected by COVID-19 can be analyzed [3]. X-ray images are grayscale images ranging from 0 to 255 with low quality in terms of intensity, contrast, and edges [4]. In order to improve the quality of image in terms of contrast and intensity, few processing steps are required. Image enhancement plays a challenging role in enhancing the image quality. Various methods exist for the image enhancement which are mainly categorized into two domains—Spatial domain and Frequency domain. Spatial techniques are directly applied to image pixels while in frequency domain, Fourier transform is used. An image is represented in a two-dimensional form, i.e., $f(x, y)$, where x and y are spatial coordinates of the image. The output image is $g(x, y) = T[f(x, y)]$, where 'T' is the transformation function [5]. The transformation function maps input image pixels into the output image.

This paper thus aims to study the various image enhancement techniques on X-ray images of the patients affected by COVID-19. The techniques mainly focused on the

Priya (✉) · R. Devi
University Institute of Engineering and Technology, Kurukshetra University, Kurukshetra, India
e-mail: priyasinghuiet@kuk.ac.in

R. Devi
e-mail: reetakuk@kuk.ac.in

N. Marriwala et al. (eds.), *Soft Computing for Intelligent Systems*, Algorithms for Intelligent Systems, https://doi.org/10.1007/978-981-16-1048-6_49

paper include Linear transformation, Logarithmic transformation, Power Law transformation, Contrast-Stretching, Histogram equalization, Contrast limited adaptive histogram equalizer. Additionally, various parameters analyzed in image enhancement and the comparative analysis of the selected image enhancement techniques are discussed. This study will be helpful in understanding pulmonary complications and other clinical imaging studies.

2 Image Enhancement Techniques

2.1 *Linear Transformation (LT)*

This transformation technique is the basic image enhancement technique which gives negative of the image. In this technique, image negative is obtained by complementing the gray values of the input image pixels. In digital image processing, the most common operation is computing the negative of an image which can be obtained by inverting the pixel gray values. For instance, if grayscale image having 256 levels is considered, then to obtain the image negative, input image pixels are subtracted from the maximum value of gray levels, i.e., 255 [6].

2.2 *Logarithmic Transformation (Log T)*

In this transformation technique, narrow range of pixels having low intensity is mapped to a wider range of output image value. This transformation is used to stretch the dark values of image pixels and compress the higher image pixel values and vice versa can be done in inverse log transformation. The log transformation of output image $g\ (x, y)$ can be obtained through $g(x, y) = c * \log(1 + r)$, where c is the constant and r is the intensity values [6].

2.3 *Power Law Transformation (PLT)*

In this transformation technique, each pixel of the input image can be varied to a static power, i.e., gamma correction denoted by γ the output image is $g(x, y) = c[f(x, y)]^\wedge\gamma$ If the gamma correction value is greater than 1, it will provide the enhanced image having high contrast and inverse operation is performed in case of gamma correction value less than 1 [7].

2.4 *Contrast-Stretching (C-S)*

In contrast stretching, the range of gray levels is expanded so that it spans the full intensity range of the display device. It improves the quality of image by increasing the contrast of the darker area more darker and bright area brighter [7]. Subjective analysis is done through image enhancement. There are no predefined values that give the required output. In contrast stretching, dynamic range of the image is expanded [8].

2.5 *Histogram Equalization (HEq)*

Image histogram represents the graphical plot between the probability of frequency of occurrence of image pixels and the number of intensity levels. Histogram equalization is the most widely used technique to improve the quality of the image. In histogram equalization, the intensity of pixels is spread automatically as per the requirement of the image. Image pixels having low intensity are mapped with high intensity. The output image obtained from this technique is of better quality and can be used in image segmentation. However, sometimes this technique creates unwanted artifacts [9].

2.6 *Contrast Limited Adaptive Histogram Equalizer (CLAHEq)*

This is the modified technique of the Histogram Equalization which reduces the unwanted noise by using variable local histogram transfer function. In this technique, pixel value of the image is divided into small parts, the mapping function of local histogram is selected and then clipping point of histogram is selected. Further, this function is applied at each part. To eliminate unwanted noise background, subtraction method can be used. But this technique is not suitable for the wide range of low contrast images [10].

Performance Analysis Parameters

The performance of the image can be analyzed on the basis of Peak Signal to Noise Ratio (PSNR) and Mean Square Error (MSE). These parameters can be calculated using the formulas as given below [7]

1. Mean Square Error (MSE): It can be calculated as the sum of the squared values differences of the input image and reconstructed image divided by number of pixels in the image. For better quality image, MSE will be low.

$$\text{MSE} = \sum\sum[f(x, y) - g(x, y)]^2/M * N \tag{1}$$

where M, N are number of rows and column in the image and summation ranges from 0 to $m - 1$ and 0 to $n - 1$.

2. Peak Signal to Noise Ratio (PSNR): It gives the quality measure of the output image. Higher the PSNR, better is the quality of the image. It is defined as

$$\text{PSNR} = 20 \log_{10}\left(I_{\text{max}}/\sqrt{}\text{MSE}\right) \tag{2}$$

where I_{max} is the maximum value of image pixel. In a grayscale, image maximum levels are 256 and the values lie from 0 to 255. So, for grayscale images, I_{max} will be 255.

Results

The results of different enhancement techniques are depicted in Fig. 1. Among all the image enhancement techniques, contrast limited adaptive histogram equalizer (CLAHEq) gives the better quality image. Table 1 shows the comparison between various image enhancement techniques as discussed above. Therefore, on the basis of selected performance analysis parameters—PSNR and MSE—it is verified that contrast limited adaptive histogram equalizer technique gives better result and best suited to enhance the X-ray images.

3 Conclusion

This paper shows the performance analysis of image enhancement techniques. The output image obtained from Contrast Limited Adaptive Histogram Equalizer gives better quality image than the other enhancement techniques. This technique can be used in pre-processing of medical images such as X-ray images. In future, quality of the image can be enhanced more by applying filters on the input image to eliminate noise. It will improve the PSNR where higher the PSNR better will be the image quality. After pre-processing, some post-processing techniques such as image segmentation and compression can also be determined.

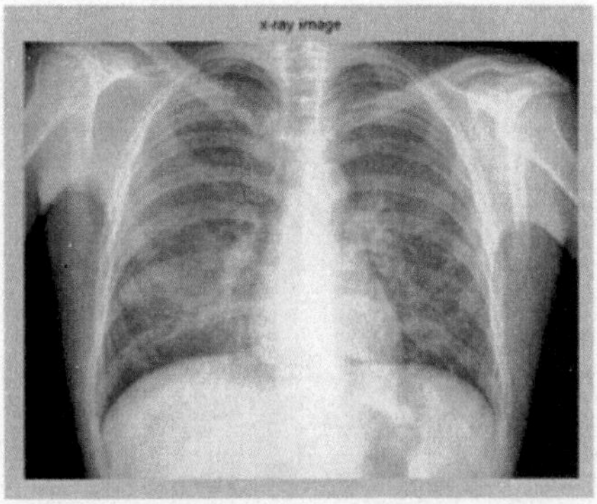

(a)

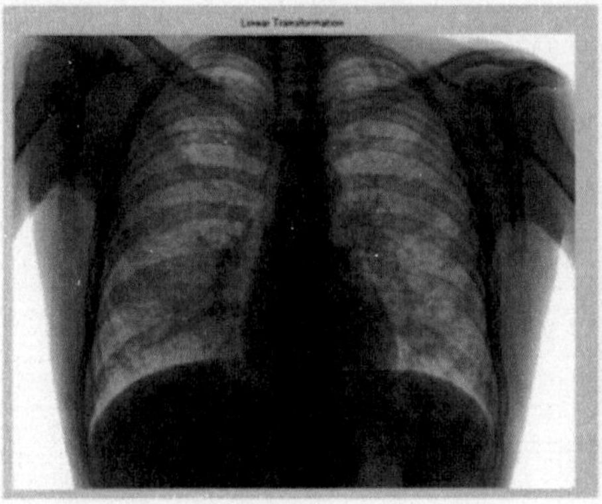

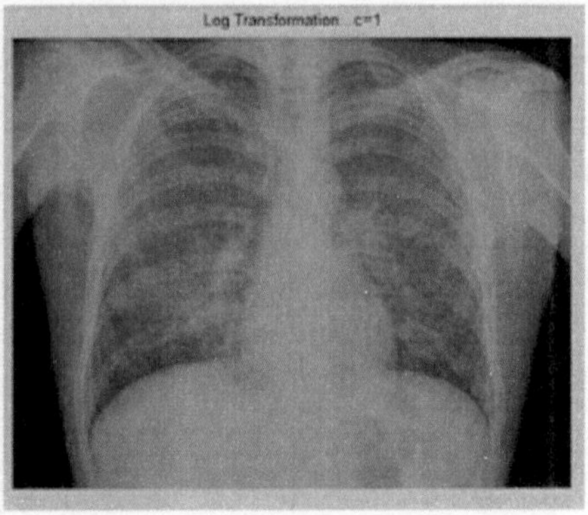

(b) (c)

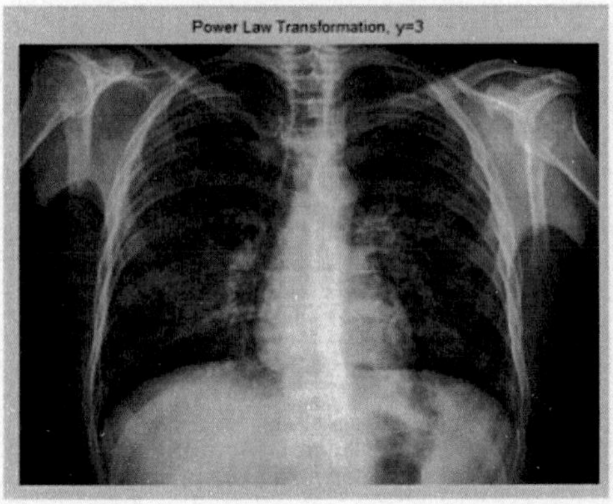

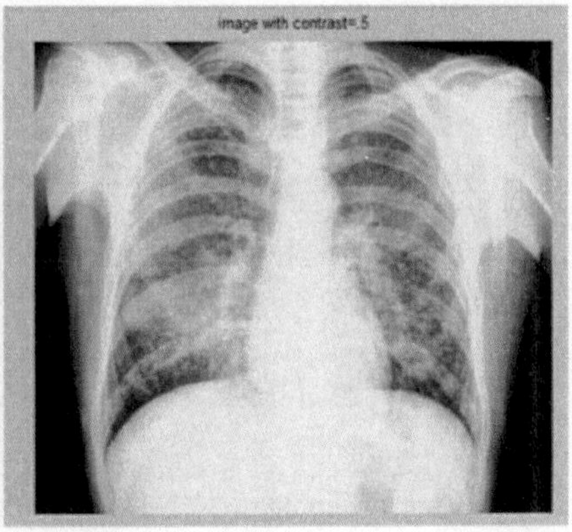

(d) (e)

Fig. 1 **a** Original Image [11], **b** output image obtained from linear transformation, **c** output image obtained from log transformation at $C = 1$, **d** output image obtained from power law transformation at $\gamma = 3$, **e** output image obtained from contrast stretching at contrast $= 0.5$, **f** output image obtained from histogram equalization, **g** output image obtained from contrast limited adaptive histogram equalizer

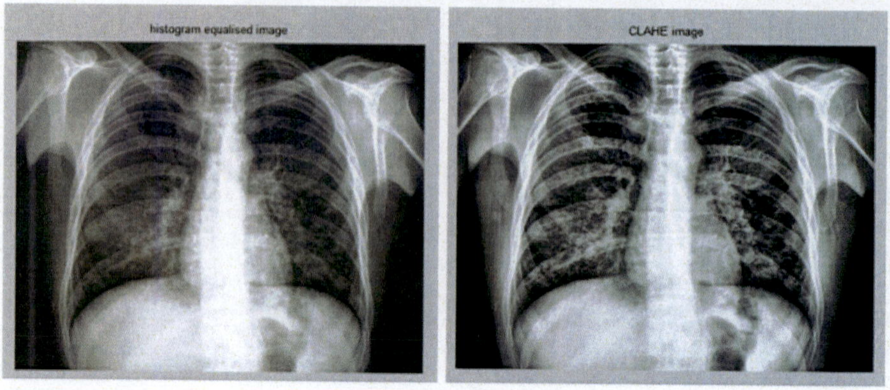

Fig. 1 (continued)

Table 1 Performance analysis of various image enhancement techniques on the basis of PSNR and MSE

Image enhancement techniques	PSNR	MSE
LT	4.563	2.27E + 04
Log T ($C = 1$)	3.1639	3.14E + 04
PLT ($\gamma = 3$)	10.6298	5.62E + 03
$C - S$ ($C = 0.5$)	3.1746	3.13E + 04
Heq	14.4389	2.34E + 03
CLAHEq	15.9888	1.64E + 03

References

1. Singhal T (2020) A review of coronavirus disease-2019 (COVID-19). Indian J Pediatr 87(4):281–286
2. Brosnahan SB, Jonkman AH, Kugler MC, Munger JS, Kaufman DA (2020) COVID-19 and respiratory system disorders: current knowledge, future clinical and translational research questions. Arterioscler Thromb Vasc Biol 40(11):2586–2597
3. Sirisha A, Venkateswararao P (2020) Image processing techniques on radiological images of human lungs effected by COVID-19. JOIV Int J Inf Vis 4(2):69–72
4. Chikhalekar AT (2016) Analysis of image processing for digital X-ray. Int Res J Eng Technol (IRJET) 3(5):1364–1368
5. Rahman S, Rahman MM, Hussain K, Khaled SM, Shoyaib M (2014) Image enhancement in spatial domain: a comprehensive study. In: 17th international conference on computer and information technology (ICCIT). IEEE, Dhaka, pp 368–373
6. Maini RA, Aggarwal H (2010) Comprehensive review of image enhancement techniques. J Comput 2(3):2151–9617
7. Meena SK, Potnis A, Mishra M, Dwivedy P, Soofi S (2017) Review and application of different contrast enhancement technique on various images. In: 1st international conference on electronics, materials engineering and nano-technology (IEMENTech). IEEE, Kolkata, India, pp 1–6

8. Gurunathan V, Bharathi S, Sudhakar R (2015) Image enhancement techniques for palm vein images. In: International conference on advanced computing and communication systems. IEEE, Coimbatore, pp 1–5
9. Yang J, Zhong W, Miao Z (2016) On the image enhancement histogram processing. In: 3rd international conference on informative and cybernetics for computational social systems (ICCSS). IEEE, Jinzhou, China, pp 252–255
10. Ikhsan IAM, Hussain A, Zulkifley MA, Tahir NM, Mustapha A (2014) An analysis of x-ray image enhancement methods for vertebral bone segmentation. In: 10th international colloquium on signal processing and its applications. IEEE, Kuala Lumpur, Malaysia, pp 208–211
11. https://github.com/ieee8023/covid-chestxray-dataset/blob/master/images/000001-10.jpg

Chapter 50
Assessment of Iris Flower Classification Using Machine Learning Algorithms

Saumya Goyal, Atul Sharma, Piyush Gupta, and Pragya Chandi

1 Introduction

The machine learning is an ambit of computer science. On the report of Arthur Samuel in 1959 told "computers are having the ability to learn without being explicitly programmed". Now from the further study of computational learning theory and pattern recognition in the area of artificial intelligence, the machine learning prospects the study and building of algorithm which can be gained from and deduce prediction on the new data. In this way such algorithms get over by the following steady program instructions by building data-driven decisions or predictions, by constructing a model from training dataset [1].

Machine learning is practiced in a scale of computing functions in which building and programming unmistakably algorithms with great results is unfeasible or tedious.

Machine learning pivot on the progress of computer programs that can make themselves how to expand and modify when disclosed to new input [2]. It is a research field at the convergence of computer science, artificial intelligence and statistics and which is termed as statistical learning or predictive analytics [3].

Machine learning is mainly categorized into two categories; one is Supervised Learning and other is Unsupervised Learning.

Supervised learning is a venture of deducing a function from labeled training dataset. In this learning, every sample is a collection of an input and preferred resulting value. Now by studying the given training dataset and by generating the concluded function, supervised learning algorithm is used to train the new samples. These problems can be further cluster into regression and classification problems.

S. Goyal (✉) · A. Sharma · P. Chandi
Department of Computer Science and Engineering, University Institute of Engineering and Technology (U.I.E.T), Kurukshetra University, Kurukshetra, Haryana, India

P. Gupta
Department of Computer Science and Engineering, J.C. Bose University of Science and Technology, Y.M.C.A Faridabad Haryana, Faridabad, India

N. Marriwala et al. (eds.), *Soft Computing for Intelligent Systems*, Algorithms for Intelligent Systems, https://doi.org/10.1007/978-981-16-1048-6_50

When the resulting variable can be categorized like "black" or "white" or "true" and "false" is known as classification problem. When the resulting variable is a real time value, like "dollar" or "height" then that is regression problem [4].

In unsupervised learning, there is no need to supervise the model. Alternatively, we need to allow the model to go on its own to uncover the information. It mainly works with unlabelled data. We don't have any objective or result variable to envision/gauge [5]. It is used in clustering the objects, which is used for exploratory object analysis, to figure out the unseen fashion and clustering in them.

And this paper particularly focuses on supervised learning.

For identification of Iris flower species, eight different classification algorithms are compared which is termed as Multiple Classifier system and accuracy and best algorithm among them is presented in this paper. This works in two stages, termed as training and testing stage. In training phase, the training dataset is uploaded into Machine Learning Models and gets assigned by labels. Further, the predictive model predicts which category of the Iris plant belongs to. Hence, the expected Iris plant species get labeled [6].

2 Existing Algorithms

Multiple classifier system is the combination of eight different linear and non-linear existing algorithms which are discussed below.

2.1 Logistic Regression (LR)

It is one of the supervised learning classification algorithms practiced to anticipate the probability of target variable. It is used to assign survey to a discrete set of classes. It is similar to Linear Regression model but the cost function used in LR is more complex and this cost function can be known as the "Sigmoid function". This sigmoid function is S-shaped curve that accepts any real-valued number and plots it into a value between 0 and 1.

2.2 Linear Discriminant Analysis (LDA)

It is dimensionally reduction technique in Machine Learning and applications of sample classification and is used as a pre-processing step. LDA is a supervised classification technique. This property of dimensionality reduction is mainly applied in areas like predictive analysis in marketing and image recognition.

2.3 K-Nearest Neighbors (KNN)

It is one of the simplest algorithms of machine learning that lies under supervised learning technique. It basically finds out the similarity among the data and on the basis of that categories it. It is also known as lazy algorithm because it does not train itself during training phase. At that phase, it just stores the dataset and when new data occurs it matches from the dataset on the basis of similarities and then places it in that category.

2.4 Classification and Regression Trees (CART)

It is only the modern term to call it, as it is also known by the most common term that is decision tree. Decision trees are being in use for a very long time and are most importantly applied for predictive purpose in machine learning. The aim of using it is to create a training model that predicts the class of the target variable by making decisions on the basis of rules inferred in old data, i.e., training data. At every node, it asks the question and on the basis of answer (yes or no) it divides the tree into subtrees.

2.5 Gaussian Naive Bayes (NB)

It is the supervised learning algorithm. It is the simple classifier with high functionality. It most importantly come in use when the dimensions of the input are high. Bayes theorem is also used to calculate conditional probability which makes it a powerful tool in machine learning. It gets trained very efficiently with the small training dataset. In this one, strong independent assumption is taken between the features. These classifiers assume that the value of a particular character is independent of the value of other characters.

2.6 Support Vector Machines (SVM)

SVM can be applied for both regression and classification tasks. N-dimensional space means having N number of features; it finds a hyperplane which distinctly finds the data point. The main objective is to find the maximum distance between the classes and data point; the more the distance more will be confident to classify data.

2.7 Random Forest Classifier (RFC)

As the name suggests, it is the forest which is made up of multiple decision tree. It is a collection of individual decision trees which are doing their task individually. RFC votes out the class prediction made by decision tree with maximum number and declares it model's final prediction. The basic concept behind the random forest classifier is simple and powerful that is the maximum vote wins. To make the correct decision, correct features should be selected while making the decision tree.

2.8 Gradient Boosting Classifier (GBC)

It is an interesting supervised learning technique. As the name suggests it boosts to do better performance for better result. Gradient boosting classifier is the collection of weaker model to make itself the strong model. Basically, GBC iteratively learns from the loop holes of the weaker model to make itself strong. That's why it lies under the category of boosting method. It is used for regression, classification, and ranking also. It has main three components—loss function, weak learner, and additive model [7–12].

3 Proposed Work

Now the following section contains the detailed description of each step of the process (Fig. 1).

Several datasets of Iris Flower are piled up. This database is well suitable for doing classification research. Dataset is well known in the field of pattern recognition. The dataset contains three categories of 50 samples each, where each category points to a type of Iris plant species that is Setosa, Versicolor, and Virginca. In this, one category is linearly separable from the other two, but last mentioned are NOT linearly separable from each other.

The gathered Iris plant datasets are uploaded onto the multiple classifier system. For the classification of the Iris plant Dataset, Scikit-learn is associated with few approved datasets. From Scikit-learn the "load_iris" function is imported. The "load_iris" function is implemented and saves the returned result in the object termed "iris".

The "iris" object is "sklearn.datasets.base.bunch" type, where bunch in scikit-learn is a specific object which stores datasets and their features. Some features of "iris" object are data_feature, data, target_names, target, etc. The type of "numpy.ndarray" is "iris.data" and is stocked in a Matrix say "M". Where "M" is two-dimensional array, which presents no. of observations on y-axis and no. of features on x-axis. The type of "numpy.ndarray" is "iris.target" and is saved in a

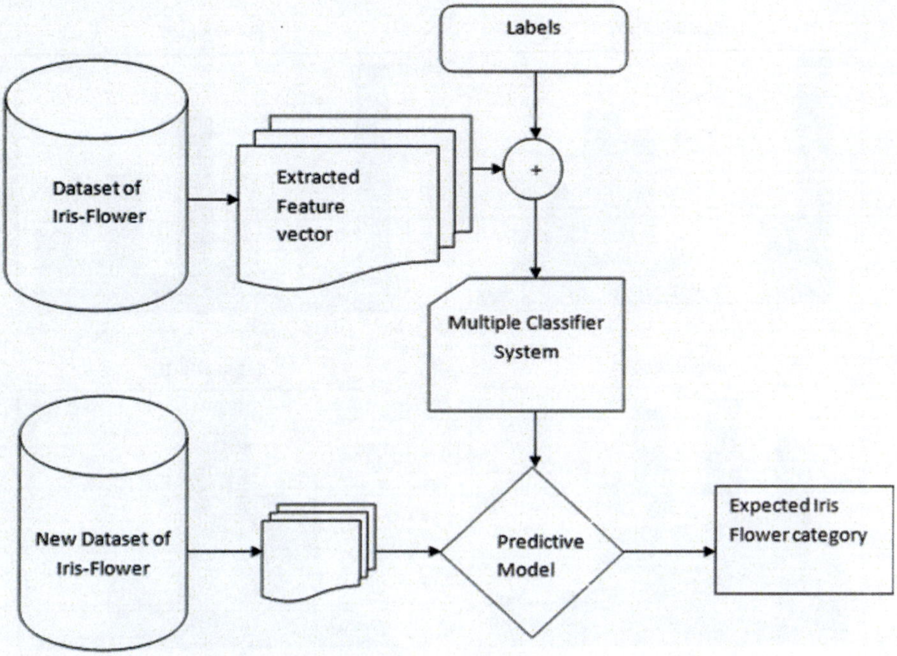

Fig. 1 Shows the working of the proposed scheme

Response Vector say "V". Where "V" is a one-dimensional array, which presents number of observations.

In Scikit-learn, every Row is an observation that is record, for example, sample, instance and every Column is a feature that is attribute, predictor, input, independent variable, Covariate, regressor.

To work with the data in Scikit-learn, there are four key requirements that should be known:

1. Responses and Features are discrete objects.
2. Responses and Features should be in number.
3. Responses and Features should be NumPy arrays.
4. Responses and Features should be of particular shapes.

The structure of M and V is (150, 4) and (150), respectively, in which 150 are observations and number of features are 4. The Target Names of Iris plant datasets are ['setosa', 'versicolor', 'virginica'] and Feature Names of Iris plant datasets are ['sepal length (cm)', 'petal width (cm)', 'petal length (cm)', 'sepal width (cm)'] [12].

Predicted attribute: class of iris plant.

Histogram Plot and Scatter Matrix plot for the Iris plant dataset are shown in Figs. 2 and 3, respectively.

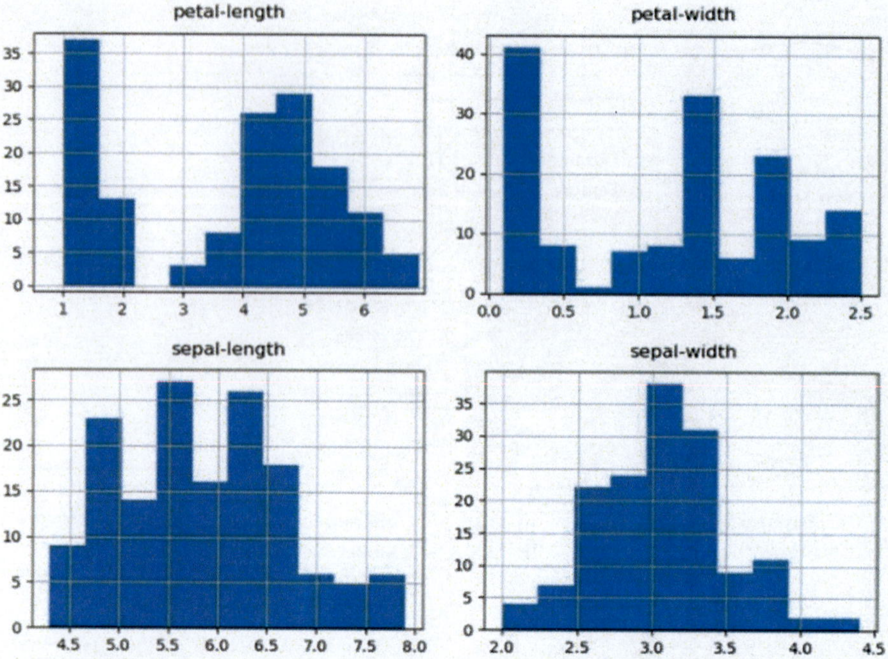

Fig. 2 Histogram plot of Iris dataset input variables

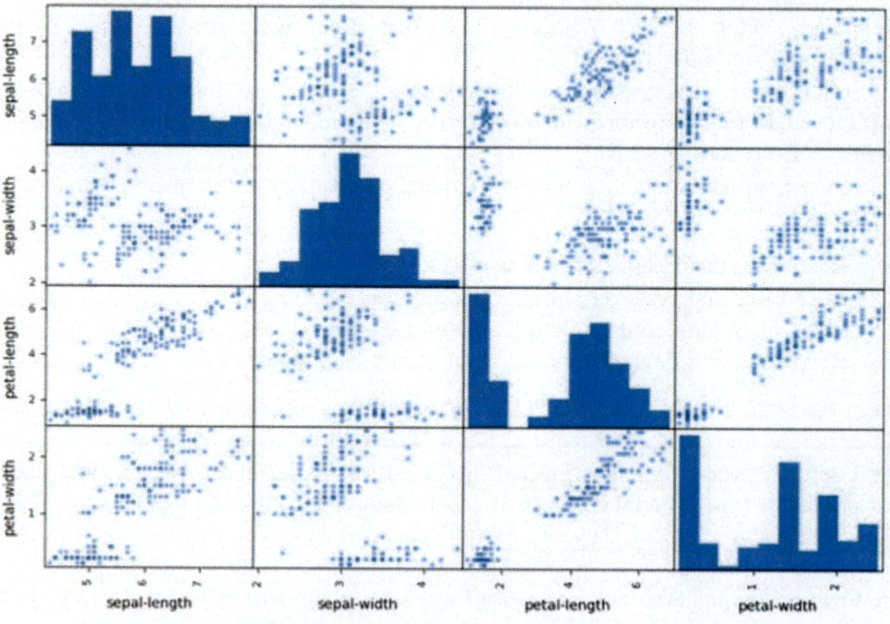

Fig. 3 Scatter matrix plot of iris dataset input variables

1. Iris dataset is divided into two sub datasets, i.e., training and validation dataset, 80% of the whole data is applied to train the model and rest 20% used as a validation dataset.
2. To estimate model accuracy we have used tenfold cross-validation.
3. We have tested our problem on 8 different linear and non-linear algorithms, which are stated below

   a. Logistic Regression (LR).
   b. K-Nearest Neighbors (KNN).
   c. Linear Discriminant Analysis (LDA).
   d. Classification and Regression Trees (CART).
   e. Support Vector Machines (SVM).
   f. Gaussian Naive Bayes (NB).
   g. Random Forest Classifier (RFM).
   h. Gradient Boosting Classifier (GBC).

These models are compared with each other, and we have selected the best among these models based on the accuracy.

4 Result and Observations

The proposed mechanism has been implemented using python. It is an open-source language. In the proposed mechanism, different algorithms are applied to Iris dataset. SVM has the best accuracy score, i.e., 0.98 or 98%. During the training phase of the model, SVM shows the best results on the dataset provided to it. After that the testing phase is performed only on the SVM algorithm.

When predictions are evaluated on the SVM with the expected results in the validation set 0.96 or 96% accuracy is observed (Fig. 4).

Above graph is the comparison graph which shows the accuracy of every algorithm of Multiple Classifier system when performed during the training stage.

In above shown graph, the box is showing the range of accuracy achieved by every algorithm and orange line shows the mean of it.

In the first graph, the box is showing the range of accuracy achieved by every algorithm and orange line shows the mean of it.

Evaluation of the trained model:

Accuracy achieved: 0.96.

Confusion Matrix:

Iris Setosa	[[11	0	0]
Iris Versicolor	[0	12	1]
Iris Verginica	[0	0	6]]

Fig. 4 Comparative analysis of different algorithm

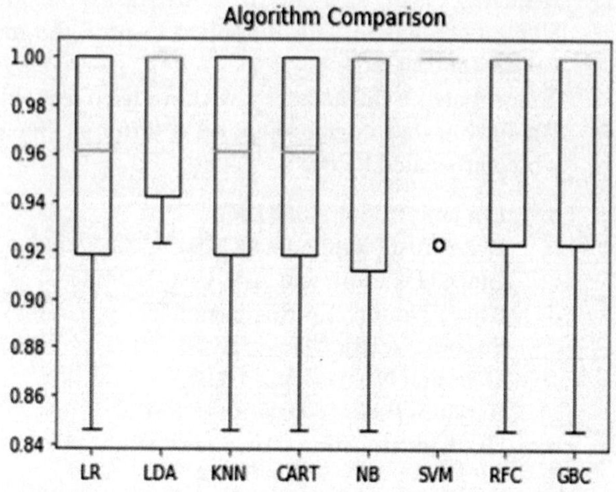

5 Conclusion

Biodiversity of earth is very rich. About 360,000 create a healthy biome within the environment of earth. Some of them are identical in physical appearance like shape, size, and color. Hence, it is difficult to recognize any species. Similarly, Irisflower species has three subspecies Setosa, Versicolor, and Virginica. We are using Iris dataset because it is frequently available. The dataset of Iris flower contains 3 classes of 50 instances each. With the help of Machine learning, Iris dataset identifies the subclasses of Iris flower. The paper focuses on how Machine Learning algorithms can automatically recognize the class of flower with the help of high degree of accuracy rather than approximately. The three phases to implement this approach are segmentation, feature extraction, and classification using Neural Network, Logistic Regression, Support Vector Machine and k-Nearest Neighbors.

References

1. Vatshayan S (2019) Performance evaluation of supervised learning for iris flower species. In: April 2019 , IJIRT 5(11). ISSN: 2349–6002
2. Shukla A, Agarwal A, Pant H, Mishra P (2020) Flower classification using supervised learning. Int J Eng Res Technol (IJERT) 9(05). ISSN: 2278–0181 IJERTV9IS050582
3. Tsoukas V, Kolomvatsos K, Chioktour V, Kakarountas A (2019) A comparative assessment of machine learning algorithms for events detection. In: 2019 4th South-East Europe design automation, computer engineering, computer networks and social media conference (SEEDA-CECNSM). IEEE, pp 1–4
4. Ahmadi N, Nilashi M, Samad S, Rashid TA, Ahmadi H (2019) An intelligent method for iris recognition using supervised machine learning techniques. Opt Laser Technol 120:105701
5. Gümüşçü A, Tenekeci ME, Bilgili AV (2019) Estimation of wheat planting date using machine learning algorithms based on available climate data. In: December 2020 under Sustain Comput

Inf Syst 100308, Elsevier
6. TaheriNejad N, Jantsch A (2019) Improved machine learning using confidence. In: 2019 IEEE Canadian conference of electrical and computer engineering (CCECE). IEEE, pp 1–5
7. Fakir Y, Lakhdoura Y, Elayachi R (2020) Comparative analysis of random forest and J48 classifiers for "IRIS" variety prediction. Glob J Comput Sci Technol: H Inf Tech 20(2) Version 1.0 Year 2020, Online ISSN: 0975–4172 & Print ISSN: 0975–4350
8. Basha SM, Rajput DS (2019) Survey on evaluating the performance of machine learning algorithms: past contributions and future roadmap. In: Deep learning and parallel computing environment for bioengineering systems. Academic Press, pp 153–164
9. Mohan L, Pant J, Suyal P, Kumar A (2020) Support vector machine accuracy improvement with classification. In: 2020 12th international conference on computational intelligence and communication networks (CICN). IEEE, pp 477–481
10. Malviya A (2020) Machine learning: an overview of classification techniques. In: Computing algorithms with applications in engineering. Springer, Singapore, pp 389–401
11. Galphade AV, Walse KH (2019) Supervised learning approach for flower images using color, shape and texture features. Int Res J Eng Technol 6(05)
12. Pico-Valencia P, Vinueza-Celi O, Holgado-Terriza JA (2020) Bringing machine learning predictive models based on machine learning closer to non-technical users. In: International conference on systems and information sciences. Springer, Cham, pp 3–15

Author Index

N. Marriwala et al. (eds.), *Soft Computing for Intelligent Systems*, Algorithms for
Intelligent Systems, https://doi.org/10.1007/978-981-16-1048-6